OBSERVATORIES IN EARTH ORBIT AND BEYOND

ASTROPHYSICS AND SPACE SCIENCE LIBRARY

A SERIES OF BOOKS ON THE RECENT DEVELOPMENTS
OF SPACE SCIENCE AND OF GENERAL GEOPHYSICS AND ASTROPHYSICS
PUBLISHED IN CONNECTION WITH THE JOURNAL
SPACE SCIENCE REVIEWS

PROCEEDINGS
VOLUME 166

OBSERVATORIES IN EARTH ORBIT AND BEYOND

PROCEEDINGS OF THE 123RD COLLOQUIUM OF THE INTERNATIONAL ASTRONOMICAL UNION, HELD IN GREENBELT, MARYLAND, U.S.A., APRIL 24–27, 1990

Edited by

Y. KONDO

NASA/Goddard Space Flight Center, Greenbelt, Maryland, U.S.A.

SPRINGER–SCIENCE+BUSINESS, MEDIA, B.V.

Library of Congress Cataloging-in-Publication Data

International Astronomical Union. Colloquium (123rd : 1990 :
  Greenbelt, Md.)
    Observatories in earth orbit and beyond : proceedings of the 123rd
  Colloquium of the International Astronomical Union, held in
  Greenbelt, Maryland, April 24-27, 1990 / edited by Y. Kondo.
      p.   cm. -- (Astrophysics and space science library ; v. 166)
    ISBN 978-94-010-5528-4   ISBN 978-94-011-3454-5 (eBook)
    DOI 10.1007/978-94-011-3454-5
    1. Orbiting astronomical observatories--Congresses. 2. Astronomy-
  -Congresses. 3. Astrophysics--Congresses.   I. Kondo, Yuji.
  II. Title. III. Series.
  QB500.267.I56 1990
  522'.1919--dc20   91-6568
                                                                   CIP

ISBN 978-94-010-5528-4

Printed on acid-free paper

TABLE OF CONTENTS

(B) ULTRAVIOLET MISSIONS

(C) INFRARED AND SUBMILLIMETER MISSIONS

(D) RADIO MISSIONS

(E) SOLAR SYSTEM & PLANETARY SYSTEMS

(F) SHUTTLE-BORNE ASTRO MISSIONS

(B) ALTERNATIVE APPROACHES

V. LONG TERM FUTURE ISSUES

Panel: Major Unsolved Problems of Astronomy

CONTRIBUTED PAPERS

FOREWORD

When I became President of International Astronomical Union Commission 44 for the triennial period 1985–1988, several members of the Organizing Committee and I agreed that it would be a good idea for our Commission to host a conference on observatories in space in view of their increasingly important role in astronomi cal research. IAU Colloquium Number 123 "Observatories in Earth Orbit and Beyond" is the first colloquium sponsored by IAU Com mission 44 on Astronomy from Space, although Commission 44 has co-sponsored numerous colloquia and symposia in the past.

The past two decades have seen a flourishing of astronomical observatories in space. Over a dozen orbiting observatories have opened up a new window on the universe, providing hitherto una vailable data in the electromagnetic spectral range from gamma-ray, X-ray, ultraviolet to infrared and radio. This has clearly demonstrated the crucial nature of astronomical observations from space.

The invited talks of present colloquium consist primarily of reviews of currently operating observatories in space, future observatories that have been approved by sponsoring government or space agencies, the launch systems of U.S.A., E.S.A., U.S.S.R. and Japan, discussions of various orbits and sites (such as the Moon), and alternate approaches in designing space observatories. Several panel discussions addressed those issues as well as the major unsolved problems of astronomy. Contributed poster papers included descriptions of space observatories that are in planning stage.

The Hubble Space Telescope was launched half an hour before the meeting opened, making it a current observatory. The X-ray satellite, ROSAT, was discussed as a future observatory at the meeting but was successfully placed in orbit a month after the colloquium. The summary of ROSAT in the proceedings was prepared several weeks after the launch and presents ROSAT as an operating satellite.

Some invited speakers and poster-paper contributors were unable to prepare in time their presentations for publication in the proceedings, quite possibly because of their involvement in various space programs or related research activities. In such cases, their abstracts are published if available. If not, their presentations are included by their titles only.

I wish to express my appreciation to the Scientific Organiz ing Committee for their assistance in organizing the conference and to the Local Organizing Committee, co-chaired by Drs. Mead and Michalitsianos, for their outstanding work. Of course, this meeting would not have been possible without the enthusiastic participation of our colleagues who came despite their busy schedule. In all, 192 people registered for the colloquium. Dr. S.A. Hall of the Library of Congress provided capable assistance in editing the Proceedings and T. Busby-Lewis of West over rendered an efficient logistic support for the conference.

Y. Kondo (ed.), Observatories in Earth Orbit and Beyond, xi–xii.
©1990 *Kluwer Academic Publishers.*

In behalf of the Organizing Committees, I would like to thank the IAU, NASA and ESA for their valuable support. It is a pleasure to acknowledge the donations received from Computer Sciences Corporation, Ball Aerospace Systems Group, Ford Aero space, Hughes Danbury Optical Systems and TRW in support of the meeting. This colloquium was also cosponsored by COSPAR.

Yoji Kondo, Editor

I. CURRENT MISSIONS

THE HUBBLE SPACE TELESCOPE

ALBERT BOGGESS

Sciences Directorate Goddard Space Flight Center

Abstract. The Hubble Space Telescope was launched from the Kennedy Space Center on April 24, 1990. Its initial check-out indicates that all sub-systems of the satellite are working very well, with two key exceptions: The line-of-sight pointing is subject to occasional jitter apparently induced by thermal stresses in the solar arrays; this is expected to be overcome. The telescope mirrors are found to contain approximately 0.5 wave rms of spherical aberration which cannot be overcome by any controls on board the satellite. This defect will limit the scientific performance of the telescope in the short run. However, the aberration can be fully corrected in the optical designs of future replacement instruments, and the delivery schedules of these instruments are being accelerated.

1. Introduction

Those of us associated with the Hubble Space Telescope wished to commemorate this IAU colloquium and to lend some sense of immediacy to its deliberations and so arranged to have the HST launched on Space Shuttle Discovery at 8:33 EDT on the morning of the opening session. We hope the astronauts' saga during the following days of deploying the telescope and establishing it as an independent free flyer provided an interesting backdrop for the meetings. The satellite is still in the midst of its engineering check-out phase and it is too soon to predict the ultimate performance of the telescope in orbit, other than to note the serious problem of spherical aberration in the telescope optics. The following paragraphs summarize what has been learned so far.

2. Launch preparations and deployment

The pre-launch preparations at the Kennedy Space Center went very smoothly for the most part. The satellite was shipped to Kennedy from the Lockheed plant in Sunnyvale, California last October, and tests quickly showed that it had arrived in good shape. The Wide Field/Planetary Camera, which had been shipped to JPL for refurbishment of some electronics boards, arrived in December and was installed without incident. Other refurbished electronics parts were installed on schedule in February and the HST was declared ready for launch. The principal remaining concern was to be sure that the flight batteries could be fully enough charged at launch to provide power to HST during the critical several-hour period from the time that supporting shuttle power would be removed in orbit until the solar arrays could be extended. The logistics necessary to allow batteries to be trickle-charged until very late in the countdown was carefully coordinated between the spacecraft and launch crews.

The initial launch attempt was on the morning of April 10. The countdown proceeded smoothly until a few seconds before lift-off, when a failed Auxiliary Power

Y. Kondo (ed.), Observatories in Earth Orbit and Beyond, 3–7.

Unit in the shuttle orbiter forced the launch to be scrubbed. Launch was rescheduled for April 24 and went flawlessly. The shuttle established a 613 km circular orbit, about 2 km higher than requested, which removes a need for reboost until the next solar cycle. On orbit, the astronauts deployed the satellite on the manipulator arm and then extended the high gain antennas and the solar arrays. One solar array was a cause for concern because monitors indicated that the tension on the array became high as it was extended. The procedure was stopped and some time spent carefully analyzing the situation to decide whether it was safe to continue. In the meantime two of the astronauts, Kathy Sullivan and Bruce McCandless, made preparations for an Extra-Vehicular Activity in order to unfurl the array manually. To everyone's relief (except perhaps for Kathy and Bruce), the array did extend fully when commanded on a subsequent attempt and the HST was ready to be released from the shuttle. A second problem soon became apparent when one of the high gain antennas, designed to be pointed toward the Tracking and Data Relay Satellites in geosynchronous orbit, encountered resistance when commanded to rotate through its full pointing range. After examining documentation photographs taken during assembly and after modeling the motions of the antenna, it was concluded that at some orientations an antenna counterweight could rub against a flexure loop in the wiring harness. These orientations proved to be easily avoided during regular operation of the antennae and the deployment sequence was allowed to proceed.

TABLE I

Hubble Space Telescope Spacecraft Characteristics

Length	13 m
Diameter (without arrays)	4.3 m
Mass	11,500 Kg
Power	2,400 w
Communications	1 Mbps through TDRSS
Tracking accuracy	0.01 arc-sec
Tracking jitter	0.007 arc-sec
Telescope aperture	2.4 m
Telescope figure	Ritchey-Chretien
Primary focal ratio	f/2.3
Secondary focal ratio	f/24
Plate scale	3.58 arc-sec/mm
Image FWHM at 540 nm	0.08 arc-sec
Radius of 70 % encircled energy	0.6 arc-sec

The final step was to back the shuttle orbiter far enough away that its emissions could not contaminate the telescope optics while the door was opened at the front end of the telescope tube. That actuation was the last task of the deployment sequence that called for back-up help from the astronauts, and so they and the shuttle were released to complete their mission and return to Earth. The engineers and ground controllers were then left with the task of learning the actual operating characteristics of HST in orbit and of developing their techniques for scheduling and using the satellite.

3. Orbital verification

The principal tasks of Orbital Verification have been to evaluate the capabilities of the various spacecraft sub-systems and learn how to operate them. These sub-systems include such areas as pointing control, telescope optics, thermal control, power, communications, command and data handling, and the scientific instruments. Of these, all are working well except the first two. The thermal control system is doing its job routinely and the passive thermal system is running somewhat warmer than intended, reducing the need for heater power. This in turn results in a total power consumption about ten percent less than expected, which can permit greater operational and observing flexibility in the early years and should delay the future date at which batteries and solar arrays will need to be replaced. The communications system is working well. Some procedural problems in establishing links through TDRSS have been identified and corrected. The link margins from the ground through TDRSS to HST and return are larger than expected, which is advantageous to operational flexibility. The command and data handling system is performing properly, although some minor problems were found and corrected in the use of the on-board tape recorders. All of the scientific instruments have been turned on and have completed their internal engineering checks. Successful exposures have been made with the two cameras, and the narrow-field instruments are awaiting the mapping of their entrance apertures in order to observe external sources.

The pointing control system is very complex and has been correspondingly difficult to make operational. The system consists of control algorithms in the spacecraft computer which accept input data from gyroscopes and star sensors and which issue pointing commands to momentum wheels. Momentum is managed through the use of magnetic torquers. The gyros control large slews and can also be used to maintain spacecraft pointing in the absence of star sensor data. There are two types of star sensors: three Fixed-Head Star Trackers pointed in various directions away from the telescope axis in order to measure approximate spacecraft orientations with respect to bright stars and three Fine Guidance Sensors mounted in the telescope focal plane to provide precise pointing information. In order to make the system useful the gyro drifts must be monitored and compensated, the angular relationships between the momentum wheel axes, the gyro axes, the Fixed-Head Star Tracker and Fine Guidance Sensor axes, the telescope axis and the positions of the entrance apertures of the various science instruments must all be measured precisely, and of course the techniques for using and integrating the data from all of these control sources must work smoothly. Further, the star trackers and guidance sensors have photometric calibrations and complex operating modes that need to be evaluated. While all of this equipment does not yet operate together in a routine way, the components do work according to their specifications and routine operations should come with experience. However, there is an unexpected disturbance source that has not yet been overcome. It appears that, during transitions from sunlight to darkness and from darkness to sunlight, thermal stresses are induced in the solar arrays which cause them to oscillate and introduce torques into the spacecraft. Although the major disturbances occur at day-night transitions, smaller disturbances occur randomly throughout the orbit and their cause is poorly understood. The oscillations are sinusoidal at the ten-second natural period of the solar arrays. Under gyro control their amplitudes are typically several arc-seconds, which is reduced to a few tenths arc-second under Fine Guidance control. In order to meet specification, which is expressed as 0.007 arc-sec rms, the pointing performance will have to be improved by more than a factor of twenty. Such an improvement appears to be possible by revising the control law algorithms specifically to accommodate torques of the frequency and amplitude that are observed. The flight control software is currently being rewritten to implement these changes.

4. Optical evaluation

Attempts to focus the telescope and to understand its optical quality have been complicated by the relative instability of the pointing system. Each of the Fine Guidance Sensors contains an interferometer designed to measure the wavefront produced by the telescope optics. The interferograms produced by these three instruments were designed to provide information on how to move the secondary mirror, which has five degrees of freedom, in order to remove focus, coma and astigmatism errors. The fringe visibility in the interferograms is greatly reduced by telescope jitter. In retrospect, it is apparent that their visibility is also low because the interferometers were not designed to work in the presence of large amounts of spherical aberration. After engineering checks were completed of the Wide Field/Planetary Camera and the Faint Object Camera, they were used to obtain stellar images at a variety of positions of the secondary mirror, and the analysis of these images made it plain that spherical aberration is present. Although the exact amount has not been finally measured as of this writing, the present data indicate it to be about 0.4 wave rms at a wavelength of 0.5 micron, with an uncertainty of about 0.1. This corresponds to an error in the profile of the primary mirror between the inner and outer masks of about 2 microns, in the sense that the curvature is too shallow. The point spread function produced by these optics consists of a sharp inner core whose full width at half maximum is about 0.08 arc-second and a faint outer skirt with a radius of about 0.7 arc-sec. Unfortunately, the inner core contains only about 15 percent of the total energy in the image.

The present plans are to set the secondary at an optimum position, proceed with instrument calibration and begin science observing as soon as possible. The objective of the focus optimization will be to remove coma and astigmatism and to maximize the energy in the central core. Pictures taken with this point spread function will be useful for many types of morphological studies and are potentially good candidates for the application of sophisticated image deconvolution techniques. Exciting scientific observations can be carried out at ultraviolet wavelengths with the cameras, spectrographs and photometer, and the astrometric programs should be able to proceed with little or no compromise. Looking to the future it appears that the surface quality of the mirrors is well within specification, so that future instruments built to compensate for the spherical aberration should achieve the full image quality that has been intended. Preliminary studies show that the three new instruments presently being developed will be able to incorporate the aberration compensation in a straightforward manner. Thus the replacement camera (WFPC II), the imaging spectrograph (STIS), and the near infrared instrument (NICMOS) can produce data that will met or exceed the original HST expectations. NASA is currently examining ways to accelerate the construction and deployment of these instruments.

EARLY RESULTS FROM THE COSMIC BACKGROUND EXPLORER (COBE)

J. C. MATHER, M. G. HAUSER, C. L. BENNETT, N. W. BOGGESS, E. S. CHENG,
R. E. EPLEE, JR., H. T. FREUDENREICH, R. B. ISAACMAN, T. KELSALL,
C. M. LISSE, S. H. MOSELEY, JR., R. A. SHAFER, R. F. SILVERBERG,
W. J. SPIESMAN, G. N. TOLLER, J. L. WEILAND

Laboratory for Astronomy and Solar Physics, Goddard Space Flight Center,
Greenbelt, MD 20771

S. GULKIS, M. JANSSEN

Jet Propulsion Laboratory, MS 169-506, 4800 Oak Grove Drive, Pasadena, CA 91109

P. M. LUBIN

UCSB, Dept. of Physics, Goleta, CA 93106

S. S. MEYER, R. WEISS

Department of Physics, Massachusetts Institute of Technology,
Cambridge, MA 02139

T. L. MURDOCK

General Research Corporation, 5 Cherry Hill Drive, Suite 220,
Danvers, MA 01923

G. F. SMOOT

Lawrence Berkeley Laboratory, 50-232, University of California,
Berkeley, CA 94720

D. T. WILKINSON

Department of Physics, Jadwin Hall, Box 708, Princeton University,
Princeton, NJ 08544

and

E. L. WRIGHT

UCLA Dept. of Astronomy, Los Angeles, CA 90024-1562

Abstract. The Cosmic Background Explorer, launched November 18, 1989, has nearly completed its first full mapping of the sky with all three of its instruments: a Far Infrared Absolute Spectrophotometer *(FIRAS)* covering 0.1 to 10 mm, a set of Differential Microwave Radiometers *(DMR)* operating at 3.3, 5.7, and 9.6 mm, and a Diffuse Infrared Background Experiment *(DIRBE)* spanning 1 to 300 μm in ten bands. A preliminary map of the sky derived from *DIRBE* data is presented. Initial cosmological implications include: a limit on the Comptonization y parameter of 10^{-3}, on the chemical potential μ parameter of 10^{-2}, a strong limit on the existence of a hot smooth intergalactic medium, and a confirmation that the dipole anisotropy has the spectrum expected from a Doppler shift of a blackbody. There are no significant anisotropies in the microwave sky detected, other than from our own galaxy and a $\cos\theta$ dipole anisotropy whose amplitude and direction agree with previous data. At shorter wavelengths, the sky spectrum and anisotropies are dominated by emission from 'local' sources of emission within our Galaxy and Solar

Y. Kondo (ed.), Observatories in Earth Orbit and Beyond, 9–18.
© 1990 *Kluwer Academic Publishers.*

System. Preliminary comparison of *IRAS* and *DIRBE* sky brightnesses toward the ecliptic poles shows the *IRAS* values to be significantly higher than found by *DIRBE* at 100 μm. We suggest the presence of gain and zero-point errors in the *IRAS* total brightness data. The spacecraft, instrument designs, and data reduction methods are described.

1. Introduction

The cosmic microwave background radiation (CMBR) from the Big Bang, and the infrared background radiation from the first objects to form after the Big Bang, are the subjects of NASA's first dedicated cosmological satellite, the *COBE*. The mission has been described by Mather (1982) and Gulkis et al. (1990). Theoretical estimates for various distortions of the microwave and infrared background radiation spectra have been given by Bond, Carr, and Hogan (1986) and by many others in this very active field.

We summarize here the key events in the expanding universe which influence these backgrounds. Processes in the early expansion of the universe such as inflation and subsequent nucleosynthesis, which take place within the first few minutes after the initial explosion, set the scale for the largest anisotropies and inhomogeneities. The number of quanta in the CMBR is essentially fixed when the time scales for their creation and destruction exceed the expansion time scale, which occurs at a redshift $z \sim 10^6$ when photons outnumber matter particles by a factor of $\sim 10^8$. After this epoch the spectrum of the CMBR can be modified to a Bose–Einstein distribution with a chemical potential μ by such processes as black hole formation, cosmic string annihilation, matter/antimatter annihilation, or the decay of exotic particles. After a redshift of $z \sim 3 \times 10^4$, the process to re-establish equilibrium becomes too slow to achieve a Bose–Einstein distribution, and any other energy release into the radiation field mediated by hot electrons produces a mixture of blackbodies at different temperatures, characterized by the parameter y. After the decoupling of matter and radiation at $z \sim 10^3$, the photon field is free to move unimpeded, and the major features of the spectrum and angular distribution of the CMBR as now observed should have been established.

The processes leading to the formation of galaxies and large scale structures, as well as the epoch when these form, are major unknowns in current cosmology. One can bracket the epoch by noting that the earliest quasars have a $z<5$ and that galaxies at their present size would overlap at z between 20 and 100. The formation of galaxies may well be accompanied by the creation of a hot intergalactic medium, which could produce anisotropies and spectral distortions in the CMBR as well as a signature in the infrared. Furthermore, it is expected that precursors to the large scale structures should have left anisotropies and possibly spectral distortions in the CMBR.

As illustrated in Fig. 1, local astrophysical sources make it difficult to measure these traces of the pregalactic universe. Interplanetary dust reflects and absorbs sunlight and reradiates at a temperature of about 250 K, while interstellar dust reflects and absorbs starlight and reradiates at a temperature of about 25 K. At longer wavelengths interstellar electrons produce synchrotron and free-free emission. The *COBE* mission was designed to produce a complete data set that would permit

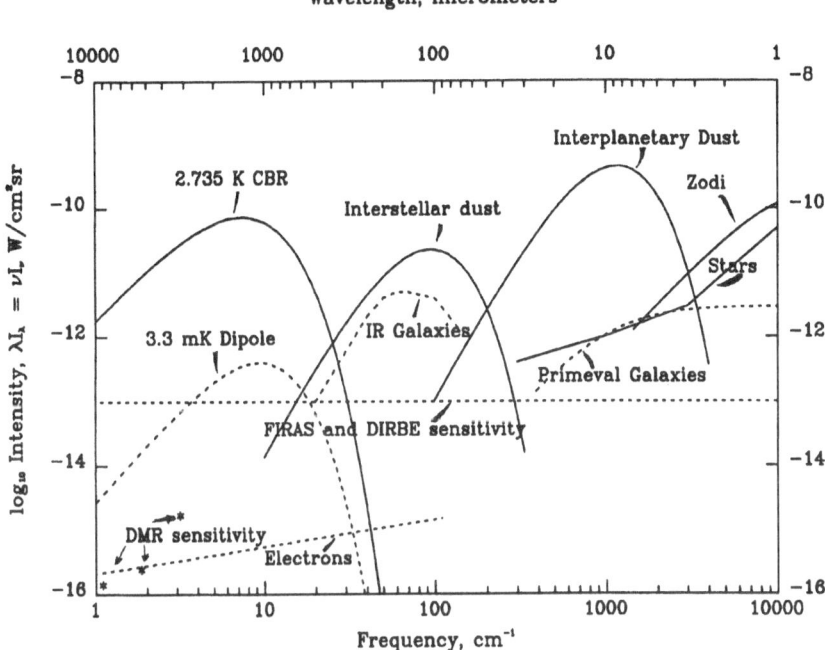

Fig. 1. Schematic of diffuse radiation fields observable by the COBE. Except for the CBR and dipole, these curves are not based on COBE measurements, which are still being analyzed.

detailed models of these local sources, and a subtraction of them to reveal the true cosmological background. The *COBE* sensitivities per beamwidth and spectral element are also shown on Fig. 1, and are well below the limitations set by local astrophysics over most of the bands.

2. Spacecraft and Orbit

The *COBE* was launched by NASA on a Delta rocket on November 18, 1989 from the Western Space and Missile Center at Lompoc, California into a nearly polar orbit over the terminator region of the Earth. With an altitude of 900 km and an orbital inclination of 99°, the plane of the orbit precesses at one revolution per year, chosen to follow the apparent motion of the Sun and keep the orbit over the terminator. In this orbit, the spacecraft can be oriented to point away from the Earth and approximately perpendicular (control range is 90° to 98°) to the Sun for the duration of the mission. For two months around the summer solstice, the satellite passes through the Earth's shadow near the South Pole for up to 18 minutes in each 103 minute orbit, and for a similar period near the North Pole the Earth limb can rise a few degrees above the top of the instrument shield and illuminate the instruments.

The three instruments are all located inside a deployable shield in the top half of the spacecraft, whose upper edge after deployment is roughly coplanar with the instrument apertures. The shield protects them from heat and light from the Earth and Sun, from terrestrial radars, and from the spacecraft telemetry antenna located at the bottom of the *COBE*. The inner surface of the shield is covered with multilayer insulation and reaches a temperature of 150 K. The *FIRAS* and *DIRBE* instruments are located inside a helium cryostat in the center of the shield, and the *DMR* receivers are located in three boxes mounted around the outside of the cryostat. The cryostat is similar to that used on the Infrared Astronomical Satellite *(IRAS)* and carried 600 liters of superfluid helium at launch, sufficient to last for one year after launch. The temperature of the helium in orbit is 1.4 K, while the *FIRAS* and *DIRBE* instruments are heated to 1.5 K by their cryogenic power dissipation of 22 mW. The helium evaporates through a porous plug which separates the liquid and gas phases, and the efflux gas exits through a vent oriented towards the Earth along the spacecraft spin axis. Because the axis is about 97° from the velocity vector, there is a component of the thrust from the helium vent along the orbit vector which lowers the orbit by $\sim$ 30 m/day.

The spacecraft spins at 0.8 rpm about its symmetry axis, to scan the *DMR* and *DIRBE* beams rapidly around the sky. Attitude control is provided by two large counter-rotating momentum wheels to create the spin, and three small momentum wheels with axes perpendicular to the spacecraft axis for three axis fine control. Sun and Earth sensors provide basic aspect information ($3\sigma < 1°$) and electromagnetic control bars exert torque against the Earth's magnetic field to keep the net angular momentum near zero. Gyroscopes provide fine resolution aspect information and are used by the attitude control servos for rate damping. Star sightings by the *DIRBE* instrument are used to calculate fine aspect solutions good to a few arcminutes. The spacecraft is powered by three solar cell panels which provide over 1000 watts of power in full sun, and Ni-Cd storage batteries provide power during the shadow passages. Data are stored on two tape recorders and played back daily to a ground station at the Wallops Flight Facility in Virginia.

Commands are loaded to the onboard memory daily through the Tracking and Data Relay Satellite, and correct operations are verified 7 times per day through the same system.

3. FIRAS

The purpose of the *FIRAS* is to compare the spectrum of the CMBR with that of a precise blackbody, to enable the measurement of small deviations from a Planckian spectrum.

The *FIRAS* instrument covers two wavelength ranges, from 0.1 to 0.5 mm and from 0.5 to 10 mm. It has a 7° diameter beamwidth, established by a non-imaging parabolic concentrator, which has a flared aperture to reduce diffractive sidelobe responses. The instrument is calibrated by a full beam, temperature-controlled external blackbody, which can be moved into the beam by command. The *FIRAS* is the first instrument to measure the background radiation and compare it to such an accurate external full beam calibrator in flight. The spectral resolution is obtained with a polarizing Michelson interferometer, with separated input and output beams to permit fully symmetrical differential operation. One input beam is the sky or full aperture calibrator while the second input beam is provided by an internal temperature controlled reference blackbody, with its own parabolic concentrator. Both input concentrators and both calibrators are temperature controlled and can be set by command to any temperature between 2 and 25 K. The standard operating condition is for the two concentrators and the internal reference body to be commanded to match the sky temperature, thereby yielding a nearly nulled interferogram which reduces almost all instrumental errors to negligible values.

The instrument's spectral resolution is limited by the maximum stroke length to 0.2 cm^{-1} in the long wavelength channels, and by the microprocessor buffer size and telemetry to 0.2 cm^{-1} in the short wavelength channel. The rms sensitivity is νI_ν $< 10^{-9}$ W/m^{2}sr per $7°$ field and per 5% spectral resolution element after one year of operation. There are four large area $(0.5$ cm$^2)$ composite bolometer detectors, two on each output of the spectrometer. The separation of long and short wavelengths is accomplished by a capacitive grid dichroic filter.

The external calibrator determines the accuracy of the instrument. It is a re-entrant cone shaped like a trumpet mute, made of Eccosorb CR-110 iron-loaded epoxy. The angles at the point and groove are $25°$, so that a ray reaching the detector has suffered 7 specular reflections from the calibrator. The calculated reflectance for this design, including diffraction and surface imperfections, is less than 10^{-3} from 0.5 to 5 mm. The instrument is calibrated by measuring spectra with the calibrator and all the other controllable sources within the instrument held at a sequence of different temperatures. The overall responsivity and the emissivity of each source can be determined relative to the external calibrator by solving a set of coupled linear equations.

The preliminary results of the FIRAS have been given by Mather et al. (1990) and may be summarized as follows. The intensity of the background sky radiation is consistent with a blackbody at 2.735 ± 0.06 K, and deviations from this blackbody at the spectral resolution of the instrument are less than 1% of the peak brightness. The quoted uncertainty in temperature is due an uncertainty in the thermometer calibration, and we expect to further refine the calibration by additional tests. The deviations can be fitted to the Sunyaev-Zel'dovitch form for Comptonization, giving a limit on the y parameter of $|y| < 10^{-3}$ $(3\ \sigma)$. They can also be fitted to a Bose-Einstein distribution with a chemical potential, giving a limit of $|\mu| < 10^{-2}$ $(3\ \sigma)$. There is a strong limit on the existence of a smooth hot intergalactic medium; it can contribute less than 3% of the X-ray background radiation even at a reheating time as recent as $z = 2$. There is no evidence of a distortion of the spectrum, such as that reported by Matsumoto et al. (1988a), and the measured temperature is consistent with previous reports.

The variation of the spectrum with position in the sky as measured by the FIRAS is dominated by the dipole pattern of the Doppler shift of the background temperature, plus a difference in the interstellar dust emission. A preliminary analysis of three weeks of data taken near the Galactic plane shows exactly such a variation. A more precise determination of the spectrum of the dipole was made by calculating the average spectra in two large circular regions of the sky each of angular diameter $60°$, one centered at $(\alpha, \delta) = (11^h.1, -6°.3)$ and the other at $(23^h.1, 6°.3)$, which lie along opposite ends of the dipole axis. The difference between these spectra can be fit extremely well by the difference of two blackbodies, and is consistent with a peak dipole amplitude of 3.3 ± 0.3 mK and the assumed dipole direction.

4. DMR

The Differential Microwave Radiometers are designed to measure the anisotropy of the cosmic background radiation at a set of wavelengths optimized to separate Galactic emissions from the CMBR. The DMR includes two radiometers at each of 3 frequencies (31.5, 53, and 90 GHz, or 9.6, 5.7 and 3.3 mm wavelength). The receivers are standard two-horn Dicke-switched mixer-preamplifier devices, with ferrite switches operated at 100 Hz. The IF bandwidths are 0.5, 0.8, and 0.8 GHz respectively, and the radiometer sensitivities are 43, 42, 15, 16, 28, and 19 mK-sec$^{1/2}$ for the six receivers. The design has been described by Smoot et al. (1990) and the calibration by Bennett et al. (1991).

Two corrugated-horn antennas feed each receiver. Each horn defines a $7°$ FWHM Gaussian beam, and the horns are oriented $30°$ from the spin axis of the spacecraft and $60°$ from each other. As the spacecraft spins and pitches in its orbit, and the orbit precesses during the year, all possible pairs of pixels on the sky $60°$ apart are compared. A least-squares fit to these differences is made to recover a relative map of the sky; the DC term is lost in the differential measurement. In addition, the least square solution is used to estimate or place limits on certain potential sources of systematic errors. Beam switching between two non-fixed points on the sky gives twice as much effective observing time as a comparable single horn receiver switching between a fixed reference and the sky.

The receivers operate within protective thermal enclosures and RF shields to provide stable operating conditions. No evidence has yet been uncovered to indicate effects due to RF interference from ground radars, other satellites, or the COBE transmitter. The thermal stability of the instrument is excellent (<0.01 K/day drift in physical temperature, <0.001 K/day drift in instrument offset). The 53 and 90 GHz receivers operate at 140 K for reduced detector noise, and the 31.5 GHz receivers operate at 300 K.

Preliminary maps of the sky made with all 6 receivers show that the expected receiver sensitivity has been achieved. The maps represent 120 days of observation, and over the two years of operations we expect more than a factor of two improvement in the random noise. No obvious structures are visible in the maps other than the expected Galactic plane emissions, and some small regions which included the moon (these will be removed in subsequent analysis). The Galactic sources show the expected variation in intensity with frequency. Analysis is underway to quantify the limits that can be placed on anisotropies and Galactic power law indices. The preliminary data are consistent with previous values from balloon and satellite work. We find the amplitude and direction of the dipole as 3.3 ± 0.2 mK towards $(\alpha, \delta) = (11^h.2 \pm 0.^h2, -6° \pm 2°)$ expressed as equivalent thermodynamic temperature variation of a blackbody spectrum. All three *DMR* frequencies and the *FIRAS* results are consistent with the Doppler shift interpretation of the dipole anisotropy.

5. DIRBE

The purpose of the Diffuse Infrared Background Experiment is to conduct a definitive search for an isotropic cosmic infrared background radiation. Such a background is expected to be produced by the cumulative emissions of pregalactic, protogalactic, and galactic systems. The experiment approach is to obtain absolute brightness maps of the full sky in 10 photometric bands (J, K, L, and M; the four *IRAS* bands at 12, 25, 60, and 100 μm, and 120–200 and 200–300 μm bands). In order to facilitate discrimination of the contribution from the bright foreground due to scattering and emission from interplanetary dust, linear polarization is also measured in the J, K, and L bands, and all celestial directions are observed hundreds of times at all accessible solar elongation angles (depending upon ecliptic latitude) in the range 64° to 124°.

The *DIRBE* instrument is an absolute radiometer, utilizing an off-axis folded Gregorian telescope with a 19 cm diameter primary mirror. The optical configuration is carefully designed for stray light immunity, utilizing both a secondary field stop and a Lyot stop, super-polished primary and secondary mirrors, a reflective forebaffle, extensive internal baffling, and a complete light-tight enclosure of the instrument. The instrument measures the absolute sky brightness by chopping between the sky and a zero-flux internal reference at 32 Hz using a tuning fork chopper. Instrumental offsets can be measured by closing a cold shutter located at the prime focus. All spectral bands view the same instantaneous field-of-view, 0.7° × 0.7°, oriented at 30° from the *COBE* spin axis. This allows the *DIRBE* to modulate solar elongation angles by 60° during each rotation, and to sample fully 50% of the celestial sphere each day. A highly-reproducible internal reference source can stimulate all detectors when the shutter is closed to monitor the stability of instrument response.

At the present time, the *DIRBE* has mapped the full celestial sphere, though all available elongations will not be covered until the end of May. The data are of very high quality, showing good sensitivity, stability, and linearity, with few artifacts other than those induced by energetic particles in the South Atlantic Anomaly. Data processing to date has been limited to quick-look checking of daily data dumps. This checking includes creation of daily sky maps. Figure 2 shows an image created by combining J, K, and L-band data from several weeks during the first 4 months of the mission. At these wavelengths the sky is dominated by Galactic starlight, with a lesser contribution from zodiacal scattered light.

To meet the cosmological objective of the *DIRBE*, the foreground light from interplanetary and galactic sources must be discriminated from the total observed sky brightness. This task will require extensive careful modelling, which is just beginning. A conservative upper limit on extragalactic light is the total observed brightness in a relatively dark direction, such as the ecliptic poles. Figure 3 presents the infrared sky brightness toward the south ecliptic pole in all 10 *DIRBE* bands

Fig. 2. DIRBE map of 90° sector of the Galaxy, combining J, K, and L bands.

based upon a preliminary calibration. As expected, the faintest foregrounds occur at 3.4 μm (L-band), in the minimum between interplanetary dust scattering of sunlight and re-emission of absorbed sunlight by the same dust, and longward of 100 μm, where interstellar dust emission begins to decrease. Through careful modelling, we hope to be able to discriminate isotropic residuals at a level as small as a percent of the foregrounds. These near-infrared and submillimeter 'windows' will allow the most sensitive search for the elusive cosmic infrared background.

The *DIRBE* total brightnesses at J, K, L, and M band are similar to those reported by Matsumoto *et al.* (1988b), though quantitative statements must await comparison of observations made in the same direction on the same day number of the year (to get the same interplanetary dust contribution). A similar conclusion is reached when our data at 100, 150, and 240 μm are compared with the submillimeter data of Matsumoto *et al.* (1988a).

We have made a more detailed comparison at several points on the sky with *IRAS* total brightness observations at 12, 25, 60, and 100 μm. By using several points, we can distinguish zero-point offsets and gain differences. We find evidence for zero-point differences, evidently not constant in time, of a few

MJy/sr at all of these wavelengths. These are most significant (as a fraction of

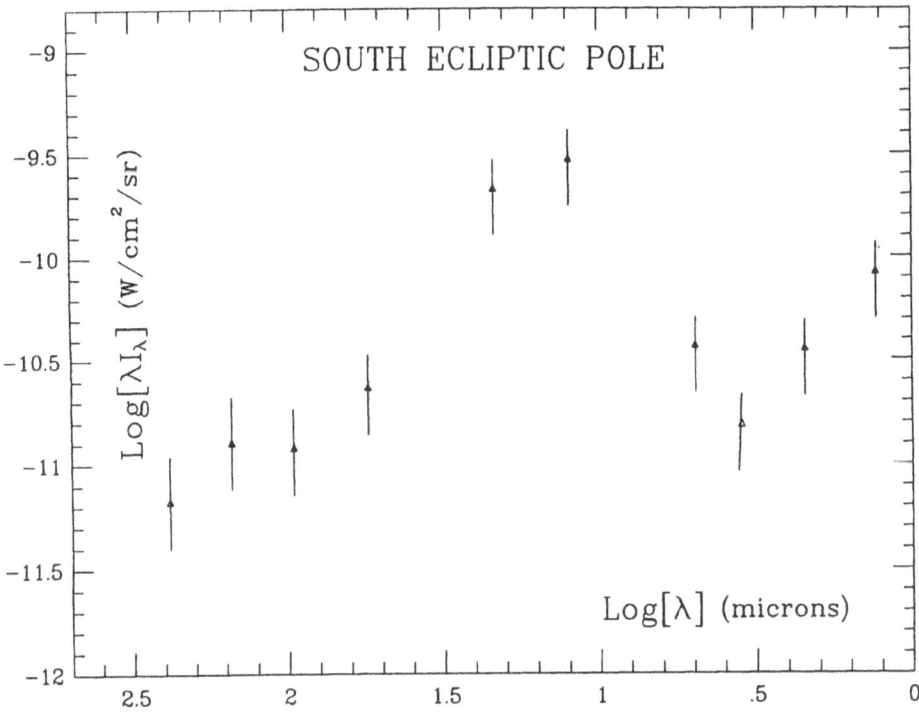

Fig. 3. South Ecliptic Pole brightness measured by DIRBE.

total brightness) at the longest wavelengths.

We also find that the *IRAS* DC gain at 60 and 100 μm is substantially too high (see the *IRAS* Explanatory Supplement, pp. IV-9 and IV-10 for a discussion of the *IRAS* DC response). The combined effect of these differences is particularly large at 100 μm toward faint sky regions: the *IRAS* ecliptic pole brightness is nearly 3 times brighter than that reported here. Since the *IRAS* gain depends on the source angular scale (scan-modulated instrument) and on the source brightness, no simple conversion between the data from the two experiments can be defined.

6. Summary and Conclusions

The *COBE* has mapped nearly the entire sky with two beam sizes at many wavelengths from 1μm to 1 cm. Its performance in orbit meets or exceeds the design goals, and the maps show the expected local sources. The detailed analysis of the data to model and remove these local foreground sources is just beginning, but limits on the cosmic $y < 10^{-3}$ and $\mu < 10^{-2}$ (3 σ) distortions of the 2.735 ± 0.06 K background spectrum have been obtained. The dipole anisotropy has the expected shape for a Doppler shifted blackbody as measured by both *DMR* and *FIRAS*, and the dipole amplitude and direction are consistent with previous measurements.

The extensive, high-quality *DIRBE* data promise to allow a very sensitive search for cosmic infrared background radiation.

Acknowledgements

It is a pleasure to acknowledge the vital contributions of all those at GSFC who devoted their efforts to making this challenging mission not only possible but enjoyable as well. The National Aeronautics and Space Administration/Goddard Space Flight Center (NASA/GSFC) is responsible for the design, development, and operation of the Cosmic Background Explorer (*COBE*). GSFC is also responsible for the development of the analysis software and for the production of the mission data sets. The *COBE* program is supported by the Astrophysics Division of NASA's Office of Space Science and Applications.

References

Bennett, C. L., *et al.* : 1991, in preparation
Bond, J. R., Carr, B. J., and Hogan, C. J.: 1986, *Ap. J.* **306**, 428-450
Gulkis, S., Lubin, P.M., Meyer, S.S., and Silverberg, R. F.: 1990, *Sci. Am.* **262**(1), 132-139
Mather, J. C.: 1982, *Opt. Eng.* **21**(4), 769-774
Mather, J.C., Cheng, E.S., Eplee, R.E., Jr., Isaacman, R. B., Meyer, S. S., Shafer, R. A., Weiss, R., Wright, E.L., Bennett, C. L, Boggess, N. B., Dwek, E., Gulkis, S., Hauser, M. G., Janssen, M., Kelsall, T., Lubin, P. M., Moseley, S. H. Jr., Murdock, T. L., Silverberg, R. F., Smoot, G. F., and Wilkinson, D. T.: 1990, *Ap. J.* **354**, L37-L41
Matsumoto, T., Hayakawa, S., Matsuo, H., Murakami, H., Sato, S., Lange, A. E., and Richards, P. L.: 1988a, *Ap. J.* **329**, 567-571
Matsumoto, T., Akiba, M., and Murakami, H.: 1988b, *Ap. J.* **332**, 575
Smoot, G. S. *et al.*: 1990, *Ap. J* **360**, TB

THE STATUS OF THE DIRBE INSTRUMENT ON THE COBE

M.G. HAUSER
NASA/Goddard Space Flight Center
T. KELSALL, H. MOSELEY, R. SILVERBERG, T.L. MURDOCK
General Research Cooperation, Danvers, Massachusetts
J.C. MATHER
NASA/Goddard Space Flight Center
G. SMOOT
University of California at Berkeley
R. WEISS
Massachusetts Institute of Technology
and
E.L. WRIGHT
University of California at Los Angeles

Abstract. The Diffuse Infrared-Background Experiment (DIRBE) on the Cosmic Back-ground Explorer (COBE) satellite is a 10-band absolute photometer covering the wave-lengths 1–300 microns using photovoltaic, photoconductive, and bolometric detectors. The input is via a 19-cm, off-axis, highly-baffled Gregorian telescope, with the detectors lo-cated at a pupil plane so they share the same field of view (0.7×0.7 degrees). The whole assembly is mounted inside a 1.4 K super-fluid, liquid-He dewar, which is shared with the Far Infrared Absolute Spectrometer (FIRAS) instrument. Each day half of the sky is surveyed, as the line-of-sight of the DIRBE is canted 30 degrees to the COBE spin axis. The whole sky is fully observed in 6 months, as the spin axis precesses at about 1 degree per day. At present each sky pixel has been observed at least once. The basic findings on the general brightness of the sky – Zodiacal light and galaxy – are provided, as well as a synopsis of the advantages and disadvantages associated with a space-borne observatory. The relationship of our experience and findings with respect to possible future missions and their scientific goals is presented.

Y. Kondo (ed.), Observatories in Earth Orbit and Beyond, 19.
©1990 *Kluwer Academic Publishers.*

THE AUTOMATIC SPACECRAFT *GRANAT*

R. SUNYAEV
Glavkosmos USSR, G.N. Babakin Center

Translated by Ben Charny, Jet Propulsion Lab, Section 331

1. Cooperation

The participants in the project are:
- USSR, by developing the *Granat* spacecraft (Scientific and Production Association S.A. Lavochkin, G.N. Babakin Research Center) and the instruments in the scientific equipment packages
- ART-P, ART-S, KONUS-V, 'Podsolnukh' (Sunflower), KS-18M (IKI of the Academy of Sciences of the USSR, Leningrad Institute of Physics and Technology of the Academy of Sciences of the USSR, Nuclear Physics Science and Research Institute of the Moscow State University);
- France (CNES), by developing and delivering the gamma ray telescope SIGMA, the high energy spectrometer Phoebus, a memory device with 128 Mbit capacity, and a system for determining the current orientation of the telescope (star sensor);
- Denmark (Danish Institute for Space Research), by developing the wide-angle X-Ray telescope VOTCh;
- Bulgaria (TsLKI Bulgarian Academy of Sciences), by developing and delivering certain control instruments for the 'Podsolnukh' apparatus.

2. Purpose and Scientific Tasks

The spacecraft *Granat* with its scientific instruments package is designed to conduct detailed studies of compact and diffuse space sources in X-Rays and soft gamma rays.

The main scientific tasks to be solved by the scientific instruments package are:
- producing images of the parts of the sky and localizing sources of space radiation in the studied parts of the sky with high accuracy;
- spectral studies in time of point sources;
- measuring linear polarization of X-Ray sources;
- patrol type monitoring of the sky in order to discover and study variable sources of radiation (bursts).

The *Granat* spacecraft will be launched on December 1, 1989 from the Baikonur (Tyuratam) launch complex.

3. Ballistic Data

The launching of the spacecraft into orbit is done using a two impulse scheme. First, the launch rocket provides for the spacecraft launch into intermediate support orbit of the Earth satellite with the following nominal parameters:
- maximum altitude, 2009 km

Y. Kondo (ed.), Observatories in Earth Orbit and Beyond, 21–25.
© 1990 *Kluwer Academic Publishers.*

- minimum altitude, 166 km
- inclination, 51.5 degrees
- rotation period, 107 min
- longitude of the ascending node, 9.3 degrees
- perigee argument, 119.7 degrees
- argument of the latitude of the insertion point, 119.7 degrees
- eccentricity is 0.123

The transfer of the spacecraft from the supporting orbit to the working Earth orbit is accomplished during the first revolution of the supporting orbit.

The working orbit has the following parameters:
- altitude in perigee, 2000 km
- altitude in apogee, 20,000 km
- inclination, 51.5 degrees
- rotation period, 4.05 days

The time the spacecraft spends in the Earth's shadow during the insertion part does not exceed one hour.

4. Main Characteristics of the Spacecraft

Mass of scientific equipment, 2,146 kg

Energy range of studied radiation sources, 2 − −40 MeV

Eight scientific experiments

Average power consumption by the scientific equipment on working parts of orbit, 350 Watt

Length of the scientific observation session, 24 hours

Orientation type: triaxal (constant solar-star)

Angular errors of the axes associated with the spacecraft coordinate system:

− constant error in angle, along the X-axis ca. 15 arcmin, along the Y-axis ca. 10 arcmin, along the Z-axis (toward the Sun) ca. 10 arcmin.

− maximum error in angular stabilization X: ca. 30 arcmin, Y: ca. 22 arcmin and Z: ca. 30 arcmin.

Main stars in orientation system: Sirius, Canopus, Vega, Rigel, Capella

Accuracy in determining current object orientation, 1 arcmin

Time the spacecraft is allowed to stay in the shadow, 3 hours

Total number of communication sessions with Earth: 200

Active lifespan of the spacecraft, 8 months

Observation Region: the entire sky

Capacity of the scientific equiment memory device, 128–200 Mbit

Maximum rate of transmission for scientific information, 65 kilobaud

5. Spacecraft Description

The *Granat* spacecraft consists of the orbital module and the scientific instruments package.

The base of the orbital module consists of the sequentially located toroidal instrument compartment and the support cylinder connected by a coneshaped spacer.

The toroidal instrument compartment is hermetically sealed. It contains functionally important spacecraft systems, providing the operations in Earth's orbit: radio complex, telemetry system, autonomous orientation and stabilization system, electric power system, elements of the thermal management system, blocks of electric automation. A block of star sensor is located on the outside of the toroidal compartment, on the side of the spacecraft that is constantly facing the Sun. It consists of photoelectric sensors for orientation by the Sun and stars.

The support cylinder is divided in two by an internal partition, perpendicular to its lengthwise axis. The top part constitutes a hermetically sealed compartment housing scientific equipment electronics.

The spacecraft's power source consists of three silicon solar panels. Two panels each having two folding sections, are located on trusses attached to the support cylinder and are symmetric in relation to the lengthwise axis of the spacecraft. Before insertion of the spacecraft into the working orbit, they are in a folded position and open up only after the separation of the IV stage of the rocket booster. The third solar panel is located in a fixed position on the support cylinder. The total area of solar panels is 8 square meters.

The narrow-angle transmit/receive antennas, working in decimeter and centimeter bands, provide stable radio communication with Earth regardless of the spacecraft's orientation. They are located on the orbital module.

Jet nozzles using compressed nitrogen provide angular positioning of the spacecraft during orientation and stabilization in space. The jet nozzles are located on the unfolding solar panels (which work using a momentum scheme) and on the toroidal equipment compartment (using a force scheme). Nitrogen is kept under high pressure in spherical tanks which are located outisde and in the lower part of the support cylinder.

Radiators are installed in the orbital module to sustain the necessary temperature in the hermetically sealed instrument compartments.

The radiator-heater is located on the toroidal instrument compartment, its surface constantly looking towards the Sun. The radiator-cooler is located on the support cylinder on the antisolar side. Both radiators are connected via gas pipes to the hermetically sealed instrument compartments, thus forming an active closed gas circulation system of thermal regulation.

On the outside, the spacecraft is covered by a multilayer thermal insulation of the screen-vacuum type. The insulation does not cover the windows which are for optical orientation sensors and the working surfaces of some scientific instruments that have special requirements.

The location of the instruments on the orbital module is based on the fact that functionally the majority of the instruments of the scientific equipment can be divided into two basic groups: narrow field-of-view telescopes (SIGMA, ART-P, ART-S) to observe stationary sources of cosmic radiation; and survey detectors (KONUS-V, Phoebus, VOTCh) to register and study variable cosmic radiation sources, bursts in particular.

The gamma telescope SIGMA is attached to the upper flange of the support cylinder through a coneshaped spacer. Its visor line coincides with the lengthwise axis of the spacecraft. Telescopes ART-P and ART-S are located on stationary bases on both sides of the SIGMA telescope, their axes coinciding with the axis of SIGMA. Such telescopes positioning allows simultaneous studies of cosmic radiation in an extremely wide energy range. A two-level movable platform, that contains scientific instruments of the 'Podsolnukh' experiment, is located on a base near the ART-S telescope. Its position is such that at any time it can point scientific instruments toward any location in the sky.

Survey detectors of the scientific equipment complex KONUS-V, Phoebus, and VOTCh are located on the orbital module providing for the possibility of full spherical survey of the sky without their field of view being obstructed by elements of the spacecraft. Some survey detectors are positioned on the stationary bases near ART-P and ART-S telescopes; some detectors are located on the toroidal instrument compartment. Besides the scientific instruments mentioned above, the charged particle monitor KS-18M is located on the toroidal instrument compartment.

The total length of the spacecraft is 6.5 m; the solar panel span is 8.5 m; the weight is 4.4 tons.

6. Complex of Scientific Instruments

The scientific equipment complex is subdivided into two parts. The first one includes with relatively narrow field of view, such as telescopes designed mainly to study almost stationary sources. The second part includes burst-type instruments designed for detailed studies of the sources of gamma-bursts and sources of X-Ray bursts. Splash instruments permit having the entire sky to be in the field of view.

6.1. TELESCOPES

Three telescopes with approximately the same information capacity form the base of the scientific instrument complex.

The French gamma telescopes SIGMA can image a 7.3 deg × 7.3 deg part of the sky and localize point sources of X- and soft-gamma radiation. It has a position-sensitive detector based on a NaI crystal, coding aperture, and is sensitive in the energy range from 30 to 2,000 keV. The telescope has a large effective surface of 1,024 cm^2, and, as a result, high sensitivity. The instruments spectral resolution is as close as theoretically possible for the given detector type.

The Soviet astronomical position-sensitive X-Ray telescope ART-P can image a part of the sky and localize point sources of X-Ray radiation with an accuracy of approximately one angular minute. It has position-sensitive detectors that are multiwire proportional high pressure chambers with coding aperture. The telescope is sensitive in the 3 to 100 keV range. The effective surface of the telescope is 2,400 cm^2. The instrument allows for the separation of sources that fall in its 1.8 deg × 1.8 deg field of view, the study of their spectra, and their behavior in time. The detector type used allows detailed spectral studies of the sources.

The Soviet astronomical spectral X-Ray telescope ART-S is equipped with detectors that are multiwire position sensitive chambers with an effective surface of 2,400 cm^2. The telescope is sensitive in the 3 to 100 keV range. Its field of view is

2 deg × 2 deg. The instrument is designed for detailed spectral and temporal behavior studies of relatively bright X-Ray sources.

These three main instruments are supposed to form an orbital observatory, sensitive in a wide spectral range from 3 keV to 2 MeV. The on-board availability of the French memory device with 128 Mbit capacity and of the Soviet memory device with 200 Mbit capacity allows 24-hour-long observations. Together with a large detector surface area, the observatory breaks records among the implemented and currently under development space projects in sensitivity in the studied range and in the breadth of scientific tasks to be solved.

6.2. BURST INSTRUMENTS

The interest in one of the most interesting and unsolved phenomena, gamma bursts, resulted in *Granat* becoming the largest goal-oriented project to study gamma bursts in the world.

The splash equipment package includes KONUS, Phoebus, Podsolnukh, and VOTCh.

The gamma burst detector KONUS consists of seven NaI(Tl) crystal-based detectors. The instrument is sensitive in 20 keV to 2 MeV range, can conduct detailed spectroscopy, study burst behavior in time, and perform rough burst localization with an accuracy of 1 deg to 2 deg.

The high energy spectrometer Phoebus consists of six detectors based on GeBi crystals and is sensitive in the range from 200 keV to 40 MeV. It has good spectral resolution in that range and is irreplacable in studying nuclear gamma lines. The opportunity to study hard tails of gamma bursts is of enormous interest.

The Podsolnukh instrument is a complex of narrov field of view telescopes conducting studies in the range from 2 to 25 keV It is located on a turnable platform together with a block for registering video information.

The turnable platform has to support real-time automated turns and the pointing of the installed instruments towards the presumed location of the source of gamma burst. Movement is based on signals from KONUS-V. The turn speed of the platform is 90 deg/sec. Furthermore, there are plans for studying the spectral contents of the gamma burst and its temporal structure, conducting precise localization of the gamma burst source. The presence of the module for registering video information allows the source in the X-Ray band to be identified with the optical radiation from a corresponding astrophysical object.

The wide angle X-Ray telescopes VOTCh consists of 4 multilayer detectors based on CSI and NaI crystals. The detectors have rotating modulating collimators. The sensitivity range of the instruments is from 5 to 150 keV.

The instrument complex VOTCh registers X-Ray bursts, localizes them with an accuracy of up to 10 arcmin and studies the energy spectrum of the bursts and their temporal structure. It also provides patrol type monitoring of the sky in the X-Ray band, while watching bright stationary X-Ray sources.

The charged particle monitor KS-18M is designed to measure the flows of charged protons and helium nuclei with energies of more than 1 MeV and electrons with energies of more 50 keV in interplanetary space.

THE HIPPARCOS MISSION:
WILL IT BE A SCIENTIFIC SUCCESS?

M.A.C. PERRYMAN

Astrophysics Division, ESTEC, Noordwijk 2200 AG, The Netherlands

Abstract. The Hipparcos astrometry satellite was launched on 8 August 1989, and after spacecraft and payload commissioning, commenced the routine data acquisition phase on 26 November 1989. Having failed to reach its planned geostationary orbit, major revisions in the mission operations were made, and the post-launch expectations of the mission were strongly degraded with respect to the original goals – principally due to the greatly reduced observational efficiency (caused by the lack of ground station coverage) and the anticipated degraded mission lifetime (as a result of the high-energy particle degradation of the solar arrays in the geostationary transfer orbit).

The final astrometric accuracies attainable by the Hipparcos mission will be influenced by the spacecraft and payload performances on the one hand, and by fraction of useful data and mission lifetime on the other. It will be shown that the elemental observational measurements correspond very closely to the predictions, and the data recovery fraction now stands at around 60–70 per cent, so that the ultimate scientific value of the Hipparcos results will be tied directly to the satellite lifetime. A measurement duration of at least 18 months is mandatory if the astrometric parameters (parallaxes and proper motions) are to be decoupled through the data reductions procedures. A somewhat longer lifetime (2.5–3 years) is necessary in order to reduce the errors on the astrometric parameters to the astrophysically-significant accuracies of around 2 milli-arcsec.

It will be shown that the present indications of the satellite performances, and the significant progress already made in the data reductions, indicate that the difficulties of the 'revised' Hipparcos mission have been largely overcome, and that these target accuracies could still be achievable.

1. Introduction

The goals of the ESA Hipparcos space astrometry mission are to acquire the astrometric parameters (positions, parallaxes and proper motions) of about 120 000 pre-defined ('programme') stars to an accuracy of about 2 milli-arcsec to a limiting magnitude of about 9–12 mag (for the main experiment), as well as positions and BV photometry, to a lower precision, for the 400 000 or so stars down to a limiting magnitude of about 10–11 mag.

Following a nominal launch by Ariane 4 on 8 August 1989, the satellite's apogee boost motor failed to ignite, and the satellite was left, not in its intended 24-hr geostationary orbit, but in a 10-hr elliptical transfer orbit, with apogee 36 000 km and a perigee of 200 km. Early estimates of the satellite's useful lifetime were considered to be limited to an approximately 6-month lifetime of the solar arrays, dictated by the high radiation background in the given orbit. The satellite operations were further complicated by numerous other considerations, new to the revised orbit, and therefore, in the early days of the mission, poorly understood.

Y. Kondo (ed.), Observatories in Earth Orbit and Beyond, 27–33.

© 1990 *Kluwer Academic Publishers.*

TABLE I

Instrument parameters

Optics:	Telescope configuration	All-reflective Schmidt
	Field of view	0°.9 × 0°.9
	Separation between fields	58°
	Diameter of primary mirror	290 mm
	Focal length	1400 mm
	Scale at focal surface	6.8μm per arcsec
	Mirror surface accuracy	$\lambda/60$ rms (at $\lambda = 550$ nm)
Primary Detection System:	Modulating grid	2688 slits
	Slit period	1.208 arcsec (8.2μm)
	Detector	Image dissector tube
	Photocathode	S20
	Scale at photocathode	3.0μm per arcsec
	Sensitive field of view	38 arcsec diameter
	Spectral range	375–750 nm
	Sampling frequency	1200 Hz
Star Mapper (Tycho) System:	Modulating grid	4 slits perpendicular to scan 4 slits at ±45° inclination
	Detectors	Photomultiplier tubes
	Photocathode	Bi-alkali
	Spectral range (B_T)	$\lambda_{eff} = 430$ nm, $\Delta\lambda = 90$ nm
	Spectral range (V_T)	$\lambda_{eff} = 530$ nm, $\Delta\lambda = 100$ nm
	Sampling frequency	600 Hz

2. Problems Posed by the Revised Orbit

The major problems encountered in the operation of the Hipparcos satellite in its revised elliptical orbit were the following:

(1) Final orbital configuration: an early decision was required to utilise the on-board hydrazine propulsion system (originally foreseen for nutation correction and station acquisition) for a possible perigee raising manoeuvre. To minimise perturbing torques at perigee, the available hydrazine was used to raise the perigee to approximately 600 km, and a corresponding orbital period of 10h 40m. The final orbit achievable nevertheless suffers from the disadvantages of poor station coverage from a single ground station as well as the continuous interception of the Van Allen radiation belts each orbit.

Following the perigee raise, the solar panels and telescope baffle covers were deployed, and the spacecraft and payload subsequently commissioned. Commissioning proceeded without major difficulties, although on a longer timescale than that originally foreseen, due to the extended periods of satellite non-visibility im-

TABLE II
The members of the ESA Hipparcos Science Team

Name	Institute	Main Responsibilities
Prof. P.L. Bernacca	Asiago (I)	Data reductions (FAST)
Dr. M. Crézé	Besançon (F)	Input Catalogue (FAST)
Prof. F. Donati	Torino (I)	Data reductions (FAST)
Dr. M. Grenon	Genève (CH)	Input Catalogue, photometry
Prof. M. Grewing	Tübingen (D)	Tycho
Prof. E. Høg	Copenhagen (DK)	NDAC and TDAC Consortia Leader
Prof. J. Kovalevsky	Grasse (F)	FAST Consortium Leader
Dr. F. van Leeuwen	Cambridge (UK)	Data reductions (NDAC)
Dr. L. Lindegren	Lund (S)	Data reductions (NDAC)
Dr. H. van der Marel	Delft (NL)	Data reductions (FAST)
Mr. C.A. Murray	RGO (UK)	Data reductions (NDAC)
Dr. M.A.C. Perryman	ESA	Project scientist, ESA
Mr. R.S. Le Poole	Leiden (NL)	Instrument performances
Dr. C. Turon	Meudon (F)	INCA Consortium Leader

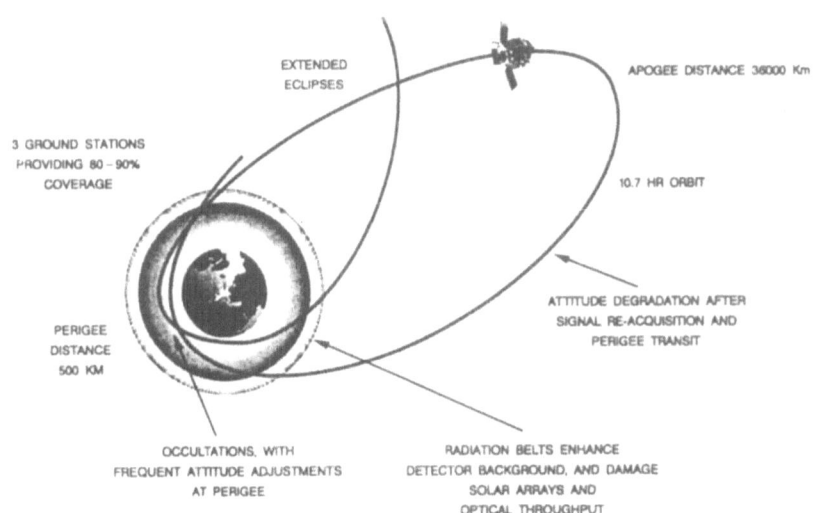

Fig. 1. The present Hipparcos orbit, and the problems associated with it.

plied by the single ground station coverage. The satellite and payload performances closely reflect the nominal, pre-launch predictions, which have been extensively reported (Perryman *et al.* 1989, *The Hipparcos Pre-Launch Status*, ESA SP-1111). In particular, the throughput of the entire optical/detection was within 10 per cent (actually better than) the budget prediction, and the instrumental chromaticity (the geometrical shift between stellar images of extreme star colours) is below 1 milli-arcsec. Signal modulation, geometrical distortion, and payload stability, were all within their respective specifications.

(2) **Solar array performance:** after the early pessimistic lifetime predictions, the on-board systems were configured to provide in-orbit measurements of the solar array voltage and current characteristics, and the orbital characteristics were used to estimate high-energy proton and electron fluencies responsible for the degradation of the solar arrays. In-orbit measurements and theoretical prediction are now in good agreement, and an estimated lifetime of approximately 3 years is now considered to be realistic. The probability of catastrophic component failure through radiation damage is less easy to quantify, and may ultimately limit the achievable lifetime.

(3) **Particle background radiation:** as well as degrading the performance of the solar arrays, systematically as a function of time, the $\sim$ 1 MeV electron fluence within the Van Allen belts also contributes to the instantaneous detector background, as a result of Cerenkov emission within the payload diotropic elements. At the level of the primary image dissector tube, the relatively small (30 arcsec diameter) instantaneous field of view limits the impact of this effect to a small loss (about 0.4 mag) in the limiting magnitude of the observations during the passage of the satellite through the radiation belts. For most purposes, this is a small effect. However, the larger field of the star mapper detector, used for the real-time attitude determination as well as for the Tycho experiment, significantly reduces the amount of data that can be used by the Tycho experiment, as well as that available for the real-time estimates of the satellite attitude (see below).

(4) **Perigee transits:** grouped into the category of problems arising during the periods around perigee transits are the higher radiation background (affecting in particular the real-time attitude determination), the extended earth occultation (when one or other of the telescope's two fields of view are occulted by the earth), the gaps in the ground station visibility, and the higher perturbing torques, resulting in modification to the precise attitude control strategy required to keep the satellite following its pre-defined scanning law.

(5) **Ground station coverage:** in the nominal mission, ground station coverage (telecommanding and telemetry) was to have been provided by a single ground station (at Odenwald, FRG) resulting in continuous satellite visibility for the ground station, and an overall useful observing time (after accounting for calibration, earth and moon occultations, and all other operational overheads) of between 90–95 per cent. In the revised orbit, the Odenwald station alone would have provided some 30 per cent useful data acquisition. Fortunately, it was possible to incorporate the ESA station at Perth, Australia, and the CNES station at Kourou, French Guiana, within weeks of the apogee motor failure. This brought the satellite observability to about 80 per cent, and the useful data fraction to about 55 per cent. Further

assistance by NASA has brought the Goldstone station, in Californis, into the network. As of mid-April 1990 the Goldstone station will bring the average satellite observability to about 93 per cent, the useful achievable data fraction to about 60-70 per cent. Very importantly, it will also reduce the present ground station coverage 'gaps' to a maximum of about 1.5 hours. This is expected to improve the preservation of the satellite real-time attitude estimation.

(6) **Eclipse passages:** unlike the geostationary orbit configuration, where maximum eclipse durations of about 5 per cent of the orbital period would habe been experienced, the present orbit results in eclipse durations of up to 15 per cent of the orbital period. This placed very severe constraints on the battery charging cycles, designed for the geostationary case. Fortunately, the largest eclipse duration expected over the first 3 years of the satellite lifetime, 105 min, was passed in mid-March with a margin of about 5 min on the battery power. This meant that no power saving procedures were necessary, and the scientific observations were able to proceed without interruption.

(7) **Real-time attitude determination:** the high detector background, the high perturbing torques at perigee and the long signal outages due to a combination of earth occultations and ground station 'gaps', have resulted in considerable difficulties in running and optimising the continuously evolving real-time attitude determination tasks. This now runs more-or-less routinely, with limited ground intervention. The cold gas propulsion system which, along with the star mapper, gyros, and on-board computer systems, is responsible for maintaining the satellite attitude, is slightly more inefficient than that foreseen for the nominal mission (due to the larger perturbing torques at perigee), although it is likely to sustain satellite operation for up to 5 years.

3. The Scientific Observations

Due to the global nature of the Hipparcos observations and reductions it has been neither desirable nor fortunately, necessary, to modify the observing programme. It is a great tribute to the work of the Input Catalogue Consortium to note that the observations have proceeded entirely nominally; not only for the 120 000 or so programme stars, but also for the 50 or so minor planets, planetary satellites (Titan and Europa) and the large amplitude variable stars. The broadband (H), Tycho bands $(B$ and $V)$ photometric measurements, and the astrometric residuals at the preliminary great-circle reduction level are entirely consistent with the ground-based compilation of their *a priori* values.

4. The Data Reductions

The operational configuration (changing ground stations, changing detector background levels, variations in the data quality as a function of satellite elevation, attitude perturbations after satellite signal re-acquisition, etc.) does mean that the burdens and complexity of the data interfaces (between ESOC and the data reduction teams) are increased over those foreseen. Although a source of frustration for the data reduction teams, these problems should not affect the eventual data

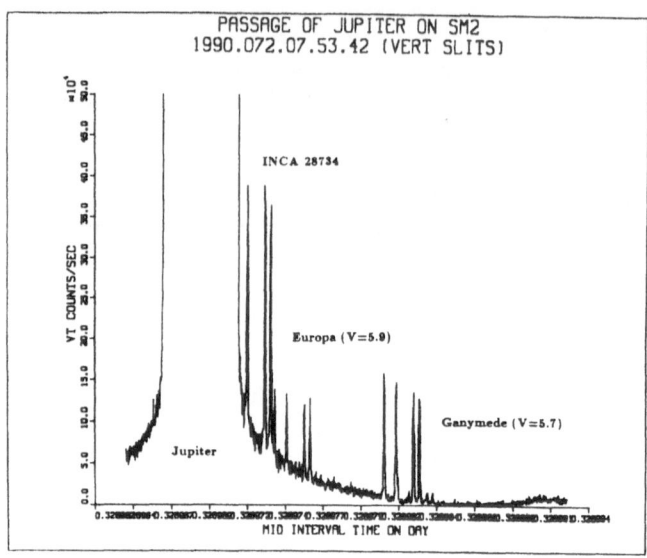

Fig. 2. Star mapper observations of Jupiter and its moons.

quality.

Although the 'mass' data treatments are not yet running routinely, the satellite data are now passing through all major elements of the data analysis chains, and dat from 12 hours of observations (approximately 1 Gbit) are now processed routinely to derive star abscissae on a reference great circle with a precision of 5–10 milli-arcsec. Instrument parameters are calibrated (as part of the ESOC real-time payload monitoring, as part of the FAST, NDAC and TDAC's routine data analysis, and within the FAST Consortium's 'first look' analysis facility) with sub-milli-arcsec precision, and are very stable. All reduction elements are in the process of becoming fully operational.

5. Final Accuracies

Simulations by the data reduction teams have shown that a mission duration of at least 18 months is necessary before the various astrometric parameters become separable. Should the satellite lifetime be less than 18 months, a significant improvement to the present astrometric reference frame will be made (individual positions will estimable to about 10 milli-arcsec) but with very limited astrophysical (parallax and proper motion) information. A progressively longer mission duration will lead to improvements in the precision of all five astrometric performance levels, the astrophysically-significant (and nominal target) accuracies of around 2 milli-arcsec should be achievable.

Present indications are therefore that the original astrometric (and hence scientific) goals could still be achievable, although there are, evidently, obvious uncertainties involved in such an extrapolation.

Acknowledgements

The rescuing of the Hipparcos mission was made possible by the skill and dedication of the many members of the operation team at ESOC, under the leadership of the Ground Segment Manager, Jozef van der Ha, and the Spacecraft Operations Manager, Dietmar Heger, as well as the project teams of ESA and the industrial prime contractor, Matra. The Agency is grateful for the continued guidance of the Hipparcos Science Team, as representatives of the scientific consortia involved in the Hipparcos project, who have worked since launch, and will continue to work, under difficult circumstances. The author acknowledges particularly the Input Catalogue Consortium and its leader, Dr. C. Turon, for providing a trouble-free operational and observing catalogue (the Input Catalogue), the FAST, NDAC and TDAC data reduction consortia, and their leaders Professors J. Kovalevsky and E. Høg, Dr. L. Lindegren for his early assessments of the accuracy implications for the revised mission, and the FAST Consortium who, through the work of Dr. J. Schrijver, established a first-look analysis facility which has greatly expedited the early understanding of the flight data.

THE INTERNATIONAL ULTRAVIOLET EXPLORER (IUE)

YOJI KONDO

Goddard Space Flight Center, Greenbelt, Maryland 20771, U.S.A.

Abstract. The International Ultraviolet Explorer (IUE) was launched into a geosyn-chronous orbit on 26 January 1978. It is equipped with a 45-cm mirror and spectrographs operating in the far-ultraviolet (1150–2000 Å) and the mid-ultraviolet (1900–3200 Å) wavelength regions. In a low-dispersion mode, the spectral resolution is some 6–7 Å. In a high-dispersion echelle mode, the resolution is about 0.1 Å at the shortest wavelength and about 0.3 Å at the longest. It is a collaborative program among NASA, ESA and the British SERC. The IUE is operated in real time 16 hours a day from NASA Goddard Space Flight Center near Washing ton, D.C. and 8 hours daily from ESA's Villafranca groundstation near Madrid, Spain. By the end of 1989, 1870 papers, using IUE observations, have been published in refereed journals. During the same period, over 1700 different astronomers from all over the world used the IUE for their research.

1. Introduction

The International Ultraviolet Explorer (IUE) was launched on a Delta rocket into an eccentric geosynchronous orbit on 26 January 1978. The IUE is a collaborative project of NASA, ESA and the British SERC (Science and Engineering Research Council). It is operated in real time 16 hours daily from NASA Goddard Space Flight Center in the suburb of Washington, D.C. and 8 hours a day from ESA's Villafranca groundstation near Madrid, Spain. The satellite observatory is equipped with a 45-cm aperture primary mirror and spectrographs operating in the far-ultraviolet (1150–2000 Å) and the mid-ultraviolet (1900–3200 Å) wavelengths. The IUE was originally designed for a 3- to 5-year mission. (Cf. Boggess et al. 1978)

From the outset, the IUE was planned as a guest observer facility. Astronomers from any country in the world may submit their proposals to use the IUE. Users have come to Goddard and Villafranca science operations centers from every continent of the world, with the inevitable exception of the Antarctica. In all, over 1700 different astronomers have come through the first twelve years of its operation. This represents a significant fraction of the astronomers of the world, particularly those in the U.S and Europe.

The total number of scientific papers from IUE observations published in refereed journals came to 1870 at the end of 1989. A list of refereed IUE papers is published from time to time in the IUE Newsletter.

Themes of research with the IUE have encompassed from solar system objects, such as planets, satellites, asteroids and comets, the interstellar and intergalactic media, and all manner of stars, including Cepheids, X-ray binaries, novae and su-perno vae, to external galaxies, including star burst galaxies, active galactic nuclei and quasars. The brightest star observed was Sirius at −1.4 magnitude and the faintest a 21st magnitude object. (Actually, Venus was observed at −4.0 magnitude

Y. Kondo (ed.), Observatories in Earth Orbit and Beyond, 35–40.

YOJI KONDO

TABLE I
SPACECRAFT

Mass	462 kg
Dimensions	
Length	417 cm
Width	145 cm
Main Body Shape	Octagonal
Orbit	
Semi-Major Axis	42162 km
Perigee	27616 km
Apogee	43953 km
Eccentricity	0.1937
Inclination	29.76 degrees
Mean Orbital Velocity	3.1 km/sec
Data Rates	
Downlink	2249.80 Mhz (S-Band); PCM Split
Phase Uplink	148.98 Mhz (UHF)
Ranging	136/86 Mhz (VHF)
Downlink Data Rates	40–12.5 Kbits/sec (uncoded or convolved)
Science Data Rates	Maximum rates 3 images per hour;
	1 image consists of 768 by 768,
	8-bit pixels (= picture elements)

but it is not very bright in the ultraviolet and is therefore not suitable for examining the dynamical range.)

2. Scientific Highlights

Highlights of research with the IUE have been reviewed in a multi-authored volume edited by Kondo et al. (1987 and 89) and in a review article by Kondo, Boggess and Maran (1989). The book contains 31 chapters but even that was not enough to cover ade quately the many astronomically significant results from the IUE. The proceedings of several IUE symposia, quoted in the latter article are also excellent sources of information. In the fol lowing, a few selected examples of IUE highlights will be briefly reviewed.

2.1. GALACTIC HALOS

IUE observations of hot stars in the Magellanic Clouds showed absorption due to C II and other neutral or low-ionization atoms and absorption due to C IV and other highly ionized atoms, which were caused by the gas outside the galactic disk. Galactic halo gases were thus discovered. Most of the gas seems to be relatively

TABLE II
INSTRUMENTS

Telescope	
Type	Ritchey-Chretian
Aperture	45 cm
Focal Ratio	f/15
Image Quality	2 arc sec
Spectrograph	
Type	echelle
Wavelength Ranges	
short	1150–2000 Å
long	1900–3200 Å
High Dispersion	resolving power = 10,000 or, 0.1 to 0.3 Å
Low Dispersion	about 7 Å
Apertures	3 arc sec and 10 × 20 arc sec

TABLE III
SPACECRAFT AND INSTRUMENT STATUS AS OF 1 JUNE 1990

Gyroscopes	originally 6* gyros, 2 operational gyros since 1985, 1-gyro control system developed
Slewing	3 reaction wheels in use; 1 back-up never used in orbit
Solar Panels	decreasing output 2–4 % a year
Batteries	2 fully functional batteries
Computers	2 onboard computers 8K standard configuration
Spectrograph Cameras	short wavelength primary – operational (SEC Vidicons)
	s.w. redundant – non-operational
	long wavelength primary – operational
	l.w. redundant – back-up at reduced sensitivity
Fine Error Sensors	FES No. 2 – operational
(or Star-Trackers)	FES No. 1 – operational, back-up

* The IUE was designed to be operated with a minimum of 3 gyro scopes; we started with 3 redundant gyroscopes for a projected 3 to 5 year mission. After the third gyroscope failure, we de veloped a 2-gyro spacecraft control system, which was placed in service after the fourth gyroscope failure in 1985. A 1-gyro control system has since been developed in case it is needed. A zero-gyro control system is a theoretical possibility that may yet be developed.

cool and exists within several hundred parsecs of the disk. The higher temperature gas exhibits different velocity distribution from that of the cooler gas and its scale height based on C IV observations is several kiloparsecs.

2.2. MASS LOSS FROM STARS

Early ultraviolet observations from rockets and previous satellites showed mass loss primarily from hot stars. The IUE has been used to observe just about all types of stars. The IUE's capability to observe much fainter stars at a high resolu tion has made it possible to extend such observations to all manner of stars; they are effectively all losing matter from its surface and that includes hot white dwarfs with relatively low surface gravity. Such universal mass loss from virtually all kinds of stars will affect the evolutionary processes in those stars and will influence the replenishment of the interstellar media.

2.3. MASS FLOW IN BINARY STARS

Until the advent of the satellite observatories that provided the ultraviolet spectra of interacting binary stars, the exact manner of mass flow in them has been unclear. Since the mass flow is governed by a hydrodynamic or a magnetohydrodynamic pro cesses further complicated by the three- body problem, theoretical models were not entirely adequate in predicting the evolutionary processes in interacting binaries. The IUE has made it possible not only to observe vastly more such binaries but, more importantly, to observe a number of binaries following its orbital phase. As a result, the wholly conservative mass flow idea that had currency in the Sixties through the mid-Seventies was shown to be wrong. According to the IUE data, some of the matter flowing out of one of the two components leaves the binary system while a fraction of the gas is apparently accreted by its companion creating a hot region. Such synoptic observations are also unraveling the physical processes involved in novae and other cataclysmic binaries.

2.4. COMETS

The IUE has been used every year to observe a number of periodic comets as well as newly discovered ones. One interest ing example of the latter type was the 1983 comet IRAS-Araki- Alcock, in which strong emission spectral features due to molecu lar sulfur were unexpectedly discovered. The comet's close approach to Earth apparently made their serendipitous detection possible. The sulfur molecular bands, never observed before, have not since been detected in any other comet. The presence of the sulfur molecules is thought to be closely tied to the proc esses in the protosolar system nebula in which comets were formed.

A prime example of periodic comets observed with the IUE was Halley's comet. The IUE was the first telescope to observe the comet from outside the Earth's atmosphere in 1985 and was also the last to do so in August 1986. Based on the OH emission lines observed near the comet's perihelion passage, Halley was dumping water into space at the rate about 10 tons per second. When the tragic accident of Space Shuttle Challenger made the Astro mission to observe the comet impossible, the IUE provided the complementary ultraviolet observations in support of the Halley flyby missions Giotto, Vega 1 and 2, Suisei.

2.5. SUPERNOVA 1987A

When the report of the discovery of SN1987A in the Large Magellanic Cloud reached Goddard on February 24, the IUE was the only telescope in the world capable of observing it immediately. When the first exposure was read down by telemetry, the supernova was still so bright in the ultraviolet that the next exposure had to be shortened substantially. It was bright enough for high resolution observations, and the spectra obtained provided first valuable information on the interstellar and intergalactic matter between the solar system and an object in the Large Magellanic Cloud.

The far-ultraviolet flux subsided by three orders of magnitude over the next four days, diminishing effectively to zero in short wavelengths and revealing the spectra of the two relatively faint hot stars in the same direction as the superno va. The in-depth analysis of the pre-supernova image of this area by Walborn et al. (1987) showed that there had actually been three stars present. The analysis of the IUE far-ultraviolet spectrum showing the two remaining stars enabled Sonneborn, Altner and Kirshner (1987) in the U.S. and Gilmozzi et al. (1987) in Europe to establish that it was the blue supergiant star SK-69°202 that exploded as a supernova. It was the first time in history that the progenitor of any supernova had been identified.

It was speculated that the blue supergiant had once been a red supergiant. After shedding its outer atmospheres it had become a blue supergiant. If this interpretation was correct, we should expect the illumination of the ejected envelope from the red supergiant. Emission lines of various ionized atoms, such as the N V lines and the forbidden lines of C III, which became observable in the summer of 1987 reached their peak intensities some four hundred days after the original brightening, placing the location of the gaseous shell previously ejected from the red supergiant at about one light year or less.

2.6. ACTIVE GALACTIC NUCLEI

Investigation of the active galactic nuclei took a step forward when it became possible to observe these objects in all wavelengths, from the X-ray to the ultraviolet, optical, infrared to radio wavelengths. Since these objects are variable, unless we obtain simultaneous synoptic observations at all wavelengths, it is impractical to formulate a viable model for the source of its immense energy. The first such observations were made in 1978 for MK 501 with HEAO-1, IUE, the 5-meter Hale Telescope at Palomar Mountain, the 2.5-meter at Mt. Wilson, the 92-meter telescope at Green Bank and the 48- meter telescope at Algonquin.

The latest such observations were made on NGC5548 with the IUE at a 4-day interval over an 8-month period. Some radio, infrared, optical and X-ray observations were also made during this period. These evenly-spaced observations yielded very good fluctuation power density spectra, which were quite steep during this epoch, being roughly proportional to the inverse square of the frequency. The largest amplitude variations were observed in the shortest wavelengths; the spectrum was harder when brighter in agreement with the thermal interpretation of the far ultravio let excess. Variability was much slower than what could be expected from the orbital period of the putative accretion disk. A close synchrony was observed between the ultraviolet and opti cal variations; the optical maximum lagged by about two days or less.

NGC 4151 was also observed synoptically over 62 days at frequent intervals. A prevailing variability timescale of 23 days was observed. The ultraviolet flux changes were as much as by a factor of 3.5, with doubling times of about 7 days and halving times of some 12 days. The shapes of the light curves were unusual.

3. Concluding Remarks

The IUE satellite has uniquely important roles to play in the era of new space missions such as the Hubble Space Telescope and ROSAT in addition to its many versatile normal roles as a guest observer facility. It is the only observatory currently operating in space suitable for systematic observations of the variabilities in astrophysically important objects, such as active galactic nuclei and novae. Its flexibility for scheduling makes it an essential component in any coordinated multi-wave length observing campaign requiring ultraviolet data.

The prospects for its continued operation over the next few years are good, thanks to the dedicated efforts of the Resi dent Astronomers, Telescope Operators, spacecraft support engi neers, especially the guidance and control system engineers at Goddard, and numerous other support personnel.

Processing of the spectra for the IUE Final Data Archive, in which its already good quality will be optimized, is scheduled to start in early 1991. This will further enhance the value of the IUE archive that contained more than 70,000 spectra in early 1990.

References

Boggess, A. et al.: 1978, *Nature* **275**, 372

Gilmozzi, R., Cassatella, A., Clavel, J., Fransson, C., Gonzales, R., Gry, C., Panagia, N., Talavera, A. and Wamsteker, W.: 1987, *Nature* **328**, 318

Kondo, Y. *et al.*: 1987, Kondo, Y. , Wamsteker, W., Boggess, A., Grewing, M., de Jager, C., Lane, A.L., Linsky, L., and Wilson, R. (eds.), *Exploring the Universe with the IUE Satellite*, D. Reidel Publishing Comp., Dordrecht

Kondo, Y., Boggess, A. and Maran, S.P.: 1989, *Ann. Reviews of Astron. & Astroph.* **27**, 397

Sonneborn, G., Altner, B. and Kirshner, R.P.: 1987, *Astroph. J.(Lett.)* **323**, L35

Walborn, N., Lasker, B.M., Laidler, V.G. and Chu, Y.-H.: 1987, *Astroph. J.(Lett.)* **320**, L41

X-RAY ASTRONOMY SATELLITE GINGA

F. MAKINO

Institute of Space and Astronautical Science, 3-1-1, Yoshinodai,
Sagamihara, Kanagawa 229, Japan.

Abstract. The X-ray astronomy satellite *Ginga* carries three scientific instruments, the Large Area proportional Counters (LAC), All Sky X-ray Monitor (ASM) and Gamma-ray Burst Detector (GBD). The LAC is the main instrument with an effective area of 4000 cm^2 giving it the highest sensitivity to hard X-rays so far achieved. *Ginga* observed about 250 targets up to the end of 1989.

1. Introduction

Ginga is the third Japanese X-ray astronomy satellite following *Hakucho* (Kondo *et al.* 1981) and *Tenma* (Tanaka *et al.* 1984). The fabrication, launching and operation of *Ginga* were carried out entirely by the Institute of Space and Astronautical Science (ISAS). General properties of *Ginga* have been described in the literature (Makino *et al.* 1987) and are summarized in Table 1. *Ginga* was launched on February 5, 1987 from Kagoshima Space Center of ISAS by the M-3S-II launch vehicle. The initial orbital parameters were a perigee height of 505.5 km, an apogee height of 673.5 km, an inclination of 31.1° and a period of 96.5 min. The altitude of *Ginga* has decreased rapidly since early 1989, because of high solar activity. The perigee and apogee height were 575.1 km and 471.0 km respectively in April 1990.

The total mass of the satellite is 420 kg of which 105 kg is for scientific instruments, and the rectangular main body measures 1 m × 1 m × 1.5 m with four deployable solar panels 1.7 m long and 0.75 m wide. A maximum power of 500 W can be generated from the solar panels and normal power consumption is about 150 W.

The scientific instruments are the Large Area proportional Counters (LAC), All Sky X-ray Monitor (ASM) and Gamma-ray Burst Detector (GBD). The LAC and GBD are prepared in collaboration with University of Leicester and Rutherford/Appelton Laboratory in the U.K. and Los Alamos National Laboratory in the U.S.A. respectively.

All the instruments have functioned normally to date, April 1990. The total number of targets observed to the end of 1989 is about 250, 51% are Galactic sources and 49% extra-Galactic sources.

2. Scientific instruments

2.1. LARGE AREA PROPORTIONAL COUNTERS (LAC)

The LAC is the primary instrument of *Ginga* and consists of eight identical counters (Turner et al., 1989). The total effective area is 4000 cm^2. Each counter is a multi-cell proportional counter which comprises 13 cell counters of four layers. The cell

Y. Kondo (ed.), Observatories in Earth Orbit and Beyond, 41–48.

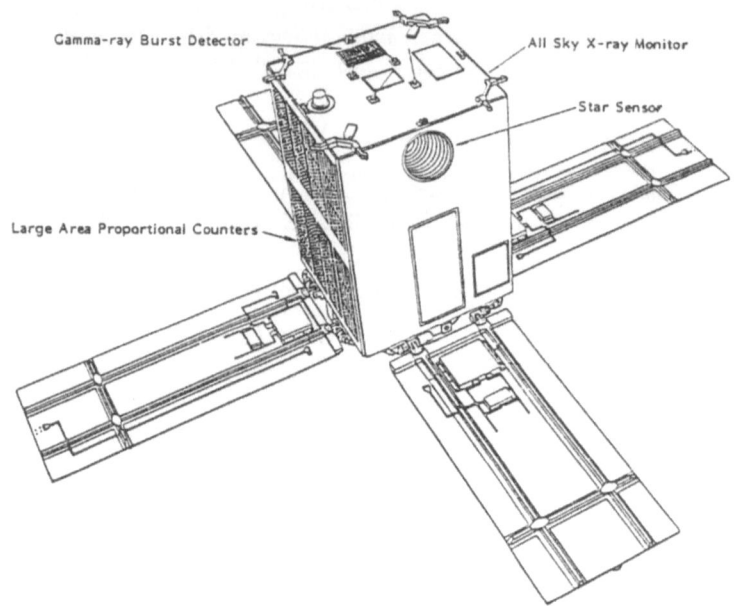

Fig. 1. The outside view of *Ginga*.

TABLE I

General characteristics of *Ginga*

Launch	February 5, 1987
Launch Vihicle	M-3SII-3
Initial Orbit	Perigee : 505.5 km, Apogee : 673.5 km
	Inclination : 31.1°, Period : 96.5 min
Mass	Total : 420 kg, Experiments : 105 kg
Experiments	Large Area Proportional Counters (LAC)
	All Sky X-ray Monitor (ASM)
	Gamm-ray Burst Detector (GBD)
Stabilization	Biased momentum three-axis control
Attitude Sensors	Two CCD Star Trackers, Sun Sensor, Four Gyroscopes
	and Geomagnetic Aspectmeters
Telemetry Bit Rate	High:16384 bps, Medium:2048 bps, Low:512 bps
Data Recorder	41.9 Mbit Bubble Memory
	Recording Time:42m40s(H), 5h41m(M), 22h44m(L)

counters sided by counter wall are connected together and used for anti-coincidence to lessen background, as well as the signals from the end plates. Because most of the background counts are produced by Compton scattering of hard X-rays and gamma-rays in the counter body. The anodes of the central two layers are connected together and those of the cell counters sided by entrance window are connected alternately. The signals from these three anode groups are analyzed by onboard data processor and are transmitted. Mutual anti-coincidence among these three anode groups is also employed for background reduction.

The entrance window is beryllium foil of 62μm thick which is supported by honeycomb type collimators made with stainless steel plates. The counters are filled with a gas mixture of argon (75%), xenon (20%) and carbon dixide (5%) at 2 atm at 293 K. The sensitive energy range is from 1.5 keV to 35 keV. The collimator field of view is elliptical of 1 degrees by two degrees (FWHM). The counting rate of X-rays from Crab Nebula is about 10000 cps, while the background rate is 70 cps of which 18 cps is due to cosmic diffuse X-ray background (CDXB). The detection limit of the LAC is confusion limit which is defined as sky to sky fluctuation of CDXB due to intensity source number relation. The 3σ limit is 2.1 cps, roughly equal to 0.2 mCrab (Hayashida et $al.$ 1989). The sensitivity is the highest so far achieved in this energy region.

The time resolution depends on the number of energy channels for pulse height analysis and the telemetry bit rate. The highest is 1 ms in the case of 2 energy channels and 16 kbps, and the lowest is 16 s in the case of 48 energy channel and 500 bps. However, a temperature dependence of the clock frequency appeared after launch, but can be compensated for using temperature data for high resolution timing analysis (Deeter and Inoue, 1990). No degradation of the LAC has been observed to the end of April 1990.

2.2. ALL SKY X-RAY MONITOR (ASM)

The ASM consists of two identical proportional counters, each containing three independent counters with veto cell counters at the back (Tsumeni et $al.$ 1989). These six counters are attached to a collimator of 1° by 45° (FWHM) of various slant angles. The monitoring of a wide sky region is conducted by slow rotation of the satellite around the z-axis. The position of the source can be determined by the time deferences of the peak position of the source for each counter. The normal frequency of scanning is once per day. The sky region which can be monitored is constrained to the equatorial region of satellite coordinate system determined by the LAC target.

The effective area of each counter is 70 cm^2. The counters are filled with xenon (96.7%) and carbon dioxide (3.3%) at 1 atm. The thickness of beryllium windows are 50μm. The detectors are sensitive to X-rays from 1 kev to 30 keV. The detection threshold is about 50 mCrab but becomes higher with increasing elevation angle from X-y plane of the satellite.

2.3. GAMMA-RAY BURST DETECTOR (GBD)

Two detectors, a NaI(Tl) scintillation counter and a Xe-filled proportional counter have been used for gamma-ray burst observation (Murakami et al. 1989). The NaI(Tl) is 8.8 cm in diameter and 1 cm in thickness, and measures hard X-rays from 13 keV to 400 keV. The proportional counter has an effective area of 63 cm^2, covering energy region from 2 keV to 30 keV. The X-ray observation of gamma-ray bursts has been conducted for the first time.

The gamma-ray bursts are unpredictable transient events shorter than 1 min. Two memory systems are employed to record bursts. The one is a fast memory triggered by the rise of the burst and is frozen until read out by command from the ground station. The sampling time is 31 ms for pulse counts and 0.5 s for pulse height distributions. The data in the time interval 16 s before and 48 s after the onset of the burst are recorded. The "time-to-spill" mode of data acquisition was used to cover wide dynamic range. The onset time are measured with accuracy of 244 μs. Only one burst can be stored in the fast memory. The other memory is a continuously sampling slow memory which sampling rate is either 125 ms, 1 s or 4 s depending on bit rate. The GBD includes small semiconductor detector to monitor radiation belt (RBM) and generate RBM flag to turn off high voltage supplies for LAC, ASM and GBD.

3. Notable results obtained with *Ginga*

3.1. CYCLOTRON FEATURES OF PULSAR SPECTRA

The merit of the LAC is high sensitivity in the energy region higher than 10 keV. Spectral structures in the hard X-ray region were detected from five X-ray pulsars, Her X-1 (Mihara *et al.* 1990), 4U 1538-52 (Clark *et al.* 1990), V 0332+53 (Makishima *et al.* 1990), 1E 2259+58 (Koyama *et al.* 1989a) and 4U 0115+634). The most pronounced spectrum was obtained from V0332+53 on October 1, 1989, as shown in Fig. 2. The spectrum can be expressed by power law with resonant absorption at 28.5 keV.

Possible interpretation of this structure is cyclotron absorption by electrons in highly magnetized plasma. The magnetic field can be estimated as 2.5×10^{12} G. Further analysis, such as pulse phase dependence of absorption energy and depth will reveal X-ray transport in pulsar plasma.

3.2. X-RAY SCATTERING BY INTERSTELLAR DUST GRAINS

The moon occulted accidentally transient X-ray source (GS 1741-28) appeared near the Galactic center. The observation was conducted on October 26, 1987 (Mitsuda *et al.* 1990). The light curves in three energy bands are shown in Fig. 4. The gradually decreasing and increasing component around the point source, which corresponds to sharp fall and rise is clearly seen at ingress and egress respectively. The light curves show that this extended component decreases with increasing X-ray energy. One can conclude that the extended component is due to scattering by cosmic dust grains from the energy dependence of relative intensity and angular

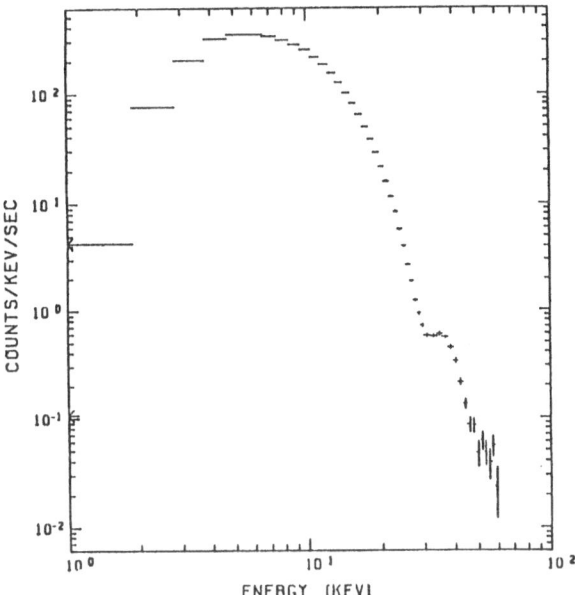

Fig. 2. The spectrum of X-ray transient pulsar V 0332+53. An absorption feature at 28 keV is clearly seen (Makishima *et al.* 1990).

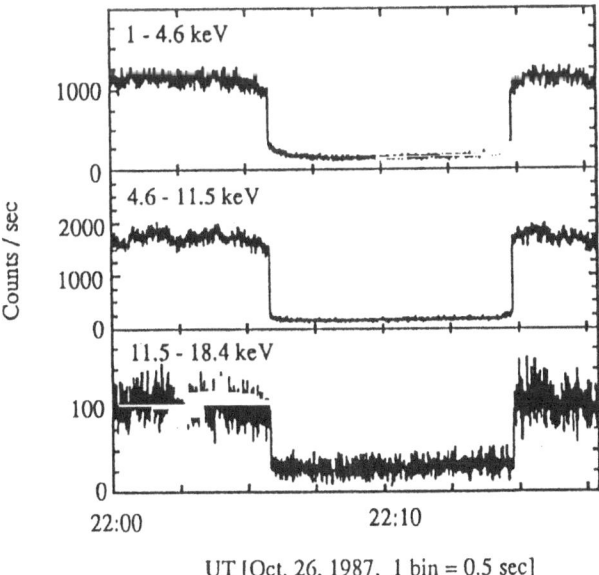

Fig. 3. Light curves of transient X-ray source GS 1741-28 in a occasion of lunar occultation. The slowly varying component is clearly visible at ingress and egress (Mitsuda *et al.* 1990).

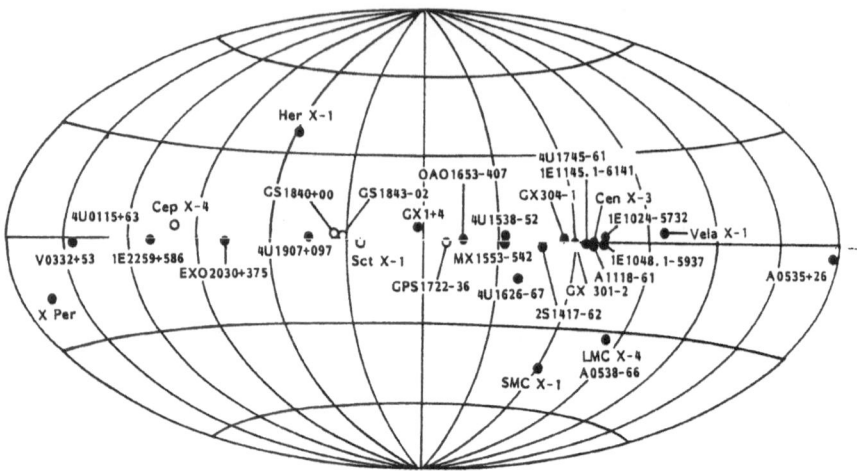

Fig. 4. Distribution of X-ray pulsars in Galactic coordinate. New pulsars discovered with *Ginga* are indicated by open circles.

diameter. The size and chemical composition responsible for X-ray scattering were estimated as about 0.1 μm in diameter and iron compound.

3.3. DISCOVERY OF NEW TRANSIENT X-RAY PULSARS

Five X-ray pulsars, Cep X-4, GS 1843+00, GPS 1722-362, GS 1843-02 and Sct X-1 have been discovered with *Ginga* (Koyama and Takeuchi, 1989). The scanning observation of Galactic plane discovered several transient X-ray sources. Most of new sources are faint and have the hard X-ray spectra characteristic of X-ray pulsars. The low energy part of the spectra is heavily absorbed by interstellar matter, and hydrogen column densities are higher than 10^{23} H-atoms cm^{-2}. These new transients which are candidate pulsars and new pulsars are concentrated in Scutum region, possibly located in 5 kpc Galactic ring. The total number of X-ray pulsars known before *Ginga* is 26 and they are located within a few kpc from solar system. *Ginga* observations suggest existence of many faint pulsars associated with inner Galactic arms.

3.4. INTENSE IRON LINE EMISSION FROM GALACTIC CENTER

The scanning observation along Galactic plane revealed diffuse emission of iron line peaked at Galactic center (Koyama *et al.* 1989b). The line energy of 6.7 keV suggests the emission is from high temperature plasma. The distribution of iron line shows sharp peak at Galactic center with angular diameter of 1.8° corresponding to 300 pc at the Galactic center. The peak flux is 1.5×10^{-7} erg s^{-1} sr^{-1} cm^{-2} and integrated luminosity of iron line is 2.3×10^{36} erg s^{-1}. The distribution of iron line extends to Galactic ridge. A possible origin of high temperature plasma

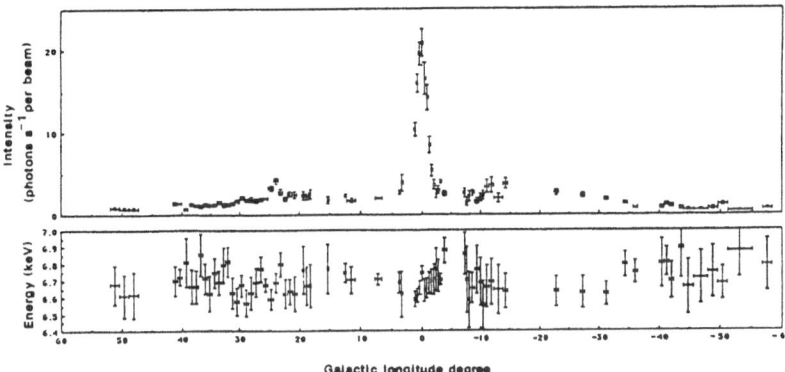

Fig. 5. Galactic longitude distribution of iron K-line. Lower panel shows line energy (Koyama *et al.* 1989b).

is supernova remnants. The distribution in such a narrow region suggests that the succesive explosion occurred at the epoch not earlier than 105 years.

3.5. SPECTRAL FEATURES AND VARIABILITY OF SEYFERT GALAXIES

Important discoveries were made on X-ray emission from Seyfert galaxies by *Ginga*. The iron line emission and K-absorption edge were observed from all Seyfert galaxies observed with *Ginga* (Pounds, 1989, Matsuoka *et al.*, 1990, Kunieda *et al.*, 1990). The line energy is 6.4 keV that is fluorescent X-rays from cold matter. The equivalent widths of iron line were about 100 cV or more. While the absorbing matter density determined from spectra below 2 keV is less than 10^{22} H-atom cm^{-2}. Therefore, the reprocessing matter is not distribute along the line of sight but it should sustend wide solid angle to central source. A possible configuration is an accretion disc (George *et al.* 1989). The variability of iron line emission from the low luminosity Seyfert 1 galaxy, NGC 6814 was observed by Kunieda *et al.* (1990). They obtained a positive correlative variation between the iron line and the continuum, and time lag between them was shorter than 250 s. This is comparable to observed variability time scale. This suggests that a geometrically thin cold accretion disc extends to the vicinity of central source. On the other hand, the continuum emission from NGC 4051 and MCG-6-30-15 varied in correlated way with spectral indices (Matsuoka *et al.* 1990). The spectra became steeper with increasing flux. This may be a further evidence for the existence of a scattered continuum which variability time scale is longer than the main component. *Ginga* observed Seyfert 2 galaxies, NGC 1068 (Koyama *et al.* 1989c), Mkn 348 (Warwick *et al.* 1989) and Mkn 3 (Awaki *et al.* 1990). Intense iron line emission at 6.4 keV was observed from NGC 1068 and Mkn 3. The equivalent widths were as high as 1 keV. The absorbing matter densities were in the order of 10^{23} H-atom cm^{-2} for Mkn 348 and Mkn 3. These facts support an idea that the Seyfert 2 galaxies are obscured Seyfert 1 galaxies. This may have an impact on the origin of the cosmic diffuse

F. MAKINO

X-ray background.

4. Concluding remark

Ginga has achieved highest sensitivity in the energy range from 1.5 keV to 40 keV and has revealed new aspects of X-ray astronomy.

Acknowledgements

Author would like to thank all the members of *Ginga* team for their preparation and operation of *Ginga* and for providing him with results of the observations. He acknowledge Dr. C. Day for careful reading of the manuscript and for his comments.

References

Awaki, H., *et al*: 1990, *Publ. Astron. Soc. Japan,* submitted

Clark, G. W., *et al*: 1990, *Astrophys.J.* **353**, 274

Deeter, J. E., and Inoue, H.: 1990, *ISAS Research Note* **No. 430**,

George, I. M., *et al*: 1989, *ESA SP-296* **2**, 945

Hayashida, K., *et al*: 1989, *Publ. Astron. Soc. Japan* **41**, 373

Kondo, I., *et al*: 1981, *Space Sci. Instr.* **5**, 211

Koyama, K., and Takeuchi, Y.: 1989, *ESA SP-296* **1**, 483

Koyama, K., *et al*: 1989a, *Publ. Astron. Soc. Japan* **41**, 461

Koyama, K., *et al*: 1989b, *Nature* **339**, 603

Koyama, K., *et al*: 1989c, *Publ. Astron. Soc. Japan* **41**, 731

Kunieda, H., *et al.*: 1990, *Nature,* in press

Makino, F., *et al*: 1987, *Astrophys. Lett. and Communication* **25**, 223

Makishima, K., *et al*: 1990, in preparation

Matsuoka, M., *et al.*: 1990, *Astrophys. J.*, in press

Mihara, T., *et al.*: 1990, *Nature,* submitted

Mitsuda, K., *et al*: 1990, *Astrophys. J.* **353**, 480

Murakami, T., *et al*: 1989, *Publ. Astron. Soc. Japan* **41**, 405

Ninomiya, K., *et al*: 1984, *Proc. of IFAC meeting,* 2915

Pounds, K. A.: 1989, *ESA SP-296* **2**, 753

Tanaka, Y., *et al*: 1984, *Publ. Astron. Soc. Japan* **36**, 641

Tsunemi, H., *et al*: 1989, *Publ. Astron. Soc. Japan* **41**, 391

Turner, M. J. L., *et al*: 1989, *Publ. Astron. Soc. Japan* **41**, 345

Warwick, R. S., *et al*: 1989, *Publ. Astron. Soc. Japan* **41**, 739

EXTREME AND FAR ULTRAVIOLET ASTRONOMY FROM VOYAGERS 1 AND 2

J. B. HOLBERG

Lunar and Planetary Laboratory, University of Arizona, Tucson, Arizona 85721,USA

Abstract. The instrumental characteristics, observational capabilities and scientific results of the Voyager 1 and 2 ultraviolet spectrometers are reviewed. These instruments provide current and ongoing access to low resolution spectra for a wide variety of astronomical sources in the 500 to 1700 Å band. Observations of the brightest OB stars and hot subluminous stars as faint as V = 15 mag. are possible. In the EUV, at wavelengths shortward of 900 Å, several new sources have been detected and a host of potential sources ruled out. In the Far UV, particularly at wavelengths between 900 and 1200 Å, Voyager is capable of observing a wide range of stellar and non-stellar sources. Such observations can often provide a valuable complement to IUE and other data sets at longer wavelengths. The Voyager spectrometers have proved remarkably stable photon counting instruments, capable of extremely long integration times. The long integration times, relatively large field of view, and location in the outer solar system also provide an ideal platform for observations of sources of faint diffuse emission, such as nebulae and the general sky background.

1. Introduction

On board the Voyager 1 and 2 spacecraft are nearly identical ultraviolet spectrometers (UVS), sensitive over the wavelength range 500 to 1700 Å. Although designed primarily to observe extreme and far UV emission from the atmospheres of the outer planets and several of their satellites, these instruments have proved exceptionally effective at conducting a wide variety of exploratory astronomical observations at wavelengths below 1200 Å. In contrast to the X-ray and longer wavelength UV, where earth orbital observations are now well established, the EUV (100 to 900 Å) and FUV (here, 900 to 1200 Å), have received scant attention. There are still wavelength ranges under active exploration, where low resolution spectroscopy or spectrophotometry can lead to important new discoveries. Over the past decade one major means of exploring these bands has been through observations conducted with the Voyager UVS. Important aspects of stellar astronomy investigated by Voyager have included: observations of luminous OB stars, a variety of active binary systems such a cataclysmic variables, hot subluminous stars including white dwarfs, subdwarfs, planetary nebulae and PG1159 objects. Other non-stellar observations have included: supernova remnants, globular clusters, active galaxies and components of the EUV and FUV sky background. Currently, the Voyager instruments are virtually the only means of conducting routine on going observations at these wavelengths.

In this paper the important instrumental, and observational characteristics of the Voyager spectrometers are reviewed. This is followed with a brief presentation

Y. Kondo (ed.), Observatories in Earth Orbit and Beyond, 49–57.
© 1990 *Kluwer Academic Publishers.*

of several different types of observations obtained with Voyager, each illustrating one or more unique aspects of these instruments.

2. Instrumentation and Observing Capabilities

Both Voyager instruments are compact, Wadsworth mounted, objective grating spectrometers (Broadfoot et al. 1977) covering the wavelength range 500 to 1700 Å. Collimation of incoming UV radiation is accomplished with a series of precession mechanical baffles which define a primary instrumental field-of-view (FOV) of 0.10 deg ×0.87 deg (FWHM). A single normal incidence reflection from a concave diffraction grating then focuses the dispersed image on to an array detector. The grating is a platinum coated replica, ruled at 540 lines/mm, blazed at 800 Å. and having a radius of curvature of 400.1 mm. Dispersion in the image plane is 93 Å/mm. In addition to the primary FOV, each UVS has a small 20 deg off-axis port which allows direct viewing of the sun, at decreased detector gain.

The open, photon counting detector consists of a dual microchannel plate (MCP) and a 128-element linear self-scanned readout array.

For wavelengths shortward of 1250 Å, the quantum efficiency of the MCP is that of the bare glass, ∼ 10%. Longward of 1250 Å, the quantum efficiency is enhanced by the use of a semi-transparent CuI photocathode deposited on a MgF_2 substrate. The MCP is normally operated at a gain of ∼ 10^6 and its output is proximity focused onto a linear array of aluminum anodes. Each of the 128 anodes (126 active plus two dead, trailing channels) is accessed every 320 μs and its output charge passed to a charge sensitive amplifier. Amplifier output for each channel is digitally converted to a 16-bit word and summed to an internal shift register memory. Memory registers are read out by the spacecraft flight data system at various defined rates, and transmitted back to earth as log compressed 10-bit words. For the astronomical observations discussed here, the two prime data rates currently are one complete spectrum every 3.84 s or every 576 s. The UVS instruments are solar blind, have no moving parts, and have operated continuously since launch in 1977. With the singular exception of a decrease in the Voyager 1 MCP gain, due to excessive radiation induced counts during passage through the inner Jovian magnetosphere, both instruments have remained photometrically stable at better than 3% level since 1977. In-flight performance of the UVS from launch through the 1979 Jupiter encounters is reviewed in Broadfoot et al. (1981).

As objective grating spectrometers, the instruments have differing spectral resolutions for point source and diffuse sources. For stellar (point) sources spectral resolution is ∼ 18 Å, while for diffuse (filled field) sources it is ∼ 30 Å. Instrumental sensitivity is optimized for the 800 to 1200 Å region. Typical limiting sensitivities at 1050 Å for stellar continua are 5×10^{-13} ergs cm^{-2} s^{-1} Å^{-1} for Voyager 2 and 1×10^{-12} ergs cm^{-2} s^{-1} Å^{-1} for Voyager 1. On-axis integration times necessary to achieve these limits are ∼ 1 day. As a practical matter, however, there is often a need to observe adjacent sky background for a comparable period of time. In addition, the average efficiency of ground station coverage must also factored in. A plot relating observed signal level at 1050 Å, the required signal-to-noise ratio, and 'total observing time' (TOT) for a stellar continuum is shown in Fig. 1. For diffuse emis-

TABLE I
Voyager Ultraviolet spectrometers

Optical Configuration
Wadsworth Mount Objective Grating
22.1 cm^{-2} Platinum replica Grating, Blazed at 800 Å
Mechanical Collimation: 0.°10 x 0.°87 Field of View

Photon Counting Detector
128 Channel Microchannel Plate, Bare for λ < 1250 Å
CuI on MgF$_2$ Filter for λ > 1250 Å
Self-Scanned Aluminum Anode Array
Integration Times of 3.84 and 576 Seconds
Dark counts 2.5 x 10^{-3} counts s^{-1} channel^{-1}

Performance
Spectral Range (Å):525 < λ < 1650
Resolution: ~ 18 Å (9.26 Å / channel)
Limiting Flux: 0.2-0.5 x 10^{-12} ergs cm^{-2} sec^{-1} Å^{-1} at 1000 Å
Stability: <2% Change in 3 Years
Absolute Calibration: ~10-15%

sion, limits of 100 photons cm^{-2} s^{-1} Å^{-1} str^{-1} and 6000 photons cm^{-2} s^{-1} str^{-1}, for line and continuum emission in the 500 to 1200 Å band have been acheived, using Voyager 2 (Holberg 1986). Very long integration times, measured in days, are a normal mode of observing with the UVS. The instruments have low backgrounds, 2.5×10^{-3} counts s^{-1} channel^{-1}, due primarily to gamma rays from the spacecraft radioisotope thermoelectric generators and scattering interplanetary H I Lyman series and He I 584 Å emission lines.

A summary of instrumental parameters is contained in Table I.

UVS data typically consist of an extended time series of a large number (1000's) of individual spectra. To obtain a representative spectrum for stellar sources, these spectra are first placed into a spatially ordered sequence. This is necessary because of objective grating nature of the spectrometers and because they are located on a spacecraft subject to small quasi-periodic attitude control motions. The net effect of these spacecraft motions is to move the instrumental FOV with respect to the target, causing off-axis spectra to experience small wavelength shifts and vignetting due to the collimator. Spacecraft attitude control motion is computed for each spectrum from the attitude control system error signals. With this information, it is possible to correct individual spectra for off-axis effects. These spectra are then summed, background subtracted, flat fielded, corrected for scattered light and photometrically calibrated. Typical extracted stellar spectra are shown in Fig. 2. A detailed discussion of the reduction of spectral data is contained in Holberg (1986) and Longo et al. (1988).

As mentioned above, two primary data modes are currently available for the

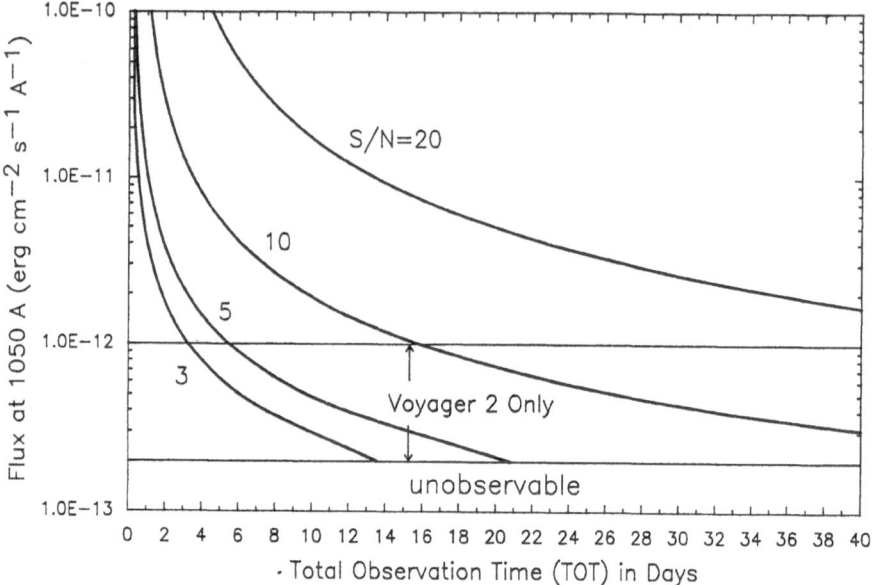

Fig. 1. Total Observing Time (TOT) vs observed flux at 1050:80 for faint objects at four signal-to-noise levels; 20, 10, 5, 3. TOT includes on-target integration time, a comparable amount of sky background integration and 50% ground station coverage.

UVS data; a high rate of 3.84 s per spectrum and a low rate of 576 s per spectrum. For observations involving stellar sources, the high data rate is preferred as it allows adequate sampling of the spacecraft limit cycle motion which has characteristic periods of $\sim$ 20 min. Stellar observations can be achieved at the low data rate but photometric accuracy often suffers as a result of the less precise knowledge of the motion of the slit with respect to the source. Low data rate spectra are useful for diffuse sources and for obtaining sky background for stellar observations.

2.1. OPERATIONS

Each UVS instrument is located, along with the other optical remote sensing experiments, on a two-axis scan platform at the end of the spacecraft science boom. Pointing the UVS at astronomical targets is accomplished by driving the scan platform to a set of predetermined actuator angles. Scan platform pointing accuracy is typically $\sim$ 2 arcmin, sufficient to place the target within the 0.10 deg wide UVS slit. Most areas of the sky are accessible from one or both spacecraft. A mutual obscuration zone in the southern hemisphere currently exists (see Fig. 3). In addition, scattered light hampers the observation of most targets with in 25 deg of the sun. These geometrical constraints change very slowly with time, as do slit orientations with respect to any specific target. All observations are preplanned as part of the overall spacecraft operating sequence resident in the spacecraft computer memory; no real-time operation is possible. In the near future, each spacecraft memory load

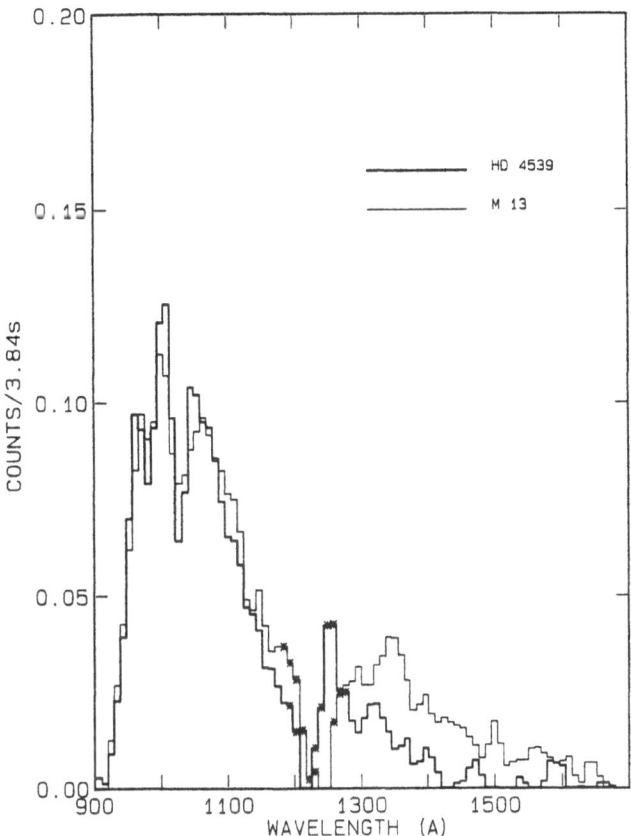

Fig. 2. A comparison of the the Voyager 2 spectrum of the integrated light from the globular cluster M 13 with a Voyager 2 spectrum of the hot sub dwarf B star, HD 4539. For this comparison the HD 4539 spectrum has been scaled to an apparent magnitude of V=11.

will operate for a 13 week period and targets will be scheduled 15 to 20 weeks in advance of the actual observation. Observations requiring simultaneous coverage of a target from the ground or another spacecraft should allow for such lead times. UVS observations, specified in simplest terms by, an initial time, right ascension, declination, and duration, are planned as part of each sequence and occur as they are clocked out by the spacecraft command and control computer. Data are received at NASA Deep Space Net ground stations, processed at the Jet Propulsion Laboratory in Pasadena and generally arrive at the Lunar and Planetary Laboratory in Tucson within one month. Ground station coverage averages between 8 to 12 hrs per day on each spacecraft.

2.2. FUTURE PLANS

Continued astronomical observations with the UVS are currently planned to be part of the Voyager Interstellar Mission (VIM). The VIM program, which officially began on January 1 1990, has as its prime objectives:

(a) to investigate the interplanetary and interstellar media, and to characterize the interaction between the two.

(b) to continue the successful Voyager program of ultraviolet astronomy.

It is anticipated that it will be practical to return UVS astronomical data at useful rates until at least 1995 when Voyagers 1 and 2 will have reached distances of 59 and 44 AU, respectively. A substantial upgrade in telemetry performance, scheduled to occur later this year, will allow the bulk of UVS data to be returned at the high (3.84 s/spectrum) data rate preferred for point source observations. This will significantly increase observing efficiency for stellar and other localized targets. Plans call for observing up to 4 targets per week per spacecraft while in this data rate; a factor of 2 to 3 increase over what has been possible in the past. In addition to astronomical observations, a parallel program of solar observations is also planned, as are continued observations of emission form the interstellar wind and interplanetary medium.

In 1988 UVS observations were opened to an initial round of guest investigator proposals. These guest investigator observations began in late 1989, following completion of the Neptune encounter activities. They are expected to be completed later this year. Currently, NASA is considering the possibility of a second round of UVS guest investigator observations.

3. Examples of Observations Possible with VOYAGER

In the remainder of this paper some of the more unique observing characteristics of the Voyager UVS are illustrated with brief accounts of several observations.

3.1. GLOBULAR CLUSTERS

In the optical, globular clusters are dominated by red giants and sub giants. In the UV, however, such stars are intrinsically faint and cluster luminosity can be dominated by a relatively small number of hot blue stars.

Obtaining integrated spectra of this stellar population has been hampered by the relatively large angular extents of most globular clusters. The UVS slit, however, is conveniently matched to the angular diameters of many clusters. At the suggestion of Dr. Horace Smith of Michigan State University, three globular clusters wer observed with Voyager 2: NGC 6752, M 13 and M 92. In each case a hot, stellar-like spectrum extending to the Lyman limit was observed. The two strongest detections were NGC 6752 and M 13. In Fig. 2 the M 13 count rate spectrum is compared with a Voyager 2 spectrum of the well known subdwarf B star, HD 4539. It is clear from IUE observations of individual cluster stars (see Castellani and Cassatella 1987) that hot subluminous stars are responsible for the UV luminosity of globular clusters. These Voyager observations have provided the first integrated spectra of globular clusters in the FUV.

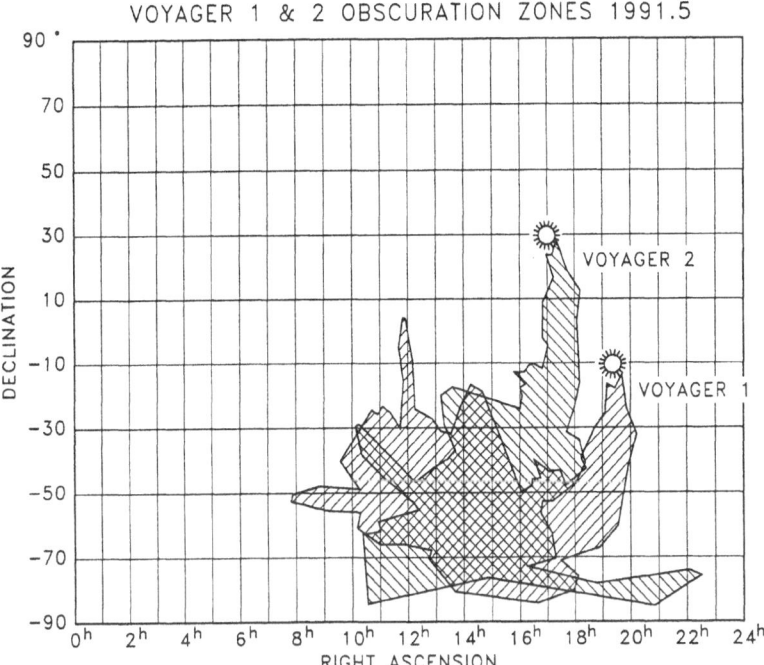

Fig. 3. Accessible regions of the sky in mid-1991, as seen from Voyagers 1 and 2. The hatched regions are portions of the sky obscured by spacecraft structure. The apparent locations of the sun as seen from each spacecraft are indicated. Scattered light becomes a problem within 25 degrees of the sun

3.2. FUV SKY BACKGROUND

The availability of extremely long integration times, the relatively low amounts of foreground emission present in the outer solar system and the relatively large field of view all contribute to the effectiveness of Voyager in observing the FUV sky background. Indeed, the best current upper limits on high galactic latitude sky backgrounds between 500 and 1200:80 have been obtained form Voyager 2 (Holberg 1986). At UV wavelengths, the principal components of the diffuse sky background are expected to be star light scattered from interstellar dust and recombination line emission from ions in a possible hot (10^5 K) component of the interstellar medium.

The first of these components is a diffuse continuum having an energy distribution characteristic of the hot, early type stars which illuminate the dust. In contrast to the upper limits on sky background found by Holberg (1986) at high galactic latitudes, several low latitude regions have now been found which exhibit diffuse stellar-like continua. The most interesting of these is associated with a 150 day long drift of the Voyager 2 FOV through the constellation of Ophiuchus. In these data a diffuse continuum extending over many degrees and having a mean 1000 Å intensity of 2500 photons cm^{-2} s^{-1} Å^{-1} str^{-1} was observed. Spectra of this emission bear a strong resemblance to a single modestly reddened O star of 11 mag. Figure 4 shows

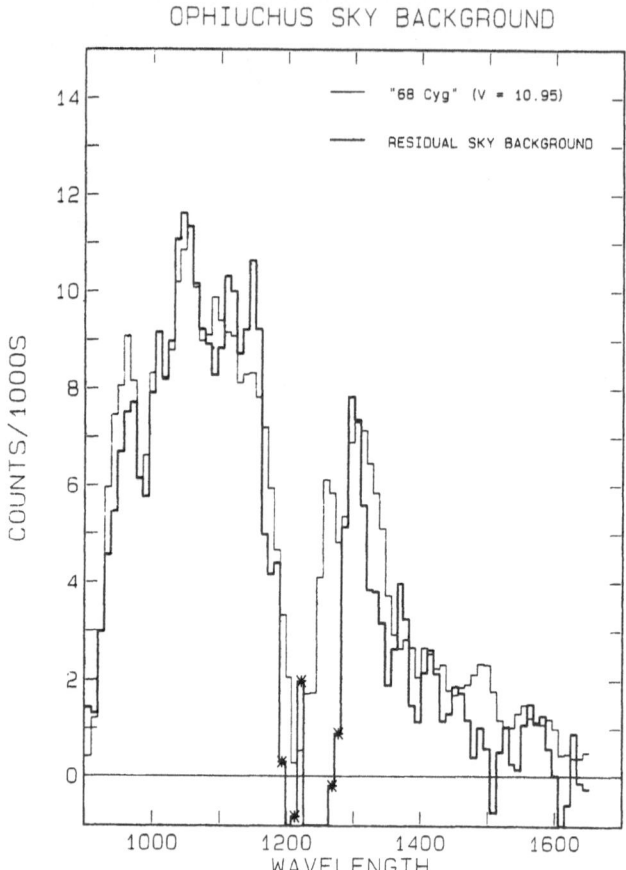

Fig. 4. A comparison of the observed spectrum of the diffuse background observed in Ophiuchus with an observation of the moderately reddened O8 star; 68 Cyg. The 68 Cyg spectrum has been uniformly reduced by a factor corresponding to 5.95 magnitudes

this spectrum compared with a Voyager 2 spectrum of 68 Cyg (O8, B–V = –0.01).

The interpretation (Holberg 1990) is that this is due to the scattering of FUV starlight from interstellar dust; in effect an extended reflection nebula is being observed.

3.3. INTERSTELLAR REDDENING

In sharp contrast to longer wavelengths, knowledge of interstellar reddening short-ward of 1200 Å has remained in rudimentary state. Virtually the only direct observations are those of York et al. (1973) who used Copernicus to obtain reddening four curves for wavelengths down to 1000 Å. Recently, however, Snow, Mason and Polidan (1990) have used Voyager UVS observations to greatly expand reddening studies to include many more reddened-unreddened pairs and to extend wavelength coverage 950 Å. This study indicates that reddening raises steeply between 1200 Å and 950 Å, as expected from theoretical studies. In addition, as is the case at longer wavelengths, considerable variation is seen among extinction curves. These studies were made possible through the use of the extensive archive of Voyager observations of luminous OB stars.

Acknowledgements

The author wishes to acknowledge the efforts of those present and past members of the UVS team who have contributed significantly to much of the work discussed here, in particular A. L. Broadfoot, R. S. Polidan, B. R. Sandel and D. S. Shemansky. Special recognition is also due S. Linick at JPL who has worked effectively over the years to obtain UVS observations. This work was supported by NASA grant NAGW-587.

References

Broadfoot, A. L. et al.: 1977, *Space Sci. Rev.* **21**, 183
Broadfoot, A. L. et al.: 1981, *J. Geophys. Rev.* **86**, 8259
Castellani, V. and Cassatella, A.: 1987, in Y. Kondo, ed(s)., *Exploring the Universe with the IUE Satellite*, Reidel, Dordrecht, 637
Holberg, J. B.: 1986, *Astrophys. J.* **311**, 969
Holberg, J. B.: 1990, S. Bowyer and C.Leinert, *IAU Symposium No. 139, Galactic and Extragalactic Background Radiation*, Kluwer Academic Publishers, Dordrecht
Longo, R., Stalio, R., Polidan, R. S., and Rossi, L.: 1989, *Astrophys. J.* **339**, 474
Snow, T. P., Mason, M. M., and Polidan, R. S.: 1990, *Astrophys. J.*, (in press)
York, D. G., Drake, J. F., Jenkins, E. B., Morton, D. C., Rogerson, J. B., and Spitzer, L.: 1973, *Astrophys. J. (Letters)* **182**, L1

II. FUTURE MISSIONS

(A) X-RAY AND GAMMA-RAY MISSIONS

RÖNTGEN SATELLITE

J. TRÜMPER

MPI für Extraterrestrische Physik, D-8046 Garching-bei-München

The ROSAT launch on June 1, 1990 happened to be after the IAU Colloquium No. 123 before the deadline for submitting manuscripts. I therefore take the liberty to deviate grossly from the manuscript of my talk and give a short up to date mission summary. A more complete description of the mission can be found in References 1 and 2.

The launch of the Delta II rocket was perfect and the orbital parameters reached are very close to nominal: height 580 km, inclination 53°. The predicted satellite's lifetime in this orbital is at least 10 years. The switch-on of the spacecraft and instrument subsystems could be completed without any losses. All systems are in good health.

ROSAT carries two instruments:
- a large Wolter telescope covering the energy range 0.1–2.4 keV with two position sensitive proportional counters (PSPC) and one high resolution imager (HRI)

and
- Wolter-Schwarzschild-telescope with two channel plate detectors covering the adjacent XUV-range.

The novel features of the X-ray telescope are:
- large collecting power
- good spectral resolution of the PSPC's: $\Delta E/E \sim 0.4$ at 1 keV; four colours
- low background of the PSPC's: 3×10^{-5} cts/arcmin^2 sec
- high angular resolution with the HRI: < 4 arcsec half power width (PSPC ~ 25 arcsec)

The ROSAT mission comprises three phases: After the initial calibration and verification phase an all sky survey is carried out during half a year, followed by a period of pointed observations.

The *calibration and verification programme* which was completed on 30 July 1990 confirmed the excellent performance of all instruments. The satellite and the ground control worked well, leading to an observational efficiency (fraction of useful time spent on targets) of $\sim 50\%$. The ROSAT "First Light" observation was performed with the PSPC and covered a field in the Large Magellanic Cloud. Highlights of this early phase are a new map of the LMC region, including a number of new sources and an upper limit of the X-ray flux from SN 1987A, the first X-ray picture of the moon, and observations of several bright supernova remnants as well as of the cluster of galaxy Abell 2252.

Y. Kondo (ed.), Observatories in Earth Orbit and Beyond, 61–62.

The *all sky survey* represents a quantum jump in sensitivity compared with previous X-ray surveys ($\sim$ factor 100 compared with HEAO-1). The limiting flux of the ROSAT survey will be 3×10^{-13} erg/cm^2 s (0.1–2.4 keV). In the XUV range ROSAT will perform the first survey ever done.

The ROSAT survey which commenced on 30 July will play a pathfinder role for further pointed X-ray missions. At the same time the survey will provide for the first time large bias-free samples of various types of sources (in particular stars, white dwarfs, galaxies, AGN's and clusters of galaxies). In order to stimulate correlated observations the ROSAT survey time line has been published in Reference 3. After the completion of the sky survey which is expected at the end of January 1991 the AO-1 pointed programme will be executed.

The *ROSAT pointed programme* will allow to carry out detailed observations of selected sources whereby the performance in terms of sensitivity, angular resolution and spectral resolution exceeds considerably that of previous X-ray telescope missions.

The announcement of opportunity for the second period (AO-2) will be issued in December 1990.

References

Trumper, J.: 1984, *Physics Scripta* **T7**, 209

ROSAT call for proposals, technical appendix, Max-Planck-Institut fur Extraterrestrische Physik, 1989.

Schmitt, J.H.M.M., IAU circular 5069, 1990.

THE GAMMA-RAY OBSERVATORY

DONALD A. KNIFFEN

NASA Goddard Space Flight Center, Greenbelt, Maryland 20771

Abstract. The Gamma-Ray Observatory (GRO), a Shuttle-launched free-flying Observatory is currently scheduled for launch in November 1990. The mission provides nearly six orders of magnitude in spectral coverage, from about 50 KeV to about 30 GeV, with a sensitivity over the entire range over an order of magnitude greater than that of previous observations. The 16,000 kilogram Observatory contains four instruments on a stabilized platform. The mission duration is expected to be from six to ten years. A Science Support Center has been established at the Goddard Space Flight Center for the purpose of supporting a vigorous Guest Investigator Program.

1. Introduction

The Gamma-Ray Observatory (GRO) was announced in 1977 as an opportunity for gamma-ray experiments to be included on a free-flying observatory to be launched aboard the space shuttle. The mission was planned as a 3-axis oriented platform containing four instruments with a planned two year lifetime on orbit. It was later decided that the enhanced science return would justify extending the duration of orbital operations. By the time such an option was adopted, the development of the instruments and spacecraft was well underway. Studies showed that the only consumable quantity clearly preventing a lifetime of from six to ten years was the propellant gas required to maintain the operational orbit against atmospheric drag. GRO was manifested on a high performance Shuttle so that it might be taken directly to its operational altitude of 450 kilometers at the beginning of the mission. This provides sufficient fuel savings to allow the operational orbit to be maintained for up to ten years and still provide fuel for a controlled reentry.

GRO is a 16,000 kg spacecraft (Figure 1) containing a complement of four instruments to make observations of the gamma-ray sky over the range from about 50 KeV to 30 MeV (Figure 2). This very wide dynamic range requires the use of different instruments with a number of detection techniques. The Z-axis of the spacecraft may be pointed to any region of the sky at any time. However, once chosen, the X-axis must be selected to satisfy sun angle constraints which affect the power from the solar panels and the thermal environment of various spacecraft components. The attitude knowledge will be maintained to an accuracy of 2 arcminutes. Absolute time will be accurate to .1 millisecond.

The extension of the mission lifetime beyond the original two years makes it possible to expand the scientific involvement beyond the original Principal Investigators (PIs) and their Co-Investigators. This not only increases the access to the data and observing opportunities, but it also enhances the scientific return by the infusion of ideas from a broader community. However, since the data analysis systems were designed for a PI class mode of operation, a program to involve those from outside the PI teams must be done in a manner that carefully considers the

Y. Kondo (ed.), Observatories in Earth Orbit and Beyond, 63–70.

© 1990 *Kluwer Academic Publishers.*

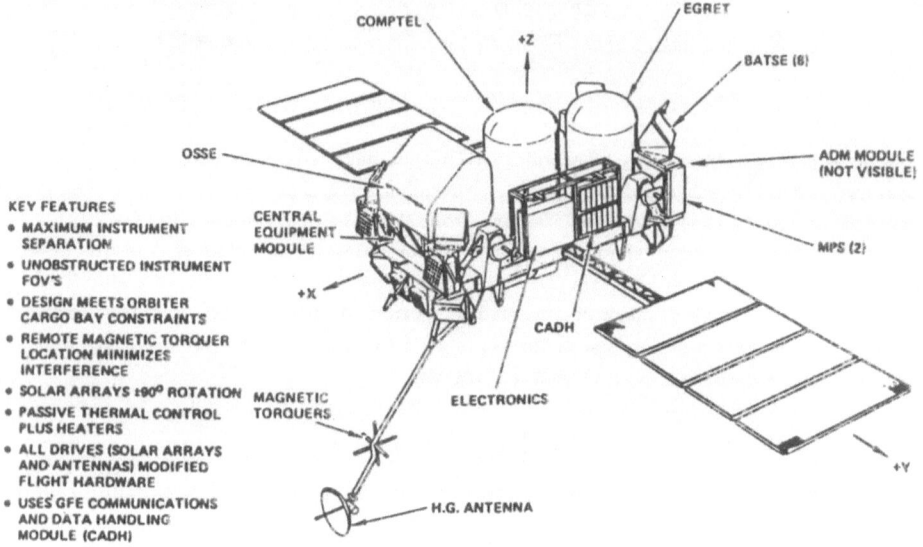

Fig. 1. The Gamma-Ray Observatory.

difficulties in analyzing and interpreting the data. The GRO Science Support Center has been established at the Goddard Space Flight Center to assist the Guest Investigators in accessing the GRO data.

2. The Instruments

Figure 2 shows the spectral coverage of the four instruments. They are the Burst and Transient Source Experiment (BATSE), the Imaging Compton Telescope (COMPTEL), the High Energy Telescope (EGRET) and the Oriented Scintillation Spectrometer Experiment (OSSE). The BATSE experiment is designed to monitor the entire sky continuously for bursts and transient gamma-ray events using eight uncollimated, wide-field detectors placed at all corners of the spacecraft, top and bottom. The OSSE experiment, with a relatively narrow field of view, will primarily be sensitive to discrete sources, the dominant feature of the low energy gamma-ray sky. The COMPTEL and EGRET experiments are wide field instruments designed to provide a high sensitivity survey of the medium and high energy gamma-ray sky.

3. BATSE

BATSE is optimized to measure brightness variations on timescales down to milliseconds. To accomplish this eight detector modules of identical configuration are used. The main detector is a large area sodium iodide scintillation crystal (20 inches in diameter by one-half inch thick). Light produced as the incident gamma rays interact in this scintillator are sensed by three large photomultiplier tubes placed behind it, while in front is a plastic scintillator used to reduce the background due

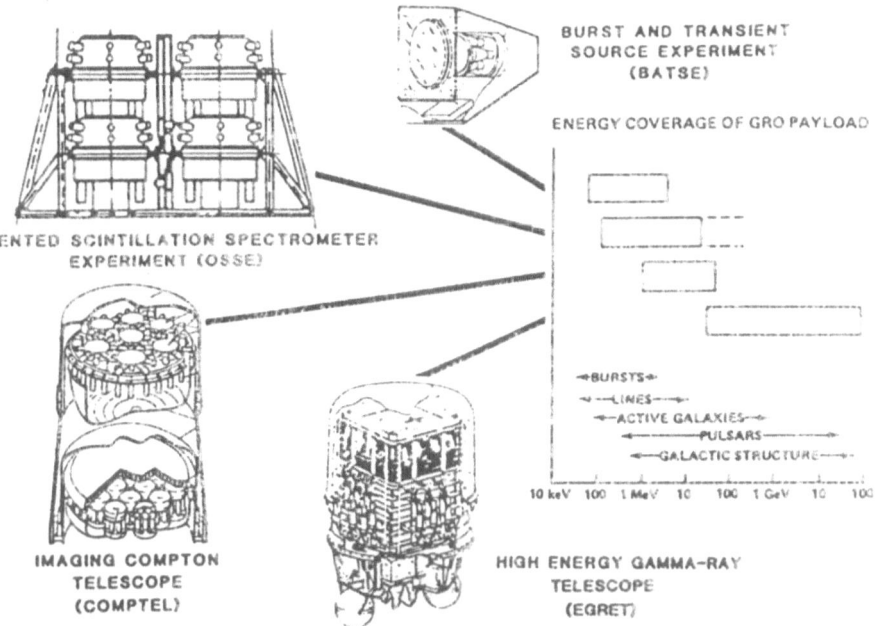

Fig. 2. Energy Coverage of the GRO Experiments.

to charged particles. Gamma-rays are inclined to pass through plastic without interaction. Upon detection of a fast, intense transient event, such as a gamma-ray burst, the BATSE processes large amounts of gamma-ray data for later transmission to the ground. It also sends a signal to the three other telescopes on the GRO that such an event has occurred so that it can be studied in greater detail over a wider energy band. A smaller spectroscopy detector, optimized for broad energy coverage and good resolution is contained with each of the 8 detectors

Dr. Gerald R. Fishman of the NASA Marshall Space Flight Center (MSFC) is the Principal Investigator of the BATSE experiment with Co-Investigators at MSFC, The University of Alabama, Huntsville, the Goddard Space Flight Center (GSFC) and the University of California, San Diego.

4. OSSE

OSSE has been designed to undertake comprehensive observations of astrophysical sources in the 0.1 to 10 MeV range, and also includes limited capability above 10 MeV, primarily for solar gamma-ray and neutron observations. The instrument consists of four identical detector systems, each of which is articulated to provide a 192-degree rotation capability about the Y-axis of the spacecraft. Normally, two detectors will observe the source, and two a nearby off-source region. The combination will be reversed at regular intervals, and the difference signal represents the net source flux. Each detector system includes a 33-cm diameter NaI-CsI phoswich assembly which provides the basic gamma-ray detection capability. Active shielding is provided by an annular NaI shield and the CsI section of the phoswich. The acceptance aperture of 3.8 × 11.4 degrees is defined by a passive tungsten alloy collimator. The offset pointing feature also allows the observation of targets away from the prime viewing axis of GRO. This permits viewing secondary sources when the main spacecraft axis is occulted by the earth, and allows for quick response observations of transient phenomena such as solar flares and novae without impacting the Observatory viewing program in a major way.

Dr. James D. Kurfess of the U.S. Naval Research Laboratory (NRL) is the Principal Investigator of the OSSE experiment with Co-Investigators at NRL, Northwestern University, Clemson University and the Royal Aircraft Establishment.

5. COMPTEL

In COMPTEL, gamma rays are detected by the occurrence of two successive interactions: the first one is a Compton collision in a detector of low-Z material (liquid scintillator) followed by a second interaction in a detector with high-z material (NaI), in which the scattered gamma-ray is totally absorbed. The sequence of collisions is confirmed by a time-of-flight measurement between the two detectors. The neutron induced background events, which may simulate double scatter gamma-ray events, are identified and rejected on the basis of the shape of the light pulse emitted in the first scatter detector. This method, in combination with effective charged particle shield detectors is very effective in suppressing the otherwise inherent instrumental background. The spatial resolution in the two detectors, together with the well defined kinematics of the Compton interaction, permits the reconstruction of discrete and extended sky images over a wide field of view with a resolution of the order of a degree. The minimum source detectability that can be achieved with COMPTEL is a few percent of the total emission from the Crab, the standard candle of gamma-ray astronomy. The first or upper detector of COMPTEL contains 7 cylindrical modules of liquid scintillator. Each module is approximately 28 cm in diameter and 8.5 cm thick and viewed by eight photomultipliers. The second detector consists of 14 cylindrical NaI blocks of 7.5 cm thickness and 28 cm diameter. Each block is viewed by seven photomultipliers from below. The distance between the first and second detectors is 1.50 meters. Each detector is surrounded by a thin anticoincidence shield of plastic scintillator. The Principal Investigator for the COMPTEL instrument is Dr. Volker Schoenfelder of the Max-Planck Institute for Extraterrestrial Physics (MPE) in Garching, West Germany. Co-Investigators are located at MPE, at the Laboratory for Space Research, Leiden, at the Space Sciences Department of the European Space Agency, and at the University of New Hampshire.

6. EGRET

The high energy instrument on GRO, EGRET covers the broadest range, from about 20 MeV to about 30,000 MeV. Like COMPTEL it is an instrument with a wide field, good angular resolution and very low background. Because it is designed for the study of higher energy gamma-rays, the detector is optimized to detect gamma-rays when they interact by the dominant high energy process which forms an electron and a positron. This unique signature allows a positive identification of the parent gamma-ray. The basic imaging portion of this instrument consists of two sections of spark chambers separated by a layer of segmented plastic scintillators. The upper section consists of 28 spark chamber modules interleaved with 27 thin foils of high-Z target material for the conversion of the gamma-rays to an electron-positron pair. After the electrons are formed their trajectory is followed through the remainder of the upper and lower spark chamber sections and into the large NaI crystal where the rest of their energy is deposited. The NaI counter will provide good precision in estimating the energy of the gamma-ray over the large sensitive range of this instrument. A central scintillator matrix, in coincidence with a similar layer below the bottom spark chamber section, identifies the presence of the electrons, applies the high voltage to the spark chambers and initiates the read-out of the digitized "picture" of the particle trajectories. A large plastic scintillator dome over the instrument identifies and rejects events containing particles which are already charged on entry into the detector.

Co-Principal Investigators of the EGRET instrument are Dr. Carl E. Fichtel of GSFC, Dr. Robert Hofstadter of Stanford University and Dr. Klaus Pinkau of the Max-Planck Institute in West Germany. Co-Investigators are located at each of these institutions and at the Grumman Aerospace Corporation.

Major characteristics and capabilities of the GRO instrument complement are summarized in Table I.

7. The Gamma-Ray Observatory Mission

The Gamma-Ray Observatory is now in launch preparations at the Kennedy Space Center in Florida. The launch is currently scheduled for the beginning of 1991, aboard the Space Shuttle Atlantis. Although the orbital altitude is degraded by atmospheric drag, it will be maintained between 440 and 450 km (Figure 3) by the on-board propulsion system. At the end of the mission the spacecraft will either be retrieved by the Space Shuttle, or will reenter in a controlled manner.

During the orbit operations data will be recorded at a 32 kilobit per second data rate, and these data will be telemetered via the Tracking and Data Relay Satellite (TDRS) once every two orbits at a 512 kilobit rate. Uplink commanding of both the experiments and the spacecraft will also be sent during these TDRS contacts. The mission operations profile are indicated in Figure 4. The Control Center will be at the Goddard Space Flight Center. Data will be received in packetized form with all ancillary information required to analyze the data already inserted on board the spacecraft. Time ordering, overlap elimination and quality checks will be done on ground before the data are forwarded, within 48 hours of receipt, to the experiment data analysis facility sites.

The viewing program for the first 15 months of operation is currently in the final process of development by the GRO Timeline Committee and will be designed to provide a nearly uniform full sky survey of two week exposures for the two wide field instruments (EGRET and COMPTEL). The narrow aperture instrument (OSSE)instrument will select 30 primary and secondary targets for discrete source

TABLE I

Summary of GRO Detector Characteristics

	OSSE	COMPTEL	EGRET	BATSE LARGE AREA	BATSE SPECTROSCOPY
ENERGY RANGE (MeV)	0.10 to 10.0	1.0 to 30.0	20 to 3×10^4	0.03 to 1.9	0.015 - 110
ENERGY RESOLUTION (FWHM)	12.5% at 0.2 MeV 8.8 " 1.0 " 4.0 " 5.0 "	8.8% at 1.27 MeV 6.5 " 2.75 " 6.3 " 4.43 "	~20% 100 to 2000 MeV	32% at 0.05 MeV 27 " 0.09 " 20 " 0.66 "	8.2% at 0.09 MeV 7.2 " 0.66 " 5.8 " 1.17 "
EFFECTIVE AREA (cm^2)	2013 at 0.2 MeV 1480 " 1.0 " 569 " 5.0 "	25.8 at 1.27 MeV 29.3 " 2.75 " 29.4 " 4.43 "	1200 at 100 MeV 1600 " 500 " 1400 " 3000 "	1000 ea. at 0.03 MeV 1800 " 0.1 " 550 " 0.66 "	100 ea. at 0.3 MeV 127 " 0.2 " 52 " 3 "
POSITION LOCALIZATION (STRONG SOURCE)	10 arc min square error box (special mode; 0.1 x Crab spectrum)	8.5 arc min (90% confidence at 2.75 MeV - 20σ source)	5 to 10 arc min (1σ radius; 0.2 x Crab spectrum)	1° (strong burst)	—
FIELD OF VIEW	3.8° x 11.4°	~ 64°	~ 0.6 sr	4π sr	4π sr
MAXIMUM EFFECTIVE GEOMETRIC FACTOR (cm^2 sr)	13	30	1050 (~ 500 MeV)	15000	5000
ESTIMATED SOURCE SENSITIVITY (10⁶ sec; off Galactic plane) — LINE	$(2-5) \times 10^{-5}$ cm^{-2} s^{-1}	3×10^{-5} to 3×10^{-6} cm^{-2} s^{-1}		0.4% equivalent width (5 sec integration)	
ESTIMATED SOURCE SENSITIVITY — CONTINUUM	2×10^{-7} cm^{-2} s^{-1} keV^{-1} (@ 1 MeV)	5×10^{-5} cm^{-2} s^{-1}	5×10^{-8} cm^{-2} s^{-1} (> 100 MeV) 1.5×10^{-9} cm^{-2} s^{-1} (> 1000 MeV)	6×10^{-8} erg cm^{-2} (10 sec - burst)	

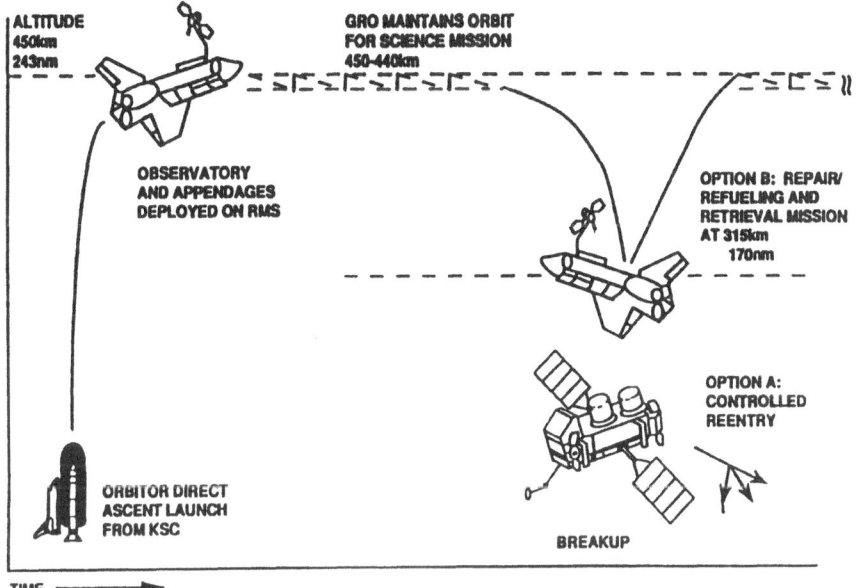

Fig. 3. The GRO Mission Profile.

studies during this 15 month period and the burst instrument (BATSE) covers the entire unocculted sky at all times. The viewing program for following years will be determined later, but will emphasize deep searches and the study of source time variations on many timescales. The timing of the NASA Research

Announcement for the Guest Investigator Participation will be chosen so that observations proposed by selected Guest Investigators will be included in the observing plan as efficiently as possible. A considerable effort has bee placed on coordinating with both space- and earth-based observations at other wavelengths to enhance the value of such observations at all wavelengths.

8. The Science Support Center

The central point of contact for Guest Investigator support will be the GRO Science Support Center (GROSSC), the Goddard Space Flight Center. In addition to its function of aiding Guest Investigators in accessing GRO data the GROSSC will be a source of information for potential users at all stages of their involvement, from the preproposal stage, through the analysis phase. The Center will provide information on the availability and timing of opportunities, and give technical support in the implementation of chosen guest investigations. It will be the source for technical information on the GRO instruments, for scientific and technical information on the GRO spacecraft and mission, for catalogs of data on GRO and other astrophysics and astronomy observations, for the status of GRO observations, for availability of useful analysis software, and other such information of use for GRO investigations.

9. Summary

The Gamma-Ray Observatory represents a dramatic step in capability at the highest energy region of the electromagnetic spectrum accessible by space platforms. Many exciting advances in our knowledge of our explosive universe can be predicted based on the pioneering work of many previous missions with instruments with narrower objectives. The Space Shuttle brings the opportunity to make these advances and the many unpredictable discoveries that result when such dramatic increases in observing capability occur. With the next decade should bring great advances in astrophysics.

THE ADVANCED X-RAY ASTROPHYSICS FACILITY

M.C. WEISSKOPF

Space Science Laboratory, NASA Marshall Space Flight Center, Alabama 35812

Abstract. The Advanced X-Ray Astrophysics Facility (AXAF) is a remarkable and unique scientific endeavor. It will be a space-based earth orbiting X-ray observatory, designed to address fundamental questions in astronomy and astrophysics. It is remarkable in that it represents enormous advances in observational capability. It is unique in that it will be the best means of obtaining high resolution, spectrally resolved X-ray images of astronomical objects for the rest of this century and beyond.

1. Introduction

Before the onset of the space age our view of the universe was restricted to those wavelength bands in which the earth's atmosphere was essentially transparent. Our current ability to place sophisticated instrumentation in orbit has opened new windows for studying the cosmos. In particular, X-ray astronomy has become a major discipline of astronomy. X-ray observations, which began by utilizing sounding rockets in the early 60's, have revealed a universe involving violent and energetic phenomena wherein particles are accelerated to relativistic velocities, wherein we see plasmas at temperatures from 10^6 to 10^9 K, and where there are magnetic fields stronger than 10^{12} Gauss. X-ray observations have discovered such phenomena in wide variety of astronomical settings ranging from normal stars to quasars and have had a profound influence on astronomy, astrophysics and cosmology.

The ubiquity of X-ray emission together with the fact that many of the fundamental physical processes give rise to copious amounts of X-ray emissions are the basis of the scientific importance of the AXAF. The growing partnership between astronomy and physics further supports this importance. The ultimate fate of the universe could be discovered by accurate and unambiguous measurements of the the Hubble constant and the deceleration parameter. The existence of exotic particles predicted by supersymmetric particle theories might be confirmed or ruled out by observing their effects on the hot X-ray emitting gas found in galaxies and clusters of galaxies. The equation of state of matter at nuclear densities can be tested by studying the X-radiation from neutron stars. These investigations, which will be performed with the AXAF, address some of the most fundamental questions in science today. The AXAF will play a crucial role in finding the answers.

AXAF is the successor to NASA's second High Energy Astronomy Observatory, the EINSTEIN, which was flown from 1978–1981, and made significant discoveries. Like EINSTEIN, AXAF will utilize grazing incidence X-ray optics capable of forming X-ray images. AXAF, however, will go far beyond EINSTEIN in capability. AXAF will have ≈10 times better angular resolution, 100 times better sensitivity for the detection of point sources, and up to 1000 times the sensitivity for high resolution spectroscopy. Using the history of science as a guide, such increases in

Y. Kondo (ed.), Observatories in Earth Orbit and Beyond, 71–79.

sensitivity guarantee that AXAF will make major advances in our understanding
of nature.

To accomplish its objectives, AXAF will be built around a large area, high resolution, grazing incidence X-ray telescope, accompanied by a complement of imaging
and spectroscopic instruments which can be maintained and/or replaced in orbit.
The optics, of Wolter-I design, are formed of parabolic and hyperbolic surfaces
of revolution. The overall characteristics of the facility and its first generation instruments are summarized in Table I. A more detailed discussion of the AXAF
characteristics may be found in the review paper by Weisskopf (1987).

It is important to note that AXAF's long life of 15 years will contribute to its
scientific potential. There will be time for the new theoretical and observational
work stimulated by unexpected discoveries to be checked against the phenomena
while the phenomena are still open to observation. This type of capability represents
an important step for space astronomy.

The AXAF will be open for use by the entire astronomical community, with the
majority of observing time and all the archival data set aside for general observers
to be selected by peer review. Based on NASA's experience with a number of
programs, this also implies substantial international participation.

2. Galaxy Clusters

AXAF is critical to progress in astrophysics because it will be the premiere facility
for observations in the X-ray region of the electromagnetic spectrum. It is no accident that this portion of the spectrum is so rich in information. This richness is a
consequence of the convolution of any extragalactic X-ray source spectrum with the
transmission of this flux in the interstellar medium in our galaxy which leads to a
peak response in low energy X-rays. The spectral sensitivity of AXAF (0.1-10 keV)
thus offers a direct means of probing the universe in a critical wavelength region.

The phenomena that give rise to X-ray emission are of paramount importance
in that they involve highly energetic cosmic events and their consequences. These
include stellar and galactic explosions, accretion of matter onto gravitational collapsed objects, and high temperature gases in variety settings including clusters
of galaxies. Phenomena such as these announce their existence primarily through
their X-ray emission and provide us with a different picture of the universe both
qualitative and quantitaive, than the one we are used to seeing. As an example,
we show in Figure 1 the optical and X-ray view of a cluster of galaxies -A1367.
The X-ray emission is dominated by the extended emission coming from a diffuse,
hot gas filling the space between the galaxies in the cluster. This gas has a characteristic temperature of about 5×10^7 K and therefore radiates primarily at X-ray
wavelengths. Visible light images gave no hint as to the existence of this gas; yet
the gas itself accounts for about half of the directly observable mass of the cluster
and provides important structural information through its density and temperature profiles. Clearly a thorough understanding of the formation, dynamics, and
evolution of such systems will be accomplished with AXAF.

AXAF spectra of the hot gas in clusters will allow examination of the heavy
element content, already known to be signficant from earlier X-ray observations,

TABLE I
AXAF characteristics

Weight	12,000 kg
Dimensions	cylindrical, 14 m long, 4 m diameter
Power	2500 watts, orbital average for entire observatory
Data Rate	32 kbps, 24 kbps science
Pointing Accuracy	30 arc sec
Pointing Stability	0.5 arc sec for 10 sec
Post Facto Aspect Accuracy	0.5 arc sec (rms diameter) relative
Post Facto Aspect Accuracy	1.0 arc sec (1σ radius) absolute
Telescope Angular Resolution	0.5 arc sec FWHM
Telescope Dimensions	1.2 m diameter, 10 m focal length, $1°$ field of view
Telescope Geometric Area	1700 cm^2
Spectral Range	0.1–10 keV
Orbit	600 km circular, 28.5 inclination
Lifetime	15 years with on-orbit servicing
Instruments	record position, energy, and arrival time for individual photons
Low Energy Transmission Gratings	cover the bandwidth from 0.1–4.0 keV with resolving power from about 2000 at low energies falling to 100 at 2 keV.
High Energy Transmission Gratings	cover the bandwidth from 0.5–9.0 keV with resolving power from 1000 at the lower energies falling to 100 at 9 keV.
CCD imager	an array of CCD devices with sub arc-second angular resolution, approximately 150 eV energy resolution and an energy-dependent quantum efficiency that averages better than 30% over most of the 0.1–10.0 keV bandwidth.
Microchannel Plate Imager	A 32' by 32' imager with 0.5 arc-second angular resolution. The device has an energy-dependent efficiency ranging from 20–50% (0.1–3 keV) and 10–20% (3–8 keV) and offers a modest spectral resolution of 1 at 1 keV.
X-ray Calorimeter	a spectrometer with 10 eV energy resolution and high quantum efficiency which subtends a 1 arc-minute square field-of-view.

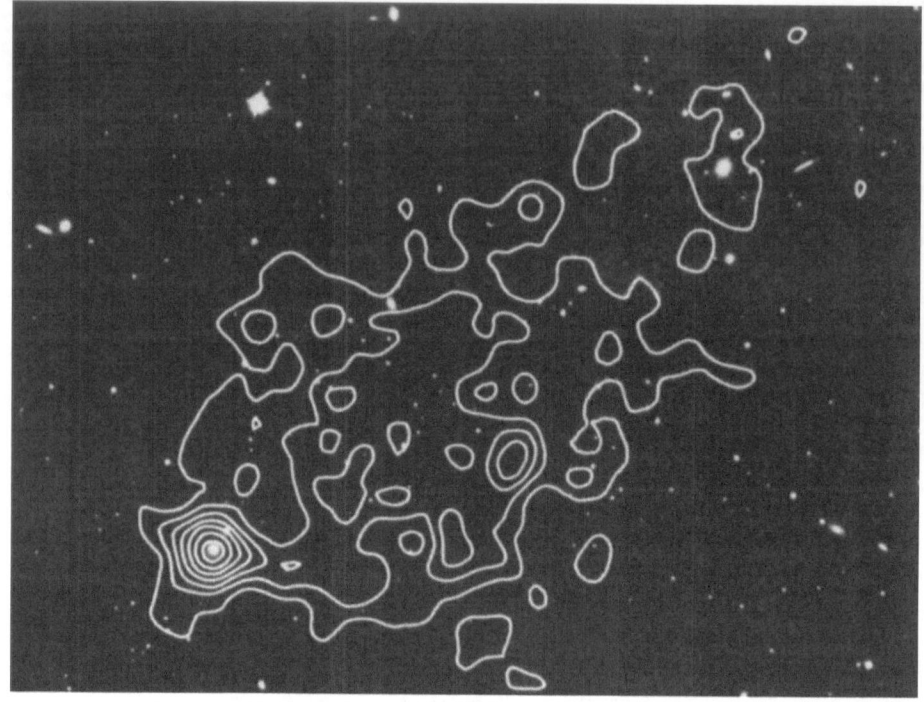

Fig. 1. Optical image and X-ray contours of A1367.

but previously impossible to study in detail due to an inability to combine high angular and high energy resolution. For the nearer clusters AXAF will be able to measure the spatial distribution, temperature and ionization state of the heavy-element content through iron. The potential of such studies is illustrated in Figure 2, the Einstein image of the galaxy M86, a member of the Virgo cluster of galaxies. The extended plume to the upper right indicates a large amount of gas being stripped from the galaxy and added to the cluster gas. AXAF will be able to measure the heavy element content of the plume and compare it to the cluster as a whole to see if the galactic gas is relatively enriched, as would be expected if it had been processed through stars. If all the heavy elements are produced in stars, AXAF studies of more distant, and therefore younger clusters, should uncover an evolutionary pattern. Even then, possible surprises could be provided by the detection of heavy elements produced by a very short-lived generation of massive stars that exploded very early in the cluster development. Such studies, including the detailed X-ray spectroscopy of supernova in our own and nearby galaxies, should cast new light on the sequence of events leading from the Big Bang to the formation of structure in galaxies, stars, and provide essential information about the seed material from which our solar system and indeed we were formed.

An extremely important question that AXAF will address involve understanding how the hot gas, that permeates individual galaxies, as well as clusters, is heated.

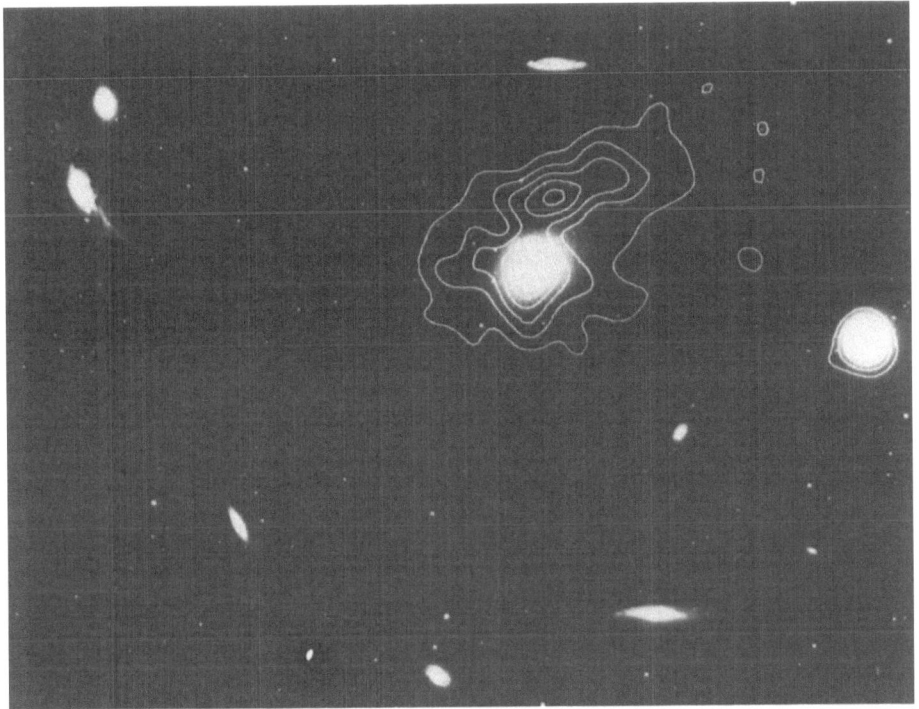

Fig. 2. Optical image and X-ray contours of M86 showing the extended plume.

Further, observations of the manner in which these gases cool through their X-ray emission have important relevance for star formation, which requires a large aggregation of gas to lose enough heat so it can collapse under its own gravity.

3. Dark Matter

Observations with AXAF will be particularly well-suited for measuring the distribution of dark matter on various size scales, from stars, to galaxies, to clusters of galaxies, to clusters of clusters or superclusters. Although theories abound, the dark matter has not to date been directly observed, it can be studied through its gravitational effects. The very hot gas (e.g., Figure 1), which can only be seen in X-rays has been observed in many large astronomical systems and is evidently being held in the gravitational grip of dark matter. Careful observation of such gas with AXAF can be used to map out the structure of this gravitational field, which depends on the temperature and mass of the individual dark matter particles. Comparisons between the observed and predicted properties of the structure of the gravitational field can help in evaluating different theories.

On a more speculative level it is feasible that AXAF may directly detect the X-ray emission associated with some of the exotic dark matter candidates such as one-dimensional discontinuities in quantum gauge fields, called cosmic strings. Some

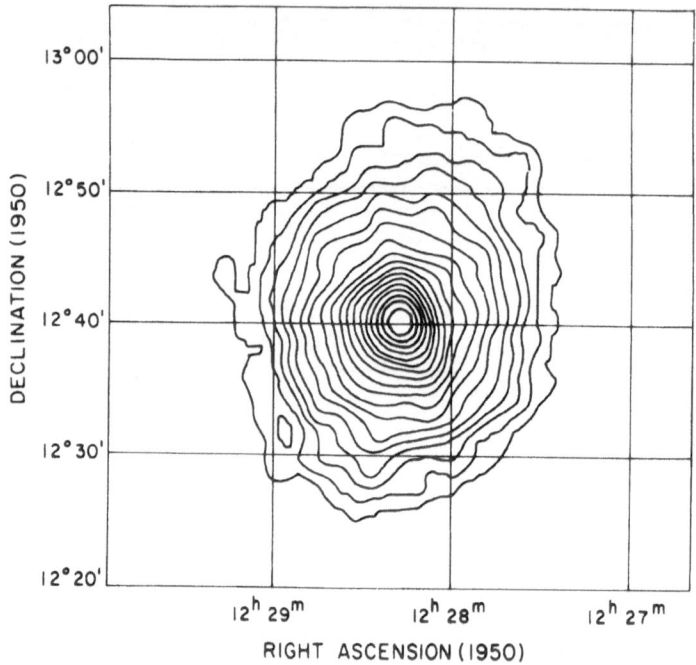

Fig. 3. X-ray contours of the giant ellptical galaxy M87.

new theories in particle physics predict that these exotic objects were created by a phase transition in the early universe and could exist over a variety of size and mass scales. If the strings are superconducting, as suggested, then their motion through a magnetized plasma could accelerate particles to high velocities and eventually produce X-radiation.

For the problem of dark matter in our galaxy, AXAF will be used to search for a variety of potential dark matter candidates. These studies include searches for the coronal X-ray emission from very low mass stars, for thermal X-ray emission, from the surfaces of isolated, cooling neutron stars, and for X-ray emission from isolated black holes accreting matter from the interstellar medium – all of which may be undetectable at other wavelengths, but are easily within the AXAF's capability.

The technique for probing dark matter in other galaxies is illustrated by the EINSTEIN observations of the giant elliptical galaxy M87 shown in Figure 3. The emission extends over a region at least three times larger than that from which visible light has been detected. Such data can be converted to a profile of the gas density by using the fact that the X-ray intensity emitted by gas at low density scales as the square of the gas density. Together with a measure of the temperature profile, and under the reasonable assumption that the bulk of the gas is in hydrostatic equilibrium with the outward directed thermal pressure balanced by the inward directed gravitational force, one can compute the total mass interior to a given radius. When this was done for M87, for example, a mass profile was determined along with a total mass estimate of between 4 and 8×10^{13} $M_\odot$. This is about 100 times more mass that determined from the visible light output of the galaxy, and also at least 30 times the mass seen in the X-ray emitting gas.

These measurements are unique in their use of the hot, X-ray emitting gas as a tracer to map the underlying gravitational potential or mass. In this manner, AXAF can measure the actual masses (including the dark matter content) of individual galaxies with hot gas in the outer regions as well as the masses of clusters of galaxies. Optical measurements to accomplish the same goals are not always feasible, often require numerous spectroscopic observations of individual tracer stars or galaxies, and may be greatly biased by local departures from isotropy and dynamical equilibrum. The X-ray data are not effected by these problems and therefore, in this particular endeavor, emphasize the uniqueness of X-ray observations. The EINSTEIN data demonstrated the feasibility of this approach, and AXAF will extend these studies to much fainter and more distant objects while also greatly improving the accuracy of surface brightness and temperature distributions. AXAF will determine masses to a precision of 10%, an improvement of a factor of 10 over EINSTEIN.

4. The Distance Scale of The Universe

Another fundamental issue which AXAF will address is the distance scale of the universe, as well as its age and ultimate fate. A particularly novel and fundamental measurement of cosmic distances can be made by combining AXAF studies of clusters of galaxies with ground-based radio or microwave observations. This measurement hinges on the so-called Sunyaev-Zel'dovich effect. The effect arises when the cosmic microwave background radiation passes through the gas in a cluster of galaxies. The energetic electrons in the hot gas cloud scatter the photons, slightly increasing their temperature. The predicted change in microwave background temperature scales as (gas density) $\times$ (radius of gas cloud) $\times$ (gas temperature). The X-ray observations will determine the luminosity [which scales as $(density^2) \times (radius^3)$], gas temperature, and angular size. The combined observations yield the cluster radius directly, and, from the angular radius, one can then determine the distance. AXAF data combined with already available microwave measurements would thus determine the distance to some clusters (e.g. A2218) to a precision of 30%. Improvements in the microwave measurements already feasible and extension to millimeter wavelengths could conceivably improve this to better that 7%

The Hubble Space Telescope will be used to measure distances with this order of precision, but on a relatively local scale. A2218, on the other hand, is at a distance comparable to that of many quasars, perhaps 50 times further away than HST will be measuring. At a minimum, AXAF will be able to apply this independent technique to determine the Hubble constant. Since the X-ray/microwave technique can, in principle, be extended to very large distances, AXAF has the potential to measure the Hubble constant as a function of distance. By looking at several clusters at distances corresponding to look-back-times of 10 billion years, AXAF could detect about a 20% change in the Hubble constant. This level of precision could thus distinguish between open and closed universes. Inflationary cosmological models involving an early phase of exponentially fast expansion, predict that the universe should be just at the boundary between open and closed. If supported by the AXAF observations, then a large amount of additional mass must be present in galaxies, in clusters, or in the voids of space.

5. Stars to Quasars

It is believed that stars, like planets, generate magnetic fields in their interiors with charged fluid motions. The primary AXAF observations in this area will involve the detection of X-ray emission from the outer coronae of thousands of stars, coupled with detailed X-ray spectroscopy for stars of varying temperatures and luminosities. The capability for performing such experiments was established by AXAF's prototype; EINSTEIN revealed the unexpected existence of hot X-ray emitting coronae around all types of stars. The conventional wisdom held that only stars with internal convectional processes like those of our Sun had the mechanisms to generate such coronae. Massive stars, too hot to have convection zones, were therefore not expected to have coronae and thus strong X-ray emission. Theorists had to go back to the drawing board and have not yet returned. Stellar dynamo theories may be tested and improved by observing the correlations of coronal X-ray emission with stellar magnetic field strength, age, rotation, rate, and composition.

AXAF will study the evolution of high mass stars and probe the nature of the supernova process. The spatially resolved spectra of supernova remnants observed with AXAF should provide the data needed to distinguish the material ejected in the supernova explosion from the interstellar matter with which it has been mixed. With its sensitivity at relatively high energies, AXAF will be able to study the critically important iron lines. Combining this data with theoretical models will then help reveal the chemical composition of the progenitor star, its mass, and details about the nuclear reactions during the supernova event itself.

AXAF also will study the thermal radiation from neutron stars that have just been produced in supernova explosions and initially radiate in the X-ray region of the spectrum. The observed rates of cooling of such neutron stars can be used as diagnostics to determine such parameters as size and thermal conductivity, which in turn depend on the equation of state of matter under extremely high densities and on general relativistic thermodynamics. Because of their enormously high density and gravity, neutron stars offer an excellent laboratory for the study of matter under extreme conditions and of the effects of general relativity. Indeed the contact with basic physics has been well established. The best current limit on the flux of magnetic monoples from space comes from EINSTEIN Observatory limits on neutron star surface temperatures which would be increased by nucleon decay catalized by magnetic monopoles, if present.

What can AXAF tell us about the formation and evolution of quasars? Due to the absence of substantial intergalactic absorption at X-ray wavelengths and the ubiquitous association of X-ray emission with quasars, AXAF can discover and study quasars at higher redshifts than any presently known or detectable with the

Hubble Space Telescope. Such studies will determine the evolution of quasars over the longest time scales.

Related observations and deep surveys with AXAF are capable of determining the composition of the all-sky, cosmic X-ray background, thereby establishing the contribution of various classes of discrete sources to this background and setting a limit for any residual diffuse component. Since currently observed, broad-band X-ray spectra of individual objects differ somewhat from the spectrum of the background, and theoretical calculations argue against a substantial diffuse component, these studies may discover new classes of astronomical objects or at least require modifications of present evolutionary scenarios.

6. Conclusions

AXAF is a member of the Great Observatories, which also include the Hubble Space Telescope, the Gamma-Ray Observatory, and the Space Infrared Telescope Facility. Operating in concert, and complementing each other in their respective wavelength capabilities, the Great Observatories should place mankind in a unique and historical position to understand the universe. It is of course presumptuous to claim that fundamental problems may be fully resolved by any new scientific mission. Indeed, the success of science is measured as much by the questions it raises as the questions it answers. In exploring such questions as those described above, AXAF has the potential to increase our knowledge at a fundamental level. With its state-of-the-art equipment for imaging and analyzing X-rays, AXAF will peer farther into space and farther back in time than any previous X-ray telescope. AXAF will help answer some of the most fundamental questions that have stirred the human mind: How far does space extend? Will the universe come to an end? What are the forces at work in the cosmos? What are the basic constituents of matter?

Acknowledgements

I wish to thank Alan Lightman, Harvey Tannanbaum, Steve O'Dell and the members of the AXAF Science Working Group. Special thanks to Paul Gorenstein, William Forman and Christine Jones for the Einstein Images.

References

Weisskopf, M.C., 1988, *Astronomy and Astrophysics with the Advanced X-Ray Astrophysics Facility*, Space Science Reviews, **47**, pp. 47-93

THE ASTRO-D MISSION

Y. TANAKA

Institute of Space and Astronautical Science
3-1-1 Yoshinodai, Sagamihara, Kanagawa-ken 229, Japan

Abstract. Astro-D, the fourth X-ray astronomy mission of ISAS following Ginga, is a high-throughput imaging and spectroscopic X-ray observatory, scheduled for launch in early 1993. It utilizes multilayer conical X-ray mirrors which provide a large effective area over a wide range energy range up to 12 keV. The focal plane instruments comprize a set of CCD cameras and Imaging gas scintillation proportional counters. Main features of Astro-D are described, and the important astrophysical objectives are discussed.

1. Introduction

The Institute of Space and Astronautical Science (ISAS) has launced three X-ray astronomy satellites so far. The first small X-ray astronomy satellite "Hakucho" was launched in 1979. "Tenma", carrying for the first time the large-area gas scintillation proportional counters, was launched four years later (1983). The third X-ray astronomy observatory "Ginga" launched in early 1987 carries a large-area (4000 cm^2) proportional counter array jointly prepared by the Japanese and the U.K. groups, and is currently operational for its fourth year in orbit.

Astro-D is the fourth X-ray astronomy mission of ISAS, following Ginga. Astro-D is designed to be a high-capability X-ray observatory from which a number of important astrophysical investigations can be carried out. The main features of Astro-D are given in Table I, and the schematic view is shown in Fig. 1. It will be equipped with multilayer thin-foil telescopes which provide a large effective area over a wide energy range from below 1 keV up to 12 keV. It is capable of imaging the X-ray sky with a spatial resolution comparable to that of the IPC of the Einstein Observatory. Two different types of detectors, imaging gas scintillation proportional counters and CCD cameras, are employed for the focal plane instruments. The CCDs will provide a spectral resolution comparable or superior to that of SSS of the Einstein Observatory but cover a much wider energy range. This is particularly powerful for the spectroscopic studies of X-ray sources, i.e. observations of the characteristic lines and absorption edges of elements from oxygen through iron. These features characterize Astro-D to be the first high-throughput imaging and spectroscopic X-ray observatory covering a wide energy band over 10 keV.

Astro-D will be launched by the ISAS launch vehicle M-3S II into an approximately circular orbit of a 550–650 km altitude. The total weight is limited to 420 kg. The spacecraft is three axis stabilized. The absolute pointing accuracy will be approximately 1 arcminute. However, the spacecraft stability and the accuracy of post facto attitude reconstruction is ecpected to be better than 0.2 arcminutes.

The key elements of Astro-D have been successfully tested during the PM phase, and activities for the flight hardware (FM) fabrication have started for the launch scheduled for early 1993.

Y. Kondo (ed.), Observatories in Earth Orbit and Beyond, 81–87.
©1990 *Kluwer Academic Publishers.*

Y. TANAKA

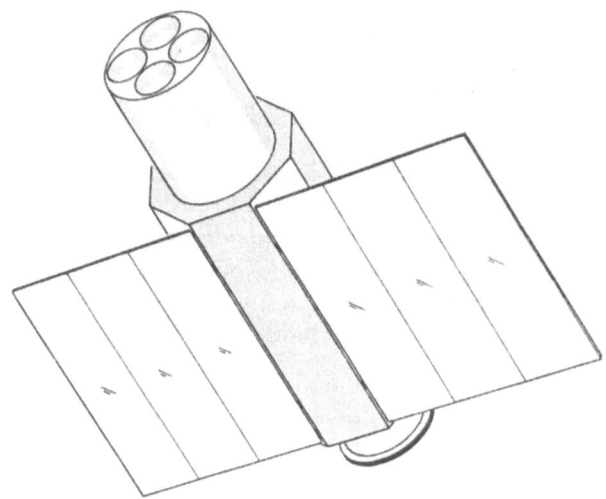

Fig. 1. Schematic view of Astro-D.

TABLE I

Main features of Astro-D

---

1. High-throughput over a wide energy range

Telescope:	4 sets of multilayer thin-foil conical mirrors (collaboration with GSFC)
Effective area:	~ 1300 cm^2 at 1 keV
	~ 600 cm^2 at 6–7 keV
Focal length:	3.5 m
Image size:	~ 2 mm half power diameter (HPD)

2. Imaging capability

Focal plane instruments:

2 sets of Imaging gas scintillation proportional counters
& 2 sets of X-ray CCD cameras (jointly with MIT/Penn.State)
Field of view: 30 × 30 arcmin^2
Point source resolution: < 1 arcmin.

3. Good energy resolution

$\leq 8\%$ (5.9 keV) with IGSPC
$\leq 2\%$ (5.9 keV) with CCD (cooled to $\sim$-70°C)

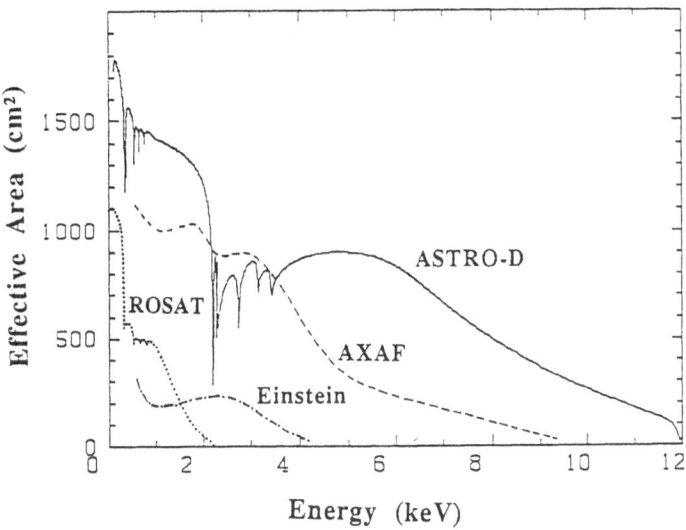

Fig. 2. On-axis effective area of four X-ray mirrors as a function of energy. Those of the Einstein Observatory, ROSAT and AXAF are also shown for comparison.

2. X-Ray Telescopes

The X-ray telescope consists of four sets of multilayer thin-foil conical mirrors. This technolgy is developed by P. Serlemitsos (1981). The mirror system of Astro-D is prepared by a collaboration between P. Serlemitsos and his collaborators at NASA Goddard Space Flight Center and the Nagoya University/ISAS group. An actual X-ray telescope consists of two stages, each 120 layers of coaxially aligned thin-foil cones in an approximate Wolter type I configuration. The radii of the outermost and the innermost cones are 17.3 cm and 5.5 cm, respectively, and each segment lenght is 10.9 cm. For such a short segment length and a long focal length, the approximation of a praboloid or a hypaboloid by a cone is fairly good. However, the image size actually achieved at GSFC is approximately 2.5 arcminutes half-power-diameter (HPD) at present due to several error sources, among which the shape error has been identified to be most important. An effort to improve the image quality for Astro-D is under way.

Although the image resolution is moderate, this technology can provide a large effective area over a wide range with a very light weight. Therefore, it is particularly advantageous for Astro-D which is severly weight limited. Astro-D carries four identical telescopes with a focal length of 3.5 m. The total on-axis effective area for four sets of mirrors Astro-D as a function of X-ray energy is shown in Fig. 2, in comparison with those of the mirrors on board the Einstein Observatory, ROSAT and AXAF, respectively.

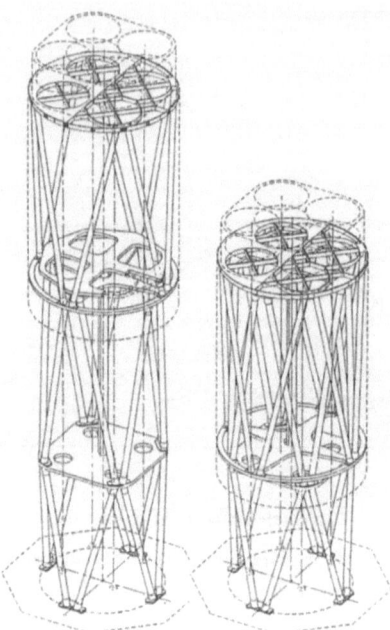

Fig. 3. Configurations of the optical bench during launch (top) and after extension (bottom).

3. Optical Bench

The nose fairing of the present ISAS launch vehicle, M-3S II, cannot accomodate an optical bench of a 3.5 m focal length. In order to achieve the required focal length, we employ an extensible optical bench which is folded short during the launch. It is extended in orbit to a 3.5 m focal length by means of a sliding mechanism, and latched precisely in position. Figure 3 shows schematically the configurations during launch and after fully extended, respectively. The optical bench is constructed by a truss structure composed of CFRP tubes in order to maintain the required accuracy against temperature changes. A prototype model of the extensible optical bench has been manufactured and verified. The entire CFRP surface will eventually be coated with aluminium in order to prevent outgassing.

4. Focal plane instruments

The focal plane instruments consist of two sets of IGSPC's and two sets of CCD cameras. These four detectors are placed at the foci of the four X-ray telescopes, respectively. These detectors will operate independently and simultaneously.

Each IGSPC is filled with Xe and sealed off with a thin berylium window. The IGSP's will cover the entire field of view of the telescope (larger than the coverage of the CCD camera). IGSPC has also a larger (and also less energy-dependent) detection efficiency for high energy photons above 5 keV than the CCD camera.

The CCD cameras are prepared jointlsy by G. Ricker and his collaborators at

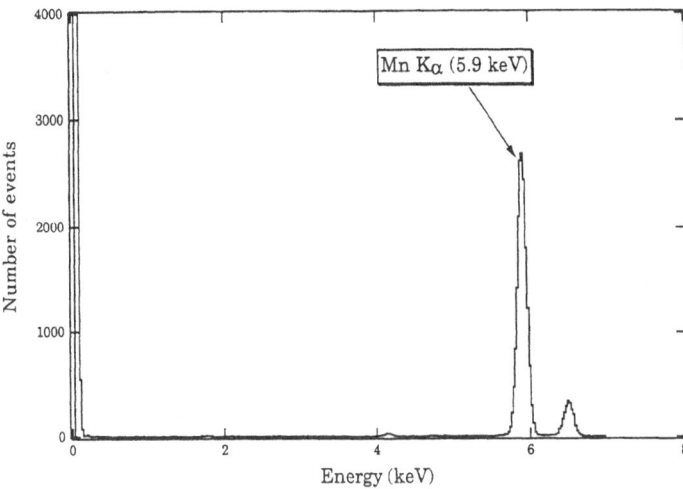

Fig. 4. Pulse-height spectrum of Mn-K lines obtained with a Lincoln CCD.

MIT and the Osaka University/ISAS team. Penn. State University group is also involved in the software development. A CCD camera head consists of the Lincoln Laboratory 840 × 840 hybrid CCD of front-side illumination (four 420 × 420 CCD chips abutted side by side, each read out by a separate preamplifier). The field of view coverage is 23 × 23 arcminutes^2. G. Ricker and his collaborators at MIT have been developing the X-ray CCD technology and demonstrated the excellent characteristics of the Lincoln lab. CCD's. An energy resolution of ∼ 2% for 5.9 keV (Mn-K) photons has been achieved at a modest temperature of ∼ -70°C. Figure 4 illustrates an example of the pulse-height spectrum of the Mn-K lines obtained with a Lincoln CCD. The K-alpha and K-beta lines are distinctly separated from each other.

The CCD camera will be a powerful instrument of Astro-D in particular for the spectroscopic studies of the lines from oxygen through nickel. It is intended to use CCD's of a ∼ 60 micron thick depletion layer, covering the entire energy range above 0.5 keV. The exptected detection efficiencies versus photon energy for IGSPC and CCD on board Astro-D are shown in Fig. 5.

Figure 6 shows the simulated energy spectra of a supernova remnant (Tycho's SNR) measured with a conventional proportional counter (PC), a gas scintillation proportional counter (GSPC), and a solid state detector (SSD) (CCD for Astro-D), respectively. IGSPC will resolve the characteristic line of each individual element unambiguously with a higher detection efficiency for high-energy X-rays. On the other hand, CCD can well resolve the K-alpha and K-beta lines of iron, and also enables measurement of the Doppler shift or broadening corresponding to a velocity of the order of 1000 km/sec which is considered to be typical in X-ray sources.

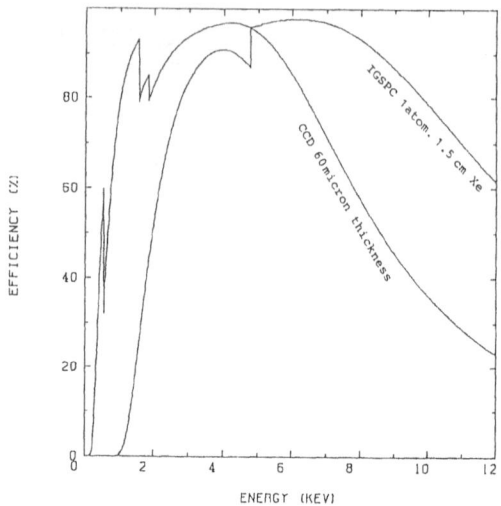

Fig. 5. Expected detection efficiencies for IGSPC and CCD.

X-ray Spectrum of Tycho Observed by

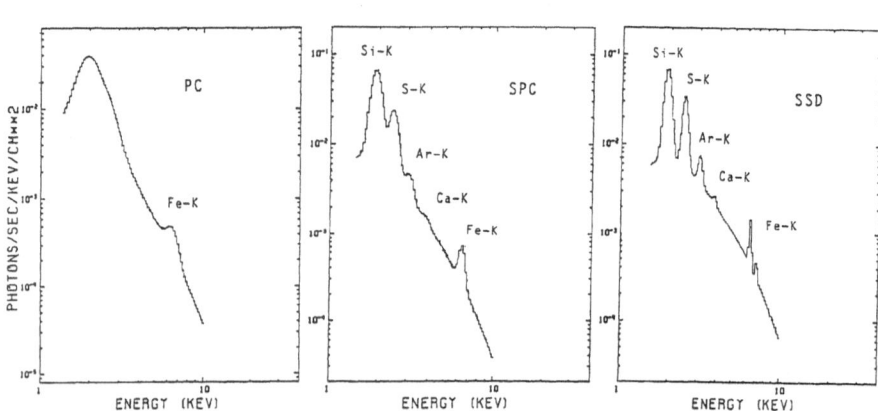

Fig. 6. Simulated spectra of Tycho's SNR measured with PC, GSPC and SSD, respectively.

5. Scientific Objectives

With the capabilities described above, a variety of important scientific objectives can be addressed. Table II lists these objectives to be investigated with Astro-D. Among others, we consider the problem of cosmic X-ray background (CXRB) to be one of the most important issues. The largest contributor to CXRB has been considered to be AGNs. However, thus far observed AGN spectra including many of those observed from Ginga are ualitatively different from that of CXRB. If discrete sources are to account for CXRB, Astro-D should be able to find new class of objects with much harder spectra than so-far known AGNs.

TABLE II
Scientific objectives of Astro-D

---

* Active galactic nuclei
        AGN's ~ 10 times fainter than the Einstein limit
        Spectra over a wide energy range
        Detection of characteristic lines

* Cluster of galaxies
        Temperature and abundance structure to z ~ 1
        Evolutionary effect
        Sunyaev-Zel'dovich effect

* Cosmic X-ray background
        Deeper than Einstein limit over a wide energy range
        Spatial fluctuation
        New population of galaxies, or diffuse?

* Galactic sources
        Spectroscopy: Emission and absorption lines,
                        Absorption edges
                        Doppler shifts and broadening
        Variability: Nature of QPO, flickering
        Individual sources in the neighboring galaxies

* Supernova & Supernova remnants
        SN 1987 A
        Spatial structure
        Temperature and abundance structure
        Search for neutron stars

* Emission along the galactic ridge
        Spatial structure
        Unidentified SNR's and/or new population sources?

References

Serlemitsos, P.J.: 1981, *Broad Band X-Ray Telescope (BBXRT)*, X-Ray Astronomy in 1980's, NASA TM 83848, p. 441

THE X-RAY TIMING EXPLORER[*]

H. V. BRADT, A. M. LEVINE, E. H. MORGAN and R. A. REMILLARD

Center for Space Research (Room 37-587), MIT, Cambridge MA 02139 USA

J. H. SWANK, B. L. DINGUS, S. S. HOLT and K. M. JAHODA

Laboratory for High Energy Astrophysics, Code 666, GSFC
NASA, Greenbelt, MD 20771 USA

R. E. ROTHSCHILD, D. E. GRUBER, P. L. HINK, AND R. M. PELLING

Center for Astrophysics and Space Science, UCSD C-011, La Jolla CA 92093 USA

and

J. G. JERNIGAN

Space Sciences Laboratory, UCB Berkeley, CA 94720, USA

Abstract. The capabilities of the X-ray Timing Explorer (XTE) are described with particular attention paid to current scientific problems it will address from galactic neutron star systems to active galactic nuclei. It features a low-background continuous 2–200 keV response with large apertures (a 0.63-m^2 proportional counter array and a 0.16-m^2 dual rocking NaI/CsI scintillation array). Rapid response (in hours) to temporal phenomena, e.g. transients, is obtained by virtue of a scanning all-sky monitor and rapid maneuverability. XTE will carry out detailed energy-resolved studies of phenomena close to neutron stars (e.g. QPO's) because of its sub-millisecond timing (to 10 μs), its high telemetry rates (to 256 kb/s), and the high throughput of its data system (to $\gtrsim 2 \times 10^5$ c s^{-1}).

1. Introduction

The X-ray Timing Explorer will carry out timing studies of compact objects, galactic and extragalactic. Its instrument complement will consist of a large-area proportional counter array (PCA; 2–60 keV; 6250 cm^2), a high-energy crystal scintillation experiment (HEXTE; 15–200 keV; 1600 cm^2), and a continuously scanning all-sky monitor (ASM; 2–10 keV; 90 cm^2). The essential parameters of XTE are given in Table I. The fields of view of the instruments are shown in Fig. 1. The effective areas of XTE and of other missions are given in Fig. 2. Expected accumulated counts for several types of sources (for differing time intervals) are given in Table II. XTE will be carried on the NASA Explorer Platform for at least 2 years. It is scheduled for launch in early 1995.

The mission was described in 1984 (McClintock and Levine 1984); the instruments have since been reduced by about 30% to reduce costs and to make them compatible with a Delta vehicle should the shuttle become unavailable (Bradt, Swank and Rothschild 1990). The science that can be addressed by the XTE and

---

[*] This article originally appeared in *Advances in Space Research* (Bradt, Swank and Rothschild 1990). It is reproduced here (with minor changes) with the permission of Pergamon Press.

Y. Kondo (ed.), Observatories in Earth Orbit and Beyond, 89–110.
© 1990 *Kluwer Academic Publishers.*

TABLE I
Parameters and Features

POINTED INSTRUMENTS
      6250 cm^2; Xe Proportional Counters; 2 - 60 keV; 'PCA';
         GSFC HEAO-A2 type; low background.
      1600 cm^2; NaI/CsI; 15 - 200 keV; 'HEXTE';
         UCSD; rocks on-off source continuously; low background

PARAMETERS OF POINTED INSTRUMENTS

Net Area	3000 cm^2 at 3 keV
	6000 cm^2 at 10 keV
	1200 cm^2 at 50 keV (NaI); 800 cm^2 at 50 keV (Xe)
	1100 cm^2 at 100 keV
	300 cm^2 at 200 keV
Field of View	1^0 FWHM circular; PCA and HEXTE coaligned
Energy Resolution	18% at 6 keV (Xe)
	18% at 60 keV (NaI)
Sensitivity*	0.1 mCrab 2-10 keV (in minutes); limit of source confusion
	1 mCrab 90 - 110 keV (3σ; 10^5 s)
Time Resolution	10 μs (system uncertainty)
Background	2 mCrab 2-10 keV
	100 mCrab (1×10^{-4} cts cm^{-2} s^{-1} keV^{-1}) at 100 keV
Telemetry	18 kb/s (PCA) and 5 kb/s (HEXTE) continuous; 256 kbs ~30 min/day (PCA)
Flight Data System (MIT)	Flexible binning criteria (microprocessor-driven)
(for PCA/ASM)	Pulsar folding and high time resolution burst searches
	Simultaneous binning with different criteria
	Real-time sub-ms resolution FFT's
	High throughput, $>2 \times 10^5$ ct s^{-1}
HEXTE Data Modes	Binned, event encoded, pulsar fold, burst trigger, optimum high-speed code

ALL-SKY MONITOR ('ASM'; MIT)

Energy Range	2-10 keV; 3 energy channels
Net Area	90 cm^2 net
Positional Resolution	0.2^0 x 2^0 (Positions to 3' x 30')
Scan time	90 min; 80% of the sky per orbit
Sensitivity	30 mCrab in 1.5 h; 10 mCrab in 1 day
Telemetry	3 kb/s
Dissemination	Routine analyses carried out immediately to assess need for pointed studies, i.e. a S/C maneuver. Results placed in public domain (computer access) to aid community in proposal writing, optical observations, etc.

SPACECRAFT and OPERATIONS

Maneuverability	25^0/min; precise to < 0.1 deg.; aspect to 0.02 deg.
	85% of sky accessible (includes anti-sun pointing for coordinated observations)
Response to Transients	Target acquisition a few hours after detection

USER (GUEST) PROGRAM

PCA/HEXTE	100% of time competitively assigned (PI's also compete)
	Single object, class studies, or contingency (i.e. transients) proposals allowed
	Observing at SOC or at PI institutions (with less support)
	Remote observing from observer's home institution possible
	Multi-wavelength coordinated observations encouraged.
ASM	Proposals for specialized analyses possible

* 1 mCrab ~ 1.06 μJy at 5.2 keV

TABLE 2: Expected Counts*

	2-10 keV	10-30 keV	>30 keV
Proportional Counter Array (PCA)**			
Background (1s)	20 cts	24 cts	16 cts
Crab Nebula (1s)	8700	1205	80
Her X-1 (1.24s) 1390	850	15	
X1728-34 (1s; burst)	10,625	3750	2
Sco X-1 (1s) 160,000	4600	4	
Cyg X-1 (1ms flare)	23	9	1
(2.5 Crab)			
SS Cyg (5 mCrab; 1s)	44	4	-
AGN (1.3 mCrab; 10 s)	113	42	4
1/2 High-Energy Experiment (HEXTE)***			
Background (1s)	-	6 cts	29 cts
Crab Nebula (1s)	-	170	130
Her X-1 (1.24s) -	40	2	
X1728-34 (1s; burst)	-	80	8
Sco X-1 (1s) -	1670	40	
Cyg X-1 (1ms flare)	-	1	3
(2.5 Crab)			
SS Cyg (5 mCrab; 1s)	-	–	–
AGN (1.3 mCrab at 5 keV; 10 s)	-	9	10
1/3 All Sky Monitor (ASM)****			
Diffuse background (1s):	40		
Crab Nebula (1s):	90		
Sco X-1 (1s) 1400			

* The assumed integration time for each source is given.

** "1 mCrab" nebula flux $\approx$ 1.06 μJy at 1.25×10^{18} Hz (5.2 keV).

*** Nominally, only 1/2 the HEXTE is on a source at a given time due to the 'rocking' collimators.

**** Counts are for one of the 3 shadow cameras with 30 cm^2 net effective area (60 cm^2 gross) for a given source. (The mask occults 1/2 of the total 60-cm^2 sensitive area.)

by the Japanese *Ginga* mission was the subject of a workshop in 1985 which was given a vivid presentation by the organizers (Epstein *et al.* 1986).

2. Unexplored measurement phase space

The power and uniqueness of the mission for timing and broad-band spectral studies derives from the synergism of its capabilities. The domains of largely unexplored phase space in 1995 will be:

Sub-millisecond timing (to 10 μs). This is made possible by high telemetry rates (32 kb/s continuous and 256 kb/s for $\sim$ 30 min. per day), large apertures, and high-rate processing capability to $\gtrsim 2 \times 10^5$ c s^{-1} (e.g. Sco X-1). The sub-millisecond regime must be explored; the natural time scale for matter near a 10-km neutron star is $\sim$ 0.1 ms (size $\div$ free fall speed $\approx$ dynamical time scale).

Rapid response to temporal phenomena. The ASM and the rapid maneuvering capability (180$°$ in 7 min) make possible early detection and rapid PCA/HEXTE

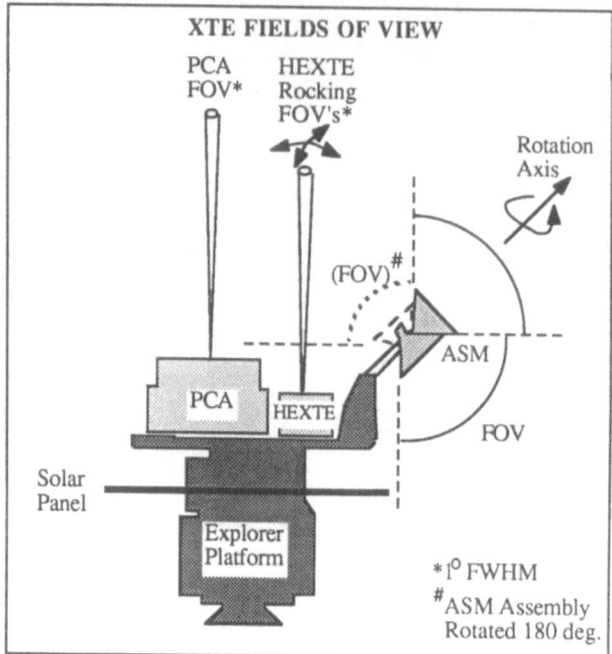

Fig. 1. Schematic view of XTE showing the fields of view of the Proportional Counter Array (PCA), the High-Energy X-ray Timing Experiment (HEXTE), the All-Sky Monitor (ASM), and elements of the Explorer Platform, i.e., the Multi-Mission Spacecraft (MMS) and the Payload Equipment Deck (PED).

acquisition of temporal phenomena (e.g. transients and state changes) within a few hours of an ASM detection.

Multi-wavelength observations. The PCA can be maneuvered rapidly (see above) and can be pointed to ∼85% of the sky any day of the year (i.e. the solar constraints are minimal). This makes possible all-night coordinated optical observations (anti-sun pointing by XTE) and source acquisition by XTE during exact times of coordinated observations.

Continuous sensitive spectra from 2 keV to 200 keV. The large effective apertures over the entire 2-200 keV band together with the low and well measured backgrounds of both the PCA and HEXTE should make XTE significantly more sensitive than other missions with comparable bandpasses.

Long-term and very-sensitive monitoring of fluxes and pulsar phases. The flexible maneuvering capability of XTE permits frequent repeated observations with the PCA. (The ASM will routinely perform such measurements for the brighter sources.)

3. The explorer platform (EP)

The Explorer Platform is essentially the same type of Multi-Mission Spacecraft (MMS) that carries the Solar Maximum mission. It is highly maneuverable (25°/min)

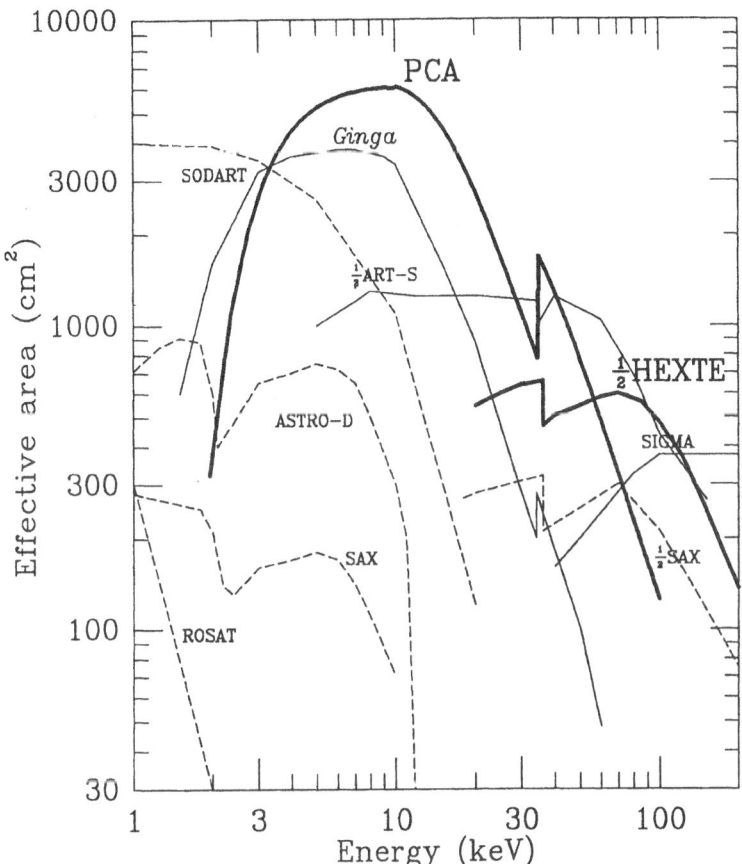

Fig. 2. Effective areas for XTE (dark lines) and other missions or instruments. The two HEXTE assemblies usually each rock on and off the source; thus nominally only 1/2 the area is on the target source at a given time. The *Granat* mission (launched 1989) carries ART-S (Sunyaev 1990) and SIGMA (Debouzy 1983). The *Spectrum* X mission (1993) will carry SODART (Schnopper 1990). The curves represent the estimated net area of the entire detection system, except for the SODART curve which shows the design goal for the geometrical area of the concentrator. *Ginga* and *Granat* are currently in orbit (Makino *et al.* 1987). The areas for *Astro-D* (1993) (Tanaka, private communication), SAX (1992) (Spada, private communication), and ROSAT (1990) (Aschenbach 1988) are also shown. The curves for future missions (dashed lines) represent our understanding of the instruments and missions; we caution that the designs for some of these instruments are still evolving. Finally, the XTE backgrounds may be somewhat less than that of other non-focusing missions. This would enhance the XTE faint-source sensitivity (in a given time) relative that of higher-background instruments.

and can point the PCA/HEXTE to any point on the sky on any day of the year provided the PCA-sun angle is > 30°. It will provide, via tape dumps through TDRSS, a continuous telemetry rate of 32 kb/s of which 26 kb/s will be available for the scientific instruments. In addition, 256 kb/s for about 30 min. a day will be available. Neither XTE nor the Platform carries expendables that would artificially limit the active lifetime in orbit. XTE will be preceded on the Platform by the Extreme Ultraviolet Explorer (EUVE). XTE will be carried into orbit on the shuttle, and the EP/EUVE will be captured and placed in the shuttle bay. XTE will be exchanged with EUVE in the bay, and then the EP/XTE will be deployed. The orbit will, of necessity, be a low-earth orbit at altitude ~ 500 km at 28° inclination. The XTE mission would terminate when another experiment replaces it on the Platform.

4. The proportional counter array (PCA)

The proportional counter array will be 5 large detectors with net area of 6250 cm² (Fig. 3). Each detector is a large version of the HEAO-1 A2 HED detectors (Rothschild *et al.* 1979) that featured low-background through efficient anti-coincidence schemes including side and rear chambers and a propane top layer. The two window (the front one and the one separating the propane and xenon/methane chambers) are each 25-μm Mylar. The xenon of the 3 detection chambers is 3.6 cm thick at 1.0 atmosphere. The PCA is effective over the range 2-60 keV with 18% energy resolution at 6 keV and at least 128-channel pulse height discrimination. The 1° FOV (FWHM) yields a source confusion limit at ~ 0.1 mCrab. The PCA is being provided by GSFC; principal investigator J. Swank.

The Crab nebula will yield 8700 c s^{-1} (2-10 keV) and 1200 c s^{-1} (10-30 keV) in the PCA. The backgrounds in these 2 bands are respectively 20 and 24 c s^{-1}, corresponding to 2 and 20 mCrab respectively. With these backgrounds, an AGN source of intensity 1.3 mCrab (2-10 keV) and energy index 0.7 will be detected at > 2σ in only 1 s at 2-10 keV and at 3σ in 10 s at 10-30 keV. Monitored anticoincidence rates will yield the background reliably to at least 10% of its value.

The microprocessor-driven flight data system for the PCA can handle high throughputs to $\gtrsim 2 \times 10^5$ c s^{-1} (Sco X-1 yields 160,000 c s^{-1}) and can time events to 10 μs. The data stream can be binned and telemetered in several modes simultaneously. Each such mode can be chosen arbitrarily for optimum tradeoffs of timing and spectral information. The binned data can be preprocessed prior to being telemetered. Features of the system in the current design are burst searches at high time resolution, pulse folding, and on-line FFT's with time resolutions to 10 μs. The data system, which will also be used for the ASM, is being provided by MIT.

5. The High-Energy X-Ray Timing Experiment (HEXTE)

The HEXTE experiment consists of two rocking clusters of NaI/CsI phoswich detectors that cover the energy range 15–200 keV (Fig. 4). The detectors are improved versions of the HEAO-1 A4 LED detectors (Matteson 1978) which attained the low-

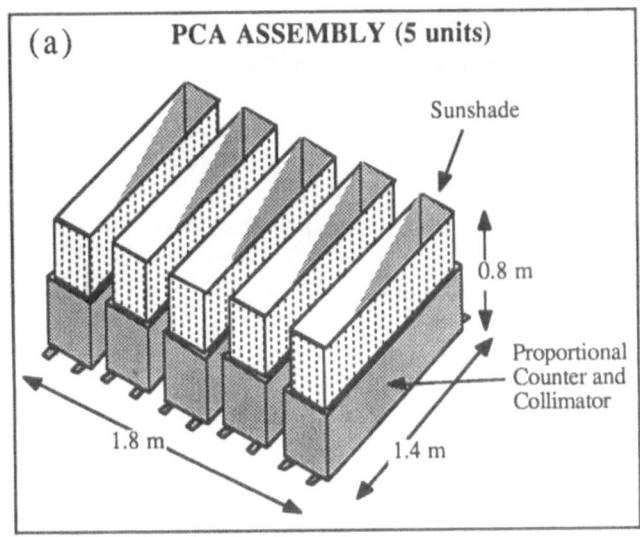

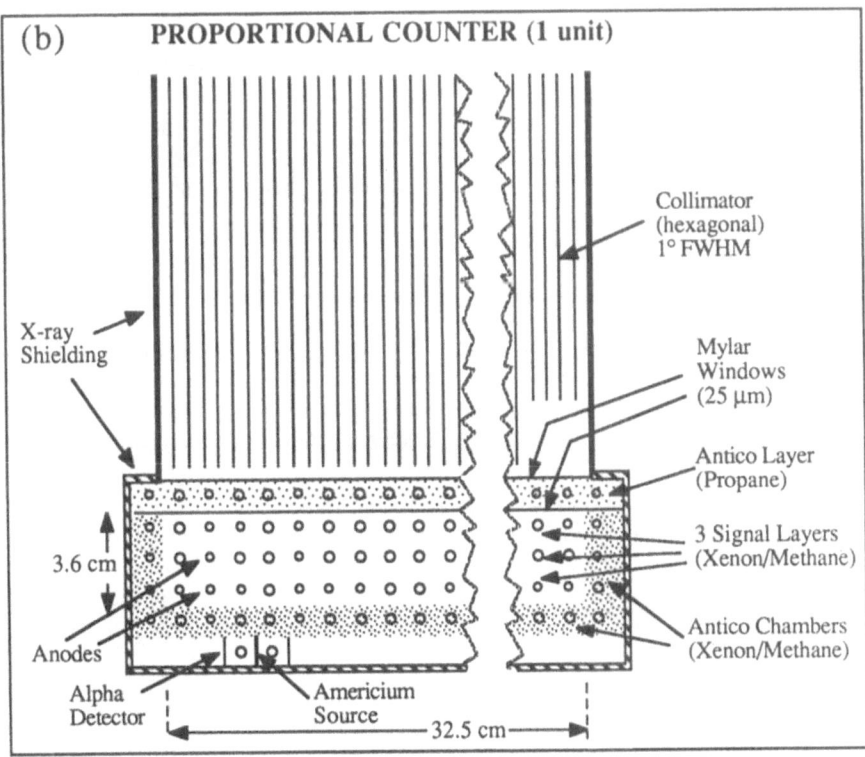

Fig. 3. Schematics of the Proportional Counter Array. (a) The 5 units of the PCA. (b) One unit of the PCA showing principal components but not engineering details (end view).

est in-orbit background for large area scintillators to date. Each detector consists of a 3-mm thick NaI primary detector coupled to a 38-mm thick CsI anticoincidence crystal that also serves as a light guide to the photomultiplier tube. Each detector has 200 cm^2 net effective area. Each cluster contains 4 detectors; the total net area of the entire system is 1600 cm^2. The 'rocking' field of view is 1° FWHM and is coaligned with the PCA when on source. The instrument will be provided by UCSD; principal investigator R. Rothschild.

The phoswich/collimators, under most circumstances, will be rotated ("rocked") on or off the source every $\sim$ 15 s to provide alternately source and background measurements. Each cluster will sample background positions on two opposing sides of the source, and the two clusters will rock in mutually perpendicular directions. Thus 4 background positions will be monitored. The rocking will be phased so that the source is continuously viewed by one of the clusters. A plastic scintillator 'box', viewed with photomultiplier tubes, serves as an anticoincidence shield for background reduction on the sides and underside of the cluster. Automatic gain control on each detector further refines the background knowledge.

The HEXTE flight data system will provide the following modes: binned, event encoded, pulsar fold, burst trigger, and an optimum high-speed code. The telemetry rate for HEXTE will be at least 3 kb/s. The Crab nebula will yield 170 c s^{-1} (15-30 keV) and 130 c s^{-1} (> 30 keV) in 1 cluster (1/2 HEXTE). The background in these energy bands will be $\sim$ 6 and $\sim$ 29 c s^{-1} respectively. At 100 keV, the background is about 100 mCrab (1×10^{-4} c cm^{-2} s^{-1} keV^{-1}). In the 90-110 keV band, the limiting sensitivity for detailed spectral analysis is expected to be about 1 mCrab (1×10^{-6} cts cm^{-2} s^{-1} keV^{-1}) or 1% of the instrument background. This sensitivity can be reached at 3 σ in 10^5 s.

6. The All-Sky Monitor (ASM)

The all-sky monitor (Fig. 5) consists of 3 'scanning shadow cameras' (SSC) on one rotating boom with a total net effective area of 90 cm^2 (180 cm^2 without masks). Each SSC is a one-dimensional 'Dicke camera' (Dicke 1968) consisting of a 1-dimensional mask and a 1-dimensional imaging proportional counter. The gross field of view of a single SSC is 6° × 90° FWHM, and the angular resolution in the narrow (imaging) direction is 0.2°. A weak source provides a single line of position of 0.2° × 90°. Two of the units view perpendicular to the rotation axis in nearly the same direction except that the detectors are rotated slightly (5° each) about the view direction. Thus they serve as 'crossed slat collimators'. The crossed fields provide a positional error region of 0.2° × 2° for a weak source and 3′ × 30′ for a brighter, $\sim$ 5σ, source. A spacecraft maneuver could reduce this to 3′ × 3′ . The third (similar) unit views along the axis of rotation. It serves in part as a 'rotation modulation collimator' and surveys one of the 2 poles not scanned by the other two. The ASM will be provided by MIT; principal investigator H. Bradt.

Each SSC detector is a sealed proportional counter filled to 1 atm with xenon-CO$_2$ and with a sensitive depth of 15 mm, in the current design. It has 9 position-sensitive anodes, a 50 μm beryllium window, a sensitive area of about 60 cm^2 of which only 1/2 can view a given celestial position at a given time, anticoincidence

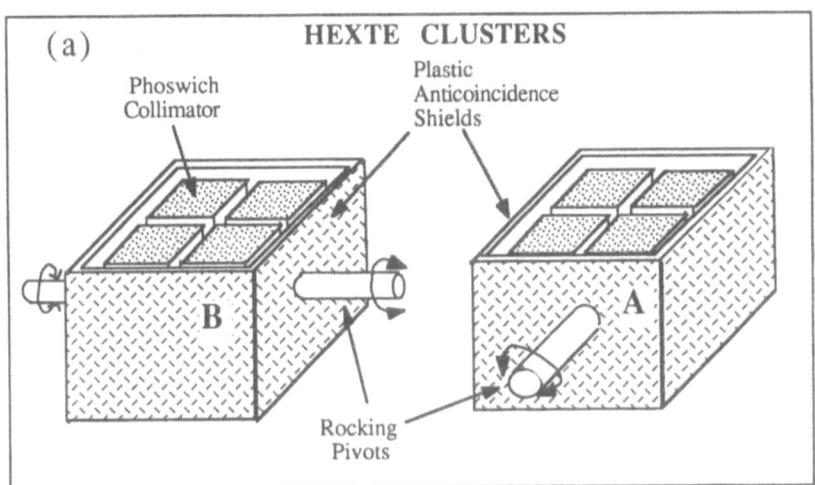

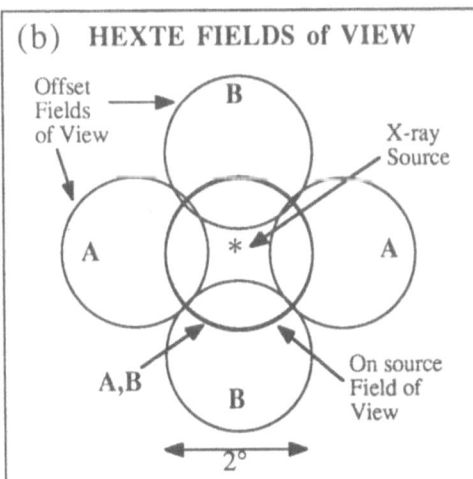

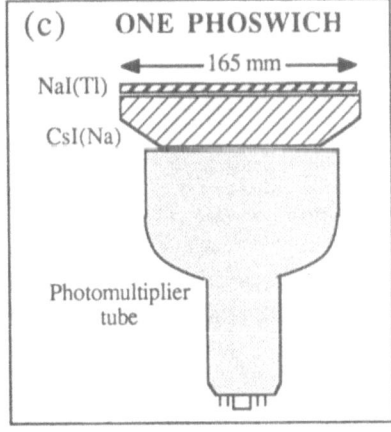

Fig. 4. Schematics of the HEXTE detector system: (a) The two rocking clusters, each with 4 phoswich detectors which lie inside a plastic scintillator anticoincidence shield ('box'). The collimators are not shown. (b) The fields of view sampled during the on-source and off-source orientations. (c) One of the 8 phoswich detectors shown without its collimator; the crystals are square when viewed from above.

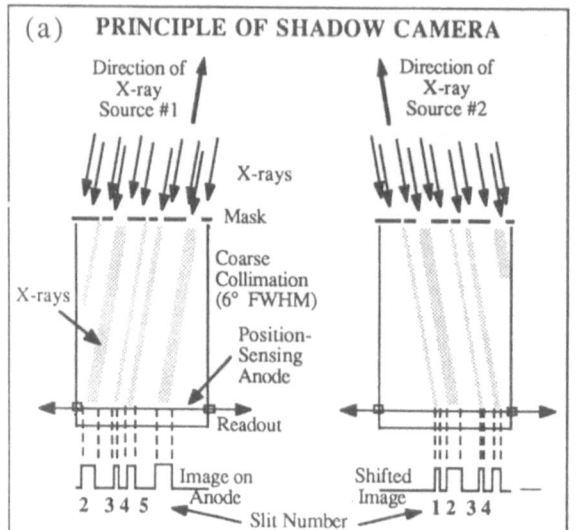

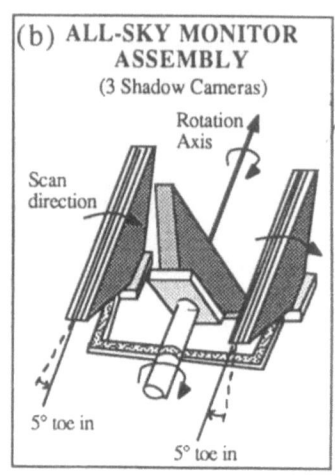

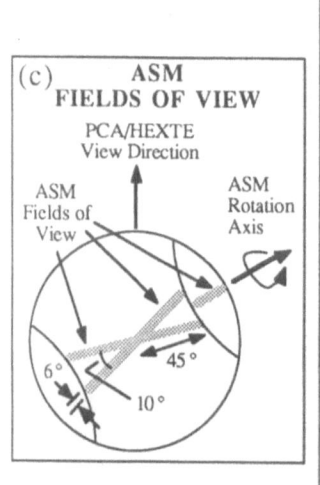

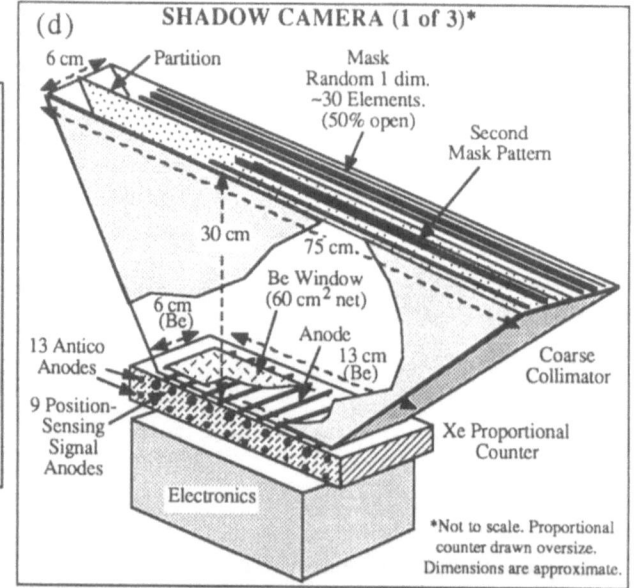

Fig. 5. Schematics of the All Sky Monitor system. (a) Principle of one-dimensional imaging showing shifted image for source #2. (b) Arrangement of the 3 shadow cameras. (c) Regions (grey) of the celestial sphere viewed by the 3 cameras at one angular position. (The rotation axis is shown oriented as in Fig. 1). (d) One of the 3 shadow cameras.

chambers on the sides and rear, and sensitivity to 2-10 keV x-rays with 3 energy channels. The one-dimensional nature of the SSC's minimizes the required telemetry rate. The scanning (or stepping) operation of the 'crossed-field detectors' ensures that each source gives rise to the entire mask pattern in the binned data, thus minimizing aliasing and side bands in the deconvolved results. The field of view in the scan direction is restricted to $\sim 6°$ to reduce source confusion and to reduce position smearing due to the finite depth of the detector. The relative merits of the SSC concept are described in some detail by Doty (1988).

The instruments will make a complete revolution once every 90 min, and $\sim 80\%$ of the sky will be surveyed to a depth of 30 mCrab (about 40 sources). Frequent spacecraft maneuvers will insure that 100% of the sky is surveyed each day. In one day, the limiting sensitivity becomes ~ 10 mCrab (~ 75 sources). The intensities derived from the data will immediately be made available in the Science Operations Center and to the community in general via computer links. The results will make possible rapid acquisition by the PCA/HEXTE of sources when they undergo a change of state, e.g. when a transient appears or when a low-mass binary system moves to the horizontal branch of the hardness-intensity plot.

7. The User (Guest) Program

The PCA/HEXTE observing program will be devoted 100% to guests ('Users') after a 30-d checkout period; proposal evaluation will be carried out by a NASA-appointed peer group. PI-institution observers will compete in this process. Proposals will be accepted for contingency observations, e.g. for a transient with particular characteristics. Results of an automated straightforward analysis of the ASM data (e.g., light curves and FFT's) will be placed in the public domain in near real time by the XTE team and made available via computer links. It is hoped the results will stimulate proposals for observations by XTE and optical/radio observatories. Specialized analyses of the raw ASM data or of preprocessed public-domain data will be possible through the User program.

Observations can be carried out at a Science Operations Center (SOC, at GSFC), at an instrumental-PI institution, or at the observer's home institution through a remote terminal. Adjustment of instrument configurations will be possible during observations to optimize scientific return. Standard analysis programs will be available at the SOC and will be made available to Users for use at their home institutions. Data will be provided promptly to Users in standard formats.

Real-time operations at the SOC will include examination of ASM data by the SOC staff and generation of commands for prompt observation of transients, etc. (Data rights will be determined by proposal; see above.) The flexible maneuvering of XTE should permit the transient observation and the previously scheduled observation to be multiplexed, thus minimizing disruption of the preplanned schedule.

The large community of U.S. observers, not limited to x-ray astronomers, is expected to be a major factor in making XTE a highly productive mission. International participation is also expected.

Multi-wavelength observations of necessity involve a wide community. The state of knowledge of the stellar systems (e.g. cataclysmic variables) is now sufficiently developed so that coordinated observations often are required for model building. Steps to facilitate such observations (e.g. by convenient scheduling) will be implemented.

8. The science objectives of XTE

XTE's capabilities allow it to address many astrophysical problems. In view of other missions expected to fly in coming years, XTE's observing program is likely to emphasize those objectives for which XTE is particularly well suited. We discuss here important XTE objectives and 1 or 2 specific observations pertaining to each.

8.1. NATURE OF THE X-RAY SOURCE IN ACTIVE GALACTIC NUCLEI

X-ray emission from quasars, Sy 1 galaxies, and BL Lac objects provides direct information about the regions close to the power source, presumably a massive black hole (10^6 to 10^8 $M_\odot$). At present, among the x-ray sources in the HEAO-1 LASS survey (Wood *et al.* 1984), 180 emission-line AGN (including $\sim$ 50 quasars) and 31 BL Lac objects are now securely identified (Remillard 1990; Bradt *et al.* 1988). These sources are all brighter than 0.7 μJy at 5 keV, well above the source confusion limit (0.1 μJy at 5 keV) of XTE. Spectra to above 100 keV may be obtained in a few days. They are also bright at optical wavelengths ($V \sim 14$–16) and thus are highly amenable to optical/x-ray studies.

Spectral studies are of utmost importance for several reasons: (1) The most elementary quantity, the x-ray luminosity, is not yet known for most objects; the energy flux per decade, νF_ν, is still rising at $\sim$ 30 keV and must cut off somewhere, possibly at 40 - 100 keV as suggested by a comparison to the diffuse x-ray background spectrum. (In a few cases, the spectrum is known to extend to $\gtrsim$ 40 MeV, e.g. Swanenburg *et al.* [1978].) (2) Extrapolations of the AGN spectra do not readily explain the diffuse x-ray background (Rothschild 1983) although AGN are a likely source of background. (3) Phenomena such as e^-e^+ pairs in the plasma or reprocessing of a high-energy beam by a cool accretion disk will influence the 2-200 keV spectrum in a manner that is discernible only in broad-band measurements.

Variability from 10^2 s to months has been detected in a number of AGN (see review by Urry [1988]), and these yield indications of the masses of the central engines. The speed-of-light crossing time for regions close to an engine of $\sim 10^7$ $M_\odot$ is $10^2 - 10^4$ s. The large-area, low-background, and high-energy response of XTE makes possible time-resolved spectral studies on these time scales which, for instance, should show hard vs soft lags if thermal Compton scattering is important. In 100 s, an AGN of 1.0 μJy (at 5 keV) yields in the PCA a 19σ detection at 2-10 keV and 6.5σ at 10-30 keV, and in the HEXTE, a 1.4σ detection at > 30 keV. Power density spectra from EXOSAT have slopes close to -1 down to $\sim$ 2000 s indicating *no characteristic time scale* (Lawrence *et al.* 1987; McHardy and Czerny 1987). XTE observations should be able to push this down to $\lesssim$ 200 s and to explore the variability at *higher* photon energies. Very low-frequency variability can be effectively monitored with daily short observations of $\sim$ 12 AGN.

BL Lac objects may be the most variable x-ray objects of all. The BL Lac object H0323+022 (Doxsey *et al.* 1983) is highly active in both HEAO-1, *Einstein* (to 60 s) and *Ginga* data (Doxsey *et al.* 1983; Feigelson *et al.* 1986; Ohashi 1988). PKS 2155-305 varies in hours (Snyder *et al.* 1980; Urry and Mushotzky 1982). The nature of variability is likely to be closely related to the relativistic jets that are believed to be operating in these objects. Spectra of the varying component above $\sim$ 10 keV should distinguish between a soft synchrotron spectrum and a hard inverse Compton spectrum. BL Lac objects are quite numerous in the HEAO-1 surveys (see above) but are deficient among the fainter *Einstein* serendipitous objects (Maccacaro 1984). The BL Lac phenomenon in the brighter HEAO-1 objects can be studied by XTE with the advantages of high statistics, a wide band pass, and a large sample of systems.

Observation: Spectra and rapid time variability. The 2-150 keV spectra of 20 bright ($\gtrsim$ 0.7 μJy at 5 keV) AGN can be measured with XTE during long observations ($\gtrsim 10^5$ s) simultaneously with the variability in the 2-10 keV band on time scales down to $\sim$ 10 s, and to $\sim$ 100 s in the 10-30 keV band. The analyzed data will bound the total x-ray luminosity, determine if AGN can provide the high-energy x-ray background, and permit a search for a characteristic time scale from the power density spectrum to $\lesssim$ 200 s.

Observation: Low-frequency variability. Low-frequency (days to years) variability of $\sim$ 10 bright AGN can be monitored with short daily PCA observations. This will yield the previously unexplored *very* low-frequency end of the PDS .

8.2. STRUCTURE OF ACCRETING NEUTRON STARS

The masses of neutron stars have been deduced from determinations of x-ray and optical mass functions through measurements of the Doppler shifts of the x-ray pulsing. (See review by Joss and Rappaport [1984]) Neutron-star radii have been determined from measures of burst temperatures and luminosities, for assumed distances. Evidence that some burst luminosities are Eddington limited have made possible direct measures of distance (Ebisuzaki 1984). Better time-resolved spectra by XTE can yield markedly improved x-ray measures of the distance to the galactic center (Joss 1988).

Fluctuations in rotation rate of x-ray pulsars (accreting neutron stars) for the most part have their origin in the torques due to infalling accreting matter. The response of the neutron-star to the accretion torques depends upon its internal structure and can be detected through the timing of x-ray pulsations (Lamb 1978). If magnetized vortices embedded in the superfluid core provide coupling to the crust, the response provides a measure of the degree to which the vortices are pinned to the crust. Studies of both x-ray and radio pulsars have established that the interior of a neutron star is not well described by a simple model with only 2 components, a crust and a loosely coupled superfluid core (F. Lamb 1985). Measurements of the angular acceleration of Vela X-1 are consistent with the neutron star being a rigid body, but with weak limits ($\lesssim$ 80%) on the amount of loosely coupled moment of inertia (Boynton *et al.* 1984).

XTE observations will provide improved instantaneous statistics that should reduce the limit on uncoupled moment of inertia to $\sim$ 10% and will extend the spectrum to higher frequencies, from time scales of several days to several hours. Frequent intermittent observations (possible with XTE's maneuverability) will yield the low-frequency power. XTE can expect to carry out measurements on a large sample of pulsars. Models of the internal structure of neutron stars will be substantially constrained.

Observations : Her X-1. (1) Study angular-acceleration fluctuations to obtain an improved limit on the loosely coupled moment of inertia of the neutron star. (2) Obtain correlations of angular acceleration with flux and pulse shape to diagnose torque and pulse-beaming mechanisms and to obtain values of the magnetic moment. (3) Search for phase shifts indicative of torque-free precession of the pulsing neutron star suggested by 35-day modulation of pulse shapes. Confirmation would imply a stiff equation of state for the neutron star.

8.3. BEHAVIOR OF MATTER CLOSE TO STELLAR BLACK HOLES

Several binary x-ray sources are thought to contain stellar black holes because of their high optical mass functions. Several of these exhibit rapid aperiodic fluctuations on time scales down to a few milliseconds, e.g. Cyg X-1 (Meekins *et al.* 1984). Intense flares on this time scale have also been reported from Cyg X-1 (Rothschild *et al.* 1977). The origin of this variability is not known; the time scales indicate strongly that it arises from the innermost region of the accretion disk. The innermost stable orbits for black holes have periods that scale linearly with mass; for $5 \ M_\odot$, they are 0.4 ms and 3.0 ms for maximally rotating and non-rotating black holes respectively.

The large area, high throughput, and sub-ms time resolution of XTE make possible definitive studies of the nature of this variability. The character of the power density spectrum (slopes and cutoffs) could show a strong quasi-periodicity due to strong Doppler beaming from the orbiting material (Sunyaev 1973). The frequency would be a function of the mass of the black hole and its degree of rotation. Temporal lags (of a few ms) in rapid variability as a function of photon energy are expected due to scattering delays for inverse Compton models of the formation of the spectrum. The timing characteristics of a number of sources could establish distinctions between black holes and neutron stars of low magnetic field which could also exhibit ms variability. At least one neutron-star system, Cir X-1, is known to exhibit $\sim$ 10-ms QPO variability (Tennant 1987).

Observation: Cyg X-1. A definitive study of millisecond variability in Cyg X-1 should be carried out. The sub-ms capability and high telemetry rates of XTE are required for this study. The character of the ms variability (aperiodic and quasi-periodic) will provide strong diagnostics of the innermost regions of accretion disks of black holes, including possible (quasi) periodicities by accreting relativistic matter in the innermost orbits. Use PCA and HEXTE burst trigger modes to study millisecond bursts in detail.

8.4. THE NATURE OF QUASI-PERIODIC OSCILLATORS IN ACCRETING SYSTEMS

The discovery of quasi-periodic oscillations (QPO's) in the 5 - 50 Hz range by EXOSAT (van der Klis 1985) has added an important, and possibly epochal, new diagnostic tool for the processes occurring in accreting x-ray binary systems. The QPO phenomenon is a major breakthrough because it exhibits the fastest (quasi-)periodic variability yet detected in accreting systems and because it occurs largely in the low-mass binary systems that have been notably difficult to diagnose because of the usual absence of total eclipses and coherent pulsing. The QPO could be a direct consequence of an underlying millisecond pulsar.

The QPO phenomenon can be a highly reproducible phenomenon that depends uniquely upon the 'state' of the source (e.g. its instantaneous position on an intensity/hardness plot). In addition, the frequency varies systematically with luminosity, and the power in a 'Low Frequency Noise' component (LFN, e.g. ~ 5 Hz) can be correlated with the QPO strength. (See reviews by Lamb 1988; Lewin, van Paradijs, and van der Klis 1988.) *These systematic features suggest strongly to us that the phenomenon is ripe for detailed understanding, similar to that of coherent periodicities.* The QPO phenomenon is clearly much more promising than that of completely aperiodic variability which appears to be much farther from detailed understanding.

A very promising model is the 'beat-frequency' model (Alpar and Shaham 1985; Lamb *et al.* 1985) wherein the periodicity is due to the interaction of the magnetosphere of a very rapidly rotating neutron star (e.g. P = 7 ms) and blobs of material in Keplerian orbits with, e.g., P = 6 ms. The observed frequency would then be the beat frequency of 24 Hz. The expected compression of the magnetosphere as the accretion flow increases naturally yields the variation of QPO frequency with intensity, I ($\nu \propto I^2$) observed in GX 5-1. This model is not universally accepted because of some counter examples in certain sources, but it exhibits the clear possibility that the QPO is driven by an underlying millisecond rotator. This is of great importance because it is likely (but not proven) that such accretion-driven rapid pulsars in low-mass x-ray binary systems are the precursors of the millisecond radio pulsars (Helfand, Ruderman, and Shaham 1983).

XTE can probe deeply the QPO phenomenon by using its sub-ms time resolution, large aperture, and high-throughput capabilities (e.g. 160,000 events per second for Sco X-1). The high sensitivity of XTE should lead to the discovery of QPO in many fainter objects thus permitting studies under a wide range of conditions.

Observation: Search for underlying coherent millisecond pulsing. XTE can search for the underlying pulsing up to and beyond the expected ~ 1 ms minimum rotation period expected from instabilities involving gravitational radiation (Friedman 1983; Wagoner 1984). There are several reasons why the pulse amplitude might be small (e.g. low beaming because of low magnetic field). A sensitive search is nevertheless of crucial importance because the existence of a fast rotator would directly demonstrate a low magnetic field ($10^9 - 10^{10}$ G) for the neutron star in an (old) low-mass x-ray system. This is to be compared to pulsing high-mass (young) x-ray systems with $\sim 10^{12}$ G. This is then direct evidence for magnetic-field decay in

neutron stars, a topic currently the subject of much discussion (Sang and Chanmugan 1987). Coherent millisecond pulsing would provide a direct connection to
millisecond radio pulsars which also exhibit low magnetic fields. The large aperture
of XTE is essential to minimize the effects of period smearing due to orbital motion
of the neutron star. XTE's sub-ms timing is essential to this study.

Observation: QPO phase-lag studies as a function of x-ray energy. Phase lags
of QPO pulsing in hard x-rays (> 5 keV) relative to soft x-rays (< 5 keV) of order
5 ms and as much as 70 ms are observed respectively in EXOSAT (Hasinger 1986)
and *Ginga* (Mitsuda *et al.* 1988) data. (*Soft* lags are seen in one unusual source, the
rapid burster.) Comptonization models can provide delays of a few ms because of
the larger number of scatterings required for higher (or lower for down scattering)
energies; however the 70-ms delay appears to be untenable. *Thus the phenomenon
is a direct diagnostic of the processes (and geometries) of the formation of the x-ray
spectrum.* Exploitation of this phenomenon requires time-resolved spectra on the
ms time scale together with the high statistics and broad bandpass that XTE can
provide.

8.5. X-RAY TRANSIENT STUDIES OF ACCRETION TORQUES, COMPACT-STAR MASSES, AND STELLAR WINDS

The sudden appearance of a very bright x-ray source where none existed previously
is one of the more dramatic occurrences in the sky. These events, called 'x-ray transients' or 'x-ray novae', are known to be the result of a major episode of accretion
onto a neutron star. They can brighten by factors up to $> 10^6$ and have decay times
of 10^1 to 10^2 days and rise times (not well observed) ranging down to hours. The
transients may be classified into 2 or 3 classes. Basically, some occur in low-mass
systems (exhibiting soft spectra and no x-ray pulsing) while others are in high-mass
systems (exhibiting hard spectra and x-ray pulsing). It is the onset and decline of
accretion in these systems that provide unique diagnostic tools.

The *high-mass* systems ('hard' transients) are often widely spaced binaries consisting of a Be star and a neutron star in a wide elliptical orbit (van den Heuvel
and Rappaport 1988). The onset of accretion presents an opportunity to diagnose
episodic accretion, both optically and in x-rays. The orbiting x-ray source acts as
a probe of the geometries, densities, and velocities of the material ejected from the
Be star. X-ray pulse timing yields the size and eccentricity of the orbit and the
position of the neutron star during the measurements. Discovery of a correlation
between the pulse period and orbital period in these systems appears to indicate
that the accreting gas in most systems arises primarily from the slowly expanding
dense circumstellar disk around the Be star (and not a high-velocity low-density
wind) (Corbet 1984, 1986). The properties of the accretion onto the neutron star
and of the x-ray beaming from the neutron star can be studied best by following
a *single* transient through a wide range of accretion rates. This was exemplified
by the serendipitous EXOSAT discovery of a hard transient (EXO 2030+375) that
showed the accretion rate (luminosity) to be strongly correlated with the neutron-
star spin-rate change and with the pulse shape (Parmar *et al.* 1989a,b). This is a
powerful diagnostic tool that XTE can use to probe other systems.

The *low-mass systems* ('soft' transients) have been more difficult to probe definitively because most do not pulse and because the optical light from the x-ray illuminated accretion disk masks the feeble optical light from the (K star) companion during outburst. The cause of the onset of accretion is not known though models exist that could possibly be distinguished through studies of the precursor and rising phases with XTE. Most important, these low-mass systems present the opportunity to study *optically* the binary systems after the x-rays have subsided. Excellent limits on the mass of the compact object (neutron star or black hole) can be obtained as exemplified by A0620-00, currently the strongest black-hole candidate (McClintock and Remillard 1986). A hard component, to ~ 100 keV, has been observed in these systems (Wilson and Rothschild 1983 and Sunyaev 1988). It probably originates in the inner accretion disk and could be an indicator of the putative black hole.

Transient events usually occur at unanticipated places on the sky; thus detection requires wide-field monitor detectors such as were carried on *Ariel*-5 (pin-hole camera) and SAS-3 (scanning slats). These satellites could only bring a low aperture (few $\times 10^2$ cm^2) to bear upon the discovered objects. More recent satellites with larger aperture lacked a monitor (EXOSAT) or had a rather modest one (*Ginga*). XTE, with its ASM and high maneuverability, is expected to detect *and* study (with high aperture) some dozen such events a year. Optical identifications of new transients will lead to fruitful optical studies. Some transients are recurrent and already identified.

Observation: Spin rate and pulse-shape changes in 'hard' transients. The spin-rate and pulse-shape changes in ~ 10 hard transients will be studied. The spin-rate change in EXO 2030+375 agreed remarkably well with the expected relation, $-dP/dt \propto \sim L^{6/7}$, except at low luminosities where the pulsar may have been spinning down. Confirmation of the spin-down phenomenon by XTE would provide a measure of neutron-star magnetic field strength from conventional accretion theory. The pulse-shape changes with luminosity permit direct modeling of beaming processes and geometries.

Observation: Masses of compact objects through detection of 'soft' transients. 'Soft' x-ray transients will be detected with the ASM, optically identified, and (later, in quiescence) studied optically to determine the optical mass functions. These yield directly strong lower limits on the mass of the compact partner. Pioneering studies of Cen X-4 and A0620-00 indicate that the former contains a neutron star and the latter a black hole. Optical identifications with the $3' \times 30'$ uncertainty region derived from the ASM data will be rapid and straightforward because the optical counterparts brighten by $\gtrsim 5$ mag.

8.6. THE MAGNETIC FIELDS OF NEUTRON STARS AND WHITE DWARFS

Magnetic cataclysmic variable systems, i.e. the DQ Her or AM Her types with white-dwarf magnetic fields of $10^6 - 10^8$ G, exhibit an exceptionally hard x-ray component, typically $kT > 20$ keV. This most likely arises from shocks above (or at) the stellar surface caused by accreting material being channeled along the converging magnetic field lines. (See reviews by Patterson 1984, Liebert and Stockman 1985, D. Lamb 1985.) Studies of the hard emission as a function of the orbital phase provide direct information about the temperatures, geometries and magnetic fields. A continuing puzzle is the 'soft x-ray problem': the hard x-ray flux from AM Her objects appears to be insufficient to produce the observed intense very soft ($\sim$ 30 eV) flux through reprocessing in the stellar surface and accretion shock (Beuermann 1988, D. Lamb 1988, Osborne 1988). Observations with good sensitivity to hard x-rays would help resolve this issue by giving better definition to the hard x-ray emission region.

Most accretion-driven pulsing neutron stars appear to have fields of $\sim 10^{12}$ G, based upon their spin-up rates and luminosities. The discovery and confirmation of a spectral feature in Her X-1 at 35 keV (if absorption) or 58 keV (if emission) is widely believed to be cyclotron emission from a $\sim 10^{12}$ G field, i.e. it is a direct indicator of the magnetic field (Trumper *et al.* 1978; Gruber *et al.* 1980). Discovery of harmonics by XTE would firmly establish the cyclotron resonance phenomenon. The ratios of line strengths would provide an estimate of the 'transverse' plasma temperature. Variation of line widths as a function of temperature and spin angle (of the neutron star) can establish the 'parallel' plasma temperature and the magnetic field direction. Beam geometries can be forthcoming from these studies also.

Observation: AM Her type cataclysmic variables. Measure the hard x-ray flux (2-30 keV) of the 6 brightest AM Her objects to determine the white-dwarf mass and the extent and height of the accretion shock.

Observation: Cyclotron features from neutron stars. Measure the hard spectrum of several accretion-driven pulsing sources out to $\sim$ 100 keV as a function of pulse phase to detect cyclotron features including harmonics. The transverse and parallel plasma temperatures and magnetic field directions would be forthcoming. XTE will be sensitive to cyclotron features an order of magnitude or more weaker than that in Her X-1.

8.7. Nature of X-ray Emission from the Galactic Plane

Results from previous x-ray observatories, e.g. EXOSAT, indicate the existence of a galactic 'ridge' of unresolved x-ray emission (Warwick *et al.* 1985). Its nature is not understood; it may consist of discrete sources of known or unknown classes. Recent *Ginga* studies of the galactic plane show the existence of a class of 'weak' (2–10 μJy; actually quite bright for XTE) and hard-spectrum transients with durations of hours to a few days in the galactic plane (Koyama *et al.* 1989). One or more of these can be found on any given scan of the plane with a 1° field of view. The nature of the transients is not known; however, their hard spectra and transient nature indicate they may be Be-star pulsing binaries.

Observation: Repeated scans of the galactic ridge. Repeated (daily to monthly) scans, each of 6-hr duration, of 1000-deg^2 of the galactic plane with the PCA should be carried out (1) to map with high sensitivity the structure on the 1-deg^2 scale of the 'galactic ridge' and (2) to determine its temporal variability. Discovered transients would be observed in the pointed mode to diagnose their nature. This could greatly increase the sample of x-ray emitting Be-star systems which are excellent stellar-wind laboratories (see above).

8.8. Search for, and Study of, Neutron Star in SN1987A

XTE will be in orbit during the period 7–9 years after the implosion that led to SN 1987A and will be able to complement observations by other observatories, including balloons and rockets. Although highly uncertain, it is not unlikely that the ejecta will become optically thin to > 20-keV x-rays at age $\sim$ 10 years and that lower-energy photons will still be absorbed or scattered (Xu *et al.* 1988). High-energy response with large aperture is thus important for the study of SN 1987A.

Successful discovery of pulsing x-rays would demonstrate unequivocally that a neutron star rather than a black hole was formed. Subsequent pulse-timing measurements would reveal binary membership and would yield measures of the braking index and response to 'glitches'. Such results address the fundamental underlying physics of SN 1987A, specifically the evolution of the progenitor, the magnetic dipole moment of the pulsar, and the structure of the neutron star.

Observation: SN1987A. A 9000-s observation with the PCA (> 20 keV) will make possible a search for pulsations in 1000 frequency bins (1–1000 Hz) with $\sim$ 99.9% confidence, for a source with the luminosity of the Crab pulsar in the LMC. If pulsing were discovered, each subsequent observation of 6×10^4 s would yield a 10-bin summed light curve with 5σ confidence in each bin.

8.9. Studies of New Prototype Objects

Specific compelling objectives for XTE also follow from the ongoing discovery of new bright (> 0.7 μJy at 5 keV) objects through optical identifications (Bradt *et al.* 1988) of cataloged x-ray sources (e.g. the HEAO-1 LASS catalog of 842 sources [Wood *et al.* 1984]). These identifications are yielding objects that extend substantially the canonical parameters of known classes. Recent examples include: (1) a QSO with unusually steep x-ray spectrum and with very strong optical FeII emission (highly pertinent to the formation of broad-line clouds and to massive accretion disks) (Remillard *et al.* 1986a, 1988), (2) a possible slightly asynchronous AM Her type object, H0538+608 (Remillard *et al.* 1986b), and (3) a BL Lac object, H1720+117, with the highest known optical polarization (17%) among x-ray selected objects (Brissenden *et al.* 1990). Such objects demand detailed studies at all wavelengths; they are likely to become prototypes of new subclasses.

The numbers of known, bright, optically identified (or x-ray classified) objects with hard (> 2 keV) emission in each of the classes (e.g. low-mass neutron-star binary, QSO's, Sy 1, AM Her type, etc.) is currently not high, from a few to $\sim$ 100. Thus each newly identified or classified object may well have important unique properties, e.g. the aforementioned examples, and hence be worthy of detailed study. *We stress that the population characteristics of the $\sim 10^3$ brightest x-ray sources in the sky are not yet fully known.*

Identifications and optical/x-ray classifications of HEAO-1 and EXOSAT sources are rapidly increasing the pool of unstudied 'bright' objects. The ROSAT survey (< 2 keV) will yield many more objects. Thus we anticipate a continual infusion of new discoveries and questions into XTE's observing program.

Acknowledgements

The authors are grateful to the entire XTE team for their past and future contributions to the mission. In addition, we are grateful to our colleagues at numerous institutions who have participated in discussions about the scientific contributions that XTE could make in the context of our current knowledge. We particularly thank P. Boynton, F. Lamb, J. McClintock, and R. Mushotzky for their efforts in this regard. Our contributions to this work are supported by NASA.

References

Alpar, M., and Shaham, J.: 1985, *Nature* **316**, 239

Aschenbach B.: 1988, *Appl. Optics* **27**, 1404

Beuermann, K.: 1988, in , ed(s)., *125*, Polarized Radiation of Circumstellar Origin, eds. G.V. Coyne, *et al.*, Vatican Observatory, Rome

Boynton, P., Deeter, J., Lamb, F., Zylstra, G., Pravdo, S., White, N., Wood, K., and Yentis, D.: 1984, *Astrophys. J. Lett.* **283**, L53

Bradt, H., Remillard, R., Tuohy, I., Buckley, D., Brissenden, R., Schwartz, D., and Roberts, W.: 1988, in , ed(s)., , The Physics of Neutron Stars and Black Holes, ed. Y.Tanaka, Universal Academic Press, Tokyo

Bradt, H., Swank, J., Rothschild, R.: 1990, *Adv. Space Res.* **10**, (2)297

Brissenden, R., Remillard, R., Tuohy, I., Schwartz, D., and Hertz, P.: 1990, *Astrophys. J.* **350**, 578

Corbet, R.: 1984, *Astron. Astrophys.* **141**, 91

Corbet, R.: 1986, in , ed(s)., *63*, The Evolution of Galactic X-ray Binaries, ed. J. Trumper *et al.*, Reidel, Dordrecht

Debouzy G.: 1983, *Adv. in Space Research* **3**, 99

Deeter, J., Lamb, F., Zylstra, G., Pravdo, S., White, N., Wood, K., and Yentis, D.: 1984, *Astrophys. J. Lett.* **283**, L53

Dicke, R.: 1968, *Astrophys. J. Lett.* **153**, L101

Doty, J.: 1988, in , ed(s)., *164*, X-ray Instrumentation in Astronomy II, ed. L. Golub, Proc. SPIE 982

Doxsey, R., Bradt, H., McClintock, J., Petro, L., Remillard, R., Ricker, G., Schwartz, D., and Wood, K.: 1983, *Astrophys. J.* **264**, L43

Ebisuzaki, T., Hanawa, T., Sugimoto, D.: 1984, *Pub. Astron. Soc. Japan* **36**, 551

Epstein R., Lamb F. and Priedhorsky, W.: 1986, in , ed(s)., , Astrophysics of Time Variability in X-Ray and Gamma-Ray Sources, Proceedings of the Taos Workshop, Los Alamos Science, No. 13

Feigelson, E. *et al.*: 1986, *Astrophys. J.* **302**, 337

Friedman, J.: 1983, *Phys. Rev. Lett.* **51**, 11

Gruber, D., Matteson, J., Nolan, P., Knight, F., Baity, W., Rothschild, R., Peterson, L., Hoffman, J., Scheepmaker, A., Wheaton, W., Primini, F., Levine, A., and Lewin, W.: 1980, *Astrophys. J. Lett.* **240**, L127

Hasinger, G.: 1986, in , ed(s)., *333*, The Origin and Evolution of Neutron Stars, eds. D. Helfand and J. Huang, Reidel, Dordrecht 1986

Helfand, D., Ruderman, M., and Shaham, J.: 1983, *Nature* **304**, 118

Joss, P. and Rappaport, S.: 1984, *Ann. Rev. Astron. Astrophys.* **22**, 537

Joss, P. 1988, private communication.

Koyama, K., Kondo, H., Makino, F., Nagase, F., Takano, S., Tawara, Y., Turner, M., and Warwick, R.: 1989, *PASJ* **41**, 483

Lamb, D.: 1985, in , ed(s)., *179*, Cataclysmic Variables and Low-Mass X-ray Binaries, ed. D.Q. Lamb and J. Patterson, Reidel, Dordrecht

Lamb, D.: 1988, in , ed(s)., *151*, Polarized Radiation of Circumstellar Origin, eds. G.V. Coyne, *et al.*, Vatican Observatory, Rome

Lamb, F.: 1988, *Advances in Space Research* 8, #2-3, 421

Lamb, F., Pines, D., and Shaham, J.: 1978, *Astrophys. J.* **224**, 969

Lamb, F., Shibazaki, N., Shaham, J., and Alpar, M.: 1985, *Nature* **317**, 681

Lamb, F.: 1985, in , ed(s)., *19*, Galactic and Extragalactic Compact X-Ray Sources, ed. Y. Tanaka and W. Lewin, Cambridge U. Press, Cambridge

Lawrence, A., Watson, M., Pounds, K., and Elvis, M.: 1987, *Nature* **325**, 694

Lewin, W., van Paradijs, J., and van der Klis, M.: 1988, *Sp. Sci. Rev.* **46**, 273

Liebert, J., and Stockman, H.: 1985, in , ed(s)., *151*, Cataclysmic Variables and Low-Mass X-ray Binaries, ed. D.Q. Lamb and J. Patterson, Reidel, Dordrecht

Maccacaro, T., Gioia, I., Maccagni, D., and Stocke, J.: 1984, *Astrophys. J.* **284**, L23

Makino F. and the Astro-C Team: 1987, *Astrophys. Lett. and Comm.* **25**, 223

Matteson, J.: 1978, *AIAA 16th Aerospace Sciences Mtg* , Paper 78-35, unpubl

McClintock J. A. and Levine A. M.: 1984, in , ed(s)., *642*, High Energy Transients in Astrophysics, ed. S. Woosley, Am. Inst. of Phys. New York

McClintock, J., and Remillard, R.: 1986, *Astrophys. J.* **308**, 371

McHardy, I. and Czerny, B.: 1987, *Nature* **325**, 696

Meekins, J., Wood, K., Hedler, R., Byram, E., Yentis, D., Chubb, T., and Friedman, H.: 1984, *Astrophys. J.* **278**, 288

Mitsuda, K., and the *Ginga* Team: 1988, in , ed(s)., *in press*, The Physics of Compact Objects: Theory Versus Observations, eds. L. Filipov and N. White, Pergamon Press, Oxford

Ohashi, T. 1988, private communication.

Osborne, J.: 1988, in , ed(s)., *in press*, X-ray Astronomy with Exosat, eds. R. Pallavicini and N. White, Memoria della Societa Astronomica Italiana, Firenze

Parmar, A., White, N., Stella, L., Izzo, C., and Ferri, P.: 1989a, *Astrophys. J.* **338**, 359

Parmar, A., White, N., and Stella, L.: 1989b, *Astrophys. J.* **338**, 373

Patterson, J.: 1984, *Astrophys. J. Suppl.* **54**, 443

Remillard, R., Bradt, H., Buckley, D., Roberts, W., Schwartz, D., Tuohy, I., and Wood, K.: 1986a, *Astrophys. J.* **301**, 742

Remillard, R., Bradt, H., McClintock, J., Patterson, J., Roberts, W., Schwartz, D., and Tapia, S.: 1986b, *Astrophys. J. Lett.* **302**, L11

Remillard, R., Schwartz, D., Brissenden, R.: 1988, in , ed(s)., *273*, A Decade of Ultraviolet Astronomy with the IUE Satellite, vol. 2, ed. E. Rolfe, ESA/ESTEC, Noordwijk

Remillard, R. 1990, private communication.

Rothschild, R., Boldt, E., Holt, S., Serlemitsos, P., Garmire, G., Agrawal, P., Riegler, G., Bowyer, S., and Lampton, M.: 1979, *Space Sci. Instr.* 4, 269

Rothschild, R., Boldt, E., Holt, S., and Serlemitsos, P.: 1977, *Astrophys. J.* **213**, 818

Rothschild, R., Mushotzky, R., Baity, W., Gruber, D., Matteson, J., and Peterson, L.: 1983, *Astrophys. J.* **269**, 423

Sang, Y., and Chanmugan, G.: 1987, *Astrophys. J. Lett.* **323**, L61

Schnopper H. 1990, this volume.

Snyder, W. *et al.*: 1980, *Astrophys. J. Lett.* **237**, L11

Spada G. 1988, private communication.

Sunyaev R. 1990, this volume.

Sunyaev, R. and the Kvant Team 1988, IAU Circ. 4606.

Sunyaev, R.: 1973, *Sov. Astron.* **16**, 941

Swanenburg, B., Bennett, K., Bignami, G., Caraveo, P., Hermsen, W., Kanbach, G., Masnou, J., Mayer-Hasselwander, H., Paul, J., Sacco, B., Scarsi, L., and Wills, R.: 1978, *Nature* **275**, 298

Tanaka Y. 1988, private communication.

Tennant, A.: 1987, *MNRAS* **226**, 971

Trumper, J., Pietsch, W., Reppin, C., Voges, W., Staubert, R., and Kendziorra, E.: 1978, *Astrophys. J.* **219**, 105

Urry, C. M.: 1988, in , ed(s)., *257*, Active Galactic Nuclei, ed. H. Miller and P. Witta, Springer-Verlag, Berlin

Urry, E. and Mushotzky, R.: 1982, *Astrophys. J.* **253**, 38

van den Heuvel, E., and Rappaport, S.: 1988, in , ed(s)., *291*, IAU Colloq. #92, Physics of Be Stars, eds. A. Slettebak and T. Snow, Cambridge U. Press, Cambridge

Van der Klis, M., Jansen, F., Van Paradijs, J., Lewin, W., Van den Heuvel, E., Trumper, J., and Sztajno, M.: 1985, *Nature* **316**, 225

Wagoner, R. V.: 1984, *Astrophys. J.* **278**, 345

Warwick, R., Turner, R., Watson, M., and Willingale, R.: 1985, *Nature* **317**, 218

Wilson, C. and Rothschild, R.: 1983, *Astrophys. J.* **274**, 717

Wood, K., Meekins, J., Yentis, D., Smathers, McNutt, D. H., Bleach, R., Byram, E., Chubb, T., Friedman, H.: 1984, *Astrophys. J. Suppl.* **56**, 507

Xu, Y., Sutherland, P., McCray, R., and Ross, R.: 1988, *Astrophys. J.* **327**, 197

THE JOINT EUROPEAN TELESCOPE FOR
X-RAY ASTRONOMY (JET-X)

K. A. POUNDS

X-ray Astronomy Group, Department of Physics and Astronomy, University of Leicester,
Leicester LE1 7RH, England

Abstract. JET-X is one of the main payload elements in the Soviet Spectrum-X Gamma mission scheduled for launch in 1993. It consists of two Wolter I X-ray telescopes feeding cooled arrays of CCD's. The combination of good angular resolution and long exposures during the four day orbit of Spectrum-X Gamma, will allow the JET-X telescopes to reach a sensitivity level ten times fainter than that of the Einstein Observatory deep survey. The CCD's will provide good spectral data for many thousands of brighter sources over an energy band ~0.3–10 keV. The JET-X project is an international collaboration involving groups in the United Kingdom, Italy, West Germany and the Soviet Union together with an input from the Space Science Department at ESTEC. A brief description is given of the design and anticipated performance of JET-X together with illustrations of the anticipated scientific results.

1. Introduction

JET-X will be assembled in Western Europe from component parts produced in the UK, Italy and the Soviet Union. Technical and scientific support is also being provided by the MPE group in Germany and by the Space Science Department at ESTEC. Project management and overall systems engineering support are the responsibility of the UK, which will also provide the focal plane assembly and electronics for the two X-ray telescopes and the JET-X attitude monitor. Full funding for the UK role in JET-X was approved by the Science and Engineering Research Council in 1988. Italy is the second major partner, being responsible for provision of the X-ray mirror assembly for each Wolter I telescope and for the co-aligned Optical Monitor. Funding of the Italian involvement in the JET-X programme is being provided by the Agenzia Spaziale Italiana. A key element in the JET-X assembly is the carbon fibre structure which will hold both X-ray telescopes, the Optical Monitor and the star tracker. This structure will act as the optical bench for the system as a whole, providing the stable platform necessary to achieve the target angular resolution. The design of the JET-X X-ray telescopes is based heavily on existing research programmes, a necessary requirement to meet the short timescale set by the proposed 1993 launch of the Spectrum-X-Gamma mission. The X-ray mirrors will be developed by the Osservatorio Astronomica di Brera in Italy, using the replication technique developed for the SAX mission and mandrels manufactured by Zeiss in West Germany (in a programme running alongside the XMM mandrel development). Design and test assistance and the unique Panter X-ray beam facility will all be made available to the JET-X project by the Max Planck Institut fur Extraterrestriche Physik in Garching, bringing to the project a wealth

Y. Kondo (ed.), Observatories in Earth Orbit and Beyond, 111–118.
© 1990 *Kluwer Academic Publishers.*

of experience gained in the ROSAT programme.

Control and data reception for Spectrum-X-Gamma will be carried out from the Soviet Union, but the complex mission planning, data reduction and scientific analysis will be a broad international effort. It is intended that the JET-X scientific programme (which will be providing observing time on an agreed formula to all participating countries) will be determined on scientific merit, in response to an international A/0, and will be integrated closely with that of SODART the other major instrument on board Spectrum-X-Gamma, (ref. Schnopper, these Proceedings).

2. Instrument Concept

Each X-ray telescope is of Wolter I design, with 12 shells of electro-formed nickel (1mm thick) and 3.5 metre focal length. A fixed array of cooled CCD's is mounted in each focal plane. The baseline optics specification calls for an on-axis resolution of 30 arc sec (HEW), with a design aim of 10 arc sec (HEW), and effective area (2 telescopes) of 360 cm^2 at 1.5 keV and 140 cm^2 at 8 keV. A pair of CCD's will provide efficient coverage of a field of view of 28 × 21 arcmin^2 for each telescope.

The chosen CCD's are derived from a development programme at GEC and Leicester University, also supported by ESA for the XMM project (Wells and Lumb, 1990). Passive cooling to space will maintain the operational temperature of the CCD's on JET-X, (170–180 K), necessary to obtain the required low noise and high energy resolution for X-ray spectroscopy and imaging of faint X-ray sources.

The two X-ray telescopes will be mounted in a common carbon fibre shell, together with the star tracker and Optical Monitor. The star tracker will provide independent attitude for JET-X to an accuracy of ~ 5 arc sec and the Optical Monitor (a Richey-Chretian design of 20 cm aperture, with integrating optical CCD readout) will yield simultaneous optical images to a magnitude limit of m_v ~ 22. Further detail of the JET-X Optical Monitor will not be given here, since its design is very similar to that described for XMM (ref. Taylor these Proceedings).

3. Design and Manufacture

The X-ray mirrors will be manufactured by the replication technique developed for the SAX programme. Mandrels, of Wolter I geometry, are being produced by Zeiss, with a super-polished finish of ~5 Å rms. A gold layer is later deposited on the mandrel surface and a nickel layer electroformed on top. On removal of the nickel replica from the mandrel, the gold layer adheres to the nickel, providing the X-ray reflecting surface. Twelve paraboloid – hyperboloid shells form each complete X-ray mirror assembly. The design characteristics of the JET-X nested optics are summarised in Table I and Figure 1 shows the total effective area as a function of energy and off-axis angle.

The CCD's to be used on JET-X are deep depletion devices of type P88930T, derived from an extended research programme at the GEC EEV company at Chelmsford. The use of high resistivity silicon (2000 ohm-cm) ensures good quantum efficiency to ~ 10 keV while thinning of the front electrode structure yields an accept-

TABLE I
Mirror Characteristics

Field of View	20 arcmin (50% vignetting)
Angular resolution	10–30 arcsec (HEW)
Focal length	3.5 m
Reflecting surface	Gold
Configuration	Wolter-I
Mirror length	2 × 300 mm
Outer mirror diameter	300 mm
Inner mirror diameter	87.5 mm
Shell thickness	2 mm
Number of shells	12
Distance between shells	3–5.8 mm
Surface finish (microroughness)	5 Å rms

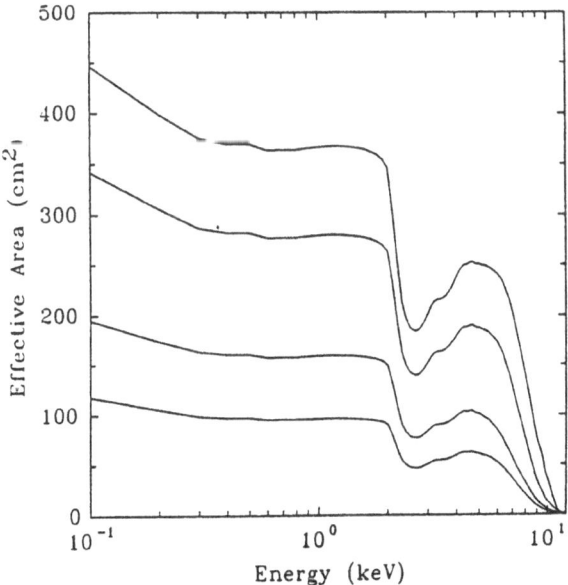

Fig. 1. Total effective area of JET-X (2 modules) as a function of photon energy on-axis, and at off-axis angles of 7.5, 15.0 and 22.0 arcminutes.

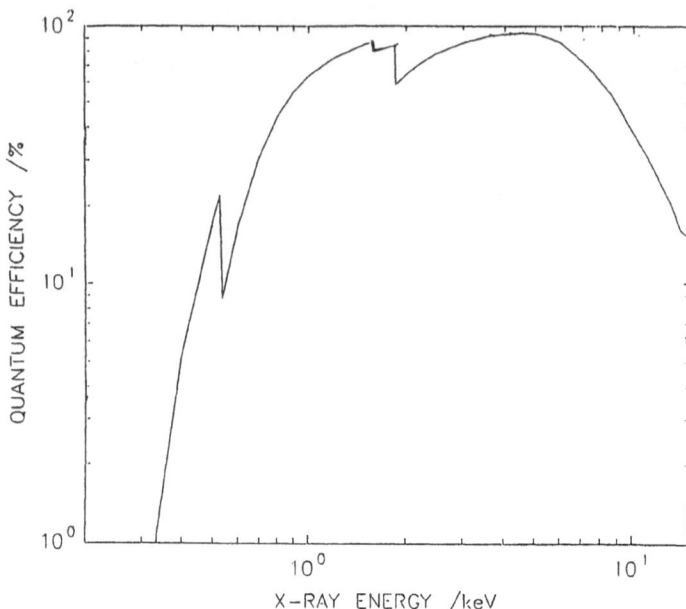

Fig. 2. Quantum efficiency of the P88930T CCD with thinned electrodes and front illumination.

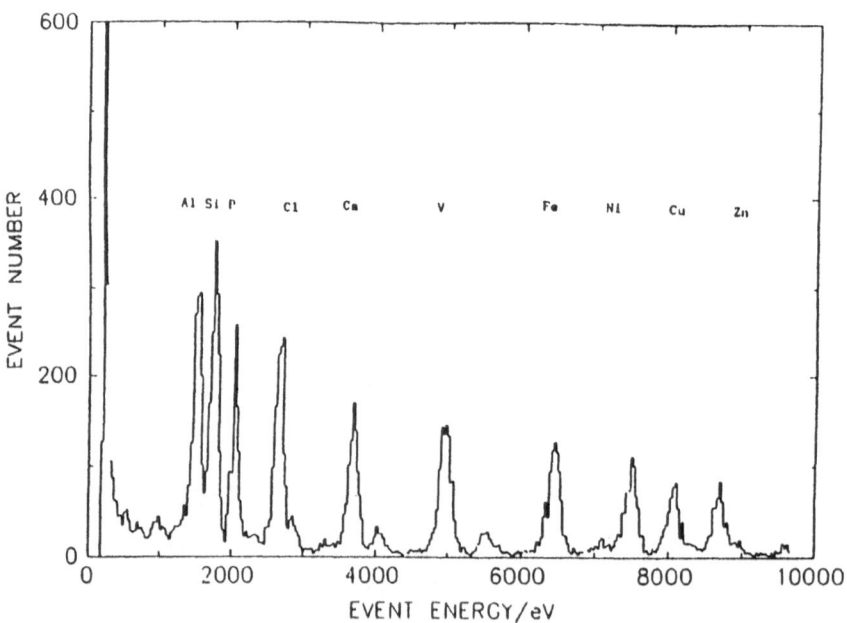

Fig. 3. Measured CCD energy response to a range of characteristic X-ray emission lines in the laboratory.

TABLE II
Performance Characteristics of type P88930T CCD

Parameter	Predicted Performance
EEV Device type	P88930T
Pixel format on chip	1024 × 768
Image area per chip	512 × 768
Pixel size	27 μm
Format of chip arrays in focal plane	2 × 1
Dead space between chips	$\lesssim 400\mu$m
Depletion depth	40 μm
Dead layer	$\lesssim 2000$ Å on 50% of pixel
Total detection depth	65 μm
Readout noise	4 electrons
Readout rate (/pixel)	12 μs
Image frame time	3 sec
Operating temperature	180–190 K
Frame store operation	4 amplifiers

able low energy response. Figure 2 shows the Q.E. of this current CCD and Table II lists the basic performance characteristics. Figure 3 shows laboratory data obtained on a number of characteristic X-ray lines to illustrate the energy resolution to be expected in JET-X.

4. Scientific Performance

The scientific potential of JET-X derives from its high angular and good energy resolution, the advantageous orbit of Spectrum-X-Gamma and its timeliness in relation to later, larger X-ray missions, such as AXAF and XMM.

Figure 4 shows the detection sensitivity of JET-X for exposures up to 105 seconds. It can be seen that the low intrinsic noise of the system allows long, photon-limited exposures, yielding:

• a sensitivity limit near 3×10^{-15} erg cm^{-2} s^{-1}, ten times fainter than the Einstein Deep Survey limit, and similar to the ROSAT Deep Survey limit.

• importantly, the higher energy response makes JET-X a valuable extension to ROSAT, for example in determining temperature structures in clusters of galaxies (Sarazin, 1988) and detecting the important Fe K-features in active galaxies (Pounds et al, 1990).

• the limit of $\sim 3 \times 10^{-15}$ erg cm^{-2}/s will remain un-confused (at 40 beams per source) provided the JET-X angular resolution achieves the specified $\sim$30 arc sec. HEW.

Figure 5 shows a second important advantage of the high angular resolution of JET-X, in resolving structure in extended sources.

For example, it can be seen that rich clusters of galaxies will be resolved, at any redshift, while cooling flows should be resolved to a high redshift, depending cru-

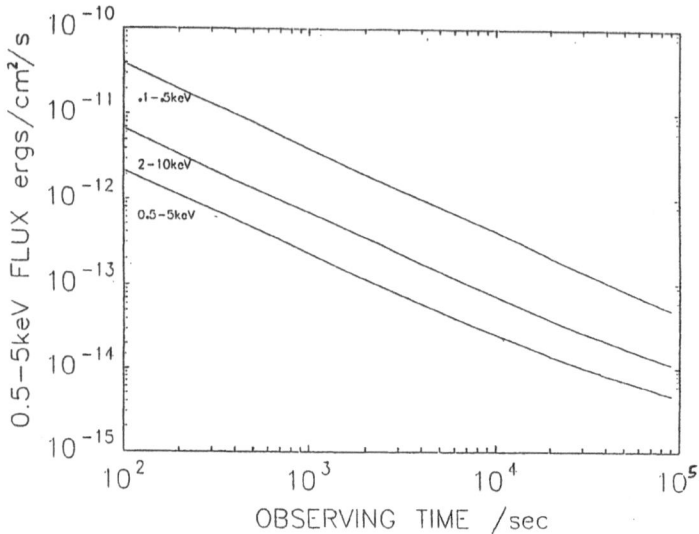

Fig. 4. JET-X point source detection (at 5σ) as a function of exposure time. Input power law of photon index 1.7 and absorbing column 4×10^{20} cm^{-2}.

cially on the resolution actually achieved. The importance of obtaining the highest possible angular resolution in JET-X is emphasised in comparing the linear size of elliptical galaxies, first shown from Einstein Observatory data to contain extended thermal X-ray emission (Fabbiano, 1990).

Figure 6 shows two examples of simulated JET-X spectral data, obtained using the latest performance figures for the optics and flight detectors. Fig. 6(a) illustrates the rich spectral detail expected from the thermal emission of a cluster of galaxies, with $L_{2-10} \sim 10^{45}$ erg/sec, a temperature of kT $\sim$ 4 keV and at a distance of 200 Mpc. The assumed exposure time is 3×10^4 seconds.

Figure 6(b) shows a 3×10^4 simulation for a Seyfert galaxy similar to NGC 4151, but ten times fainter. Here, complex partial covering of the X-ray nucleus and Fe K-fluorescent line and absorption edge features are easily seen.

5. A Five Year Mission Profile

The major scientific goal of JET-X is a study of faint X-ray sources, by spectroscopy, imaging and timing. As noted earlier, the high resolution optics, low noise detectors and 4-day orbit, combine to underpin the value of long exposures for JET-X. With these characteristics in mind a model 5-year mission profile has been constructed (Table III), assuming JET-X will be the prime instrument (i.e. determining the pointing axis) for 180 spacecraft orbits, each with an effective observing time of 2.7×10^5 seconds.

Among the key scientific objectives of each class of observation will be:

deep fields: direct measurement of source counts, to a level corresponding to $\gtrsim 0.5$ of the X-ray background flux, hence identifying the major constituent(s) of the XRB in the 2–10 keV band (and possibly revealing an entirely new class of

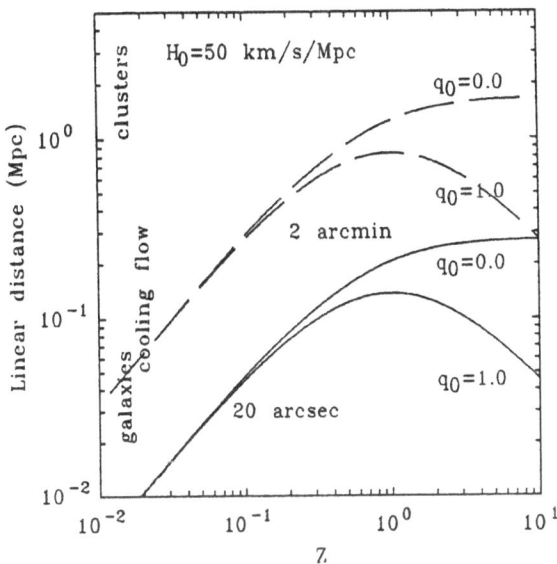

Fig. 5. Linear distance corresponding to a resolution element of 20 arc seconds as a function of redshift. Also indicated are typical sizes for galaxies, clusters of galaxies and cooling flows in clusters.

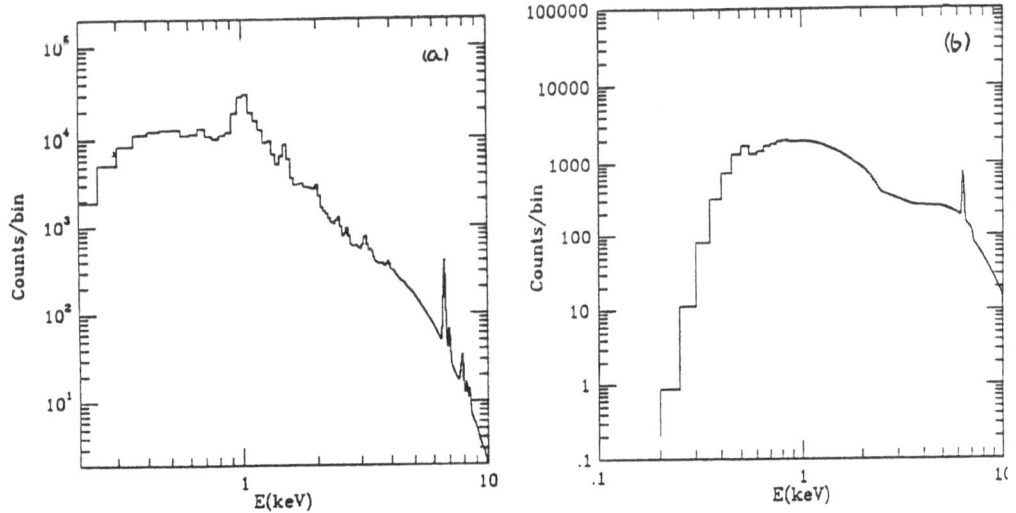

Fig. 6. Simulated JET-X spectra for a 3×10^4 second exposure on (a) a rich cluster of galaxies of $L_x \sim 10^{45}$ erg/sec and kT $\sim$ 4 keV, at 200 Mpc, and (b) a Seyfert galaxy of one tenth the brightness of NGC 4151 with partial covering and reprocessing in cold matter.

TABLE III
JET-X Prime Mission Profile (5 year mission)

Observation	Cts/source (Percentage)	Flux limit (erg cm^{-2}/s^{-1})	Percentage Exposure	Number	of Tin
Deep survey (source counts, XRB)	20	10^{-15}	2.7×10^5	14	8
Medium survey (Spectra of ROSAT Sources)	10^3	10^{-13}	10^5	200	40
Extended source spectra	10^5	$\leq 10^{-11}$	10^5	200	40
Bright sources (in window or timing modes)	$\geq 10^5$	$\geq 10^{-11}$	$\sim 2 \times 10^4$	300	12

faint, strongly evolving sources);

medium survey: broad-band spectroscopy of many X-ray sources detected in the ROSAT survey, yielding information on the possible evolution of the X-ray emission and absorption in AGN, etc.;

extended source spectra: imaging and spectroscopy of rich clusters, elliptical galaxies, SNR, etc. to determine their temperature, density and chemical abundance distributions; hence their mass and evolution;

bright source observations: to study, in particular, the variability in flux and spectrum and hence the geometry of individual emission and absorption features.

Note 1: Total time corresponds to $\sim$ 180 Spectrum X-Gamma orbits

Note 2: 20–100 serendipitous sources are expected in each 105 sec exposure

References

Fabbiano G.: 1989, *Ann. Rev. Astron. Astrophys.* **27**, 87

Pounds K.A., Nandra K., Stewart G.C., George I.M. and Fabian A.C.: 1990, *Nature* **344**, 132

Sarazin C.L.: 1988, *Rev. Mod. Phys.* **58**, 1

ells A. and Lumb D.H., 1990, *Proc. SPIE*, 1159

SODART TELESCOPE ON SPECTRUM-RÖNTGEN-GAMMA
AND ITS INSTRUMENTATION

HERBERT W. SCHNOPPER

Danish Space Research Institute, Gl. Lundtoftevej 7, DK-2800 Lyngby, Denmark

Abstract. SPECTRUM-RÖNTGEN-GAMMA (SRG) is one of a new series of large astronomical missions being planned by the Soviet Union and is scheduled for launch in
mid-1993. The Space Research Institute (IKI) of the Academy of Sciences of the USSR
and the Babakin Center (BC) are responsible for the scientific supervision and spacecraft
construction, respectively. Mission objectives include broad and narrow band imaging
spectroscopy over a wide range of energies from EUV through gamma rays with particular emphasis on extragalactic objects. The design of the Soviet Danish Röntgen Telescope
(SODART) consists of two thin foil, conical shell approximations to Wolter 1 geometry.
The reflectors are rolled aluminum foils which have been dipped in acrylic lacquer and
coated with gold resulting in a super smooth surface. Each telescope has an aperture of
60 cm, a focal length of 8 m, a field of view of 1 deg and is designed to have a halfpower
width of ≤ 2 arcmin. The conical geometry contributes 15 arcsec and manufacturing tolerances in the support structure and the quality of the figure of the foil the rest. The
contribution from X-ray scattering is insignificant. Focal plane slides can position one of
four instruments at the focus of each telescope. Images and spectra will be recorded with
position sensitive proportional counters with spectral resolution as good as 13% at 6 keV.
Spectral resolution of 2.5% at 6 keV is provided by an array of 19 cooled silicon detectors. A broad band polarimeter will be sensitive to residual polarization as low as 1%. An
objective Bragg crystal panel, placed in front of one of the telescopes, will be capable of
high resolution spectroscopic studies ($(E/\Delta E)) \sim 1000$) of point- and extended sources.

1. Introduction

SRG is to be the first of a series of new astronomical observatories to be launched
by the Soviet Union under the sponsorship of the Academy of Sciences of the
USSR. The expected launch date is mid-1993 and the observatory is expected to
be operated for a period of 5 years. The satellite will be launched into a deep,
highly eccentric orbit with a period of about four days from which long duration
observations can be made. With all the operational constraints taken into account,
there is access to about 80 per cent of the celestial sphere at any time. The major
facility on board is the SODART X-ray telescope and its associated focal plane
instruments. Instrumentation in addition to SODART will cover the energy range
from EUV to gamma rays. A complete description of the SRG system (Figure 1)
will be found in the paper presented by R. Sunyaev (1990) at this Colloquium.

2. SODART

The system consists of X-ray optics, thermal and structural elements, focal plane
instruments, a focal plane transport assembly and an objective crystal spectrometer

Y. Kondo (ed.), Observatories in Earth Orbit and Beyond, 119–128.

SPECTRUM RÖNTGEN–GAMMA

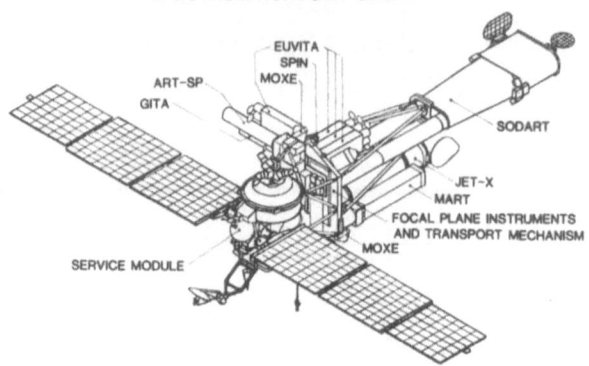

Fig. 1. SPECTRUM-RÖNTGEN-GAMMA (R.S. Kremnev, Babakin Center.)

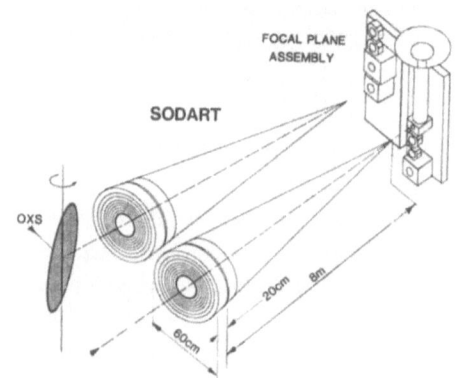

Fig. 2. The telescopes and instruments of SODART.

(Figure 2). The elements of SODART provided by DSRI are called XSPECT. Brief descriptions of these elements follow.

2.1. X-RAY OPTICS

Serlemitsos, Petri, Glasser, and Birsa (1984) pointed out the advantages of using thin foil, conical approximations to Wolter 1 optics. A large fraction of the available aperture can be filled by densely nesting many shells of thin foils yielding a large collecting area and an extended energy range. The design goal for the XSPECT telescope is a Half-Power Width (HPW) of <2 arcmin of which <15 arcsec comes from the conical geometry. Sheets of commercially available rolled aluminum foil, 0.3 mm thick are the raw material for the telescopes. After an initial inspection, the foils are cut to size, chemically cleaned, rolled to an approximate radius of curvature, coated with a thin uniform layer of acrylic lacquer and, finally, a thin layer of gold is evaporated onto the surface. The parameters chosen for the XSPECT telescope are listed in Table I (Westergaard et al., 1990). Two such telescopes are to be provided

TABLE I
XSPECT conical approximation telescopes

Field of view	60 arc min
Half-power width	<2 arcmin
Focal length	8 m
Diameter	60 cm
Reflecting surface	500 Å gold
Effective area	1700 cm^2 @ 2 keV
	1200 cm^2 @ 8 keV
	60 cm^2 @ 20 keV
Shell material	0.3 mm aluminum
Outer shell diameter	60 cm
Inner shell diameter	16 cm
Shell separation (min)	0.5 mm
Shell length	20 cm
Number of shells	154
Mass per telescope	90 kg
Total mirrored surface	75 m^2

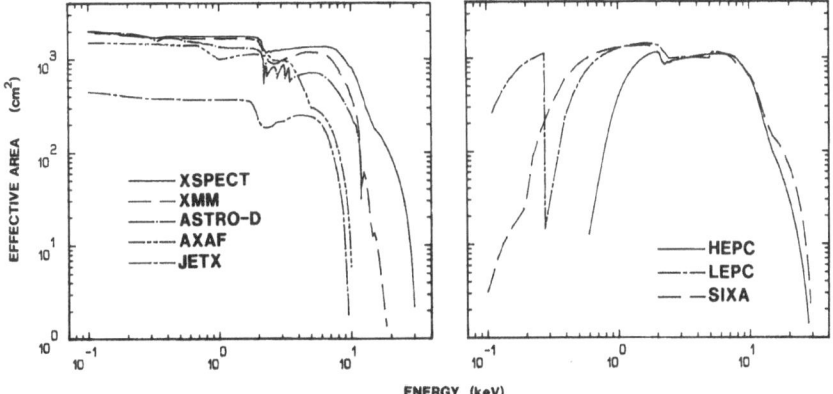

Fig. 3. Left: On-axis collecting area for various telescopes. Right: XSPECT's effective area for various detectors (N.J. Westergaard, DSRI).

for SODART. Figure 3 shows the collecting area which can be expected for a single telescope. Curves for other telescopes are shown for comparison. Measurements of reflection efficiencies at various X-ray energies for foil samples prepared at DSRI have been found to agree with the theoretical values which have been used to construct the XSPECT curve.

Each XSPECT telescope consists of two sections, 1 alpha and 3 alpha, each of which is divided into quadrants (Figure 4). After assembly, each quadrant is mounted to a reference hub whose axis is made accurately parallel to the axis of a well collimated optical beam 25 cm in diameter. The reflected beam can be recorded by a CCD camera which can be placed at any distance along the converging beam. These optical test are used to control the adjustments of the ribs which guide the foils into the optimum position. Diffraction makes it impossible to measure the HPW directly by this method. When, however, a slit is used to scan the quadrant it is possible to measure the centroid of the image focused by the small portion of the illuminated foils. The distribution of these centroids is a good indication of the contribution to the HPW which comes from manufacturing tolerances and misalignment. Recent measurements on an assembled, but not aligned quadrant show that mechanical errors are well within the goal of HPW < 2 arc min.

Once the alignment procedure is completed, the X-ray reflecting properties, including the HPW, will be measured in a specially constructed facility. A pencil beam from a high power, rotating anode X-ray tube is made highly parallel and monochromatic by perfect silicon crystals. The quadrant under test (and later the completed telescope) is mounted on a translation and rotation stage (rotation around the optical axis). An imaging proportional counter is placed just after the telescope and a raster scan of the beam is made for each offset angle to simulate the on-and off-axis behavior of the telescope. The synthesized full beam image at the correct focal distance can be constructed by computer processing. The anode material is replaceable to allow tests at several energies. Work is in progress to prepare a full aperture beam test using the synchrotron radiation facility at Daresbury (Lewis et al., 1990). Final calibration will take place in orbit.

2.2. FOCAL PLANE DETECTORS

2.2.1. XSPECT MicroStrip Proportional Counters (MSPC)

In conventional proportional counters the electron avalanche takes place in the high field region defined by anode and cathode grids made from finely spaced wires. These designs usually lead to gain and position non-linearity, count rate restrictions and high operating voltages. The MSPC is a novel approach, in which the wire grids are replaced by narrowly spaced, conducting microstrips (Oed, 1988; Budtz-Jörgensen et al., 1989) which are accurately deposited ($\pm 2\mu$m) on an insulating substrate. The XSPECT MSPC detector based on these principles is shown in Figure 5. A high- and a low-energy (HEPC and LEPC) detector will be provided for each telescope. The parameters chosen for them are listed in Table II.

X-direction positions are obtained from the charge induced on the wire grid which is suspended above the microstrip. The wires are connected with resistors and the charge division is read out by suitable spaced preamplifiers that insure the appropriate position sensitivity. Y-direction positions are obtained from the cathode strips on the microstrip which are connected by a resistive strip on the substrate. Energy, event risetime, and anti-coincidence discrimination are used to reject background. Results from energy- and position-resolution, and pulse-shape background rejection measurements are shown in Figure 6.

2.2.2. Other focal plane detectors

Two imaging Focal Plane Röntgen Detectors (FRD) described by Sunyaev (1990), a Silicon X-ray Array (SIXA) described by Vilhu (1990) and a Stellar X-Ray Polarimeter (SXRP) (Kaaret et al., 1989) complete the complement of focal plane instruments. They are shown in place on the focal plane transport assembly (Figure 7).

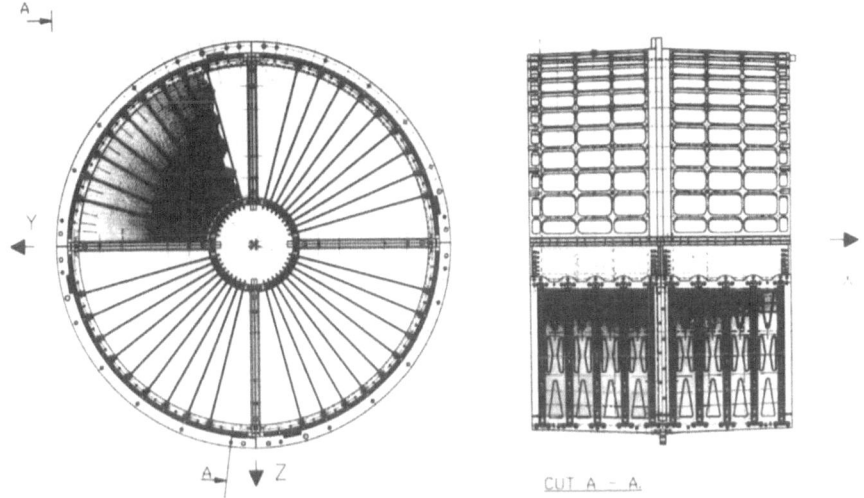

Fig. 4. Structure of the XSPECT telescope (J. Polny, DSRI).

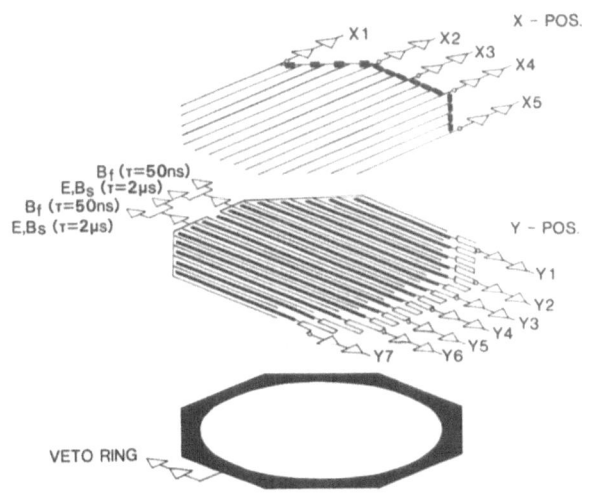

Fig. 5. The XSPECT MSPC (C. Budtz-Jörgensen).

TABLE II
Proportional Counter Parameters

Field of view	60 arcmin
Diameter of active area	160 mm
Diameter of window	140 mm
Position resolution	<1 mm
Time resolution	<5 μs
Background rejection	>99 per cent
Total memory	128 Mbytes
Gas	Xe 90%, Propylene 10%
Energy resolution	0.32E$^{\frac{1}{2}}$ keV

	HEPC	LEPC
Energy range	2 - 25	0.5 - 8 keV
Dynamic range	625	800
Efficiency	>0.7, 2<E<15	>0.7, 1<E<8 keV
Gas thickness	4	2 cm
Gas pressure	1	0.5 atm
Window	13 μm Be	0.5 μm plastic
		0.2 μm diamond
		0.02 μm Al

2.3. XSPECT Objective Crystal Spectrometer (OXS)

The OXS concept separates the processes of energy dispersion and imaging (Schnopper and Byrnak, 1987). The large mosaic panel of flat crystals in front of the telescope acts as a narrow pass filter and a mirror. Each pixel in the reflected field of view of the telescope satisfies a specific Bragg angle on the crystal and, therefore, each pixel in the detector can be identified with a particular energy. Scans which involve repositioning of the telescope axis and the angle between the crystal panel and the telescope axis (45+15/−10 deg) yield either the spectrum of a point source or energy resolved images of an extended source. Parameters chosen for OXS are given in Table III. It is intended that MultiLayer (ML) structures be deposited on the surfaces of the LiF and Ge crystals to allow simultaneous measurements of the Fe, S and O energy ranges. The use of both 1st and 2nd order ML reflections will allow studies of line emission at energies below the C K-edge cut-off imposed by the entrance window (Christensen et al., 1990).

2.4. Thermal and structural elements

SODART is launched in a folded configuration which just fits inside the shroud attached to the PROTON launcher. SODART is then deployed in orbit and locks into position for observing within the carbon fiber epoxy structure shown in Figure 1. Doors at the lower end of the telescope thermal enclosure open and telescoping tubes are lowered to make contact with the focal plane transport assembly. Heaters and thermal baffles maintain the appropriate temperature conditions for the telescopes and the crystal panel.

Each of the telescope assemblies will be provided with a laser ranging system which will measure displacements in x, y and z with respect to the intersection of the telescope axis with the ideal focal plane. The tilt of the telescope axis with

TABLE III
Crystals for the XSPECT OXS

	Crystal	Plane	2d (Å)	Range (keV)	R_c (rad)	E/ΔE (35<Θ<60 deg)
Side 1	LiF	(220)	2.848	5.00-7.40	1.7×10^{-4}	1375-2300
	ML Ni/C on LiF (220)	(001) (002)	67 67	0.21-0.31 0.43-0.63	4.0×10^{-3} 1.6×10^{-4}	70 300
or,	ML Ni/C or W/B_4C on LiF (220)	(001) (002)	39.2 39.2	0.42-0.62 0.84-1.23	4.0×10^{-4} 1.0×10^{-4}	300 400
Side 2	Ge	(111)	6.532	2.19-3.23	2.9×10^{-4}	2575
	RAP	(001)	26.121	0.55-0.81	5.0×10^{-5}	770
	ML Ni/C on on Ge (111)	(001) (002)	56 56	0.26-0.38 0.51-0.75	4.3×10^{-3} 2.0×10^{-3}	80 300

respect to the line perpendicular to the focal plane will also be measured. These measurements will be correlated with 3-axis aspect information derived from star sensors mounted on the fixed portion of the focal plane structure. The possibility of a star sensor mounted on the telescope structure is being considered. BC is responsible for these elements of SODART. They also provide the spacecraft and integrate the payload. IKI provides the overall project management.

3. Sensitivities

The sensitivity to emission lines of a single module in combination with a gas detector has been calculated with the parameters given in the Table IV. The iron line at ~6.7 keV is a blend of relatively few lines whereas the oxygen lines around 0.7 keV are surrounded by many other lines which makes the continuum determination difficult.

The points in Figure 8 are AGN from the Turner and Punds (1989) EXOSAT sample. The equivalent widths are assumed and used to indicate the typical times required to observe the lines. The continuum flux has been evaluated from the reported luminosity and power-law slope.

Acknowledgements

The staff at DSRI, IKI and BC have made significant contributions to this work and I thank them warmly for their assistance in preparing the manuscript.

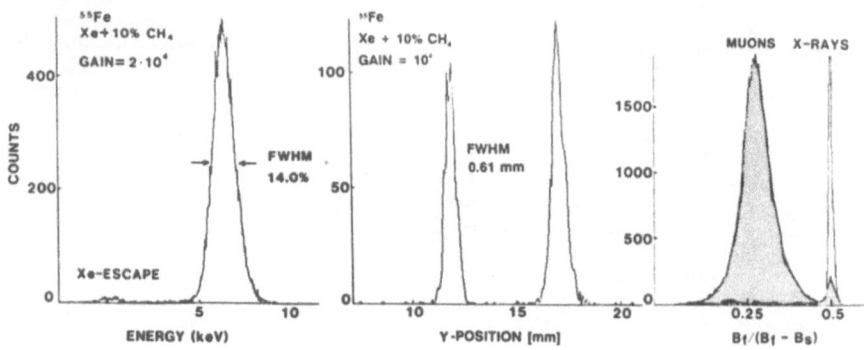

Fig. 6. Energy, position and pulse shape background rejection resolution with a breadboard MSPC (C. Budtz-Jörgensen, DSRI).

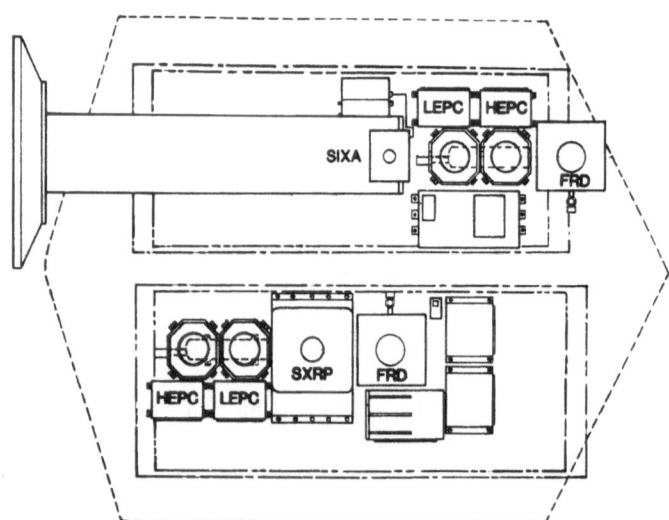

Fig. 7. The SODART focal plane assembly.

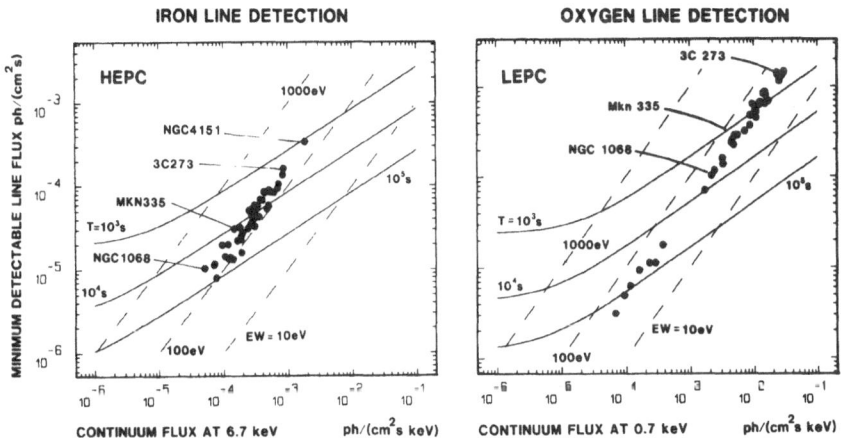

Fig. 8. Minimum detectable iron and oxygen line fluxes for various observing times and line equivalent widths (EW). The AGNs shown have all been assigned the same EW (150 eV iron and 50 eV for oxygen).

TABLE IV
Parameters used for Line Sensitivity Calculations

	Oxygen in LEPC	Iron in HEPC	
Line energy	0.7	6.7	keV
Telescope eff. area	1766	1356	cm^2
Detector quant. eff.	0.54	0.80	
Energy resolution	0.27	0.80	keV
Particle background	10^{-3}	10^{-3}	cts cm^{-2} s^{-1} keV^{-1}
Search cell size	0.22	0.22	cm^2
Fraction in cell	0.5	0.5	

The points in Figure 8 are AGN from the Turner and Punds (1989) EXOSAT sample. The equivalent widths are assumed and used to indicate the typical times required to observe the lines. The continuum flux has been evaluated from the reported luminosity and power-law slope.

References

Budtz-Joergensen, C., Madsen, M.M., Jonasson, P., Schnopper, H.W., and Oed, A.: 1989, *SPIE Proceedings*, 982

Christensen, F.E.: 1990, *SPIE Proceedings*, in press

Kaaret, P. *et al.*: 1989, *SPIE Proceedings*, 1160

Lewis, R., Bordas, J., and Christensen, F.E.: 1990, *Opt. Eng.*, in press

Oed, A.: 1988, *Nucl. Instr. Meth.* **A263**, 351–359

Schnopper, H.W., and Byrnak, B.P.: 1987, *Appl. Opt.* **26**, 2871

Serlemitsos, P.J., Petre, R., Glasser, C., and Birsa, F.: 1984, *IEEE Trans. Nucl. Sci.* NS-31, 786

Sunyaev, R.: 1991, *IAU Colloquium 123*, Kluwer Academic Publishers, Dordrecht

Turner, T.J. and Pounds, K.A.: 1989, *Mon. Not. R. Astr. Soc.* **240**, 833

Vilhu, O.: 1991, *IAU Colloquium 123*, Kluwer Academic Publishers, Dordrecht

Westergaard, N.-J., *et al.*: 1990, *Opt. Eng.*, in press

ESA's X-RAY ASTRONOMY MISSION, XMM

B.G. TAYLOR and A. PEACOCK

Space Science Department, ESTEC-ESA, Noordwijk, The Netherlands

Abstract. ESA's X-ray Astronomy Mission, XMM, scheduled for launch in 1998, is the second of four cornerstones of ESA's long term science program Horizon 2000. Covering the range from about 0.1 to 10 keV, it will provide a high throughput of 5000 cm^2 at 7 keV with three independant telescopes, and have a spatial resolution better than 30 arcsec. Broadband spectrophotometry is provided by CCD cameras while reflection gratings provide medium resolution spectroscopy (resolving power of about 400) in the range 0.3–3 keV. Long uninterrupted observations will be made from the 24 hr period, highly eccentric orbit, reaching a sensitivity approaching 10^{-15} erg cm^{-2} s^{-1} in one orbit. A 30 cm UV/optical telescope is bore-sighted with the x-ray telescopes to provide simultaneous optical counterparts to the numerous serendipitous X-ray sources which will be detected during every observation.

1. Introduction

The concept of a high-throughput. medium spatial resolution, X-ray astronomy mission with spectroscopic capability over a wide energy band was studied extensively within ESA in the mid-80's.

A set of scientific requirements, in order of priority was established as outlined below.

1. Broad band spectroscopy at full grasp:
  a) Area goals of 5000 cm^2 at 8 kev, 10,000 cm^2 at 2 keV.
  b) Energy band 0.2–10 kcV.
  c) Spatial resolution specified at < 30 arcsec.

2. Medium resolution spectroscopy below 3 keV:
  a) Resolving power > 250.
  b) At maximised grasp.

3. High resolution spectroscopy at some energies:
  a) Oxygen-K and Iron-K lines priority.
  b) Resolving power > 1000

Within ESA's long term scientific programme Horizon 2000, a ceiling of 400 MAU at 1985 economic conditions, was set for cornerstone missions of which XMM is the second. This is equal to about US$ 500 million today. As in all ESA missions, this sum includes spacecraft development, the launch vehicle and orbital operations for two years (though XMM is expected to have a lifetime of 10 years). In the case of XMM it also includes the procurement of X-ray optics, but not the cost of the instruments, which are provided by national funding.

Y. Kondo (ed.), Observatories in Earth Orbit and Beyond, 129–140.
©1990 *Kluwer Academic Publishers.*

B.G. TAYLOR AND A. PEACOCK

TABLE I
XMM mirror module – optical design

Focal length	7500 mm
Mirror radius	159-350 mm
Mirror wall thickness	0.66-1.26 mm
Axial mirror length	600 mm
Minimum packing distance	1 mm
Reflective surface	iridium/gold
Number of mirrors	58
Mirror module mass	170 kg

System level studies carried out in industry and supported by scientific working groups led to the present telescope/instrument concept of XMM as indicated schematically in Figure 1. Three independant telescopes are to be flown each with a CCD camera in the focal plane. On the two telescopes, fixed reflection gratings intercept 50% of the beam to feed strip CCD, secondary focus cameras to provide simultaneous spectroscopic coverage. A practical arrangement within the technical and financial constraints could not be found for a high resolution, dispersive spectrometer. However, in the light of XMM's X-ray sensitivity, a UV/optical telescope to give a simultaneous optical view of the X-ray field is included.

2. System Constraints and Confoguration

Experience with the EXOSAT demonstrated the virtues of an X-ray observatory in a highly eccentric orbit (HEO), e.g. the ease of operation and long uninterrupted observing times. Thus while a HEO would be natural for XMM, the choise was further enforced by the aim to use purely passive cooling for CCD's (–100 degC required) as a means to achieve a long-lived mission. Economic considerations led to the selection of the Ariane 4-2L, capable of lifting a 2.4 tonne spacecraft into a 24 hour 70,000 km apogee, 60deg inclination orbit. These orbital parameters permit the use of a single ground station which can view XMM without interruption when it is above 40,000 km, i.e. beyond the background of the radiation belts. Seventeen hours per day (70%) are thus availble for observations.

3. X-Ray Optics

Within the overall mass budget, up to 220 kg has been allocated on one mirror module, implying the use of a lightweight replication process for mirror manufacture. The principal characteristics of a mirror module are given in Table I. In the

TABLE II
Characteristics of XMM's prime focus CCDs

	pn CCD	MOS CCD
Number of CCD's/camera	12	8
Depletion Depth (μm)	280	30
Quantum efficiency (%)		
0.5 keV	60	60
1.0 keV	90	90
10.0 keV	90	25
Spatial resolution (μm)	150	20
Spectral resolution (eV)		
0.5 keV	50	50
6.7 keV	130	130
Time resolution mode:		
Full frame transfer (s)	0.024	50
High time (μs)	20	2000
Electon Radiation Tolerance (krad)	1000	14
Operating Temperature (K)	175	175
X-ray CTE (per pixel)	0.99999	0.99998
Background rejection (%)	99.999	99

currently baselined process, each Wolter I shell carrier (paraboloid and hyperboloid in one piece) of CRFP is laid upon a steel mandrel, having a shape tolerance of 10 micron and a surface roughness below 0.25 micron rms. The gold (or iridium) layer is evaporated on to a glass mandrel polished to a surface acurracy of beteer than 5 Å. The coated mandrel and the CFRP carrier are moulded together using a 50 micron epoxy layer and after curing the shell is removed from the mandrel.

In the technology development programme three mirror modules have been produced in this way and tested at the Panter facility in Munich. With 'full beam' illumination, half energy widths of 67, 70 and 94 arcseconds have been achieved at 0.93, 1.49 and 8.05 keV respectively. Pencil beam measurements along a meridian showed a best resolution of 11 arcsec (HEW) with 22 arcsec (HEW) average. This indicates some global deformation of the carrier during the replication/demoulding process.

The nickel electro-forming process is also being pursued, which has demonstrated good performance in the SAX double cone approximation and JET-X programmes. The main difficulty with this process for XMM is of course the mass limitation. However, it is faster and less expensive than CFRP replication.

The decision in mirror technology should be taken in April 1991. A mirror development model containing three of the 58 shells produced to full X-ray standard, together with the 'electro-optical' models of the instruments are scheduled for testing in the Panter facility in October 1992.

The effective area of the XMM optics as a function of energy and comparisons with other missions shown in Figure 2, while the off-axis performance is shown in Figure 3.

TABLE III
Characteristics of XMMs reflection grating spectrometers

Spectrometer readout	MOS CCD's
Number of CCDs/readout camera	10
Grating mean line density (1/mm)	641
Grating configuration	In-Plane
Number of grating plates/module	240
Grating technology	Beryllium/replication
Waveband (Å)	
Order = - 1	5-35
- 2	5-20
- 3	5-10
Maximum resolving power 5 Å	200
10 Å	360
15 Å	500
25 Å	430
35 Å	560

CCD Prime Focus Camera's – PI G. Bignami.
A number of options for the CCD devices are currently under investigation, including the pn CCD and the MOS CCD (see Figure 2, 4 and 5 for characteristics).

Reflection Grating Spectrometer – PI A.C. Brinkman
The characteristics of this instrument are shown in Table III and in Figure 4.

Optical Monitor – PI K.O. Mason
This 30 cm aperture Cassegrain telescope will span the UV/Optical range from 1500 to 10,000 Å using two detectors, a microchannel plate for the 'blue' beam and a CCD for the 'red' beam. The latter will cover the same FOV as the X-ray telescopes. Characteristics are shown in Figure 4.

4. Sensitivity

The sensitivity of XMM is crucially dependant on the collecting area of the X-ray optics (58 shells per telescope), the achieved angular resolution (< 30 arcsec) and the quantum efficiency and energy range coverage of the CCD's. Assuming nominal performance, then the minimum detectable source strength as a function of time as in Figure 6 should be obtained. In one XMM orbit of 60,000 sec the minimum detectable source would be approaching 10^{-15} erg cm^{-2} s^{-1}. For every typical exposure of 10^4 sec a limiting sensitivity some 20 times fainter than the Einstein Deep Survey will be achieved.

The minimum detectable line strength is indicated in Figure 7. For a full orbit's observing an Oxygen line with an equivalent width of 90 and 200 eV will be detected from a source with a flux of 10^{-14} erg cm^{-2} s^{-1} by the cameras and reflection gratings respectively.

TABLE IV
Characteristics of XMMs optical monitor

| Telescope Type | | Cassegrain |
Primary mirror diameter (mm)		30
Characteristics	Blue Beam	Red Beam
Angular field (arcmin)	8	30
Plate scale (μm/arcsec)	37	6
Spatial resolution (arcsec)	1	4
Detector	MCP	CCD
Spectral range (Å)	1500-6500	4000-10000
Magnitude limit in 1000 s	24.5	21
Magnitude of brightest star	14.5	6(in 0.25 s)

5. XMM As An Observatory

Figure 8 shows the allocation of guaranteed observing time granted to PIs, mission scientists and the observatory team and open time for the general observer. The first year of observations is largely taken up by guarantedd time but by the second year, 75% of the time is open. For the third and subsequent years, all observing time will be open to general observers.

References

Peacock, A., Taylor, B.G., Ellwood, J.: 1990, *Adv. in Space Res.* **10-2**, 273–285

XMM – OPTICAL PATH DIAGRAM

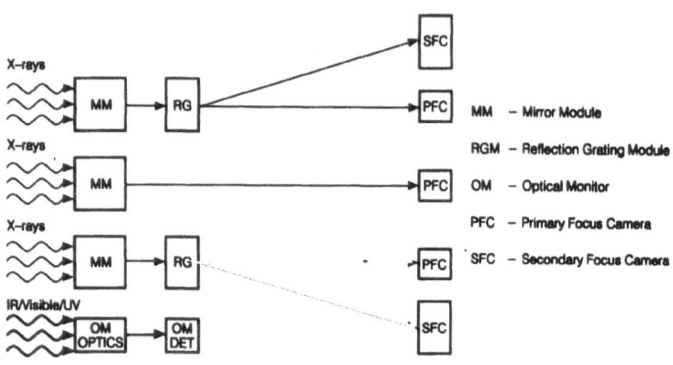

Fig. 1. Schematic of XMM's instruments.

XMM MM PERFORMANCE CHARACTERISTICS

Effective Area: Resolution:

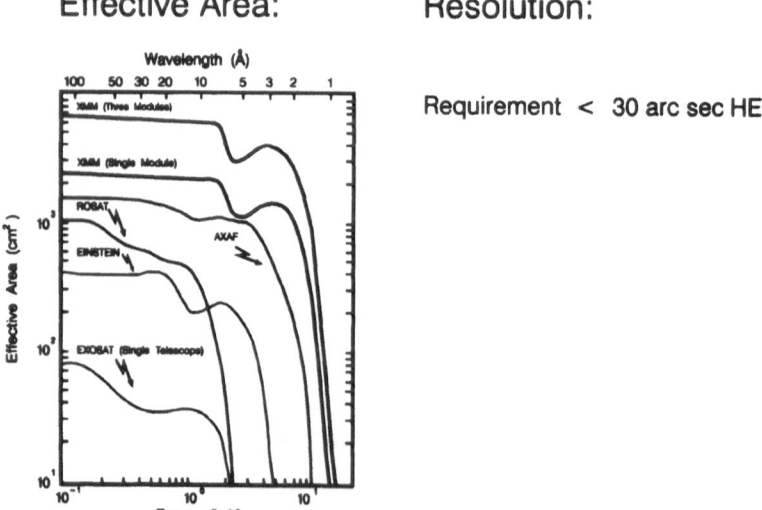

Requirement < 30 arc sec HEW (TBV)

Fig. 2. Effective area of the XMM X-ray optics as a function of photon energy compared with other missions.

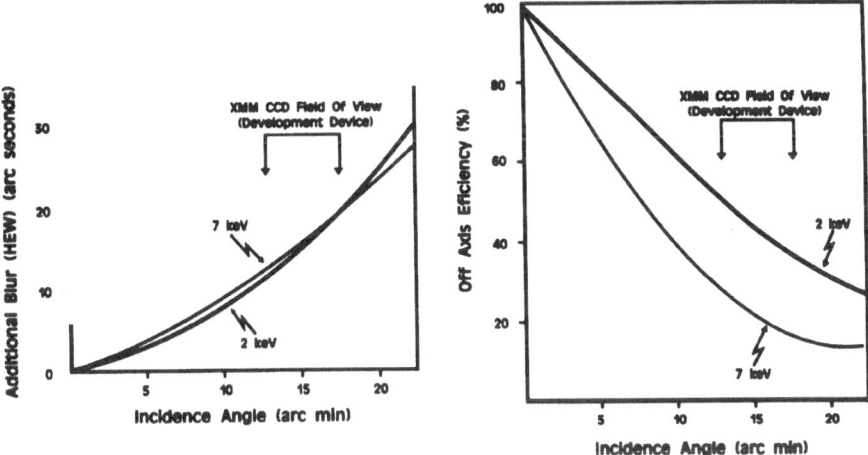

Fig. 3. Variation in the additional blur contribution to the spatial resolution and off-axis telescope efficiency for two X-ray photon energies as a function of angle of incidence.

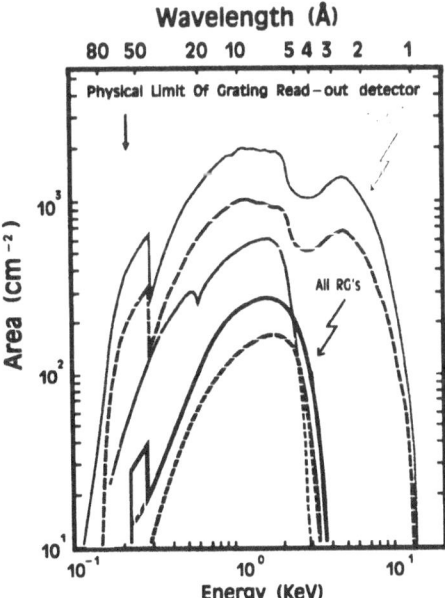

Fig. 4. Effective on-axis area of the XMM CCD camera fitted to all three telescope modules (upper curves) and to a single module (dashed curve). Note that the effect of the loss of ˙ area due to the reflection gratings behind two telescopes has been included in the total area (full curve). The effective area of the gratings is also shown.

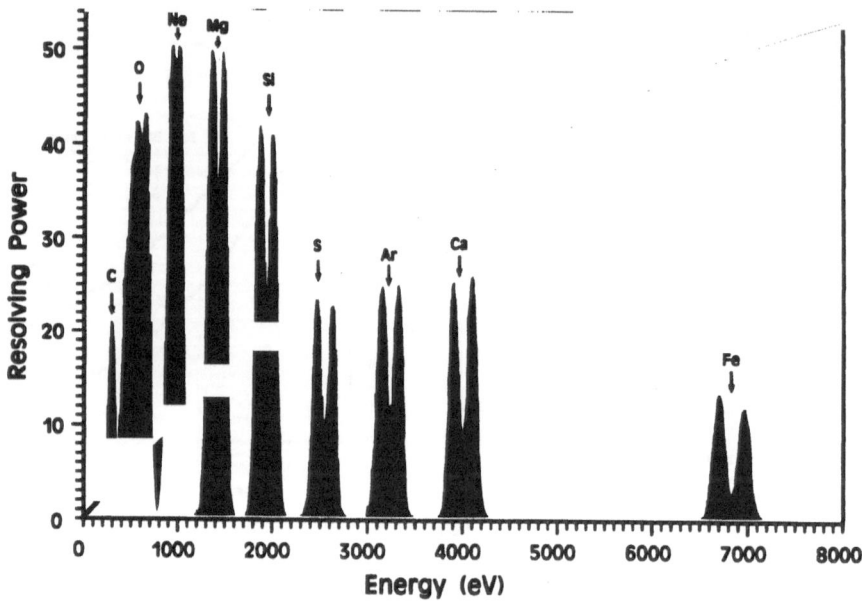

Fig. 5. Resolving power of the XMM CCD as a function of photon energy. The CCD and mirror responses have been combined and all transitions set at equal strength.

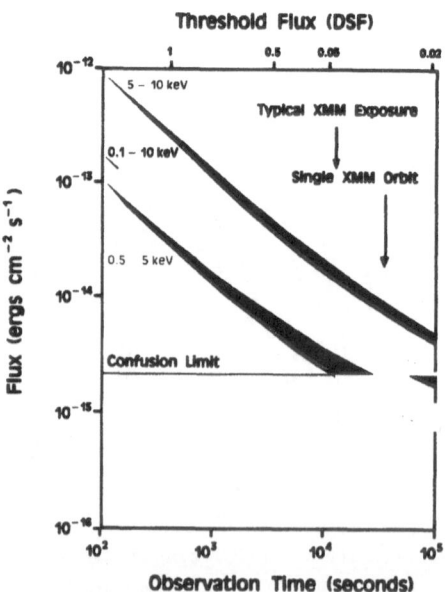

Fig. 6. Minimum detectable source strength as a function of observing time. The equivalent threshold flux in units of the Einstein Deep Survey (DSF) is indicated.

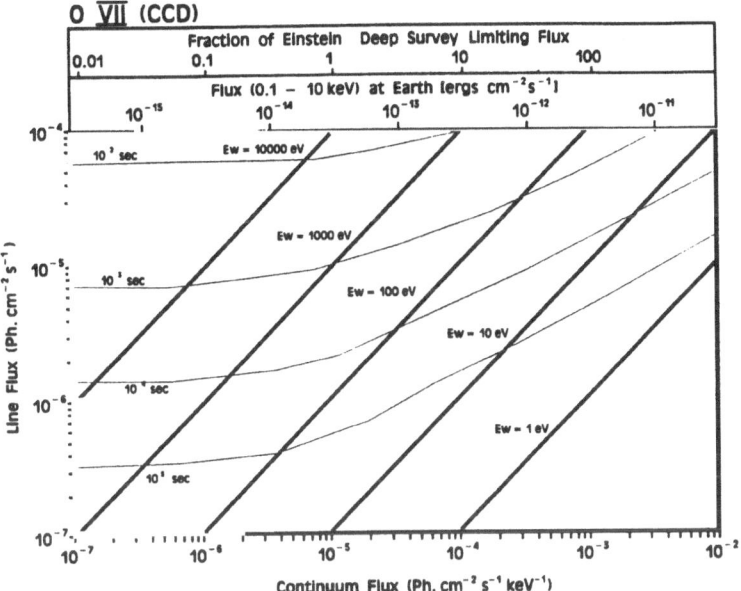

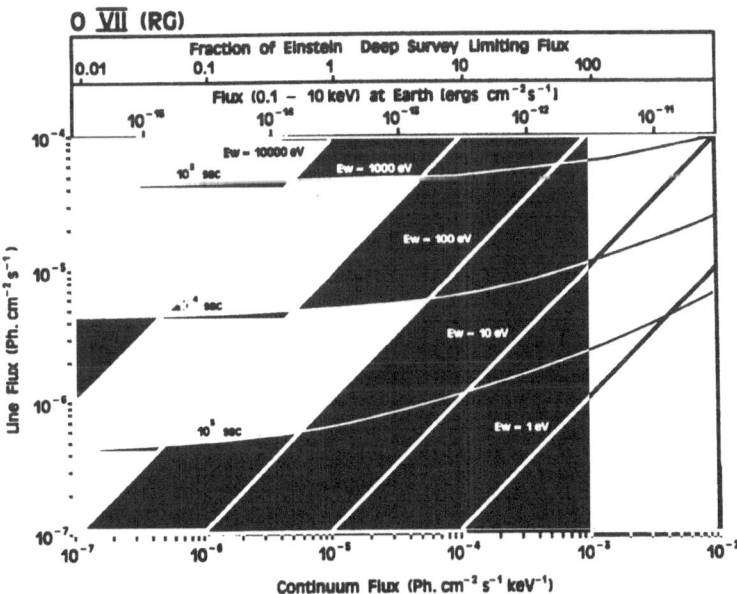

Fig. 7. (a)–(c). Minimum detectable line strength as a function of continuum flux at the He-like line energies of O (a), Si (b) and Fe (c), for both the CCD and grating modules.

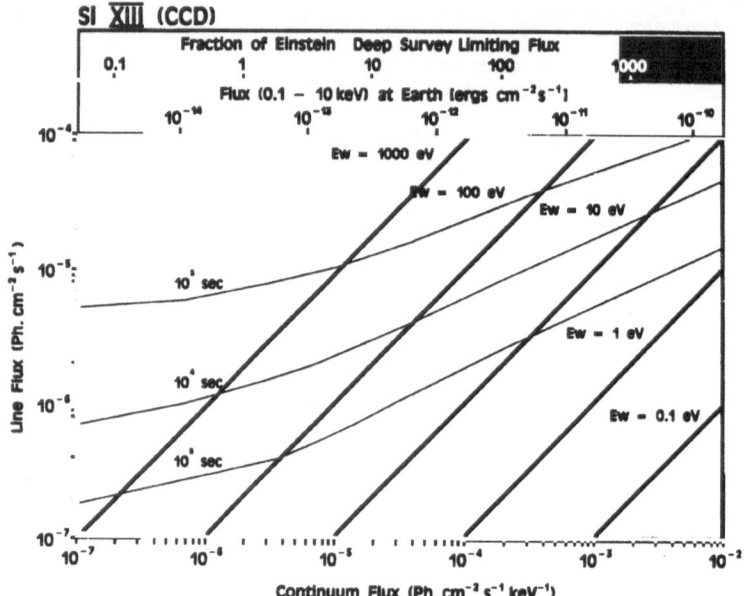

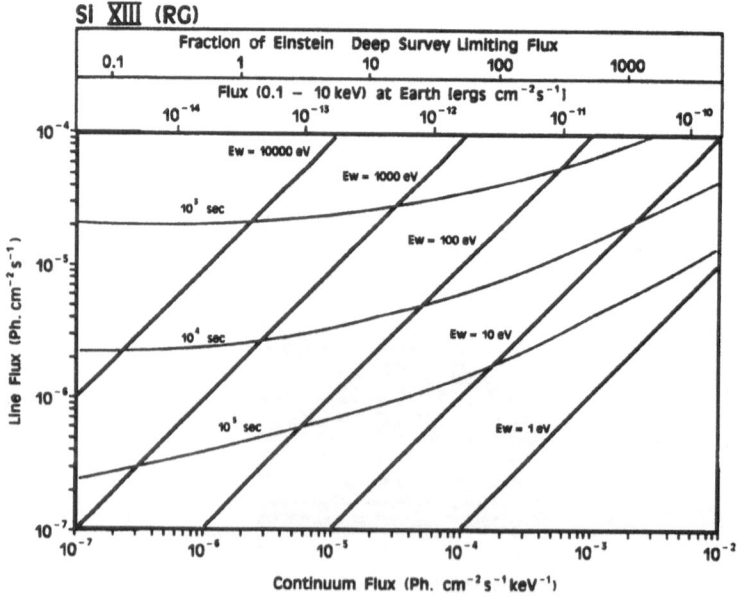

Fig. 7b.

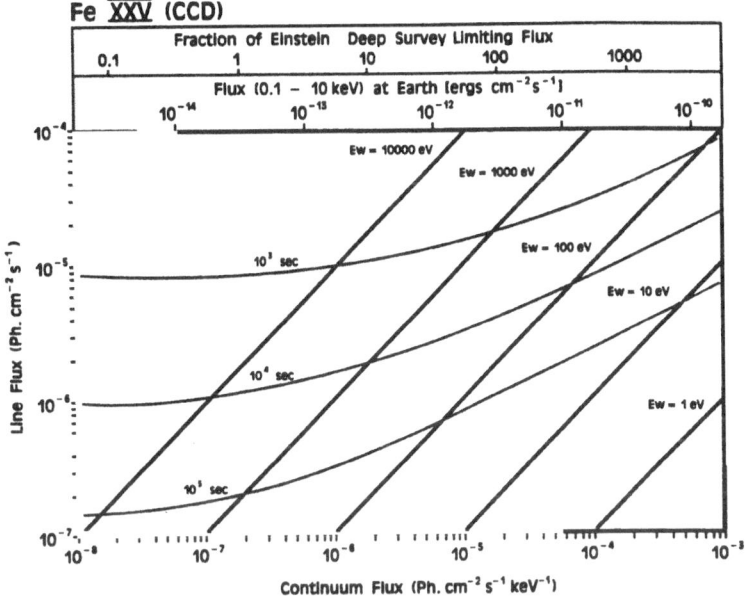

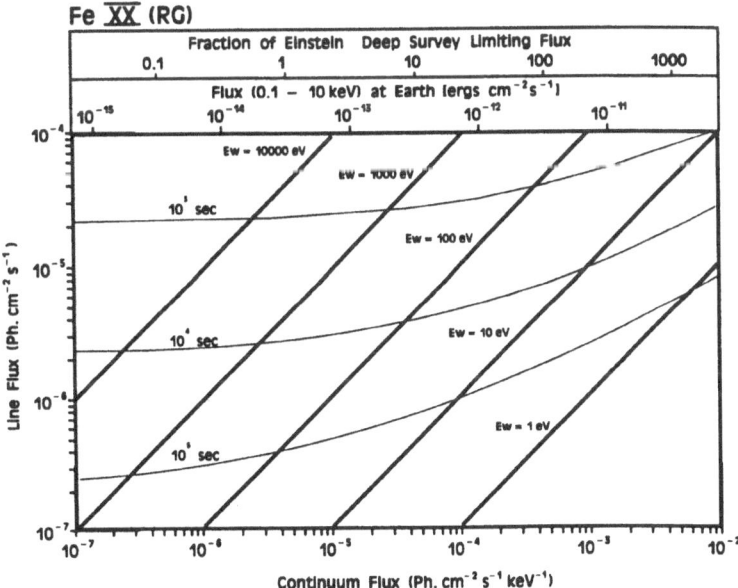

Fig. 7c.

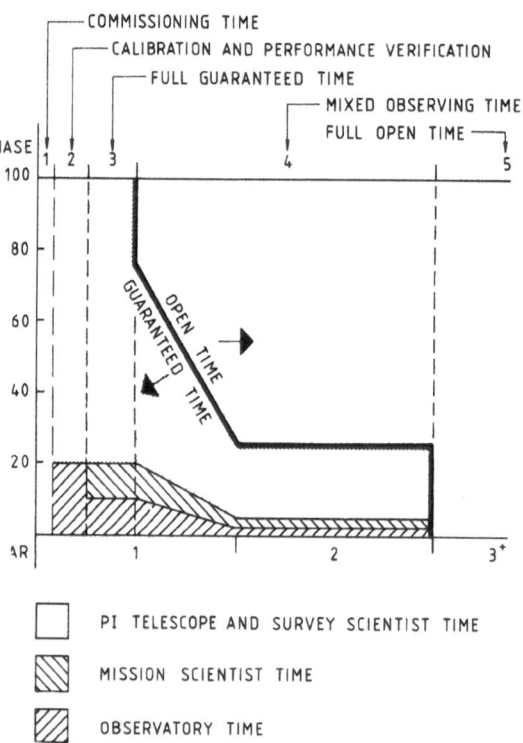

Fig. 8. The allocation of XMM observing time.

THE SAX MISSION FOR X-RAY ASTRONOMY

R.C. BUTLER

Agenzia Spaziale Italiana, Roma, Italy

and

L. SCARSI

Istituto di Fisica Cosmica/CNR and Dipartimento di Energetica,
Univ. of Palermo, Palermo, Italy

Abstract. The satellite for X-ray astronomy SAX, to be launched at the end of 1993, is devoted to systematic, integrated and comprehensive, studies of galactic and extra-galactic sources in the energy band 0.1–200 keV, and is under joint development by the Italian Space Agency (ASI) and the Netherlands Agency for Aerospace programs (NIVR), with the participation of SRU/SRON and SSD/ESTEC. The basic scientific objectives can be summarized as follows:

– Broad band spectroscopy (E/Δ E=12) from 0.1–10 keV with imaging resolution of 1 arcmin.
– Continuum and line spectroscopy (E/Δ E=5–20) in the energy range 3–200 keV.
– Variability studies of bright source energy spectra on timescales from milliseconds to days and months.
– Systematic long term variability studies over the entire sky down to a source intensity of 1 mCrab.

The payload consists of four concentrator/spectrometers (3 units 1–10 keV, 1 unit 0.1–10 keV, for a total effective area of 200 cm^2 at 7 keV), a high pressure gas scintillation counter (3–120 keV, effective area of 300 cm^2 at 6 keV), a phoswich scintillation counter (15–200 keV, effective area of 600 cm^2 at 60 keV) which are all coaligned, and is completed by two wide field cameras (2–30 keV, 20 deg ×20 deg FWHM, and effective area of 250 cm^2, pointing orthogonally with respect to the other instruments.

The three axis stabilised satellite will be placed into a circular orbit at 600 km with inclinatiobn of two degrees by an Atlas G-Centaur. The minimum mission lifetime will be two years but extendable upto four years.

1. The Sax Mission Overview

SAX, 'Satellite for Astronomy in X-rays', is a major joint program of the Italian Space Agency (ASI) and the Netherlands Agency for Space Programs (NIVR). SAX finished its Phase B activities in 1988 and entered its Phase C/D (a continuous phase) in April 1989. The satellite will be launched at the end of 1993. The chief characteristics of the mission, spacecraft, and payload instruments, are given in Table 1. In the past two decades a succession of X-ray astronomy satellites, from Uhuru to the Einstein Observatory, EXOSAT and Ginga, have led to the discovery of more than 1000 individual X-ray sources and after the completion of the first all-sky survey by an imaging X-ray telescope on board Rosat, this number can be expected to increase by nearly an order of magnitude. The results so far have

Y. Kondo (ed.), Observatories in Earth Orbit and Beyond, 141–150.

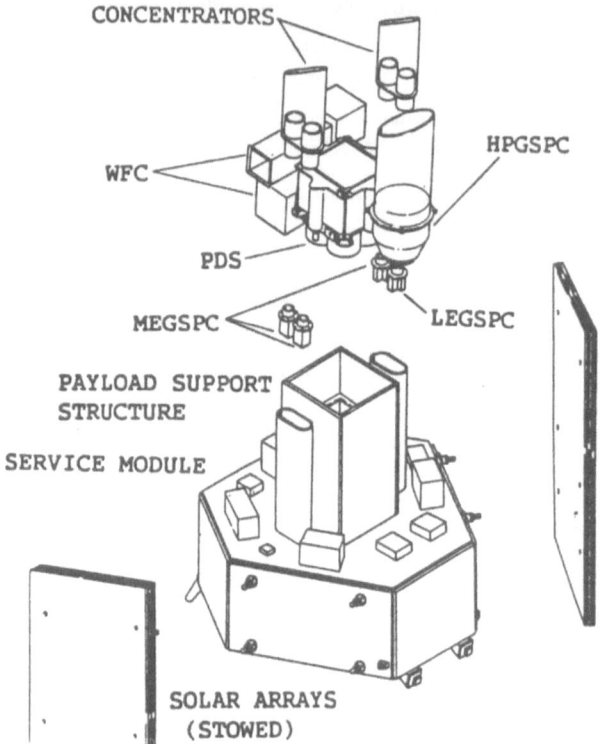

Fig. 1. SAX spacecraft and payload.

demonstrated the importance of extending X-ray observations with simultaneous spectroscopic, spectral and time variability studies over a broadened energy range, and thus the scientific objectives of SAX and the choice of its payload instruments have been made to carry out systematic and comprehensive observations in the 0.1-200 keV energy range with special emphasis on spectral and timing measurements.

The payload complement includes a medium energy (1-10 keV) concentrator/spectrometer, MECS, consisting of three units, a low energy (0.15-10 keV) concentrator/spectrometer, LECS, a high pressure gas scintillation proportional counter (3-120 keV), HPGSPC, and a phoswich detector system (15-200 keV), PDS, all of which have narrow fields of view and point in the same direction. Two wide field cameras (2-30 keV), WFC, which point in diametrically opposed directions perpendicular to the narrow field instruments complete the payload. The spacecraft and payload are shown in Fig. 1.

SAX will be launched by an Atlas G-Centaur directly into a 600 km orbit at 2 degrees inclination. The payload will thus nearly avoid the South Atlantic Anomaly, and take full advantage of the screening effect of the Earth's magnetic field in reducing the cosmic ray induced background (proton fluxes typically a factor of 20 lower than in the Exosat orbit above the radiation belts), while undergoing the minimum

TABLE I

The Main Characteristics of the SAX Mission, Spacecraft and Payload Instruments

Orbit

Type	Circular at 2 deg inclination to the equator
Height	600 km (BOL)/450 km (EOL)
Lifetime	2 years extendable to 4 years
Period	97 min (Sun eclipse 37 min)
Ground contact	11 min per orbit
Launch Date/Vehicle	End of 1993/Atlas G-Centaur

Spacecraft

Total Mass/Power	1200 kg / 800W (Sunlight), 790W (Eclipse)
Payload Mass/Power	385 Kg/290 W

Pointing

Observation time	Typically 10^4 to 10^5 sec
Abs. pointing error	3 arcmin (Three axis stabilised)
Abs. pointing stability	1 arcmin
Pointing Reconstruction	1 arcmin

Telemetry

Mass Memory Capacity	450 Mbits
Average effective rate	70 Kbps
Maximum effective rate	100 Kbps
Dump rate to ground	1 Mbps

Payload Instruments

	Aperture FWHM	Ang. Res.	Total Eff. Area	Energy Res. FWHM
Concentrator/Spectrometer (3 units, 1–10 KeV, MECS) (1 unit, 0.1–10 KeV, LECS)	30′	1′ 200 cm² 7 KeV	8% 6 KeV 56 cm² 0.2 KeV	28% 0.27 KeV
High Pressure Gas Scint. Proportional Counter (3–120 KeV, HPGSPC)	60′	–	280 cm² 60 KeV 300 cm² 6 KeV	3% 60 KeV 10% 6 KeV
Phoswich Detector System (15–200 KeV, PDS)	87′	–	140 cm² 200 KeV 680 cm² 20 KeV	17% 60 KeV
Wide Field Cameras	20° × 20°	5′	250 cm² (/unit through mask)	20% 6 KeV

Sensitivity	Photons/cm² s KeV at 5 sigma in 10^4 s	Emission lines at 5 sigma Pht/cm² s in 10^4 s
Conc/spectr (4 units)	4.4×10^{-6} (1–10 KeV)	1.5×10^{-5} at 6.7 KeV
HPGSPC	3×10^{-5} (10–30 KeV) 2×10^{-5} (35–120 KeV)	1×10^{-3} at 51 KeV
PDS	1.2×10^{-5} (15–30 KeV) 5.3×10^{-6} (30–80 KeV) 3.8×10^{-6} (80–200 KeV)	2×10^{-4} at 51 KeV
WFC/unit	3 mCrab (2–30 KeV)	
PDS (Gamma-bursts)	10^{-6} erg/cm² s (100–600 KeV)	

change in magnetic cut-off. This choice of orbit is particularly important in order to minimise the background and systematic effects caused by spallation and changes in incident charged particle fluxes around the orbit to achieve the necessary sensitivity of the PDS for weak source observations, estimated at 1% of the background in this orbit, essential if the broad band sensitivity of SAX is to be ensured. SAX has an official design lifetime of two years but is expected to remain in operation for up to four years. The satellite will achieve one arcminute pointing stability continuously for a typical maximum single observation time of 10^5 seconds, with a postfacto pointing reconstruction accuracy of 1 arcmin. The chief attitude constraint derives from the need to retain the normal to the solar arrays within 30 degrees of the Sun (occasional excursions to 45 degrees will be made for WFC) to ensure proper battery maintenance and allow observation to be continued throughout the Sun eclipse periods whith all the instruments. Thus while the whole sky will be available during one year, during a single orbit, and subject to eclipse by the Earth (which has a diameter of about 130 degrees at 600 km) the narrow field instruments will have a band in the sky 60 degrees wide available for observations which includes about 50% of the sky, and the WFC a slightly larger band commensurate with their field of view. In the mission plan the narrow field instruments will normally have the first choice of observation direction while the WFC will observe the regions of the sky available to them. Periodically the WFC will be given priority to perform long term studies of centain sky regions e.g. the galactic centre and along the galactic plane. During each orbit upto 450 Mbits of information will be stored on board and relayed to the ground during station passage. The average data rate available to the instruments will be approximately 70 kbits, but peak rates of upto 100 kbps will be catered for. The ground station for telecommand uplink and telemetry retrieval will be situated near the Equator (at the San Marco Base, Malindi, Kenya), while the Operations Control Centre connected by a communications relay satellite link will be in Italy. The SAX mission development is supported by a consortium of Institutes in Italy together with Institutes in The Netherlands and the Space Science Department of the ESA. A collaboration with the Max Planck Institute for Extraterrestrial Physics also exists for X-ray mirror testing and the calibration of the concentrator mirrors. The composition of the consortium is as follows:

- Istituto per le Tecnologie e lo Studio delle Radiazioni Extraterrestri, ITESRE / CNR, Bologna

- Istituto di Astrofisica Spaziale, IAS/CNR, Frascati

- Istituto di Fisica Cosmica e Tecnologie Relative, IFCTR/CNR and Unit GIFCO, Milano

- Istituto di Fisica Cosmica d Applicazioni dell'Informatica, IFCAI/CNR and Unit GIFCO, Palermo

- Istituto dell'Osservatorio Astronomico, Universit di Roma *La Sapienza*

- Space Research Utrecht, SRON, The Netherlands

- Space Science Department, SSD, of ESA, Noordwijk, The Netherlands

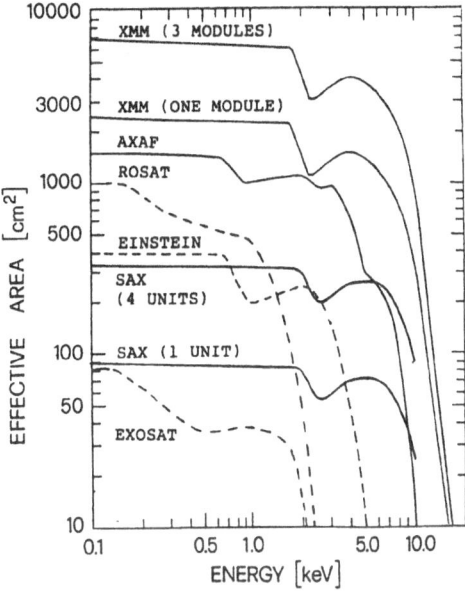

Fig. 2. Effective area of several X-Ray satellites.

2. Brief Description of the SAX Payload Instruments

The main characteristists of the SAX payload instruments, described also by Spada (1983) and Perola (1990), are given in Table I.

2.1. THE CONCENTRATOR/SPECTROMETER

The instrumentation consists of four separate concentrator mirror assemblies each with a focal length of 185 cm and a position sensitive, Xenon filled, gas scintillation proportional counter (GSPC) in the focal plane. The mirror assemblies, for which the prototype development and testing has been performed by IFCTR and IFCAI, are composed of 30 nested gold coated, confocal, double cone approximations to a Woltjer 1 paraboloid/hyperboloid configuration with a focal length of 185 cm. They will be produced by electroforming nickel (0.2–0.4 mm) onto the gold coated surface of super polished mandrils. X-ray tests on the mirrors have shown that the design goal of less than 10 angstroms of surface roughness has been reached, and that an angular resolution of slightly less than one arc minute HPR will be achieved, Citterio et al (1988, 1990). The concentrators are designed to maximize their effective area around 7 keV for iron line studies as shown in Fig. 2. The sum of the effective areas of the 4 units compared with AXAF is particularly notable in this respect.

Three of the GSPC's will have 50 micron beryllium windows which are essentially opaque below about 1.2 keV, and be read out by crossed wire anode Hamamatsu photomultipliers. Laboratory tests on the prototype by IFCAI have demonstrated an energy resolution of 8% FWHM at 6 keV and a position resolution of 0.75 mm FWHM at the same energy, equivalent to 0.76 arcminute HPR. Their energy resolution will be comparable with that of the solid state spectrometer of the Einstein Observatory and better by a factor of more than two than that of previously flown proportional counters at the iron K-line at approximately 7 keV. The fourth GSPC developed by SSD/ESTEC will have a 1.5 micron thick polypropylene window to extend its energy range down to 0.1 keV, and be viewed by a nine anode readout Hamamatsu photomultiplier. Laboratory tests on the prototype have produced an energy resolution of 28% FWHM at 0.27 keV and a performance at higher energies equal to that of the other GSPC's. This detector is described in detail by Favata and Smith (1989). The source confusion limit of the Concentrator/Spectrometers reached in 4×10^4 s will be approximately three times below the lower limit Rosat all-sky survey,, and thus they will be fully capable of exploiting the survey in the selection of representative samples of faint objects for detailed studies upto 10 keV.

The High Pressure Gas Scintillation Proportional Counter The xenon filled gas cell of the HPGSPC, is viewed by an Anger camera arrangement of seven photomultiplier tubes, and surrounded at the sides and from below by a graded lead/tin shield. The X-ray event arrival position is used to correct the event energy with an overall improvement of about a factor of two in energy resolution when compared with proportional counters. Special care has been taken in the structural design of the beryllium entrance window (1300 microns) to reduce its thickness to a minimum for the low energy response while allowing the gas cell to be filled to 5 atmospheres to increase the sensitivity at higher X-ray energies. The escape gate technique is used above the xenon K-edge at 34.5 keV to further enhance the instrument's energy resolution and background rejection. A rocking collimator is used for alternatively sampling the flux in the source direction and the background. The prototype instrument is under development at IFCAI and described in detail by Giarrusso et al (1989). The very good energy resolution (about 5 times better than that of the PDS) will be particularly important in the detailed study of narrow cyclotron lines, both in emission and absorption, as illustrated by the simulation of the Her X-1 cyclotron line in absorption in Fig. 3. It will also complement the iron line studies of the concentrator/spectrometers and the source continuum observations of these and the PDS.

2.2. THE PHOSWICH DETECTOR SYSTEM

The instrument is based on a central group of four phoswich units (3 mm of NaI(Tl) above 50 mm of CsI(Na), surrounded at the sides by an anticoincidence shield of CsI(Na) and with a plastic scintillator particle shield over the aperture. The source direction is viewed trough two collimators that can be offset to sample the background. The instrument, developed by ITESRE and IAS using the Lapex balloon borne telescope as test bed is described in Frontera et al (1990). Extrapolating the background from the Lapex results to the SAX orbit has confirmed the sensitivity of the PDS. The energy resolution, demonstrated on the prototype to be better than 17% FWHM at 60 keV, combined with its high source flux sensitivity over a

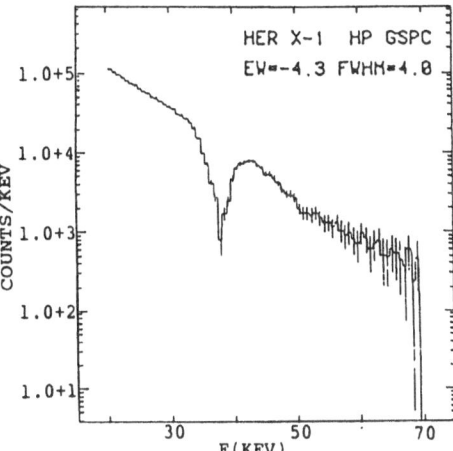

Fig. 3. Simulated spectrum of Her X-1].

broad energy range make the instrument particularly useful in the detailed studies of the continuum in galactic and extragalactic sources and their time variability. It is also well suited to the study of cyclotron line features in know sources and the search for these features in other binary where it will extend the possible measurements of the HPGSPC particularly for broad line features. Additionally, the active shielding around its sides will be used for all-sky monitoring of gamma-ray bursts with a limiting sensitivity of 10^{-6} erg/cm^2 sec and a timing capability of 0.5 to 10 ms.

2.3. THE WIDE FIELD CAMERAS

Each WFC contains a position sensitive proportional counter filled with two atmospheres of xenon, which views the sky through a random mask aperture. These instruments, described by Jager et al (1989), are under construction by SRU/SRON and are a development from their COMIS experiment actually in operation on the Russian MIR space station. The energy range of the WFC's will complement those of the narrow field instruments, while their observation program will include the long term monitoring of both galactic and extragalactic variable sources and the detection/localisation of transients on timescales from 2 ms upwards. Their typical source detection sensitivity is illustrated as a function of time in Fig. 4.

The overall broad band sensitivity of the narrow field instruments can best be illustrated by the simulated results of on a typical AGN in Fig. 5. using all of the narrowfield instruments, from which it is possible to determine the spectral index with an uncertainty approximately one order of magnitude smaller than in previous experiments (Exosat, HEAO-1). Deviations from ther power law (soft excesses or breaks at high energies), if present, shoud be clearly visible.

The narrow field instruments will also be used for time variability studies on time scales from milliseconds to days and months. Their ability is demonstrated in Fig. 6.

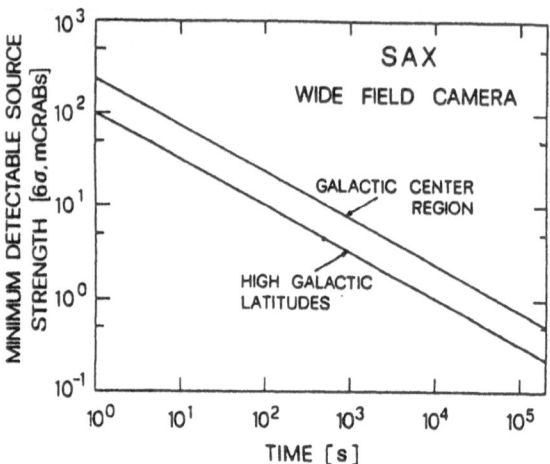

Fig. 4. Simulated spectrum of an AGN.

where the sensitivity of the concentrators/spectrometers is compared with that of the PDS for variability studies on NGC4151 and Cyg X-1. The figure illustrates the balance which has been achieved over such a wide X-ray energy range.

3. SAX Scientific Objectives

The scientific objectives of SAX have been described in detail by Perola (1983, 1990). In the mission life of a least two years SAX will perform between 2000 and 3000 separate observations. These will be based on a core program chiefly devoted to systematic studies of various classes of objects, and a guest observer program allocated about 20% of the time. A selection of the areas in which SAX is expected to make its most significant contribution is given below:

Compact galactic sources: the shape and variability of their continuum and temporal studies of features such as iron fluorescence lines, cyclotron lines and absorption effects as a function of orbital phase and rotation, transient detection and light curve studies.

Supernova remnants: spatially resolved spectra of extended ($\ll$ 1 arcmin) galactic SNRs, and the spectra of the Magellanic Clouds remnants

Stars: coronal emission spectra with a sensitivity comparable to that of the Einstein Observatory upto 10 keV

Active galactic nuclei: spectral and temporal variability studies of their continuum upto 200 keV, the spectra upto 10 keV of very distant sources ($z = 3.2$ for sources equivalent to 3C273) and soft X-ray excess, photoelectric absorption and iron fluorescence line studies.

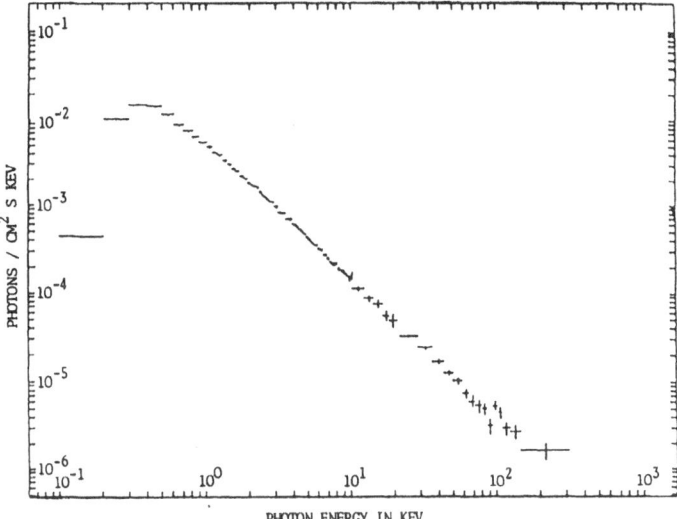

Fig. 5. Variability studies on a large range of timescales.

Clusters of galaxies: spatially resolved spectra upto 10 keV of the nearby clusters with iron fluorescence line and temperature gradient studies and high energy spectra for $z < 0.1$ and temperature measurements out to $z = 1$

Normal galaxies: spectral studies of their extended emission. The SAX payload has been chosen to cover the energy range 0.1–200 keV with a series of instruments in a coordinated fashion to notably extend the spectroscopic and time variability studies performed to date. Its launch date at the end of 1993 will give SAX a first opportunity to take advantage in a systematic way of the many new results that will become available from the all-sky imaging survey of ROSAT, and it will precede the large observatory type missions due for the second half of the 1990's which will concentrate on X-ray astronomy upto 10 keV only.

Acknowledgements

The data presented here is the result of the work of many people who are participating in SAX. We wish to acknowledge the contribution of G.C. Perola, G. Boella and G. Di Cocco, and in particular O. Citterio, G. Conti, and B. Sacco for the concentrator mirrors of the concentrator/spectrometer, A. Smith for its low energy focal plane detector, G. Manzo and S. Re for its medium energy focal plane detector and the HPGSPC, F. Frontera and E. Costa for the PDS, and J. Bleeker, B. Bringkman, R. Jager and P. Ubertini for the WFC. Finally we wish to thank from the SAX team at ASI, G. Manarini, M. Casciola and B. Negri.

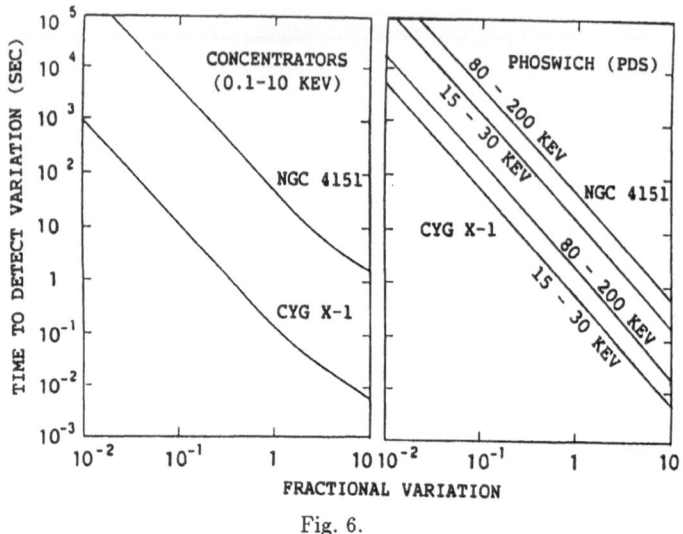

Fig. 6.

References

Citterio, O, et al. : 1988, *Applied Optics* **27**, 1470

Citterio, O, et al.: 1990, *Il Nuovo Cimento* **13C N.2**, 375

Favata, F., and Smith, A.: 1989, *SPIE* **1159**, 488

Frontera, F., et al., 1990, Cospar XXVII, The Hague, in preparation.

Giarrusso, S, et al. : 1989, *SPIE* **1159**, 514

Jager, R., et al. : 1989, *SPIE* **1159**, 2

Perola, G.C.: 1990, *Adv. Space Res.* **10N.2**, 287

Perola G.C.: 1983, in Perola G.C. and Salvati, M., ed(s)., *Proc. of Workshop on Non-thermal and Very High Temperature Phenomena in X-Ray Astronomy*, Rome, 10-20 December 1983, 175

Spada, G., 1983, ibid.

(B) ULTRAVIOLET MISSIONS

THE EXTREME ULTRAVIOLET EXPLORER MISSION

STUART BOWYER

Center for Extreme Ultraviolet Astrophysics,
University of California, Berkeley, California 94720

Abstract. The Extreme Ultraviolet Explorer mission is described. The satellite is now scheduled to be launched in September, 1991. For the first six months, an all sky survey will be carried out covering 90 to 820 Å, or essentially the entire EUV bandpass. This EUV survey will be made in four bands, or colors, $\lambda\lambda 90$–150 Å; 170–250 Å, 400–600 Å and 550–750 Å. A portion of the sky which is free from the normally intense 304 Å geocoronal helium background will be surveyed at greater sensitivity; the wavelength coverage of this band is from 90 to 400 Å. Following the sky survey portion of the mission, spectroscopy of individual sources will be carried out. Three spectrometers employing novel variable line spaced gratings will provide spectra with ~ 1 Å resolution over the band from 90 to 800 Å. This spectroscopy will be carried out by guest observers chosen by NASA in a manner roughly analogous to the IUE guest observer program.

1. Introduction

The Extreme Ultraviolet Explorer (EUVE) mission, currently scheduled for launch in September 1991, is the culmination of some 25 years of effort at the University of California at Berkeley to develop the field of extreme ultraviolet (EUV) astronomy (defined here as astronomy from roughly 100 to 1000 Å). When I started this effort in the mid 1960's, there was universal agreement by the astronomical community that stellar observations at these wavelengths would not be possible because of the high opacity of the interstellar medium. Furthermore, astronomical instrumentation for the EUV was primitive to nonexistent. Solar EUV astronomy was carried out with photographic film as the detector of choice, and spark discharge tubes were the standard laboratory sources for EUV radiation. There were no calibrated standards for use at these wavelengths. We began by developing astronomical instrumentation for these wavelengths, which were then flown as part of my sounding rocket program (Henry et al. 1975a; Henry et al. 1975b; Henry et al. 1975c). We also developed laboratory calibration instrumentation. As an adjunct to this work we developed and deployed in space instrumentation to characterize the diffuse EUV background since this background was unknown and might limit astronomical measurements (Bowyer et al. 1974; Bowyer et al. 1977; Bowyer et al. 1981). It is interesting to note that this background has turned out to be of intrinsic interest (Chakrabarti et al. 1982; Paresce et al. 1983; Chakrabarti et al. 1983; Bowyer et al. 1983; Kumar et al. 1983).

This initial work led to and culminated in the flight of a Berkeley flux collector on the Apollo-Soyuz Mission in 1975. Although the instrumentation employed was crude by current standards and the observational program was quite limited, four astronomical sources of EUV radiation were discovered (Lampton et al. 1976; Margon et al. 1976; Haisch et al. 1977; Margon et al. 1978). While the results obtained

Y. Kondo (ed.), Observatories in Earth Orbit and Beyond, 153–169.
©1990 *Kluwer Academic Publishers.*

were of intrinsic scientific interest, their major impact was that they provided a vivid demonstration that EUV astronomy was, in fact, possible. Within a year or so thereafter, a NASA review panel identified the Berkeley EUVE mission as a prime mission for development and NASA chose this mission for implementation.

2. EUVE Mission and Instrumentation

The primary goal of the EUVE mission will be to carry out an all sky survey over the entire EUV band. This survey will be conducted in four subbands, or colors. It was originally expected that this would be the first survey in this band, but in the passage of time other EUV missions have been developed which will carry out at least partial surveys in this band. Most noteworthy in this regard is the British Wide Field Camera to be flown as part of the ROSAT mission. This mission (described elsewhere in this volume) will carry out a survey at the shortest EUV wavelengths. While it is certain that the majority of EUV sources will be observed at shorter wavelengths, it is clear that the sky should be mapped over the entire EUV band at least once. In addition, our experience with x-ray astronomy, where at least five surveys of roughly equal sensitivity were conducted, has shown that multiple surveys bring out new and unexpected results.

The EUVE mission will also carry out a deeper survey over a limited portion of the sky. This data will provide insights into the types of sources which would be discovered in a more sensitive all sky survey. In specific would fainter sources be similar in character to the sources observed at higher intensities, or would an entirely new class of sources be evident?

What kinds of objects will EUVE detect? The answer is that no one knows. This is a true "explorer" mission; no surveys have been carried out and we must wait and see what nature has provided. It is generally agreed, however, that at least two classes of objects will certainly be detected: hot white dwarfs and coronally active stars. Estimates of the number of these objects which will be observed vary widely but generally fall in the range of 200 to 2000 objects. Hence it seems likely that the survey will at least detect a number of EUV sources which is equivalent to that detected by the first x-ray surveys.

Finally, EUVE will carry out spectroscopy of the brighter sources discovered in the survey phase of the mission. This spectroscopy will be carried out exclusively by guest observers, similar in manner to the IUE guest observer program.

In Figure 1 I show the sensitivity of the instrumentation on the EUVE satellite. For comparison I show the spectrum of the hot white dwarf HZ43, the most intense EUV source now known. The all sky, all band surveys will be capable of detecting sources roughly one-hundredth as intense as HZ43. The deep survey is much more sensitive but covers only part of the EUV band and only a part of the celestial sphere. The sensitivity of the spectrometer depends on the integration time and the character of the source; I show in Figure 1 the sensitivity for a continuum source and one day's integration.

The telescopes for EUVE are shown in Figure 2. The total complement of telescopes consists of two Wolter-Schwarzschild Type I telescopes used to carry out the short wavelength bands of the survey and one Wolter-Schwarzschild Type II Tele-

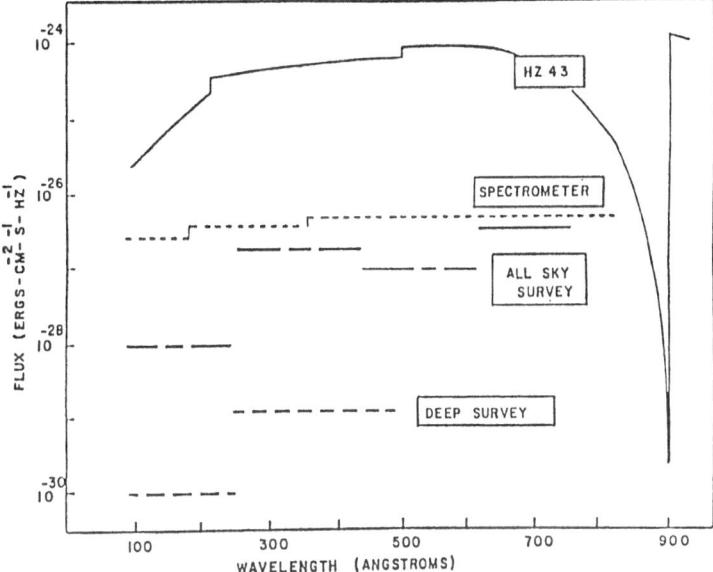

Fig. 1. The overall sensitivity of the EUVE instruments when combined with the anticipated sky integration times in the survey phase. The dotted line that is marked spectrometer shows the sensitivity for a continuum source and one day's integration time.

scope for the longer wavelength band passes. The large graze angles in this design
(Finley et al. 1988) prevents the (presumably more numerous) shorter wavelength
sources from registering in the long wavelength band and compromising this part
of the survey. A standard Wolter-Schwarzschild telescope is the collector for the 3
spectrometers and the deep survey.

The mirrors are made of metal and represent some fifteen years of development
at Berkeley that began with mirrors initially developed for my sounding rocket
program (Bowyer et al. 1985). The fabrication of these mirrors starts with aluminum
forgings which undergo extensive pre- and post-machining thermal stabilization.
The optical figure is cut utilizing the precision machining facilities of the Lawrence
Livermore National Laboratories, and a final surface polish is employed to remove
microscopic tool marks. The performance of these mirrors is exceptional. (Green
et al. 1986). In Figure 3 I show the encircled energy of a point source as a function
of encircling diameter for one of these mirrors. As can be seen, the half energy
diameter is about 4 arcseconds, which is more than an order of magnitude better
than any other metal mirror reported in the literature, and is almost a factor of two
better than the Einstein glass mirror which cost well over an order of magnitude
more to fabricate.

The band pass separation for the sky survey is provided by thin (300 to
3000 Å thick) organic and metallic filters working in combination with the characteristics of the telescopes. We began development of these filters as part of my
sounding rocket program (Chakrabarti et al. 1982; Jelinsky et al. 1983; Labov,
Bowyer and Steele 1985; Hurwitz, Labov and Chakrabarti 1985). On the basis of

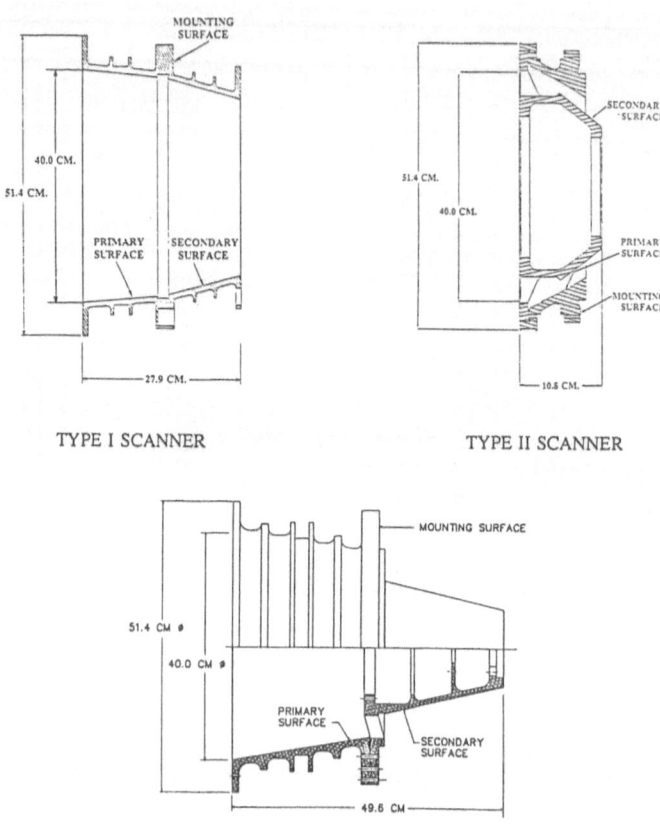

Fig. 2. Cross sectional schematic diagrams of the three types of mirrors employed.

what was known in 1975 we proposed filters of (a) parylene (a carbon/hydrogen polymer), (b) aluminum overcoated with carbon, (c) antimony, and (d) tin. Upon initiation of the EUVE Program we intensified our efforts in this area both in the range of filter materials explored and in testing (Vedder et al. 1989). A summary of some of our efforts in this regard is shown in Table I.

The response of the filter complement employed in EUVE is shown in Figure 4. These filters are rugged and stable, separate the EUV band into reasonable sub-bands, and minimize (for the most part) the background due to non-stellar sources.

The detectors employed are photon counting multichannel plate intensifiers with wedge-and-strip encoding. This type of detector was invented in our laboratory by Dr. Michael Lampton and co-workers (Martin et al. 1981) and was extensively improved and space qualified by Dr. Oswald Siegmund and co-workers (Siegmund et al. 1986). The detectors in the EUVE provide 1680 × 1680 RMS independent resolution elements, are linear to better than 1/2%, and possess a variety of additional attributes (they are stable, rugged, and solar blind; they have graceful degradation with high counting rate, plus more).

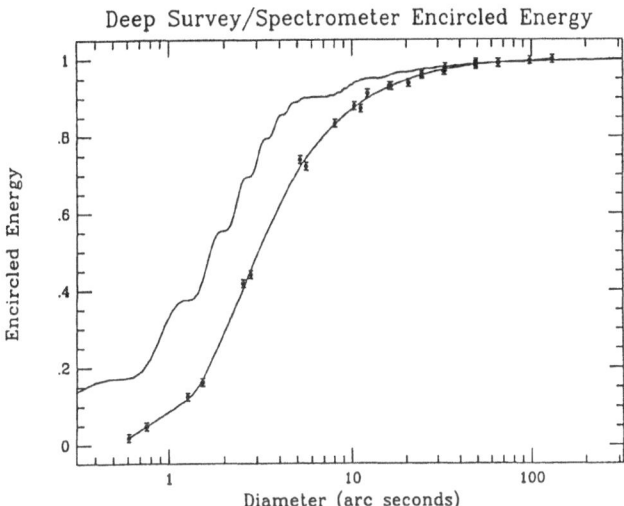

Fig. 3. Encircled energy for the deep survey/spectrometer telescope. This is well over an order of magnitude better than other metal mirrors and is a factor of two better than the Einstein mirrors.

TABLE I

Filter	Problem	Solution
100Å Bandpass:		
Parylene	FUV leak	Lexan/C substitution
Lexan/C	Atomic O attack in Orbit	Replace carbon with boron
200Å Bandpass:		
Al/C	Atomic O attack of C	Replace carbon with boron
Al/B	Boron flakes off	Revert to carbon + reverse so that aluminum on front surface
450Å Bandpass:		
Antimony	Oxidizes, not stable	Add titanium on both sides
Ti/Sb/Ti	Short wavelength leak	Redesign telescope
Ti/Sb/Ti	Pinholes during vibration	Change thermal profile during deposition
Ti/Sb/Ti	Pinholes during vibration	Add 1000Å of Al
Al/Ti/Sb/Ti	Pinholes during vibration	Add more Al
600Å Bandpass:		
Sn	Non Optimum Bandpass	Try Indium
Indium	FUV leak	Add Sn
In/Sn	Intermetallic Compound Bandpass not stable	Add SiO
In/SiO/Sn	Whisker growth/Pinholes	Remove from substrate quickly
In/SiO/Sn	Insensitive	Revert to Sn

Normalized Effective Area of EUVE Scanners

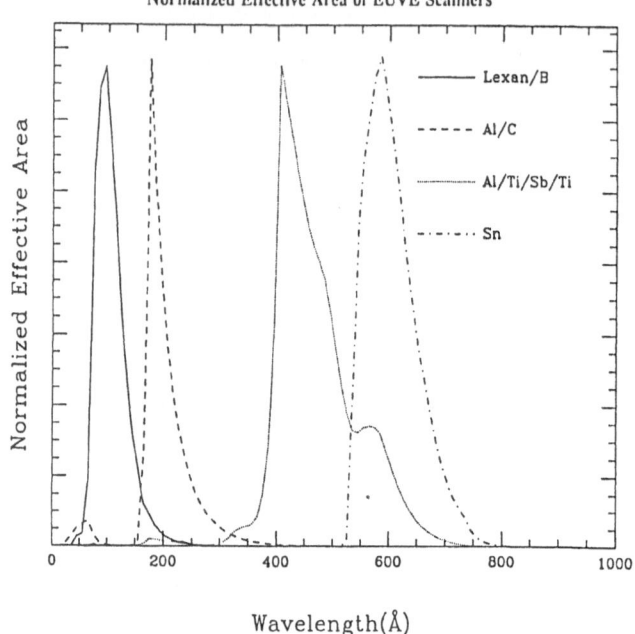

Wavelength(Å)

Fig. 4. The filter complement employed in defining the subbands of the all sky survey.

We have carried out an extensive effort to develop stable, high quantum efficiency photocathodes for the EUV, and to optimize their incorporation in the detectors (Martin and Bowyer 1982; Everman et al. 1987; Siegmund et al. 1986; Siegmund et al. 1988). In Figures 5a and 5b I show the quantum efficiency of two of the materials which have been developed and characterized in this effort.

The telescopes have been extensively parameterized in our calibration facility (Jelinsky and Malina 1988). In Figure 6 I show one of the images obtained in the calibration sequence. The linear enhanced areas are an image of a laboratory source as the articulated calibration mounting replicates the movement of an astronomical source as it will be observed in the normal observing mode. The images are offset as the satellite spin vector moves to follow the change in the sun-earth vector. The lower portion of the image plane appears blank because the filter in this half of the detector is (almost) opaque to the radiation employed.

In Figure 7 I show the source after it has been reconstructed from the data in the upper half of the image shown in Figure 5. Not surprisingly, the source is quite obvious. In Figure 8 I show the image reconstructed from the data in the lower half of the image plane. Although the source was certainly not obvious in the raw data shown in Figure 6, it stands out clearly in this reconstruction.

The spectrometer is an entirely new design invented at Berkeley (Hettrick and Bowyer 1983). This spectrometer uses variable line spaced gratings and provides a substantial number of advantages for spectroscopy with grazing incidence optics. First, it is highly efficient in that it requires a minimum number of grazing incidence reflections. Second, the image plane is nearly normal to the direction of the principal

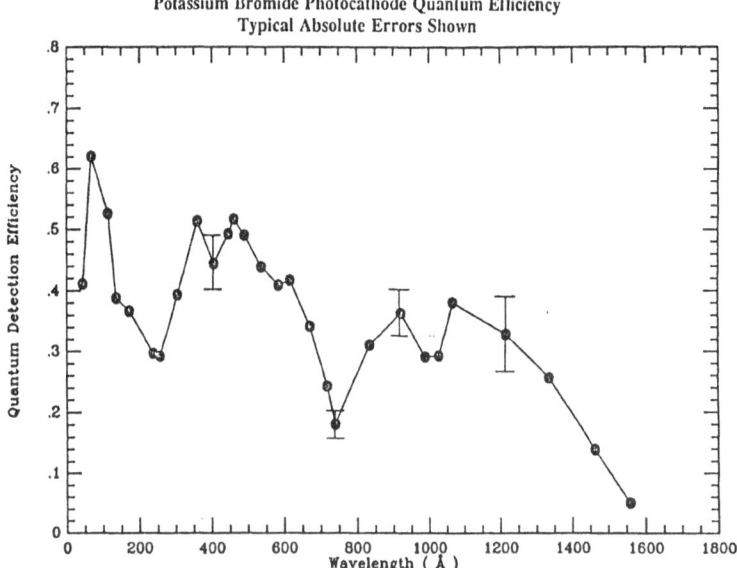

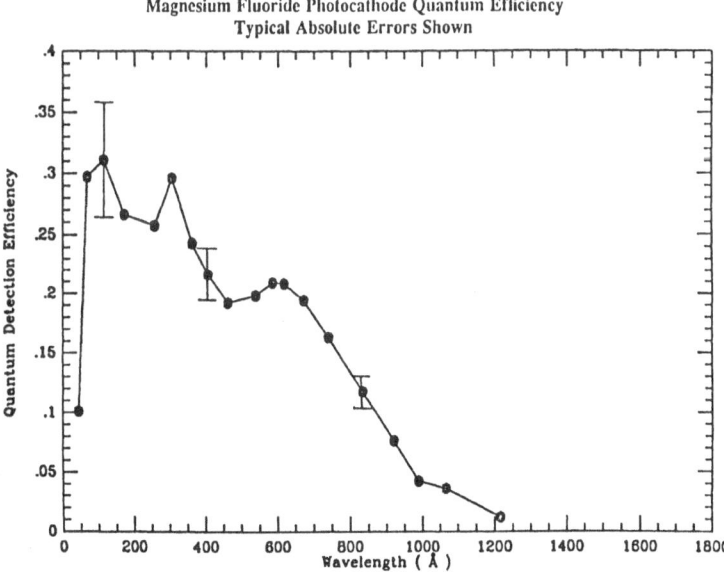

Fig. 5. Quantum efficiency of potassium bromide and magnesium fluoride as a function of wavelength.

TABLE II
Grating Characteristics

	Short	Medium	Long
Line Density (mm^{-1})	1675 - 3550	830 - 1750	415 - 875
Average Plate Scale (Å/mm)	2.4	4.8	9.6
Blaze Angle	3°	3°	3°
Coating	Rhodium	Platinum	Platinum
Filters	Lex/B	Al and Lex/B	Al and Al/C

ray. Finally, it is readily adaptable to grazing incidence optics and provides the opportunity for optimizing compactness, plate scale, resolution, etc. The concept of this spectrometer is illustrated in Figure 9.

This concept has been actualized in the EUVE deep survey/spectrometer assembly (Martin et al. 1985). This assembly is shown in Figure 10. Three spectrometers are employed so the gratings will be reasonably on-blaze over the wavelength range from 80 to 800 Å. The characteristics of the gratings used in these spectrometers are listed in Table II.

The spectrometers on EUVE have been extensively calibrated. The results of one of the calibrations are shown in Figure 11. This data was obtained at a synchrotron facility and characterizes the continuum response of the grating (Jelinsky et al. 1988). The structure exhibited in the response of the grating is due to the thin metallic overcoat on the grating surface.

We have generated simulated spectra of the response of the telescope/spectrometer to astronomical objects using models of the objects folded through the calibrated response of our instrument. In Figure 12 I show the raw spectrum of a continuum source, the hot white dwarf G191-B2B, and in Figure 13 I show the spectrum of Capella assuming its chromospheric emission is enhanced in intensity by the ratio x-ray$|_{Capella}$/x-ray$|_{Sun}$. Both simulations include the effects of interstellar absorption and detector background and assume a 40,000 second observation.

The spectrum of G191-B2B shows the high photometric accuracy expected to be obtained with brighter continuum sources in reasonable observation times. The spectrum of Capella suggest the variety of lines which we might hope to observe in coronally active stars. This is especially noteworthy since extensive work on the sun has demonstrated the power of EUV emission line studies as diagnostics of chromospheric plasmas.

Finally, in Figure 14, I show the simulated spectrum of a continuum source with 1/20 the intensity of HZ43 and an intervening column of 10^{18} hydrogen atoms and a neutral hydrogen to helium ratio as indicated.

The absorption edge at 504 Å due to neutral helium in the interstellar medium is clearly visible even in the weak source modeled here. These simulations demonstrate the power of the EUVE spectrometer in determining this parameter. Green, Jelinsky and Bowyer (1990) have carried out a measurement of this type with a rocket borne payload to provide constraints on the ionization state of the nearby interstellar medium in one view direction. With EUVE we can expect to determine this parameter in a significant number of view directions.

Acknowledgements

The development of the Extreme Ultraviolet Explorer has been a cooperative effort involving a substantial number of people at NASA and UCB. I especially acknowledge the contributions of Roger F. Malina, Michael Lampton, Oswald Siegmund, Pat Jelinsky, Herman Marshall, John Vallerga, David Finley, and Peter Vedder in the development of these instruments and in the preparation of this manuscript. This work has been supported by NASA contract NAS5-29298.

References

Bowyer, S., Freeman, J., Paresce, F., and Lampton, M.: 1977, *Appl. Opt.* **16**, 756

Bowyer, S., Green, J., Finley, D., and Malina, R.: 1985, in Diamond Turned Grazing Incidence Mirrors for the Extreme Ultraviolet: Ten Years of Fabrication and Performance Results, ed(s)., , Proceedings of the NASA Grazing Incidence Optics Workshop, Annapolis, Maryland

Bowyer, S., Kimble, R., Paresce, F., Lampton, M., and Penegor, G.: 1981, *Appl. Opt.* **20**, 477

Bowyer, S., Kumar, S., Paresce, F., and Lampton, M.: 1974, *Appl. Opt.* **13**, 575

Bowyer, S., Paresce, F., Chakrabarti, S., and Kimble, R. A.: 1983, *J. Geophys. Res.* **88**, 10247

Chakrabarti, S., Bowyer, S., Paresce, F., Franke, J., and Christensen, A.: 1982, *Appl. Opt.* **21**, 3417

Chakrabarti, S., Paresce, F., Bowyer, S., Chiu, Y., and Aikin, A.: 1982, *Geophys. Res. Letters* **9**, 151

Chakrabarti, S., Paresce, F., Bowyer, S., Kimble, R., and Kumar, S.: 1983, *J. Geophys. Res.* **88**, 4898

Finley, D. S., Jelinsky, P., Bowyer, S., and Malina, R.F.: 1988, *Appl. Opt.* **27**, 1476

Green, J., Finley, D., Bowyer, S., and Malina, R. F.: 1986, *The Extreme Ultraviolet Explorer: Optics Fabrication and Performance* **628**, Proc. SPIE

Green, J., Jelinsky, P., and Bowyer, S.: 1990, *Ap.J.* **359**, to appear

Haisch, B. M., Linsky, J. L., Lampton, M., Paresce, F., Margon, B., and Stern, R.: 1977, *Ap. J.* **213**, L119

Henry, P., Cruddace, R., Paresce, F., Bowyer, S., and Lampton, M.: 1975*A*, *Ap. J.* **195**, 107

Henry, P., Cruddace, R., Paresce, F., Lampton, M., and Bowyer, S.: 1975*B*, *Rev. Sci. Instrum.*, **46**, 355

Henry, P., Cruddace, R., Lampton, M., Paresce, F., and Bowyer, S.: 1975*C*, *Ap. J. Letters*, **197**, L117

Hettrick, M. C. and Bowyer, S.: 1983, *Appl. Opt.* **22**, 3921

Hurwitz, M., Labov, S., and Chakrabarti, S.: 1985, *Appl. Opt.* **24**, 1735

Jelinsky, P., Jelinsky, S., Miller, A., Vallerga, J., and Malina, R.F.: 1988, in L. Golub, ed(s)., *Synchrotron Radiation Calibration of the EUVE Variable Line Spaced Diffraction Gratings at the NBS SURF II Facility*, 356,
X-Ray Instrumentation in Astronomy II, Proc. SPIE 982

Jelinsky, P., Martin, C., Kimble, R., Bowyer, S., and Steele, G.: 1983, *Appl. Opt.* **22**, 1227

Kumar, S., Chakrabarti, S., Paresce, F., and Bowyer, S.: 1983, *J. Geophys. Res.* **88**, 9271

Labov, S., Bowyer, S., and Steele, G.: 1985, *Appl. Opt.* **24**, 576

Lampton, M., Margon, B., Paresce, F., Stern, R., and Bowyer, S.: 1976, *Ap. J. Letters* **203**, L71

Margon, B., Lampton, M., Bowyer, S., Stern, R., and Paresce, F.: 1976, *Ap. J. Letters* **210**, L79

Margon, B., Szkody, P., Bowyer, S., Lampton, M., and Paresce, F.: 1978, *Ap. J.* **224**, 167

Martin, C. and Bowyer, S.: 1982, *Appl. Opt.* **21**, 4206

Martin, C., Jelinsky, P., Lampton, M., Malina, R. F., and Anger, H. O.: 1981, *Rev. Sci. Instrum.* **52**, 1967

Martin, C., Mrowka, S., Bowyer, S., Malina, R.F.: 1985, in The Extreme Ultraviolet Explorer Spectrometer: Performance Characteristics Based on Development of the Collimator, Variable Line Space Gratings, Telescope and Detectors, ed(s)., *284*, Proc. SPIE 597,

Paresce, F., Chakrabarti, S., Bowyer, S., Kimble, R., and Kumar, R.: 1983, *J. Geophys. Res.* **88**, 4905

Siegmund, O.H.W., Everman, E., Vallerga, J.V., and Lampton, M.: 1988, *Appl. Opt.* **27**, 1568

Siegmund, O.H.W., Everman, E., Vallerga, J.V., Sokolowski, J., and Lampton, M.: 1987, *Appl. Opt.* **26**, 3607

Siegmund, O.H.W., Lampton, M., Bixler, J., Chakrabarti, S., Vallerga, J., Bowyer, S., and Malina, R.F.: 1986, *J. Opt. Soc. Am.-A* **3**, 2139

Siegmund, O.H.W., Vallerga, J.V., Everman, E., Labov, S., Bixler, J.: 1986, in High Quantum Efficiency Opaque CsI Photocathodes for the Extreme and Far Ultraviolet, ed(s)., *117*, Proc. SPIE 687,

Vedder, P. W., Vallerga, J. V., Siegmund, O.H.W., Gibson, J., and Hull, J.: 1989, in The Filters for the Extreme Ultraviolet Explorer: Calibration and Lifetesting Results, ed(s)., *392*, Proc. SPIE 1159,

B. Welsh, Jelinsky, P. and Malina, R. F.: 1988, in L. Golub, ed(s)., *The Berkeley Extreme Ultraviolet Calibration Facility*, X-Ray Instrumentation in Astronomy II, Proc. SPIE 982, 335

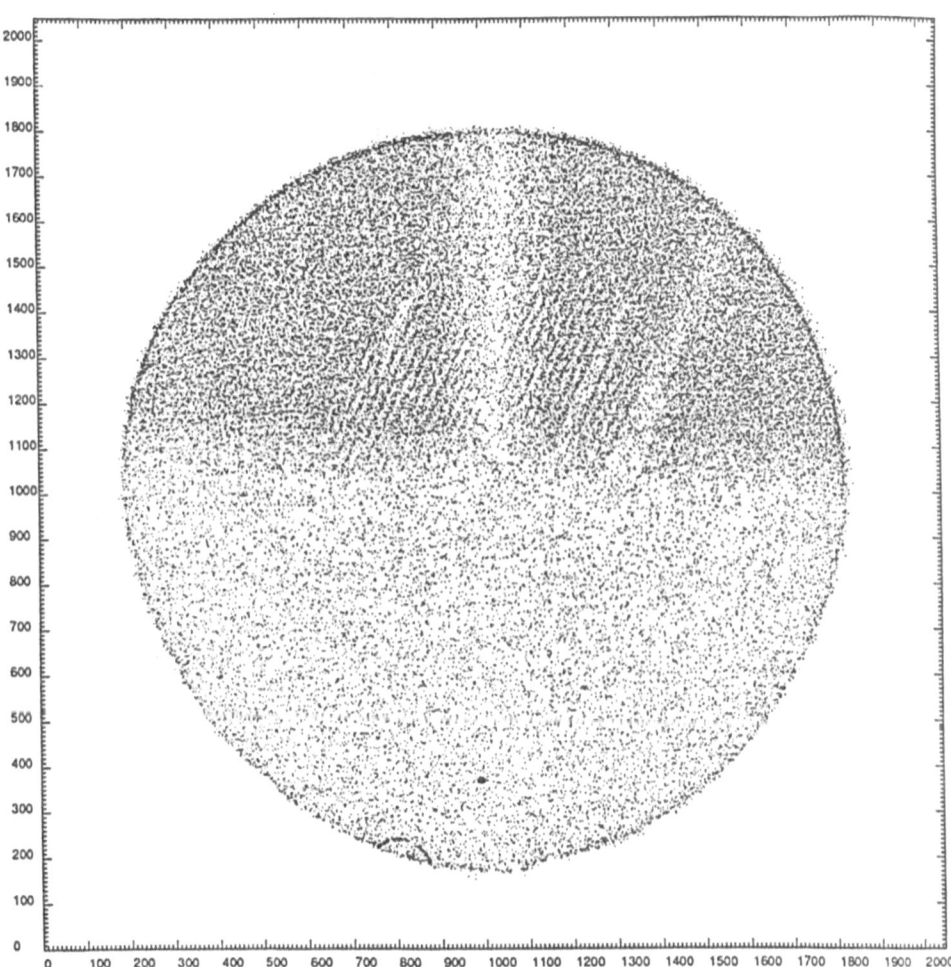

Fig. 6. Laboratory simulation of a source as it will appear in the normal operation mode in space. The thin linear structures are the source traversing the image plane as the telescope scans the sky. These are offset each orbit because of the progressive change in the sun-earth vector with time. The lower half of the image plane appears blank because of the effect of the filter which blocks radiation at this wavelength.

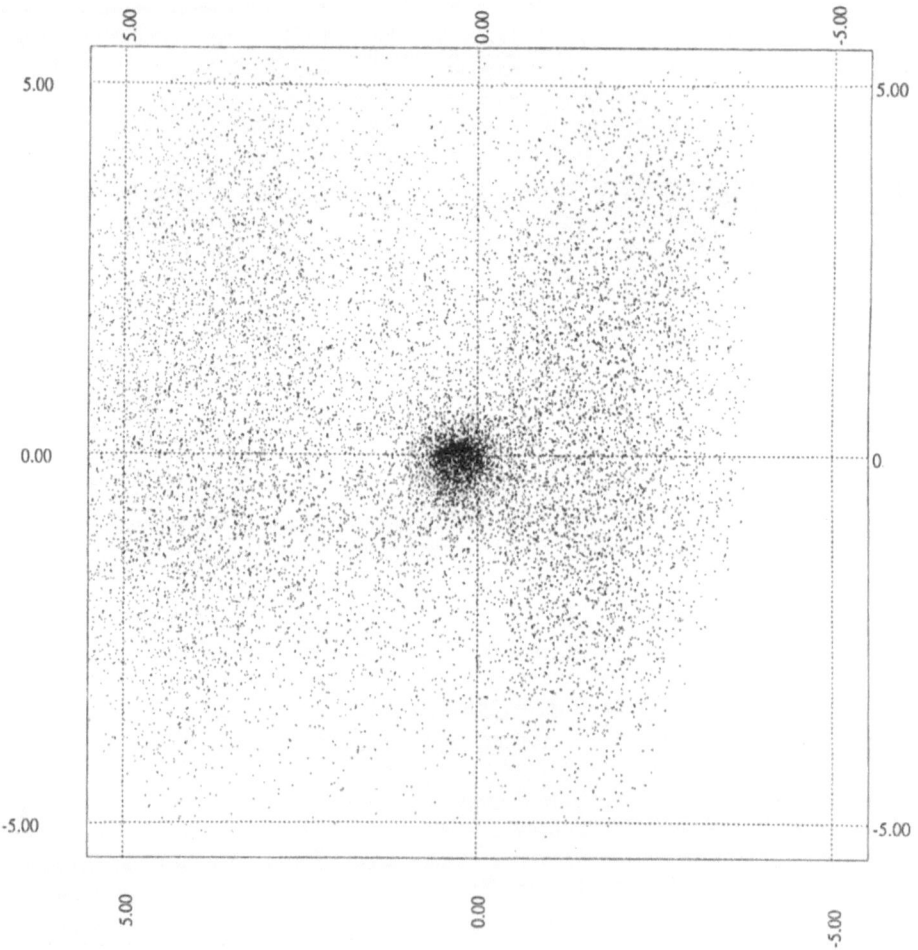

Fig. 7. A reconstruction of the data obtained from the top half of the image shown in Figure 6.

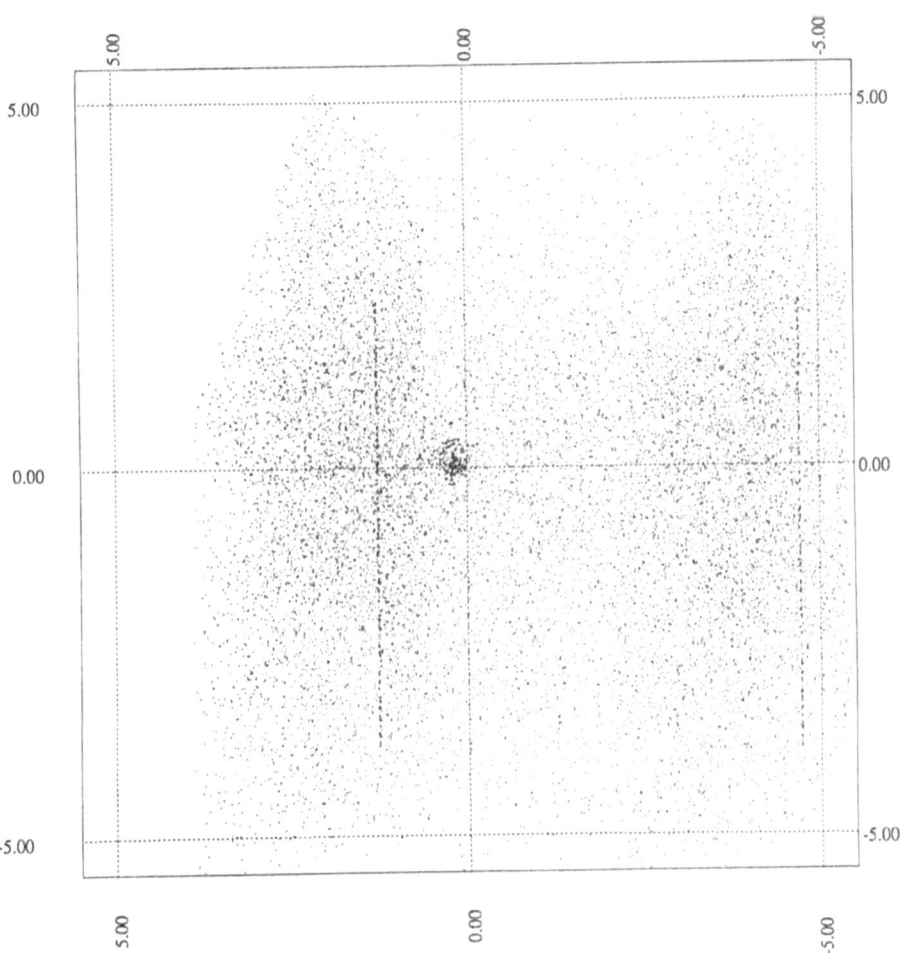

Fig. 8. A reconstruction of the data obtained from the bottom half of the image shown in Figure 6.

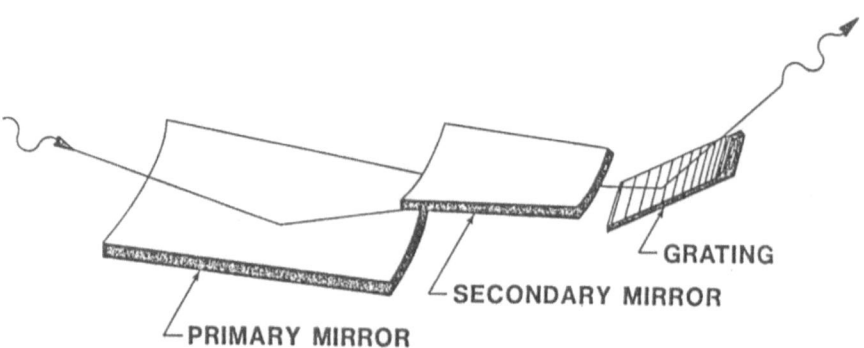

Fig. 9. A schematic illustration of the concept of the variable line space grazing incidence spectrometer.

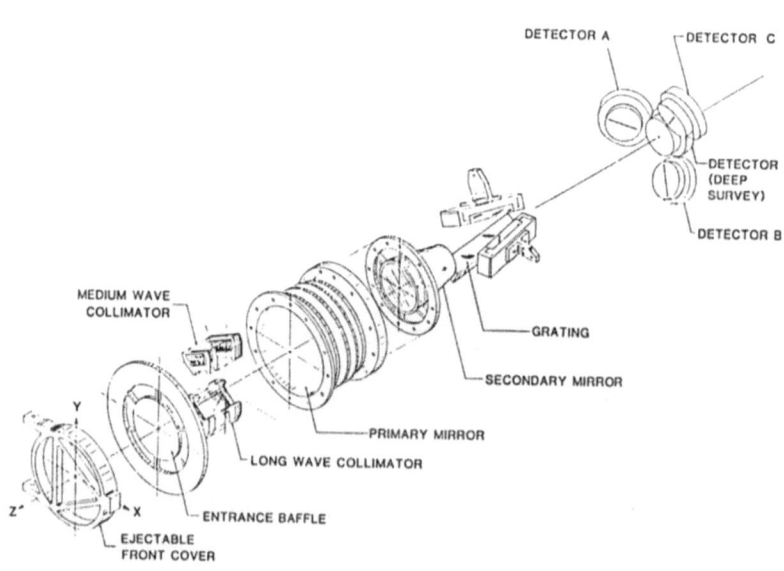

Fig. 10. An exploded schematic of the spectrometer on EUVE.

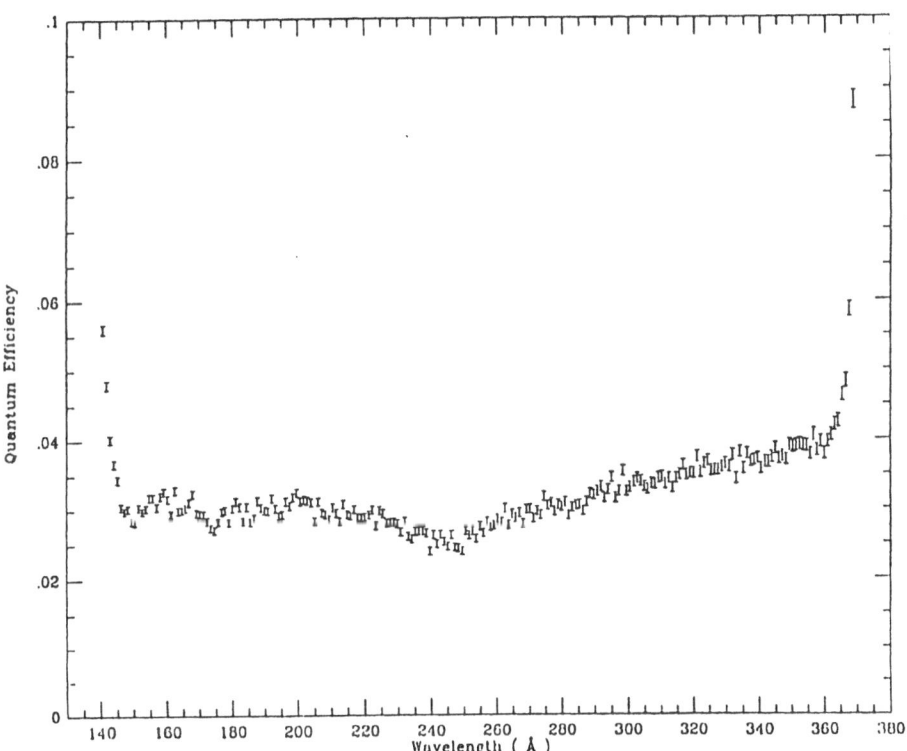

Fig. 11. The response of one of the gratings to continuum radiation.

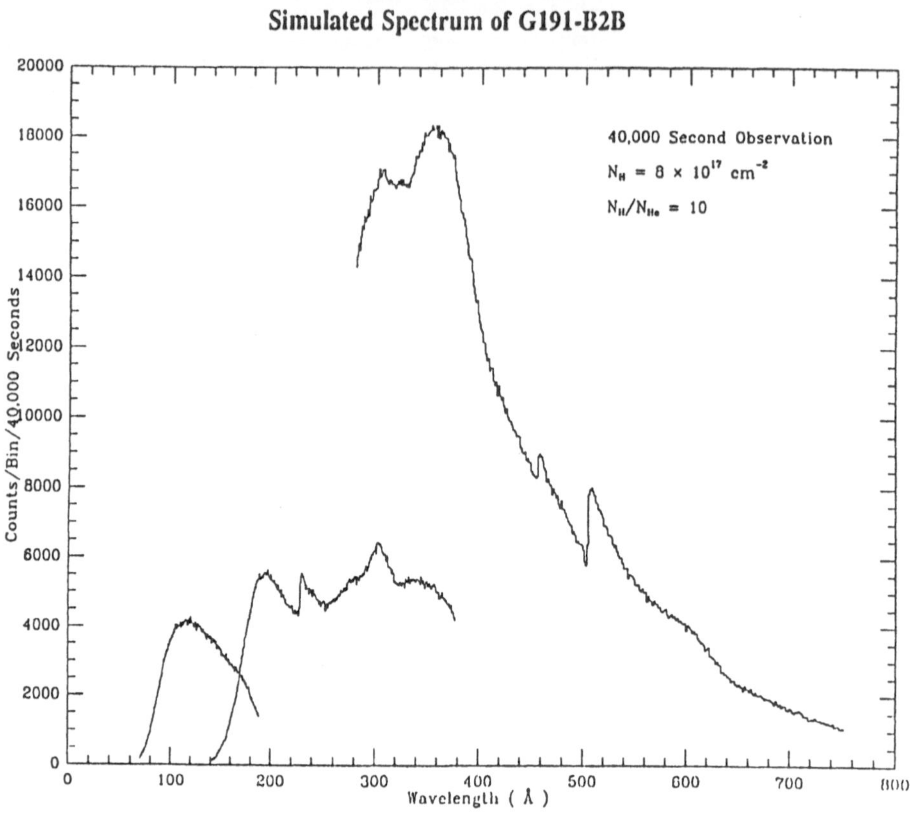

Simulated Spectrum of G191-B2B

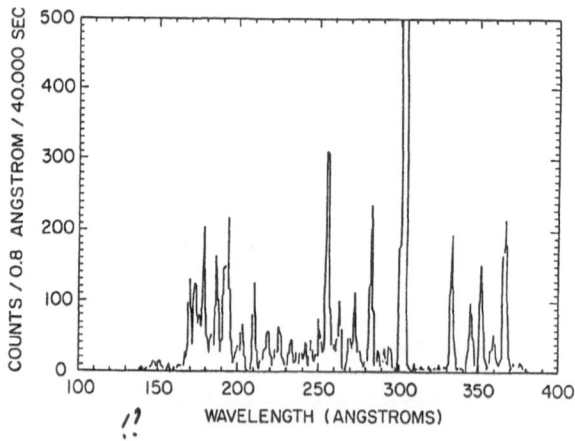

Fig. 12. Simulated spectrum of the hot white dwarf G19B2B as observed for 40,000 seconds. The effects of absorption of the ISM are included.

Fig. 13. Model of Capella's spectrum as observed for 40,000 seconds.

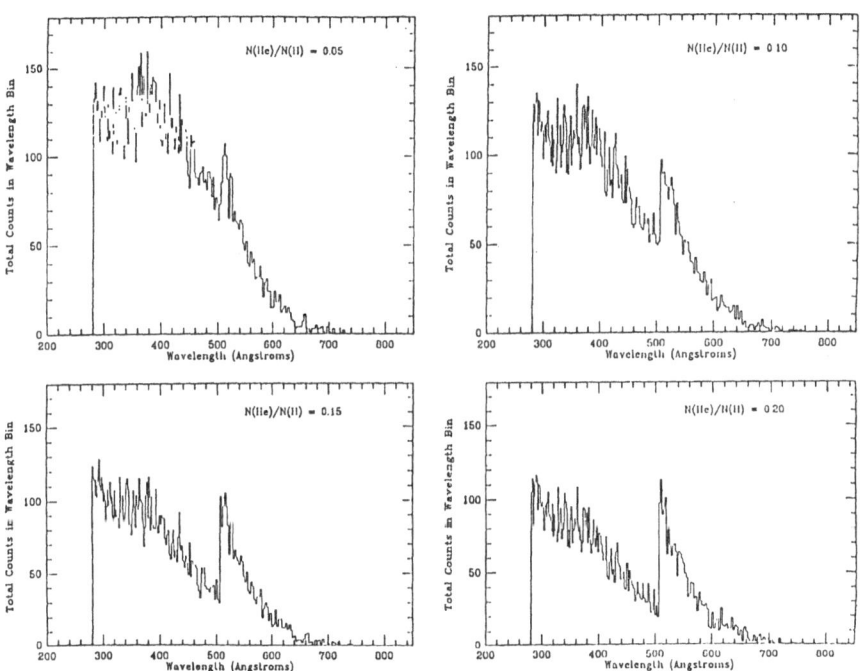

Fig. 14. The simulated response of a source with 1/20 the intensity of HZ43 with an intervening column of 10^{18} hydrogen atoms and a neutral hydrogen to helium ratio as indicated.

LYMAN THE FAR ULTRAVIOLET SPECTROSCOPIC EXPLORER

WARREN MOOS

The Johns Hopkins University, Department of Physics and Astronomy,
Homewood Campus, Baltimore, Maryland 21218

Abstract. The Lyman Far Ultraviolet Spectroscopic Explorer mission will obtain high resolution spectra (R 30,000) in the 912–1200 Å region with sufficient sensitivity to study faint sources throughout the galaxy and at large extragalactic distances. This spectral region provides unique tools for solving problems in cosmology (e.g. deuterium), galactic structure and evolution (e.g. O VI and H_2), stellar evolution (e.g. O VI) and planetary science (e.g. H_2). Recent advances in optical and detector technology, which make this goal possible with a moderate size experimental package, also enable the spectral coverage to extend down to 100 Å with good sensitivity and only a minimal increase in complexity. Thus a secondary goal is to cover the 100–912 Å region with moderate spectral resolution. In 1989, following a Phase A study of the mission concept, NASA selected the mission for Phase B study in 1989. Both Canada and the United Kingdom are participating in the definition and development of the mission.

1. Mission Overview

The Lyman Far Ultraviolet Spectroscopic Explorer (FUSE) completed its Phase A study in June of 1989. The Phase B Definition Study will begin in the summer of 1990. Phases C/D are expected to start in July of 1992 and the launch is planned for 1997. This paper is based on the Phase A study and Table I lists the scientific personnel associated with the study. Table II is a brief summary of the mission. Additional details are given in the final report for the study (Moos et al. 1989).

The FUSE instrument consists of a 70 cm Wolter type II glancing incidence telescope with 1 arcsec image quality feeding a spectrograph. This spectrograph has several channels which provide high resolving power on the order of 30,000 from the Hubble Space Telescope limit at near 1200 Å down to the limit set by the photoionization continuum of atomic hydrogen at 912 Å. In addition other channels in the spectrograph will provide moderate resolving power down to 100 Å. Most of the channels on the instrument are astigmatic and permit imaging along the slit to remove background or study extended objects. The sensitivity is quite high and as a result is appropriate for bridging the gap between the Hubble Space Telescope (HST) with a cut-off near 1200 Å and the Advanced X-ray Astrophysics Facility (AXAF) with an upper limit near 100 Å. Figure 1 shows the estimated limiting sensitivity of the FUSE instrument in comparison with several other missions.

The Canadian Space Agency and the British National Space Center are cooperating with NASA in the definition and development of the FUSE mission. Canada will supply the baffle system for the telescope and cooperate with the United Kingdom in developing the fine error sensor. In addition to development of the fine error sensor, the United Kingdom will also supply the focal plane assembly, the main electronics box and the associated software.

Y. Kondo (ed.), Observatories in Earth Orbit and Beyond, 171–176.

TABLE I
Scientific Personnel for Phase A Study

Co-Investigators

The Johns Hopkins University	*University of Hawaii*
Warren Moos, Principal Investigator	L. L. Cowie
P. D. Feldman	
	Smithsonian Astrophysical Obs.
Goddard Space Flight Center	A. K. Dupree
A. Boggess	L. P. Van Speybroeck
A. G. Michalitsianos	
J. Osantowski	*Kitt Peak National Obs.*
G. Sonneborn	R. F. Green
B. Woodgate	
	Princeton University
University of California	E. B. Jenkins
C. S. Bowyer	
	Royal Greenwich Obs.
University of Colorado	M. V. Penston
W. C. Cash, Jr.	
J. L. Linsky	*University of Wisconsin*
M. Shull	B. Savage
Stanford University	*ESA Villafranca Satellite*
J. G. Timothy	*Tracking Station*
	W. Wamsteker
University of Chicago	
D. G. York	

International Advisory Panel

Max-Planck-Institut fur	*Space Research Laboratory*
Physik und Astrophysik	*Holland*
B. Aschenbach	H. J. Lamers
Dominion Astrophysical Obs.	*Universität Tubingen*
J. B. Hutchings	M. Grewing
The Australian National University	*Institut d'Astrophysique*
M. A. Dopita	A. Vidal-Madjar

Consultants	**Mission Study Scientist**
University of California	G. Sonneborn
R. Malina	
O. Siegmund	**JHU Phase A Study Scientist**
	S. D. Friedman
University of London	
A. Willis	

2. Scientific Goals

The Lyman FUSE mission has a powerful and unique set of tools for understanding astrophysical problems. A number of astrophysically important species such as deuterium, molecular hydrogen, helium, O^+ to O^{5+}, Ne to Ne^{5+}, A, S^{3+} to S^{12+} and many others have their strongest transitions in this spectral region. The number of interstellar absorption transitions increases dramatically below 1200 Å. Emission and absorption transitions in this spectral region come from species at very different temperatures ranging from molecules to highly ionized species such as Fe XXIV. Ratios of emission intensities provide electron densities over the range of 10^6 to 10^{14} electrons/cm^3. The high spectral resolution will permit the measurement of gas velocities as small as a kilometer per second. Finally the high sensitivity permits the study of new classes of objects which were not accessible to the Copernicus mission or other missions which studied this spectral region. The ability to study white dwarfs will permit accurate measurements of interstellar species in the near interstellar medium. Extra-galactic sources such as active galactic nuclei will be used to study the gas at the very edge in the halo of the galaxy and between galaxies out to large distances. It is worth noting that the Copernicus mission was able

Bridging the Gap

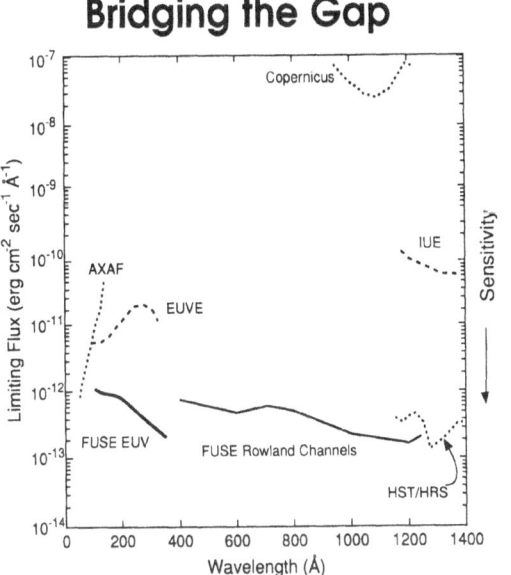

Fig. 1. The sensitivity of Lyman FUSE compared with other UV and EUV missions. The most sensitive instruments are at the bottom of the graph. An integration time of 1800 seconds and 100 detected photons per resolution element are assumed. The HST/HRS flux limits were computed for coverage of a 120 Å section of the spectrum, equal to the spectral range of each of the FUSE high-resolution Rowland channels. The Copernicus sensitivity is off the top of the graph and is displayed only to show the shape of the response.

to measure the deuterium to hydrogen ratio in approximately a dozen hot stars within about 1000 parsecs of the sun. In addition a similar number of stars have been measured using the chromospheric Lyα line using both the Copernicus and IUE satellites (Boesgaard and Steigman 1985). As a result, reliable measurements of this cosmologically important number exist only for gas clouds in a small vicinity of our spiral arm.

The Phase A Study team examined the areas where the FUSE mission would make the most important advances and concluded that there were three major areas. First, it will expand our understanding of the early universe by providing reliable measurements of big bang light element nuclear synthesis and also by measuring the temperature of the gas between galaxies. Second, the mission will also be an important tool for understanding the evolution and fundamental processes in galaxies. Measurements will range from the hot gas in disks and halos of galaxies to examining supernova and their remnants which feed much of this hot gas into the galaxy. Third, the mission will provide fundamental insight into a number of problems associated with the evolution of stars and planetary systems. In particular it is expected that accretion processes, winds and magnetic activity in solar type stars, studies of primordial abundances in the solar system, and planetary atmospheric excitation processes will be important.

Figure 2 shows a simulated spectrum for a hot object behind a cloud with a

Determining the Deuterium Abundance

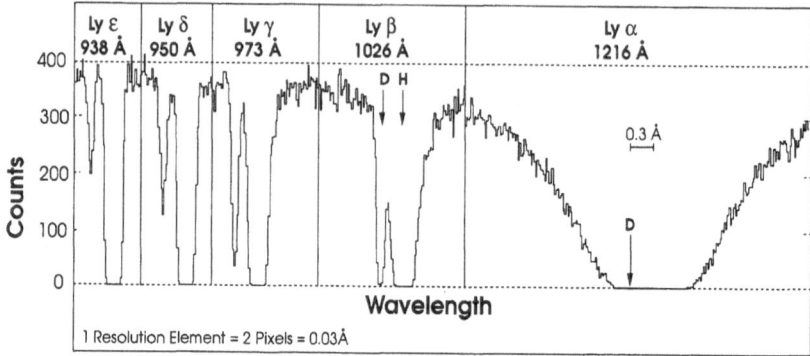

Fig. 2. Absorption line spectrum showing the splitting of the H (broad) and D (narrow) lines. This simulated spectrum corresponds to 10^5 sec integration using an approximately 16th mag QSO as a background source. An abundance ratio of $D/H = 2 \times 10^{-5}$ and a hydrogen column density of $N_H = 3 \times 10^{19}$ cm^{-2} has been assumed.

deuterium abundance of 2 parts in 10^5. It is clear that near 1200 Å it is not possible to make reliable measurements. In fact one must go shortward of even Lyβ and well into the Lyman series before one could obtain lines which are not heavily saturated. Thus, the full range of the FUSE mission is necessary to provide reliable measurements of deuterium. Deuterium is thought to be a critical indicator of the density at about 100 seconds in the Big Bang. As a consequence, it will be an important diagnostic of the baryon density then and now and will help in understanding the missing mass issue. However, in order to determine the primordial value, it is important to understand the effect of possible processing of the primordial material. The sensitivity of the instrument will enable FUSE to study the deuterium abundance in many environments with different evolutionary histories both at a number of locations in our galaxy, in nearby galaxies, and in the gas between galaxies out to very large distances.

The FUSE mission will provide new information on the location of the hot gas in the disks and halos of galaxies. The very important lithium-like oxygen ion O VI can be studied only in this spectral region. As mentioned previously, the sensitivity of FUSE will make it possible to examine the halo gas in absorption using objects outside the galaxy as light sources. It is important to note that this mission will be important for studying supernova. It is quite possible that the interaction of supernova 1987a as it expands into the interstellar medium with that medium will lead to a sharp increase in the ultraviolet emissions in the late 90s. In addition, the sensitivity of the instrument will be such that several extragalactic supernova that are discovered each year using routine patrols can be studied without disrupting the regular observing program.

Because the ions which emit in this spectral region exist over a wide range

TABLE II
Lyman FUSE Mission Summary

Telescope: 70 cm diameter and 7 meter focal length. Wolter type II glancing incidence with gold overcoat. One arcsec image diameter image.

Spectroscopic Capabilities: High spectral resolution. Three channels cover 910-1250 Å with R=30,000 at 1 arcsec width. Imaging along 45 arcsec long slits divided into 15 arcsec segments with widths of 0.5, 1.0 and 3.0 arcsec.

Planetary and survey grating. 400-1600 Å. Imaging along 45 arcsec long slit divided into a 30 arcsec by 4 arcsec segment (R=1000) and a 15 arcsec by 1 arcsec segment (R=4000).

EUV grating. 100-350 Å with R=300 and 800. Limited imaging along slit.

Spacecraft and Orbit: Explorer Bus (MMS spacecraft) in ~500 km orbit with 28° inclination. Three year mission.

Tracking: Explorer Bus Inertial Reference Units and Star Trackers with updates by Fine Error Sensor in telescope focal plane. Pointing uncertainty ~0.3 arcsec. FES field of view is 7 arcmin by 7 arcmin. Dynamic range is 17m to -2m for point (1 arcsec) sources. Solar avoidance angle is 43°.

Guest Observer Program: Most of the observing time available to the scientific community through a Guest Observer program similar to those of IUE and HST.

of temperatures, this mission will be extremely important for studying the outer atmospheres of late type stars. The fundamental problems to be addressed include whether stellar coronae exist at all temperatures, and in fact how do such solar type stars lose mass. The wavelength region covered by the FUSE instrument includes features from species which exist at very different temperatures. Thus, emission measures will be determined for the stellar plasma from a few thousand degrees to several tens of millions of degrees.

3. Status of the Technology

As part of the Phase A study, the Science Team constructed a detailed candidate observing plan. For the instrument parameters of Table II, the observing plan would require about one year of observing time for successful completion. Assuming a typical efficiency for lower earth orbit, a three-year mission will be necessary to accomplish the baseline science goals. The instrumentation technology is now at a point where this mission is practical. The telescope performance and optical fabrication errors and related manufacturing problems have been analyzed in detail by industrial firms. A design study has demonstrated the feasibility of a fine error sensor with a 7 × 7 arcminute field of view. The development of silicon carbide coatings with normal incidence reflectivities greater than 30% below 1000 Å will give high transmission down to well below 900 Å. A prototype of the distorted ellipsoid surface necessary for astigmatic imaging in the Rowland mount has been constructed (Cash 1989). The EUV gratings and detectors are based on the technology developed for the EUV mission. The FUV detectors are based on detectors presently under development for STIS, SOHO and EUVE. Finally, the very careful and detailed modelling of the mechanical and thermal stability of the structure has shown that it can be achieved without heroic efforts. The Explorer platform which is being constructed for the EUVE and XTE mission is baselined for this mission. The instrument has been designed so that it will be dual adaptable to either a shuttle or expendable launch vehicle, thus providing two paths into space.

Mission operations will build on the EUVE and XTE experience and software and will use the facilities developed for those missions. The Science Operations Center will be located at the Goddard Space Flight Center and a Guest Investigator program will be allocated about 90% of the observing time.

4. Summary

The Lyman FUSE mission will have a major impact on both astrophysics and science as a whole by providing new spectroscopic measurements of unparalled sensitivity in the 100 Å to 1200 Å region. Understanding of the scientific problems, necessary data and the required technical capabilities of the instrumentation are at a point where large steps forward are to be expected for many important scientific problems. Finally, of extreme importance, the technology is mature and ready for application in a space environment.

References

Boesgaard, A.M. and Steigman, G.: 1985, *Palo Alto: Annual Reviews* **23**, 319

Cash, W.C., Jr., 1989, private communication

Moos, H.W. *et al.*: 1989, Lyman The Far Ultraviolet Spectroscopic Explorer: Phase A Study Final Report

FAR AND EXTREME ULTRAVIOLET ASTRONOMY WITH
ORFEUS

G. KRÄMER, J. BARNSTEDT, N. EBERHARD, M. GREWING, W. GRINGEL,
C. HAAS, A. KAELBLE, N. KAPPELMANN, J. PETRIK

Astronomisches Institut, Waldhäuserstr. 64, 7400 Tübingen, F.R.G.

I. APPENZELLER, J. KRAUTTER, H. MANDEL, R. ÖSTREICHER

Landessternwarte Heidelberg, Königstuhl, 6900 Heidelberg, F.R.G.

and

S. BOWYER, M. HURWITZ

Space Sciences Laboratory, University of California, Berkeley, CA, 94720, U.S.A.

Abstract. ORFEUS (Orbiting and Retrievable Far and Extreme Ultraviolet Spectrometer) is a 1 m normal incidence telescope for spectroscopic investigations of cosmic sources in the far and extreme ultraviolet spectral range. The instrument will be integrated into the freeflyer platform ASTRO-SPAS. ORFEUS-SPAS is scheduled with STS ENDEAVOUR in September 1992. We describe the telescope with its two spectrometers and their capabilities i.e. spectral range, resolution and overall sensitivity. The main classes of objects to be observed with the instrument are discussed and two examples of simulated spectra for the white dwarf HZ43 and an O9-star in the LMC are shown.

1. Introduction

ORFEUS is a collaborative project between the Federal Republic of Germany represented by the BMFT and the United States of America represented by NASA. The scientific institutes participating in the development of the instruments and the scientific preparation of the mission are the Astronomical Institute of the University of Tübingen (AIT), the Landessternwarte Heidelberg (LSW) and the Space Sciences Laboratory of the University of California at Berkeley (SSL).

The ORFEUS telescope will be the first scientific payload (out of four) which will fly on the ASTRO-SPAS satellite. The platform will be launched and retrieved by the Shuttle. The mission duration will be several days.

2. The Space Platform ASTRO-SPAS

Fig. 1 shows the ASTRO-SPAS platform (Shuttle Pallet Satellite) with the 1 m ORFEUS telescope integrated into its modular structure. The satellite is designed to operate autonomously in low earth orbit in the vicinity of the Space Shuttle for a limited period of 7 to 10 days. The ASTRO-SPAS has its own power supply (batteries), its own attitude measurement and control system and facilities for data handling and storage. A telemetry/telecommand link through the Space Shuttle provides for short periods the possibility to dump specific quick-look data to the ground station and to send commands for online control of the SPAS and the

Y. Kondo (ed.), Observatories in Earth Orbit and Beyond, 177–184.

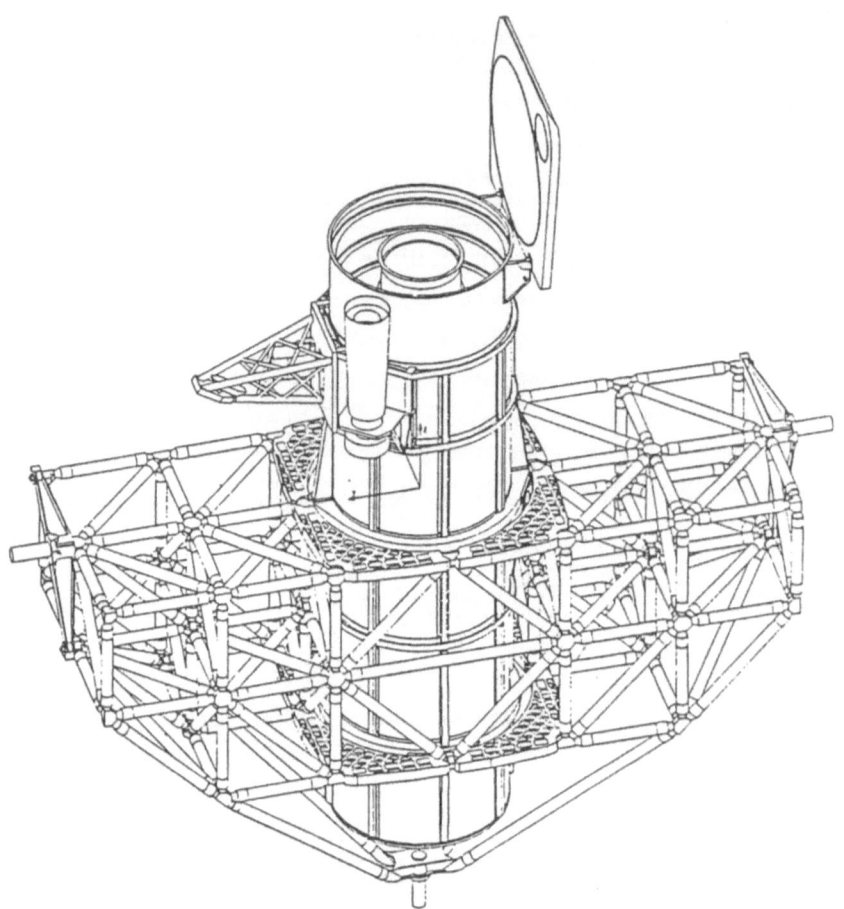

Fig. 1. The Space platform ASTRO-SPAS with the ORFEUS telescope.

experiments. Furthermore a preprogrammed target list for the next autonomous phase will be loaded into the satellite data processing unit during these contact times.

The pointing capability of the ASTRO-SPAS is of particular importance for OR-FEUS since high resolution spectroscopy requires a pointing stability in the range of arcseconds or better. The ASTRO-SPAS offers an absolute pointing accuracy of ±7 arcsecs and a pointing stability of ±5 arcsecs continuously for more than 25 mins.

Manoeuvering capability is achieved through a cold gas system which allows the aquisition of a target in less than 20 minutes and a total number of more than 100 major slews (typically 180 degrees).

3. The ORFEUS Instrumentation

The ORFEUS payload comprises a 1 m normal incidence telescope feeding light into two spectrographs which can be operated alternatively (Grewing et al. 1990; Krämer et al. 1988). The overall payload configuration is shown in Fig. 2. In the following sections we briefly describe the main characteristics of the telescope and the two spectrometers.

3.1. THE TELESCOPE

The telescope is built around a 1000 mm clear aperture f/2.4 light weight (90 kg) Zerodur primary mirror. The optical bench, a graphite-epoxy cylinder, supports the primary focus instrumentation including the Rowland spectrograph. The total weight of the ORFEUS telescope will be about 800 kg.

A wheel mechanism with 3 diaphragms of angular diameters 10 arcsecs, 20 arcsecs and 10 arcmins is located in the focal plane. The 10 arcmin hole will be used for in orbit alignment of the optical axes of the telescope and the star tracker which is part of the attitude control system of the ASTRO-SPAS. The relative position of the optical axis of the telescope is controlled by means of an image system which is integrated into the Berkeley spectrometer. This system, which is illuminated by a small part of the telescope aperture, re-images the target within the entrance aperture onto a sealed MCP detector.

The undeflected light beam from the primary mirror will enter the modified Rowland spectrometer designed and constructed by the SSL. Alternatively, a pick off mirror can be moved into the beam to feed the light into the Echelle spectrometer developed and designed by LSW and AIT.

3.2. THE ECHELLE SPECTROMETER

This spectrometer covers the wavelength range from 90 nm to 125 nm and offers a spectral resolution of approximatly 10,000 (Appenzeller et al. 1988). The Echelle is equipped with a parabolic off axis collimator which collimates the light and reflects it towards the Echelle grating which is used in the diffraction orders 45 to 62. The different orders are subsequently separated by a cross disperser which also acts as a correctorless Schmidt camera. This camera images the twodimensional spectrum onto a microchannelplate detector with a wedge and strip readout system. The image format is 1024 × 512 pixels.

With conventional coatings for the optical elements like iridium, gold and ARC 1000 the overall effective area of the instrument ranges between 1 cm^2 and 12 cm^2 for the 90 to 125 nm wavelength region. Especially close to 90 nm the sensitivity might be improved substantially if the collimator and the gratings could be coated successfully with SiC. With this Echelle spectrometer ORFEUS essentially extends the IUE capabilities into the FUV range, i.e. we will be able to observe unreddened early type stars in high resolution down to 10th magnitude. As the instrument will be photon noise limited, rebinning photon events offers the possibility to reach even much fainter objects.

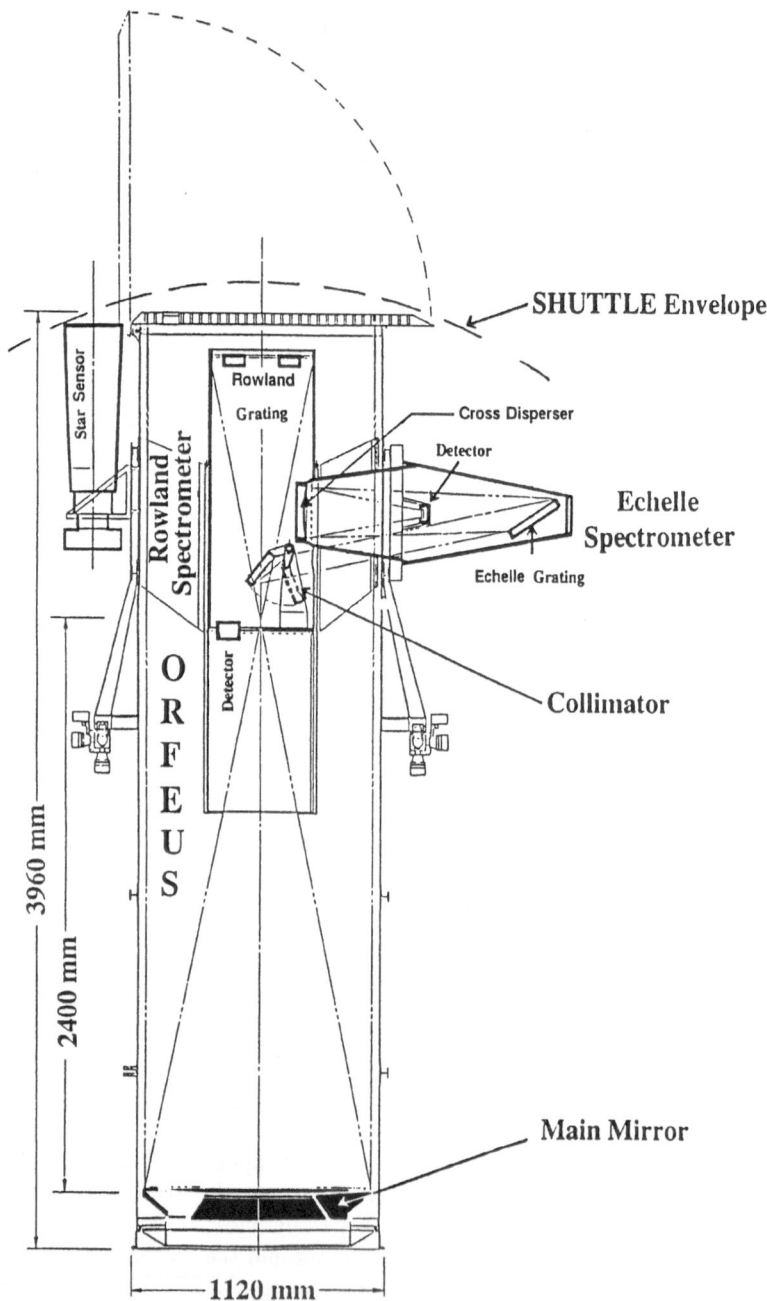

Fig. 2. Cross Section of the ORFEUS telescope system.

For brighter targets the spectral resolution may be improved above 10,000 by post flight correction of the telescope jitter: The star tracker has an intrinsic pointing accuracy of 2 arcsecs. With these data available in 1 sec intervals and the time tagged individually recorded photon events the movement of the spectrum on the detector induced by the jitter may be corrected.

3.3. THE BERKELEY SPECTROMETER

The Berkeley spectrometer covers the wavelength range from 39 to 120 nm. Its spectral resolution will be about 5000. It consists of four concave diffraction gratings, each of which is illuminated by approximately 20% of the telescope aperture. Each grating has a unique central line density, so that each disperses a sub-bandpass across a spectral detector. The line density on each grating is not constant, but is instead a fourth order function of position on the optic. This reduces astigmatism and improves spectral resolution substantially over what could be obtained with conventional uniform line-space gratings. The two short-wavelength and the two long-wavelength gratings respectively share a common detector. These two spectral detectors contain curved microchannel plates coupled to a delay line/charge division readout system, which yields very high pixel resolution in the dispersion direction and moderate resolution in the perpendicular direction.

4. Scientific Objects

a. *Chromospheres and coronae of cool stars* with photospheric temperatures like the sun show strong emissions at FUV and EUV wavelengths. Since ions from different ionization stages probe the temperature and density at different heights within the atmospheres, ions like CIII having all its resonance lines at these short wavelengths will add important information on the structure of the outer atmospheric layers of cool stars.

b. For *hot stars* it is, by contrast, mainly the understanding of the photospheric processes that will be enhanced by measurements at FUV and EUV wavelengths. Observations of continua and absorption edges will allow detailed tests of the theoretical predictions from atmosphere calculations. Studies of FUV spectra will enhance our knowledge about the ionization in stellar winds and improve our knowledge about stellar mass loss rates.

c. Obvious targets for observations at FUV and EUV wavelengths are *highly evolved* stars such as the subdwarfs, the nuclei of planetary nebulae, and hot white dwarfs because they have surface temperatures well in excess of 10^4K. Both the slope of their continua and their line spectrum will be of interest for very much the same reason as in the case of young early type stars, i.e. to study their atmospheric properties, especially the chemical composition of their atmospheres.

d. As shown by the COPERNICUS satellite the FUV range offers unique information about *the interstellar medium*. For example OVI absorption being seen

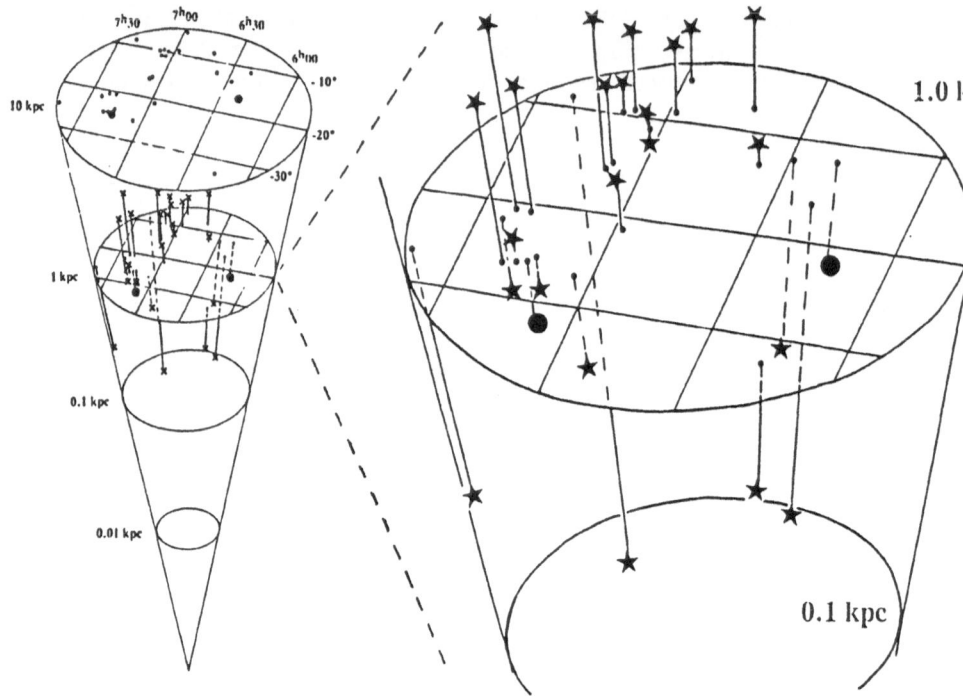

Fig. 3. Selected Field of Stars of Spectral Type O6 to B1.

along many lines of sight was a totally unexpected discovery. The large amounts of molecular hydrogen that were discovered through the FUV band spectrum totally changed the picture about the molecules in the interstellar mass budget and stimulated a large number of theoretical studies on their origin. ORFEUS, more sensitive by a factor of 100 will allow to expand these studies for a number of lines of sight by using far more distant background targets. Fig. 3 shows a field of stars of spectral type O6 to B1. Two of them could be observed with COPERNICUS, whereas ORFEUS can observe all of them with full spectral resolution. Consequently the ISM can be traced in a much narrower network at least in a selected area of special interest.

e. Last but not least gas streams and accretion discs in *close binary systems* (cataclysmic variables, X-ray binaries etc.) are to be mentioned as further classes of highly interesting objects which have never been observed in the FUV/EUV-region.

5. Model Spectra

In order to demonstrate the power of ORFEUS a model spectrum of an O9 star with $M_v = -6.95^m$ in the LMC was simulated. The spectrum in Fig. 4 is based on the COPERNICUS spectrum of μCol and an adopted integration time of 1000

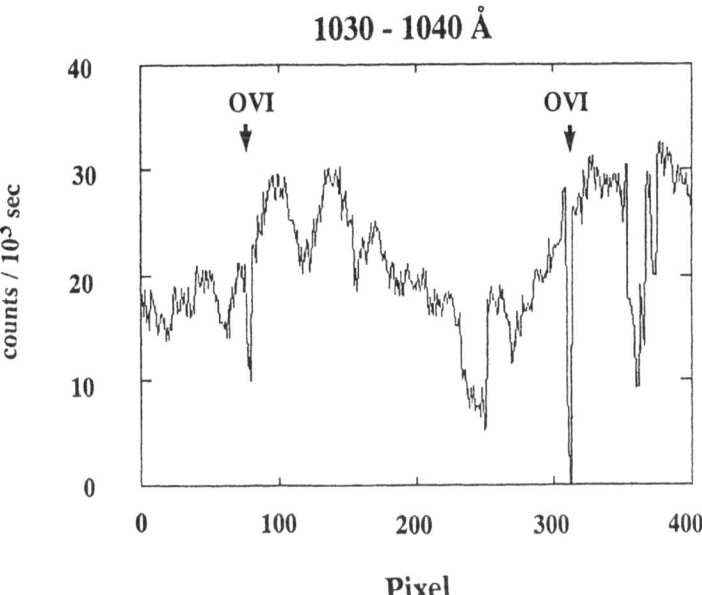

Fig. 4. Section of a Simulated ORFEUS Echelle Spectrum of an O9 Star in the LMC.

secs. It shows the details in the spectral range of the OVI doublet at 103.19 and 103.76 nm. The spectral resolution achieved corresponds to a pointing jitter of the telescope of 10 arcsecs peak to peak. As discussed before the spectral resolution may be improved by almost a factor of two by post flight restauration.

Fig. 5 shows model spectra of the hot white dwarf HZ43 calculated for both spectrometers: The Echelle spectrum is simulated for different coating combinations for the optical elements of the instrument. The coatings are itemized as follows: Primary mirror, collimator, Echelle grating, cross disperser. The abbreviations mean Ir (iridium), SiC (silicon carbide), ARC 1000 and 1200 are special coatings developed by Acton Research. No line spectra are included in this model. The Berkeley model spectrum includes the first few resonance lines of hydrogen and helium. Both spectra are based on calculations for the active area of the instruments including reflectivities and diffraction efficiencies of mirrors and gratings and the quantum yield of the MCP detectors.

Acknowledgements

This research was funded by BMFT Grant 01 OS 8501 8 and NASA Grants NASA/NGR-05-003-805 and NGT-50185.

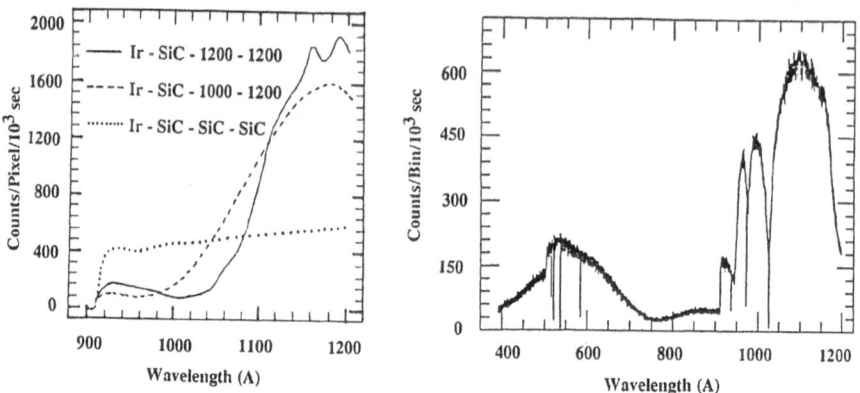

Fig. 5. Model Spectra of HZ43 calculated (A) for the Echelle and (B) the Berkeley spectrometer with an adopted integration time of 1,000 secs. For details see text.

References

Grewing M., Krämer G. et al.: 1990, in R.F. Malina and S. Bowyer, ed(s)., *Extreme Ultraviolet Astronomy*, Pergamon Press, New York,

Hurwitz M., Bowyer S.: 1990, in R.F. Malina and S. Bowyer, ed(s)., *Extreme Ultraviolet Astronomy*, Pergamon Press N.Y.,

Krämer, G. *et al.*, 1988, ORFEUS: A 1 m EUV/FUV Telescope on the Space Platform ASTRO-SPAS, in *A Decade of UV Astronomy with IUE* ESA SP 281, Vol. 2, p. 333

Hurwitz, M., Bowyer, S.: 1988, in , ed(s)., *A Decade of UV Astronomy with IUE*, ESA SP 281, Vol. 2, 329

Appenzeller I. et al.: 1988, *A Decade of UV Astronomy with IUE*, ESA-SP-281, Vol. 2, 337

THE SPECTRUM-UV PROJECT

THE SPECTRUM-UV TEAM

Abstract. The Ultra-Violet Space Telescope "SPECTRUM-UV" consists of a 170-cm diffraction limited telescope for spectroscopy and imaging at $\lambda \geq 912$ Å and a co-aligned assembly of two 50-cm telescopes for imaging and spectroscopy in the 400 to 1200 Å range and of four 20-cm, multilayer coated telescopes for narrow band imaging in the 100 to 400 Å range. The Observatory is an international facility to be launched in a highly eccentric orbit in the mid-1990s.

1. Project Outline

The SPECTRUM-UV mission, recently approved by the Academy of Sciences of the U.S.S.R., is an international orbiting observatory aimed for spectroscopy and imaging in the UV, EUV and XUV domains. It will be launched in the mid-1990s in a highly eccentric orbit with period 4 to 7 days, as to allow long (up to 24 hours), uninterrupted exposures.

The instrument complement will consist of:

— a 170-cm telescope (T-170) for both high and low dispersion spectroscopy and for wide field, broad band imaging in the 1150 to 3500 Å range. Medium dispersion spectroscopy will be extended to the 912 to 1150 Å region where, in spite of the low normal-incidence reflectivity, an accurate control of the thickness of the overcoating layers and the large collecting surface can provide a sizable effective area.

— two iridium (or SiC) coated 50-cm telescopes for, respectively, imaging and low resolution ($\Re \approx 20$) spectroscopy in the 400 to 1200 Å range.

— four multilayer coated 20-cm telescopes (T-20) for narrow band imaging in the 100 to 400 Å interval.

A $\simeq 2.5$ arcsec stability of the spacecraft will be provided on three axis. For the main T-170 telescope, a $\simeq 0.1$ arcsec pointing and tracking performance will be obtained by tilting the secondary mirror around the neutral point. A Fine Guidance System (FGS) will control the actuators of the secondary mirror.

A *real time observatory* concept is adopted as a baseline. A backup automated operating mode is also provided in case of unavailability of the link through the Deep Space Network. The ground segment will be kept as close as possible to that of other missions like the NASA/ESA/SRC International Ultraviolet Explorer and the INTERCOSMOS Astron: a Mission Operation Center will bear the overall responsibility of all spacecraft related tasks, while a Science Operation Center, under direct control of the USSR Academy of Sciences, will be responsible for the scientific exploitation of the mission, and, in particular, of the operational mode, scientific data flow, pointing and tracking.

Y. Kondo (ed.), Observatories in Earth Orbit and Beyond, 185–190.

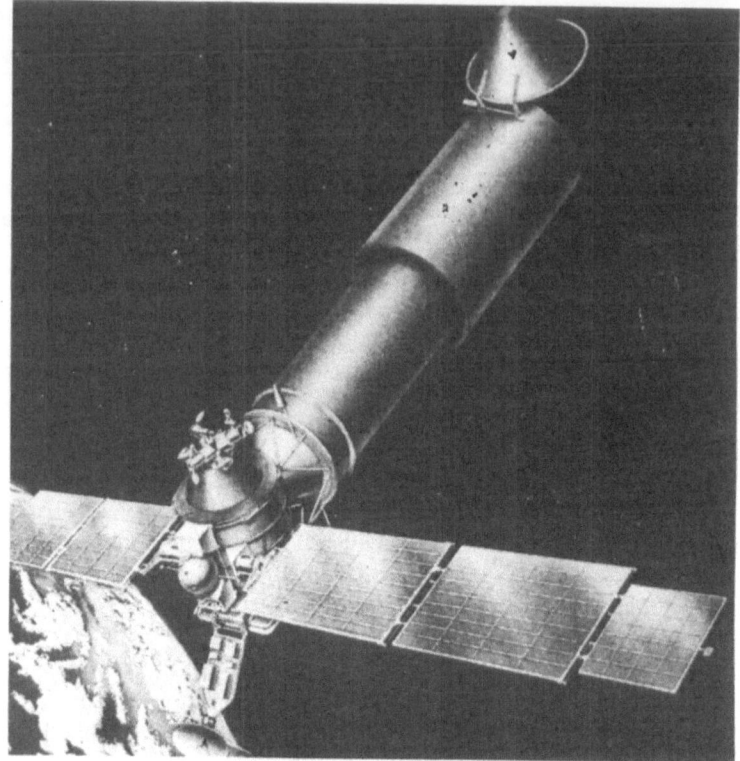

Fig. 1. SPECTRUM-UV concept.

A feasibility study of the Project is being carried out by an international Team which includes scientists from Bulgaria, Canada, Czechoslovakia, East Germany, Italy and the USSR.

2. The T-170 telescope

The T-170 telescope is in the Ritchey-Chrétien configuration with a primary mirror diameter of 170 cm and an equivalent focal length of 17.0 m. The main parameters are given in Table I.

The primary mirror consists of a thin (10 cm) meniscus of equal thickness, with a 58 cm central hole. The hyperbolic secondary is at 3.5 m from the primary apex and can be rotated on two axis around the neutral point as to achieve the fine pointing and tracking required, while keeping to a minimum the aberrations due to decentering. Both optical elements will be polished with the ion etching technique, now being implemented at the Crimea Observatory for mirrors up to 3 m diameter.

The on axis overall image quality will be kept close to the diffraction limit over the whole wavelength range with at least 70% of the energy encircled within 1 arcsec over the whole (40 arcmin) f.o.v.

A dual echelle spectrograph ("A" and "B" of Figs. 3 and 5), similar to that

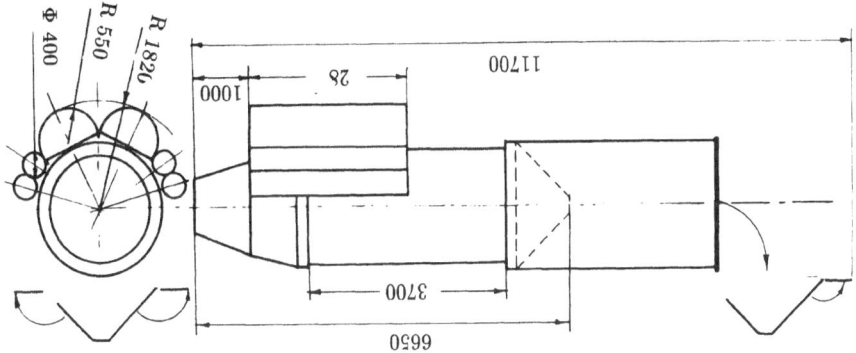

Fig. 2. Telescope assembly layout.

TABLE I
T-170 Telescope Parameters

Primary Mirror

Diameter	1.7 m
Focal ratio	f/2.8
Curvature	$0.10589 \cdot 10^{-3}$ mm^{-1}
Asphericity	1.0539

Secondary Mirror

Diameter	0.5 m
Curvature	$-0.29545 \cdot 10^{-3}$ mm^{-1}
Asphericity	3.6829

Overall Parameters

Equivalent focal length	17 m
Plate scale	12.13 *arcsec*· mm^{-1}
Field of view (2β)	40 arcmin
PSF on axis	diffraction limited
PSF 20 *arcmin* off-axis	1.0 arcsec (70%EE)
Mirrors separation	3500 mm
Focal extraction	900 mm
F.P. Curvature	1270 mm

aboard of IUE, is being designed to achieve high resolving power ($\Re \simeq 4 \times 10^4$) for the 120–190 nm and 190–350 nm intervals.

A low resolution ($\Re \simeq 10^3$) mode can be obtained by inserting a flat mirror in front of the echelle. A second spectrograph ("C" and "D" of Figs. 3 and 56), with turnable gratings in the Rowland mounting, will provide a resolving power $\Re \simeq 10^4$ in the Lyman region and $\Re \simeq 3 \times 10^3$ over the 120–320 nm interval. In both spectrographs the entrance slits open into the flat mirrors of the FGS, to minimize acquisition and tracking problems.

A nebular spectrograph ("E" of Figs. 3 and 5), with a 200 × 2 arcsec^2 slit, provides $\Re \simeq 10^2$ in the 400 to 900 nm range.

A direct imaging camera with broad band filters in the region 120–300 nm and f.o.v. $\sim$ 6 arcmin will be mounted on-axis. A $\sim$ 16 arcmin f.o.v., off-axis focal reducer in the 190–350 nm range will feed a low resolution imaging camera.

To ensure the full exploitation of the the instrument complement and of the expected mission profile, state-of-the-art detectors will be adopted. Large format CCD with enhanced UV response and with pixel size adequate for optimal sampling of the expected PSF are now becoming available. The low read-out noise (2 to 3 electrons rms) and the high quantum efficiency achievable make them eligible for the imaging cameras. Thermal control to maintain the detector system at the operating temperature ($\sim$ -100 °C) will consist in a thermoelectric cooler connected to large size radiator plates having direct view to space. The detectors in the main spectrographs will all be two-dimensional photon-counting devices, based on microchannel plate (MCP) technology with a serial read out system (e.g. wedge and strip anodes). Detector quantum efficiency as high as 40% can be obtained in the UV and FUV region with KBr and CsI photocathodes.

3. The T-50 and T-20 Telescopes

The following ancillary telescopes will be mounted co-aligned with the T-170 telescope (see Figs. 2 and 4):

— two off-axis paraboloids (T-50), each consisting of a half 80-cm mirror (collecting surface equivalent to a 50 cm diameter mirror) with iridium (or SiC) coating to enhance reflectivity in the 400 ÷ 1200 Å range. The "T-50 I" telescope will be equipped with a direct imaging camera to achieve a few arcsec angular resolution over the whole 4 arcmin f.o.v.. The "T-50 S" telescope will be equipped with a very low resolution ($\Re \simeq 20$) grating objective.

— four 20-cm off-axis, f/12.5 paraboloids (T-20) with multilayer coatings to cover four narrow bands $\Delta\lambda/\lambda = 0.05$ in the 100÷400 Å region.

Windowless MCP detectors, in the photon counting mode will be used in the focal plane of these telescopes.

4. The Spacecraft

The design of the spacecraft for the SPECTRUM-UV mission is under the joint responsibility of the Babakin Research Center and the Lavochkin Science and Industry Corporation. It is a common design platform to be used also for the RADIO-

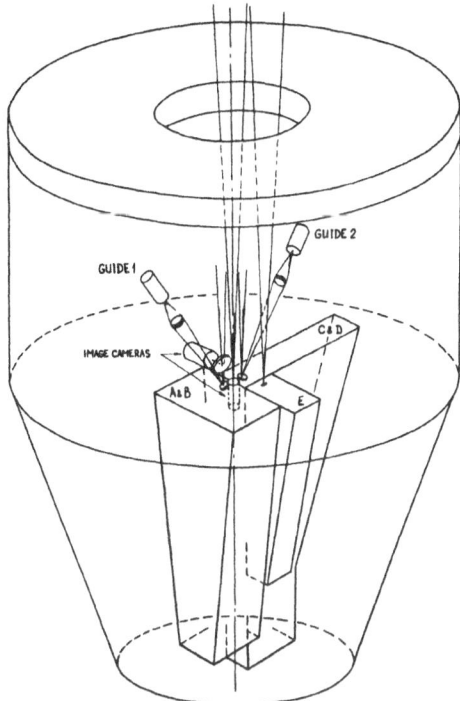

Fig. 3. T-170 telescope: focal instruments accommodation study.

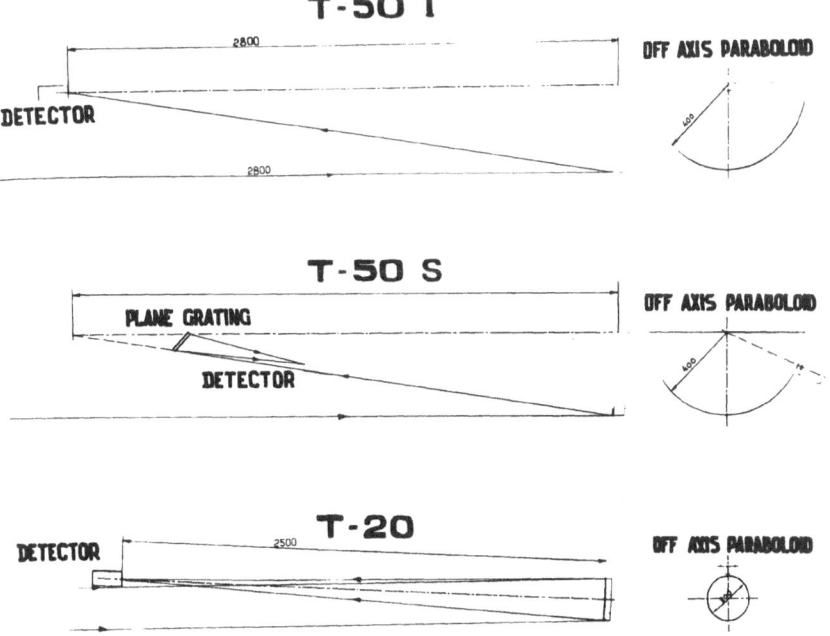

Fig. 4. Optical scheme of the T-50 and T-20 telescopes.

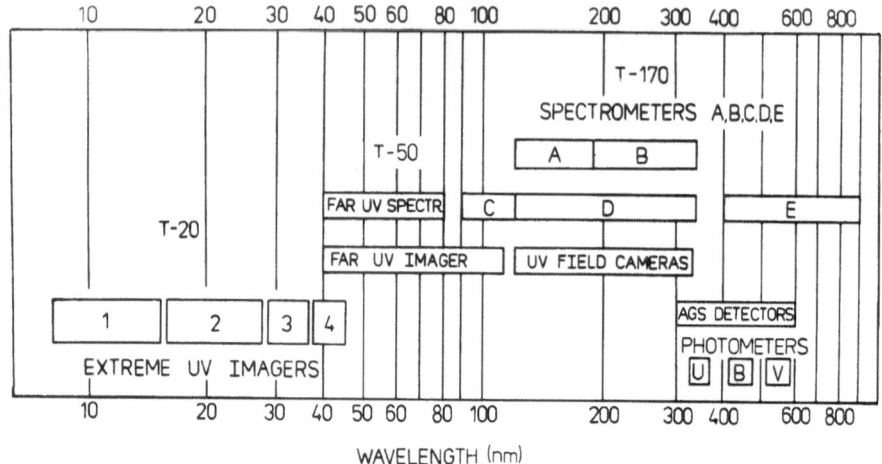

Fig. 5. Overview of the wavelength coverage of the instrument complement.

ASTRON Project (a 10-m radio telescope) and the SPECTRUM-X mission (a high energy orbiting observatory).

The platform will host scientific payloads up to 2.5 tons, providing an operational life in excess of 3 years. The telemetry bit rate is of 64 kbit sec^{-1} up to a distance of the satellite of 5×10^5 km. The satellite will be equipped with an on-board timing system of synchrofrequency generation from 1 Hz to 1 mHz.

The basic attitude determination system of the spacecraft consists of high precision gyros. In order to derive absolute pointing direction and to correct for drifts, the spacecraft relies on sun and star sensors. The actuators are reaction wheels periodically unloaded by gas jets. The spacecraft will be 3-axis stabilized in an inertial coordinate system. Its stability under the control of the gyros will be 1 to 3 arcsec on 1 minute time scale and 30 to 40 arcsec on a time scale of 24 hours, due to temperature instabilities.

For fine pointing and stabilization, the signal from the star under investigation or from a nearby offset star will be used. This signal will be obtained from the main optical system of the telescope. Stars brighter than $m_V = 14$ can be used as guiding star. Two guiding stars are needed to ensure stability during long exposures (up to 24 hours). The change of instrument configuration or mode of operation can be performed automatically without any command from the ground, except the change of the entrance diaphragm or the spectrometer slit, which should be made under control by the ground during the communication section. Such a session to change either the target object or an entrance slit of an instrument can take about 0.5 hours, whereas the data dump session and software uplink can take 2–3 hours.

The spacecraft will be launched by a PROTON booster on a high apogee orbit with initial inclination of 51°. Both solar and chemical batteries will supply 1 kW for the scientific payload for any attitude of the spacecraft.

(C) INFRARED AND SUBMILLIMETER MISSIONS

THE SPACE INFRARED TELESCOPE FACILITY (SIRTF)

GIOVANNI G. FAZIO

Smithsonian Astrophysical Observatory, Cambridge, MA 02138 USA

and

PETER EISENHARDT

Jet Propulsion Laboratory, California Institute of Technology, Pasadena, CA 91109 USA

Abstract. The Space Infrared Telescope Facility (SIRTF) is a one-meter class observatory for infrared astronomy that will be launched into high earth orbit by NASA in the late 1990's. SIRTF's three focal plane instruments will permit imaging and spectroscopy over most of the infrared spectrum with sensitivities of 100 to 10,000 times their predecessors. This paper briefly reviews SIRTF's capabilities, science objectives, and current status.

1. Introduction

The Space Infrared Telescope Facility (SIRTF) is a cryogenically-cooled one-meter class observatory for infrared astronomy, which will orbit the earth at an altitude of 100,000 km (Figure 1).

Along with the Hubble Space Telescope, The Advanced X-ray Telescope, and the Gamma-Ray Observatory, SIRTF will be part of NASA's family of Great Observatories. The SIRTF mission will be launched in the late 1990's and will have a lifetime of 5 years or longer. SIRTF will be operated as a facility for the entire scientific community, with over 80the mission available to General Observers. Members of the SIRTF Science Working Group are listed in Table I.

The outstanding success of previous infrared space missions such as the Infrared Astronomy Satellite (IRAS), the first space telescope to survey the entire sky, and the Cosmic Background Explorer (COBE), which is exploring the large-scale cosmic background radiation, has demonstrated that both technically and scientifically, the next step in our exploration of the infrared universe is a facility such as SIRTF. The hoped-for success of ESA's Infrared Space Observatory (ISO) will just increase the need for a facility with the capabilities planned for SIRTF. SIRTF, with its excellent optics, high pointing accuracy, use of the best available detector arrays, wide wavelength coverage, and long, uninterrupted observing periods, will provide the best possible follow-on to these missions.

SIRTF's instruments will permit imaging at all infrared wavelengths from 700 microns and spectroscopy from 2 to 200 microns. Dramatic advances in infrared detector technology will enable SIRTF to gain by factors of 100 to 10,000 in sensitivity over its predecessors; an additional enormous increase in capability results from the availability of these highly sensitive detectors in large-format arrays containing tens of thousands of pixels. The full potential of these devices will be realized through cooling SIRTF below 4K with super-fluid helium and by the use of diffraction-limited optics.

Y. Kondo (ed.), Observatories in Earth Orbit and Beyond, 193–203.

Fig. 1. The Space Infrared Telescope Facility (SIRTF).

TABLE I
Members of the SIRTF Science Working Group

DR. MICHAEL W. WERNER JPL	CHAIRMAN OF THE SWG PROJECT SCIENTIST
DR. DALE CRUIKSHANK NASA-AMES	INTERDISCIPLINARY SCIENTIST
DR. GIOVANNI FAZIO SAO	PI, INFRARED ARRAY CAMERA (IRAC)
DR. FRED GILLETT NASA HQ/NOAO	PROGRAM SCIENTIST
DR. JAMES R. HOUCK CORNELL UNIVERSITY	PI, INFRARED SPECTROMETER (IRS)
DR. MICHAEL JURA UCLA	INTERDISCIPLINARY SCIENTIST
DR. FRANK LOW UNIVERSITY OF ARIZONA	FACILITY SCIENTIST
DR. GEORGE RIEKE UNIVERSITY OF ARIZONA	PI, MULTIBAND IMAGING PHOTOMETER (MIPS)
DR. B. THOMAS SOIFER CALTECH/JPL	DEPUTY PROJECT SCIENTIST
DR. EDWARD L. WRIGHT UCLA	INTERDISCIPLINARY SCIENTIST

TABLE II

The SIRTF system parameters and their comparison with the Infrared Astronomy Satellite (IRAS)

SIRTF

JPL   **SIRTF/IRAS COMPARISON**

	SIRTF	IRAS
MIRROR DIAMETER	>90 CM	60 CM
WAVELENGTH COVERAGE	2–700 µm	8–120 µm
DIFFRACTION-LIMITED WAVELENGTH	3 µm	~15 µm
ANGULAR RESOLUTION	(λ/4 µm) ARCSEC	15 ARCSEC at 12 µm
POINTING STABILITY/ACCURACY	0.15/0.25 (ARCSEC)	2 ARCSEC
MODULATION	ARTICULATED SECONDARY	SCAN
FIELD OF VIEW	15.7 ARCMIN	1 DEGREE
SENSITIVITY		
10 µm	6 µJy	70 mJy
60 µm	100 µJy	70 mJy
NUMBER OF DETECTORS	>200,000	~60
SPECTRAL RESOLVING POWER	$\geq$2000	20
MODE	OBSERVATORY	SURVEY
LIFETIME	5 YR	10 MONTHS
ORBIT	100,000 KM CIRCULAR	900 KM, POLAR SUN-SYNCHRONOUS

2. SIRTF in high earth orbit

The SIRTF described in this paper is quite different from the 1988 version of SIRTF, which was designed for operation in low earth orbit at 900 km altitude (Werner et al. 1988). The high earth orbit (HEO) mission, at an altitude of 100,000 km, will will have greatly improved scientific performance (Werner et al. 1989). With an expected lifetime in excess of 5 years and a greater than 80equivalent to 10-15 years of operation in low earth orbit. Improved response at far-infrared and submillimeter wavelengths will result because of reduced telescope background radiation, longer integration times will be possible, and continuous viewing zones at high galactic latitudes will be available which are ideal for deep surveys. A comparison of the relative obscuration by the Earth in the two orbits is demonstrated in Figure 2.

Because SIRTF will orbit beyond the trapped radiation belts, the effects of the South Atlantic Anomaly on SIRTFs detectors will no longer be a problem. However, the cosmic ray flux in HEO will increase and the detectors will now be exposed to occasional solar flare effects.

The high earth orbit SIRTF will also lower the mission risks: there will be no deployable solar panels, safe hold modes of operation will exist, and there will be no atmospheric effects, such as spacecraft glow and atomic and molecular gas contamination.

Spacecraft operations will also be considerably streamlined by the 4-day orbit, excellent sky access, and use of the Deep Space Network for telemetry.

TABLE III

A summary of SIRTF's science instruments

JPL

SIRTF
SCIENCE INSTRUMENTS
OVERVIEW

INSTRUMENT	PRINCIPAL INVESTIGATOR	CHARACTERISTICS
INFRARED ARRAY CAMERA (IRAC)	G. FAZIO, SMITHSONIAN ASTROPHYSICAL OBSERVATORY (SAO)	WIDE FIELD AND DIFFRACTION LIMITED IMAGING, 1.8-30μm, USING ARRAYS WITH UP TO 256 x 256 PIXELS. POLARIMETRIC CAPABILITY. BROAD AND NARROW FILTERS.
INFRARED SPECTROGRAPH (IRS)	J. HOUCK, CORNELL UNIVERSITY	LONG SLIT AND ECHELLE SPECTROGRAPHS, 2.5-200μm USING DETECTOR ARRAYS UP TO 256 x 256 PIXELS. RESOLVING POWER FROM 100 TO >2000.
MULTIBAND IMAGING PHOTOMETER (MIPS)	G RIEKE, UNIVERSITY OF ARIZONA	BACKGROUND-LIMITED IMAGING AND PHOTOMETRY, 30-200μm, WITH PIXELS SIZED FOR COMPLETE SAMPLING OF AIRY DISK. ARRAY SIZES UP TO 32 x 32 PIXELS. BROADBAND PHOTOMETRY, 200 TO >700μm. POLARIMETRIC CAPABILITY.

3. The SIRTF facility

The SIRTF system parameters and their comparison with IRAS are shown in Table II. Although SIRTFs mirror is larger than IRAS, SIRTFs greatly improved sensitivity will be achieved primarily through its higher angular resolution, improved pointing stability, long integration time, and the availability of large-area infrared array detectors with vastly improved sensitivity. SIRTFs sensitivity will be limited only by the natural astrophysical background radiation over most of its operating band.

SIRTF will cover the entire spectral region from 2 microns to 700 microns wavelength and thus uniquely encompasses two important cosmic windows, minima in the natural background radiation located at approximately 3.5 microns and 300 microns, which permit deep views of the early Universe. SIRTFs optical system will provide diffraction-limited images at wavelengths longward of 3 microns over a 15.7 arcmin field of view, and the pointing accuracy and stability are matched to the 0.73 arcsec image diameter at 3 microns.

Figure 3 shows a cross-sectional view of the current concept for the SIRTF telescope.

The optical design is that of a standard Cassegrain telescope with Ritchey-Chretian optics. A 95-cm primary mirror collects the light and reflects it onto a secondary mirror that forms an image plane behind the primary mirror. Also located behind the primary mirror is the Multiple Instrument Chamber (MIC).

Within the MIC a pyramidal-type, multifaceted dichroic mirror (the tertiary mirror) divides the field of view into sectors, directing each sector to one of the instruments. Behind the tertiary mirror will be located the Fine Guidance Sensor, which will used with the spacecraft gyroscopes to provide pointing and stabilization. In a manner similar to IRAS and COBE, the entire telescope system, including the MIC, is suspended in an annular cryogenic tank, containing superfluid helium, which cools the optics and instruments to 1.6 K. The secondary mirror will also be used for conventional chopping at far-infrared and submillimeter wavelengths and for image

SIRTF
ORBIT GEOMETRY

900 km ALTITUDE   **100,000 km ALTITUDE**

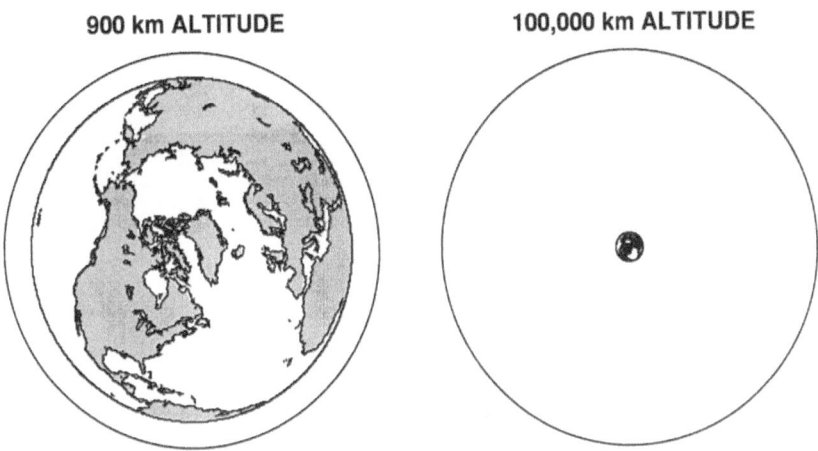

Fig. 2. Orbital geometry of SIRTF in the 100,000 km orbit in comparison to the former 900 km low earth orbit.

SIRTF

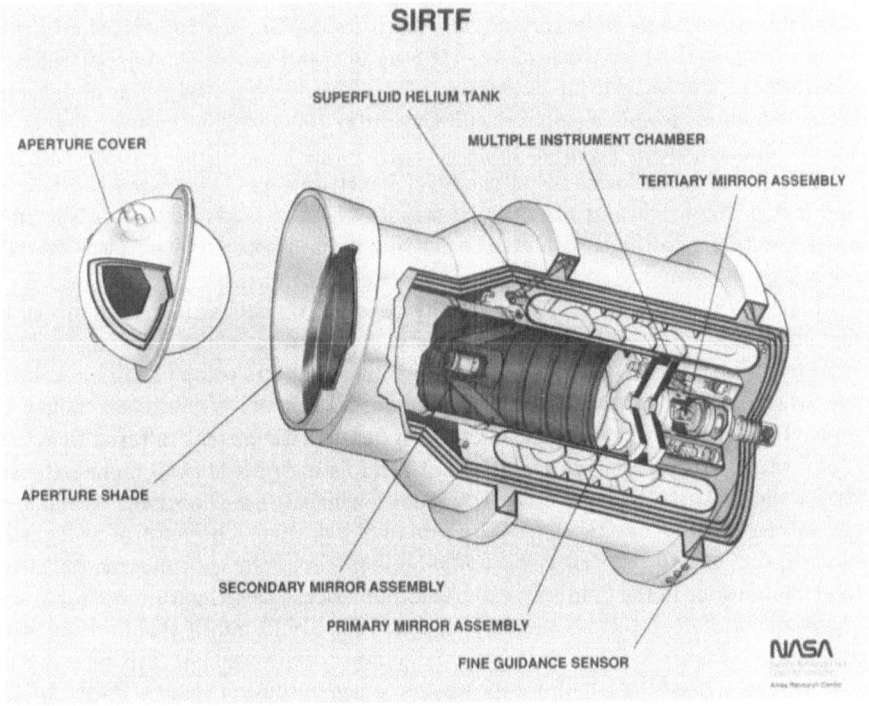

Fig. 3. A cross-sectional view of the current concept for SIRTF.

scanning at the shorter wavelengths. A truncated sun shade shields the telescope interior from stray radiation, allowing the telescope to point within 80 degrees of the Earth and Sun limb without putting an undue load on the cryogenic system. The aperture cover is ejected once the telescope is in orbit and the spacecraft has outgassed.

4. Focal plane instruments

The three focal plane instruments were selected in 1985 and are presently under definition study.

The Infrared Array Camera (IRAC) will use large area, two dimensional infrared array detectors to provide wide-field (5 arcmin) and diffraction-limited imaging over the spectral region from 2 to 27 microns. The Multiband Imaging Photometer for SIRTF (MIPS) will provide background-limited imaging and photometry over the wavelength range from 30 to 200 microns, using arrays with pixels sized for complete sampling of the Airy disk; wide-field, high resolution imaging from 50 to 120 microns; and broad band photometry and mapping from 200 to 700 microns, with a possible extension to 1.2 mm wavelength. The very high sensitivity and complete spatial sampling of IRAC and MIPS, combined with the high quality, stable images to be provided by SIRTF, will permit the use of super-resolution techniques to improve the areal angular resolution of the images over that implied by the Rayleigh criterion by factors of five or more. IRAC and MIPS will also have the capability to measure polarization. The third focal plane instrument, the Infrared Spectrograph (IRS), consists of several long slit and echelle-mode spectrographs covering the wavelength interval from 2.5 to 200 microns using two dimensional array detectors. Resolving power will vary from 100 to 2000. An overview of the science instruments is given in Table III.

The expected performance of the SIRTF instruments is compared in Figures 4 and 5 with the brightness of potential targets, the flux levels reached in the IRAS all-sky survey, and the capabilities of existing ground-based and airborne facilities for infrared astronomy.

Figure 4 highlights SIRTF's capability to study objects at large redshift in the distant universe, while Figure 5 features examples from the solar neighborhood. Note from Figure 4 that SIRTF will easily be able to obtain complete high resolution spectra of even the faintest IRAS survey sources – many of which are below the limit of detectability even in broad spectral bands with current infrared facilities.

Of utmost importance is the fact that SIRTFs instruments will make extensive use of the infrared detector array technology which is now becoming available to the astronomical community. SIRTFs sensitivity goal is to be natural-background limited; that is, SIRTFs detectors should be limited only by the fundamental statistical fluctuations in the faint infrared background of the earths natural astrophysical environment. The detectors currently in hand for SIRTF reach this limit at wavelengths between 5 and 200 microns, and further improvements can be expected over the next several years. On a per-pixel or per-resolution element basis, SIRTF thus realizes a full 100 to 10,000-fold increase in sensitivity over the achievements of IRAS and over the best current capabilities at infrared wavelengths (Figures 4 and 5). SIRTFs focal plane instruments will incorporate arrays with tens of thousands of pixels, each operating at this extremely high sensitivity level. In contrast, the IRAS focal plane incorporated 62 discrete detectors, and ISO relies primarily on small arrays (32 × 32 pixels) and on many discrete detectors. SIRTF will thus be the first mission to combine the intrinsic sensitivity of a cryogenically-cooled telescope for infrared astronomy with the tremendous imaging and spectroscopic power of large-format detector arrays. SIRTF represents a truly enormous increase in our capability to explore the Universe at infrared wavelengths.

5. SIRTF science objectives

SIRTF's long lifetime will permit scientists from all disciplines to use the facility to carry out a wide variety of astrophysical programs. In the paragraphs below we will discuss key areas of astrophysics where SIRTF's contributions are critical, i.e., where SIRTF's can substantially advance key areas of astrophysics during the coming decade. These principle scientific objectives include understanding how our solar system and the planets formed, the search for other solar systems, the birth and evolution of stars, searching for dark matter in the Universe, probing the earliest stages of the Universe to understand how galaxies formed and evolved, and understanding cosmic energy sources in active galaxies and quasars (Rieke *et al.* 1986).

5.1. OUR SOLAR SYSTEM

SIRTF will allow us to measure the chemical composition and the temperature of the most primitive objects in the solar system, such as planetary satellites, asteroids, and the nuclei of distant comets, and thus study what the solar system was like when it formed. SIRTF's measurements of the changing appearance of comets as they approach and recede from the Sun will be particularly useful for determinations of the structure and composition of cometary nuclei. Complete infrared spectra of Pluto, Chiron, and Triton will give definitive information on the atmospheres and icy surfaces of these small and remote bodies.

5.2. SEARCH FOR OTHER SOLAR SYSTEMS

IRAS found disks of matter around several nearby stars that suggest the formation of planetary systems. By applying SIRTF's high resolution and sensitivity to the study of circumstellar material around nearby stars, we can learn how commonly such disks occur, as well as their dimensions, structure, and chemical composition. Zodiacal dust clouds like the Sun's can be imaged by SIRTF around the nearest solar type stars, while disks like those found by IRAS can be studied around stars more distant than 1 kpc. For the most prominent systems, such as Vega and Fomalhaut, SIRTF's images will show the orientation, structural features and detail morphology of the disks. The effects of companions, both stars and massive planets, on the dust disks should be discernible using SIRTF's high precision images and super-resolution techniques.

5.3. BIRTH AND EVOLUTION OF STARS

SIRTF is ideally suited for the study of star formation. Stars are born within dense clouds of interstellar gas and dust. These dust clouds absorb any visible light from stars forming within them, but infrared radiation from these stars in formation can escape from even the densest clouds. SIRTF will produce images and spectra of these regions which will greatly increase of understanding of the chemical, structural, and dynamical evolution of stars in the earliest stages of their evolution. In particular, SIRTF will be a very sensitive facility for the detection of low luminosity protostellar objects. SIRTF's maps of dense, dark clouds will search for the earliest stages of star formation and will be so sensitive that all stars that are formed within these complexes will be detected and characterized through their spectra. These surveys will permit the determination of the entire initial mass and luminosity functions in these clouds.

5.4. BROWN DWARFS

Objects more massive than Jupiter and less massive than the 0.08 solar masses required for a star to sustain nuclear burning are expected to be visible in the infrared as they radiate the heat generated by their gravitational contraction. Searches to date have been generally unsuccessful in discovering these objects, called "brown dwarfs", which have been proposed as a form of dark matter. SIRTF will be capable of carrying out substantial searches, both in deep surveys and around nearby stars, over a wide range of luminosities and temperatures for these objects. Their abundance would also provide an important challenge to our understanding of the formation of stars and planets.

5.5. FORMATION AND EVOLUTION OF GALAXIES

The identification of the epoch of first star formation in galaxies is crucial to our understanding of the process of galaxy formation. The quest for this "protogalaxy" epoch has spurred observers for decades. SIRTF can detect a young galaxy of average mass at a redshift of 5, seen when the Universe was less than 10ultra-deep surveys in several broad bands to search for these objects. Observations of nearby galaxies and galaxies at intermediate distances will allow us to describe the time evolution of the physical and chemical properties of galaxies and guide our understanding of how galaxies develop from massive clouds of gas into highly organized systems of stars, gas, and dust.

5.6. ULTRALUMINOUS GALAXIES AND QUASARS

The IRAS survey has suggested that violent galaxy collisions may trigger activity in the nuclei of galaxies, perhaps providing an explanation for the origin of quasars. Quasars and other active galactic nuclei (AGN), are thought to be powered by massive black holes at their centers, but much of this activity is heavily obscured by gas and dust. SIRTF's imaging surveys will trace the evolution of quasars and ultraluminous infrared objects to redshifts well in excess of 3 (Figure 4) and SIRTF's

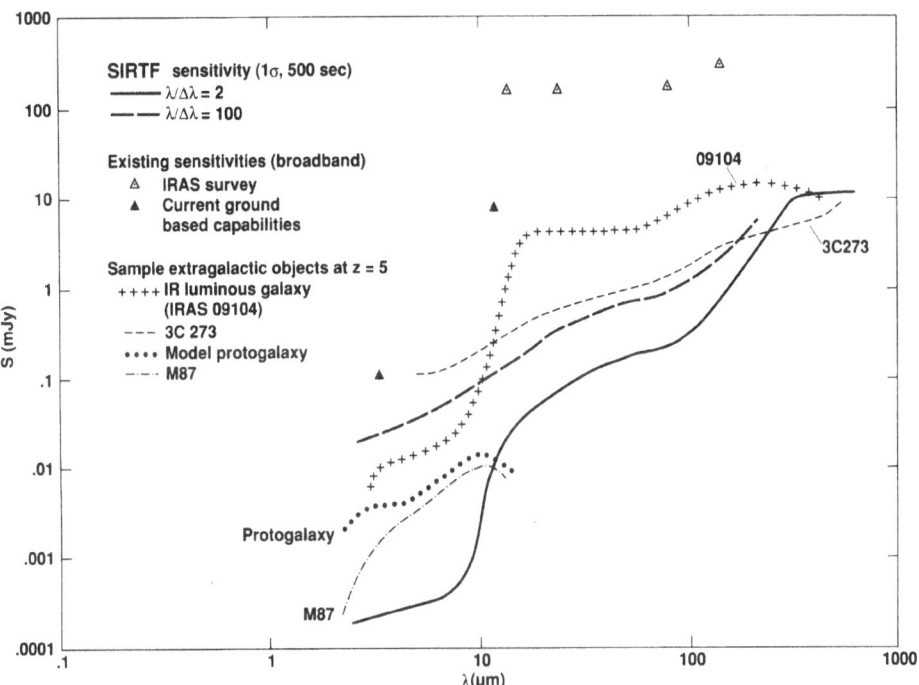

Fig. 4. SIRTF's sensitivity for objects in the distant Universe. SIRTF's sensitivity (1 sigma in 500 seconds) in photometric and low resolution spectroscopic modes is compared with that of IRAS (survey mode) and of large ground-based telescopes. The limiting sensitivity in the far infrared of current airborne telescopes is about that shown for IRAS. The SIRTF predictions are based on demonstrated detector performance or current expectations and also include both natural background and confusion limits. Superimposed on these sensitivity limits are the fluxes of known extragalactic objects of the types to be studied by SIRTF, but moved to cosmological redshifts assuming a Hubble constant of 50 km/sec/Mpc and $q_0 = 0.5$: the giant elliptical galaxy M87, the luminous IRAS galaxy 09104+4109, and the quasar 3C273. The protogalaxy model assumes 10^{11} solar masses of stars are formed at at constant rate for 0.8 Gyr prior to observation at the redshift of 5 (90universe at that epoch), and that no dust is present.

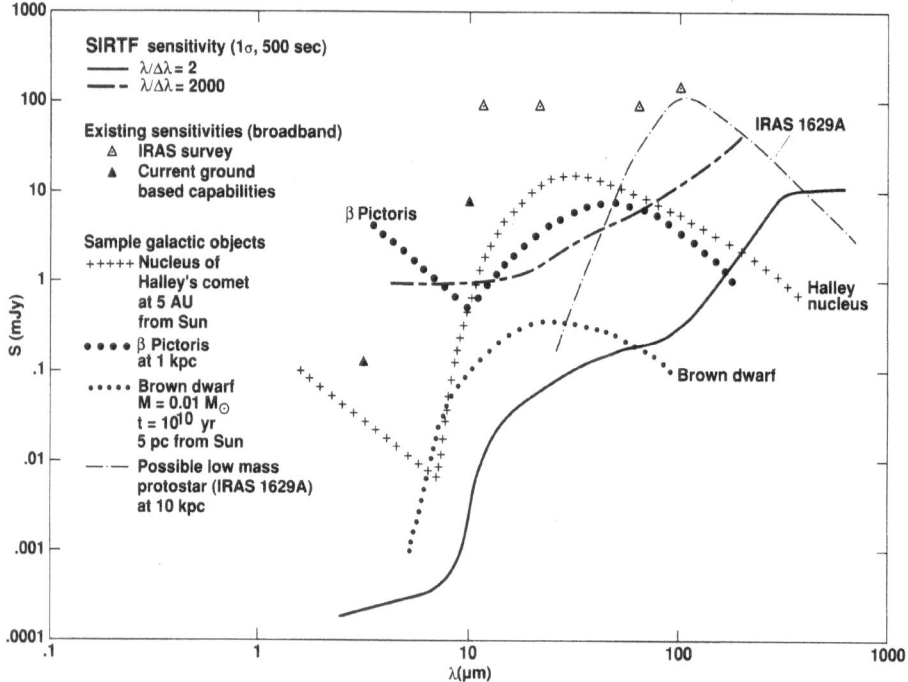

Fig. 5. SIRTF's sensitivity for objects in the solar neighborhood. SIRTF's sensitivity (1 sigma in 500 seconds) in photometric and high resolution spectroscopic modes is compared with that of IRAS (survey mode) and of large ground-based telescopes. The limiting sensitivity in the far infrared of current airborne telescopes is about that shown for IRAS. The SIRTF predictions are based on demonstrated detector performance or current expectations and also include both natural background and confusion limits. Also shown are the fluxes of known and predicted galactic and solar system objects of the types to be studied by SIRTF: Halley's comet, the star beta Pictoris with its associated planetary debris disk, a faint brown dwarf model, and a possible protostar.

spectrogaraphs will identify spectral features which will determine the redshift, and hence the luminosity, of these infrared-bright objects, as well as determine the ionization and excitation conditions in the vicinity of the luminosity sources. Complete flux-limited surveys will test the cosmic evolution of ultraluminous infrared galaxies and infrared-selected quasars by determining their number density evolution with redshift. SIRTF has the unique capabilities to make major breakthroughs in our understanding of these objects.

6. SIRTF'S future

In January, 1990, NASA Headquarters transferred the responsibility for the management of SIRTF to the Jet Propulsion Laboratory (JPL). Since that time the Request for Proposals for Phase B for the facility have been released and SIRTF's Phase B program will begin in October, 1990. The new start date (Phase C/D) is anticipated for FY1993, with a launch date in 1999.

References

Werner, M.W., and Eisenhardt, P.: 1988, *Astro. Lett. and Communications* **27**, 89

Rieke, G. H., Werner, M. W., Thompson, R. I., Becklin, E. E., Hoffmann, W. F., Houck, J. R., Low, F. J., Stein, W. A., and Witteborn, F. C.: 1986, *Science* **231**, 807

Werner, M. W., Brooks, W. F., Manning, L. A., and Eisenhardt, P.: 1989, *Proc. Third Infrared Detector Technology Workshop*, ed. by C. McCreight, p. 387, NASA TM-102209

THE INFRARED SPACE OBSERVATORY

C.J. CESARSKY

Service d'Astrophysique, DPhG, CEN Saclay, France

and

M.F. KESSLER

Space Science Department of ESA, Noordwijk, The Netherlands

Abstract. The Infrared Space Observatory (ISO), a fully approved and funded project of the European Space Agency (ESA), is an astronomical satellite, which will operate at wavelengths from 3–200μm. ISO will provide astronomers with a unique facility of unprecedented sensitivity for a detailed exploration of the universe ranging from objects in the solar system right out to distant extragalactic sources. The satellite essentially consists of a large cryostat containing at launch about 2300 litres of superfluid helium to maintain the Ritchey-Chrétien telescope, the scientific instruments and the optical baffles at temperatures between 2K and 8K. The telescope has a 60-cm diameter primary mirror and is diffraction-limited at a wavelength of 5μm. A pointing accuracy of a few arc seconds is provided by a three-axis-stabilisation system consisting of reaction wheels, gyros and optical sensors. ISO's instrument complement consists of four instruments, namely: a photo-polarimeter (3–200μm), a camera (3–17μm), a short wavelength spectrometer (3–45μm) and a long wavelength spectrometer (45–180μm). These instruments are being built by international consortia of scientific institutes and will be delivered to ESA for in-orbit operations. ISO will be launched in 1993 by an Ariane 4 into an elliptical orbit (apogee 70000km and perigee 1000km) and will be operational for at least 18 months. In keeping with ISO's role as an observatory, two-thirds of its observing time will be made available to the european and american astronomical community.

1. Introduction

The Infrared Space Observatory (ISO) is a fully approved mission, currently under development by the European Space Agency. ISO was selected in 1983, just as the first data sent by the IRAS satellite was reaching the earth. IRAS has demonstrated the great advantage of space cryogenic missions for infrared astronomy: by getting rid of the absorption and emission of the earth atmosphere, and by limiting the emission from the telescope and the instruments, cooled down to superfluid helium temperature, it becomes possible to explore the infrared universe in unprecedented ways. Throughout its 10 months lifetime, IRAS has surveyed the whole sky in four wide bands around 12, 25, 60 and 100 μm. IRAS has revealed the richness of the infrared sky, and shown the relevance of infrared observations for the study of virtually all branches of astrophysics, from comets and interplanetary phenomena to galaxies and cosmology. Many unexpected discoveries were made : cometary trails, dust disks around main sequence stars, large number of bright IR stars in the galactic bulge, diffuse emission of the interstellar medium at 12 and 25 μm, ultrabright

Y. Kondo (ed.), Observatories in Earth Orbit and Beyond, 205–214.

infrared galaxies... Exciting as they are, the IRAS results suffer from many limitations , which most often make it difficult to interpret them unambiguously through
physical models. We are left wondering about the emissivities of IRAS sources in
the wavelength ranges outside the IRAS bands ; we would like to know in more
detail the spatial and spectral distribution of the radiation, in some cases its degree
of polarization ; and of course we would welcome an increment in sensitivity to be
able to detect remote bright sources , such as far away starburst galaxies, and to
explore better the realm of intrinsically faint sources, such as brown dwarfs.

Hence the necessity of a versatile mission such as the Infrared Space Observatory,
with a wider wavelength coverage (2.5 to 200 μm), improved angular resolution (by
a factor of about 10), and enhanced sensitivity (by 2 to 3 orders of magnitude), a
variety of spectral resolutions and polarimetric capabilities. With ISO, it will be
possible for the first time to image the infrared sky from space with array detectors
at various spatial resolutions, and to draw medium ($\Delta\lambda/\lambda \sim 200$ to 1000) and
high resolution ($\Delta\lambda/\lambda \sim 10^4$ to 3.10^4) spectra over the entire range 3 μm to
180 μm. ISO will help to elucidate many of the problems revealed by IRAS, and
the potential for discovery is clearly high as well.

2. Satellite and mission design

The satellite, consisting of a payload module and a service module, is 5.3 m high,
2.3 m wide and will weigh around 2400 kg at launch. The basic spacecraft functions
are provided by the service module. These include the structure and the load path
to the launcher, the solar array mounted on the sunshield, and sub-systems for
thermal control, data handling, power conditioning, telemetry and telecommand
(using two antennas), and attitude and orbit control. The last provides the three-
axis stabilisation to an accuracy of a few arc seconds and also the raster pointing
facilities needed for the mission. It consists of sun and earth sensors, star trackers,
a quadrant star sensor on the telescope axis, gyros, reaction wheels and uses a
hydrazine reaction control system. The nominal down-link bit rate is 33 kbps of
which about 24 kbps are dedicated to the scientific instruments.

The payload module (Figure 1) is essentially a large cryostat. Inside the vacuum
vessel is a toroidal tank filled with about 2300 litres of superfluid helium, which will
provide an in-orbit lifetime of at least 18 months. Some of the infrared detectors
are directly coupled to this helium tank and are at a temperature of around 2 K.
Apart from these, all other units are cooled using the cold boil-off gas from the
liquid helium. This gas is first routed through the optical support structure, where
it cools the telescope and the scientific instruments to temperatures of 3–4 K. It is
then passed along the baffles and radiation shields before being vented to space. A
small auxiliary tank, containing about 60 litres of normal liquid helium, fulfills all
of ISO's cooling needs for the last 72 hours before launch. Mounted on the outside
of the vacuum vessel is a sunshield, which prevents the sun from shining directly
on the cryostat. The solar cells are carried by this sunshield.

Suspended in the middle of the main helium tank is the telescope, which has a
Ritchey-Chrétien configuration with an effective aperture of 60 cm and an overall
f/ratio of 15. A weight-relieved fused-silica primary mirror and a solid fused-silica

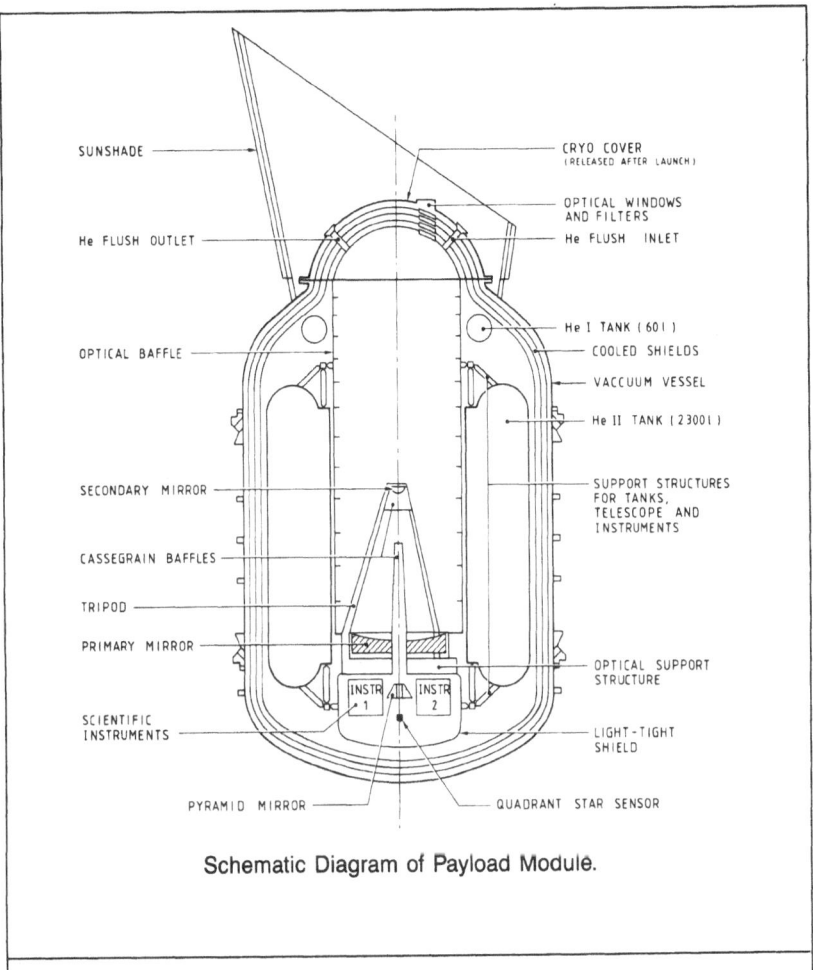

SUNSHADE

CRYO COVER
(RELEASED AFTER LAUNCH)

OPTICAL WINDOWS
AND FILTERS

He FLUSH OUTLET

He FLUSH INLET

OPTICAL BAFFLE

He I TANK (60 l)

COOLED SHIELDS

VACCUUM VESSEL

He II TANK (2300 l)

SECONDARY MIRROR

SUPPORT STRUCTURES
FOR TANKS,
TELESCOPE AND
INSTRUMENTS

CASSEGRAIN BAFFLES

TRIPOD

PRIMARY MIRROR

OPTICAL SUPPORT
STRUCTURE

INSTR 1   INSTR 2

SCIENTIFIC
INSTRUMENTS

LIGHT-TIGHT
SHIELD

PYRAMID MIRROR

QUADRANT STAR SENSOR

Schematic Diagram of Payload Module.

Fig. 1.

secondary mirror have been selected as the telescope optics. The optical quality of these mirrors is adequate for diffraction-limited performance at a wavelength of 5 μm. Stringent control of stray light, particularly from bright infrared sources outside the telescope's field of view, is necessary in order to ensure that the system sensitivity is not degraded. This control is accomplished by imposition of viewing constraints and by means of the sunshade, the cassegrain and main baffles, and an additional light-tight shield around the instruments.

The scientific instruments are mounted on the opposite side of the optical support structure to the primary mirror, each one occupying an 80° segment of the cylindrical volume available. The 20 arc minute total unvignetted field of view of the telescope is split up between the four instruments by a pyramidal mirror. Thus, each instrument simultaneously receives a 3 arc minute unvignetted field centred on an axis at an angle of 8.5 arc minutes to the telescope optical axis. To view the same target with different instruments, the satellite has to be repointed.

An Ariane-4 vehicle will launch ISO into a transfer orbit with a perigee height of around 200 km and an apogee height of around 70000 km. ISO's hydrazine reaction control system will then be used to raise the perigee to attain the operational orbit (24-hour period, perigee 1000 km and apogee 70000 km). The inclination to the equator will be between 5° and 20° (to be finalised later). ESA plans to supply only one ground station, enabling ISO to make astronomical observations during the best 14 hours of each orbit. The mission's scientific return, however, could be greatly increased by the addition of a second ground station, which would permit ISO to be operated for the entire time that it spends outside the main part of the Earth's radiation belts. An international collaboration is being sought by ESA to provide this second ground station.

3. Scientific instruments

The ISO scientific payload consists of four instruments which, although being developed separately, have been designed as a package to offer complementary facilities to the observers. Each instrument is being built by a consortium of scientific institutes using national non-ESA funding and will be delivered to ESA for in-orbit operation. Details of the individual instruments and of their capabilities are given in Table I. The four instruments view different adjacent patches of the sky, but, in principle, only one will be operational at a time. However, when the camera is not the prime instrument, it can be operated in a so-called parallel mode, either to gather additional astronomical data or to assist another instrument in acquiring and tracking its target. In order to maximise the scientific return of the mission, the ISOPHOT instrument will be operated during as many satellite slews as possible so as to make a partial sky survey at a wavelength of 200 μm, a region not explored by IRAS.

The ISOCAM (Figure 2) instrument[1] consists of two optical channels, each with a 32×32 element detector array, operating in the wavelength ranges 2.5–5.5 μm and 4–17 μm. The short wavelength array uses an InSb detector with a CID readout and the long wavelength detector is made of Si:Ga with a direct read out (DRO). Each channel contains a wheel for selecting various filters (including circular variable filters, CVF with a resolution of 45) and a second wheel for choosing a pixel field of view of 1.5, 3, 6, or 12 arc secs. Polarisers are mounted on an entrance wheel common to both channels. A sixth wheel carries mirrors for selecting between the channels, of which only one is operational at a time.

The LWS (Figure 3) instrument[2] consists of a reflection diffraction grating used in 1st and 2nd order with an array of 10 discrete detectors to provide a spectral resolving power of ~200 over the wavelength range from 45 μm to 180 μm. The detectors are made of Ge:Be (to be confirmed) and Ge:Ga (stressed and unstressed) material. Two Fabry-Pérot interferometers are mounted in a wheel and either can be rotated into the beam to increase the resolving power to ~10^4 across the entire wavelength range.

The SWS (Figure 4) instrument[3] provides a resolving power of between 1000 and 2000 across the wavelength range from 2.4 μm to 45 μm by means of two reflec-

TABLE I
Characteristics of the Instruments

Instrument and Principal Investigator	Main Function	Wavelength (Microns)	Spectral Resolution	Spatial Resolution	Outline Description
ISOCAM (C. Cesarsky, CEN-Saclay, F)	Camera and Polarimetry	3 - 17	Broad-band, Narrow-band, and Circular Variable Filters	Pixel f.o.v.'s of 1.5, 3, 6 and 12 arc seconds	Two channels each with a 32x32 element detector array
ISOPHOT (D. Lemke, MPI für Astronomie, Heidelberg, D)	Imaging Photo-polarimeter	3 - 200	Broad-band and Narrow-band Filters. Near IR Grating Spectrometer with R≈100	Variable from diffraction - limited to wide beam	· sub-systems: i) Multi-band, Multi-aperture photo-polarimeter (3-110 μm) ii) Far-Infrared Camera (30-200 μm) iii) Spectrophotometer (2.5-12 μm)
SWS (Th. de Graauw, Lab. for Space Research, Groningen, NL)	Short-wavelength Spectrometer	3 - 45	1000 across wavelength range and 3×10^4 from 15 - 30 μm	7.5x20 and 12x30 arc seconds	Two gratings and two Fabry-Pérot Interferometers
LWS (P. Clegg, Queen Mary College, London, GB)	Long-wavelength Spectrometer	45-180	200 and 10^4 across wave-length range	1.65 arc minutes	Grating and two Fabry-Pérot Interferometers

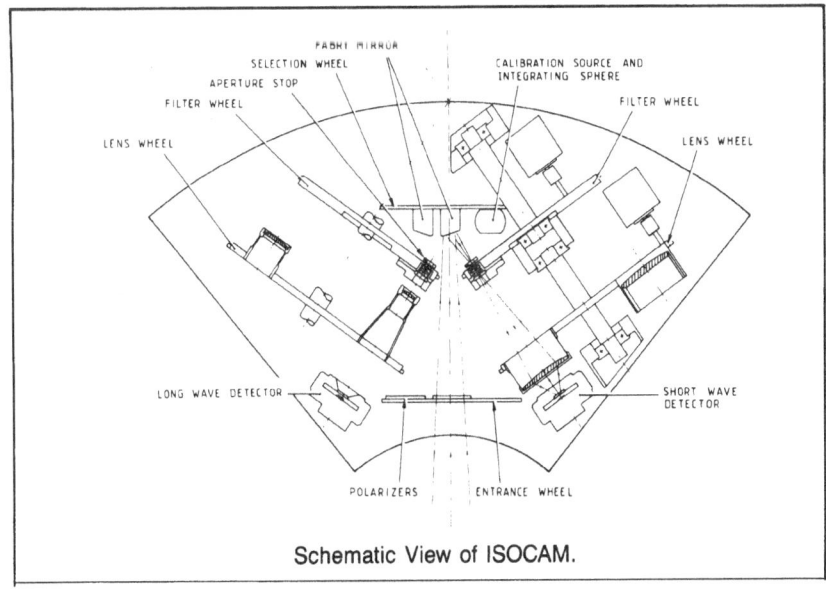

Schematic View of ISOCAM.

Fig. 2.

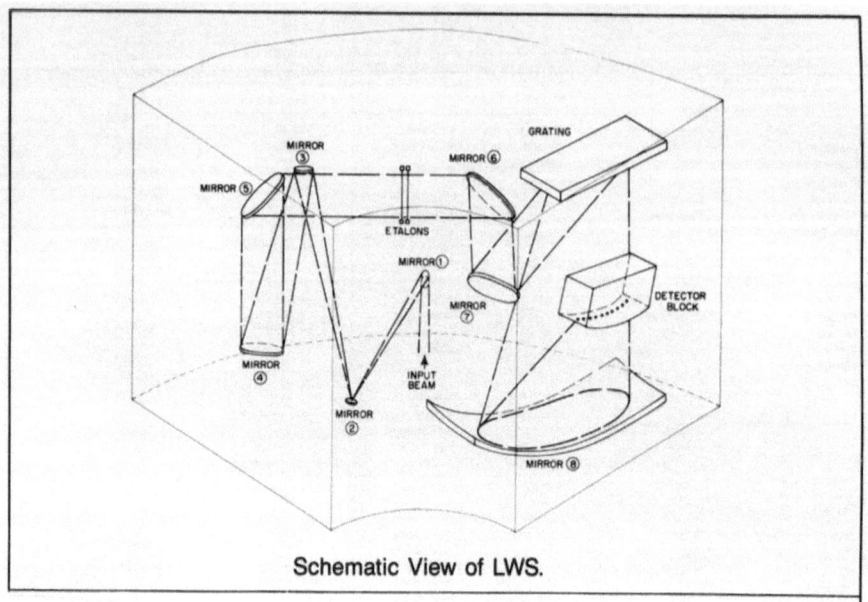

Fig. 3.

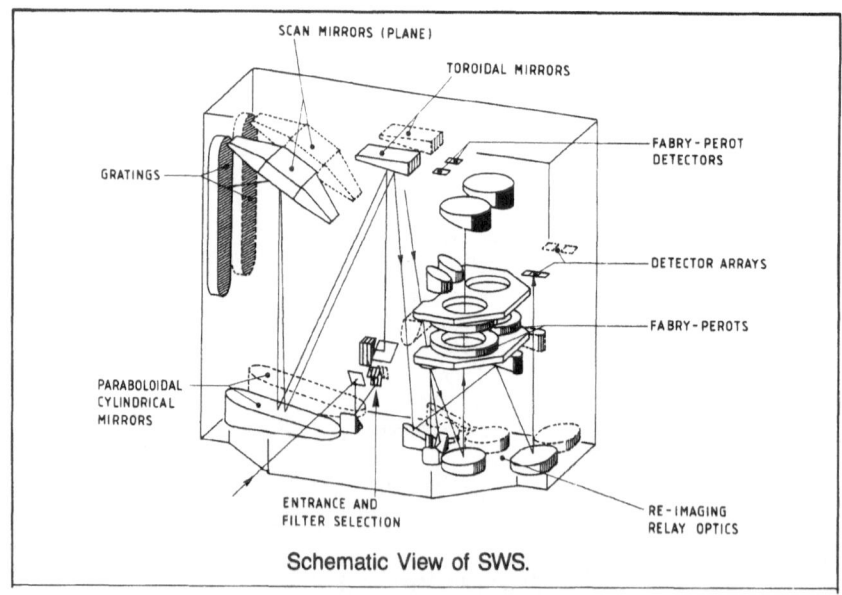

Fig. 4.

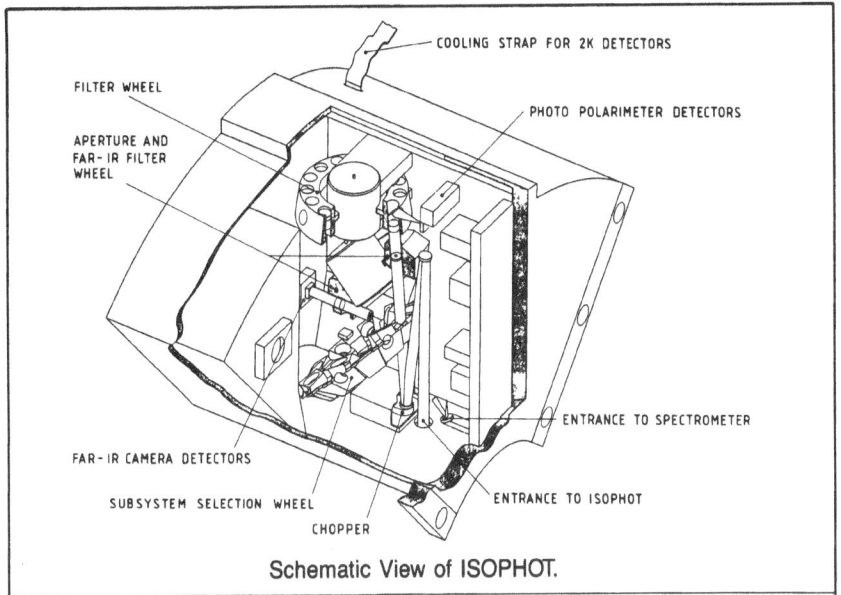

Schematic View of ISOPHOT.

Fig. 5.

tion diffraction gratings used in 1st, 2nd and 3rd orders. Filters for order-sorting
are placed at the instrument's various entrance apertures. Detectors made from
InSb, Si:Ga, Si:P and Ge:Ga material are used. Over a part (14–30 μm) of the
SWS's operating range, the resolution can be increased to ~2 x 10^4 by directing
the incident radiation through either of two Fabry-Pérot interferometers.

The ISOPHOT (Figure 5) instrument[5] consists of three sub-systems:

— ISOPHOT-C: a photopolarimeter which also provides imaging capability at
close to the diffraction limit in the wavelength range from 40 μmto 200 μm.

— ISOPHOT-P: a multi-band, multi-aperture photopolarimeter for the wavelength range from 30 μmto 110 μm.

— ISOPHOT-S: a dual grating spectrophotometer which provides a resolving
power of ~90 in two wavelength bands simultaneously (2.5–5 μmand 6–12 μm)

A focal plane chopper with a beam throw of up to 3′ is also included in
ISOPHOT. Selection between the different modes of the various sub-systems is
achieved with appropriate setting of three ratchet wheels. ISOPHOT contains several types of infrared detectors made of Si:Ga and Ge:Ga (stressed and unstressed).
These detectors are read out by specially designed cryogenic electronics, which exists in both multiplexed and 'un-multiplexed' versions.

4. Observing time

The two thirds of ISO's observing time will be available to the scientific community via the submission and selection (by peer review) of proposals. In addition to
this *Open Time*, there will also be *Guaranteed Time* for the groups who provide
the instruments, for the five Mission Scientists (T.Encrenaz, H.Habing, M.Harwit,

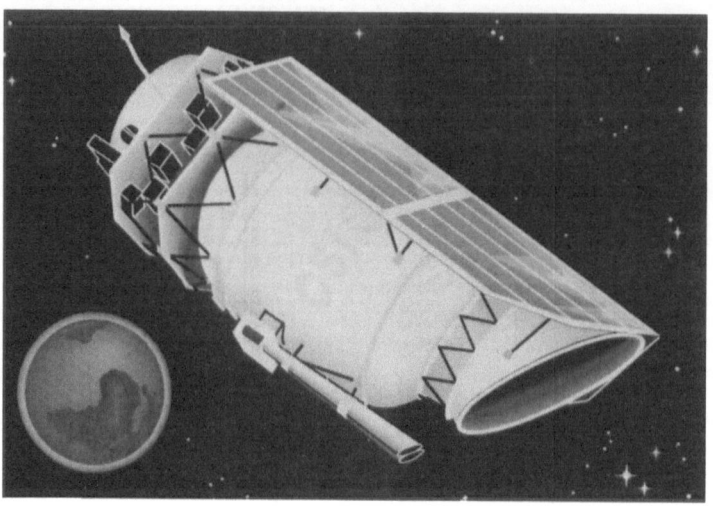

Fig. 6.

A.Moorwood, J.L.Puget) and for the Observatory Team, who will be responsible for all scientific operations. The division of time between these two categories will vary as the mission progresses. After launch, it is anticipated that there will be a period of up to 8 weeks during which the operational orbit will be attained, the spacecraft sub-systems switched on and checked out and the scientific performance of the instruments established. Following this, there will be a 1-month period, consisting of 50% *open time* and 50% *guaranteed time*, during which astronomical observations, designated by the *Observing Time Allocation Committee* as being of the highest priority, will be carried out. For the rest of the mission (at least 15 months), 65% of the time will be *open time*.

The programme to be carried on during the guaranteed time is currently being defined by the instrument groups, the mission scientists and the ESA Project Scientist (M. Kessler). It will include a number of fundamental obervations which may require long observing times, but which are indispensable outputs of a mission such as ISO. For example, complete spectra of a number of bright objects will be drawn over the whole range available to ISO ; these will allow to explore unknown wavelength domains and will be used as templates for more reduced observations of fainter sources. Low resolution spectra of the diffuse interstellar medium at wavelength below 15 μmwill be obtained with the ISOCAM CVF and/or with ISOPHOT-S; in this way, it will be possible to test the hypothesis that this emission is due to the same kind of very small grains or large molecules (possibly polyaromatic hydrocarbons) that emit bright "unidentified" lines in reflexion or planetary nebulae. Star formation regions will be mapped with various ISOCAM filters and in fine structure lines (CII and OI) with the long wavelength spectrometer, to study both star formation processes, and the structure of molecular clouds. The distribution of the main constituents of comets will be determined, and the deuterium abundance of Uranus and Neptune will be measured precisely, with the consequent cosmogonical implications. There will be a variety of programmes on normal galaxies, studying the dust in them through their far infrared emission, the distribution and excitation of very small grains and PAH, molecules wich emit mid infrared photons, the cooling of the gas through the emission of far IR fine structure lines, the chemical abundances, the regions of star formation, the nuclei at various wavelengths... Also, ISOCAM and ISOPHOT will perform deep surveys in chosen regions of the sky, devoid of cirrus. In this way, discoveries will be made or at least meaningful limits will be established on the number of bright, perhaps interacting, galaxies in the past, and of brown dwarfs in the solar neighborhood.

The first *Call for Observing Proposals* will be issued 18 months before launch. It will contain details of expected instrument performances and the program for the guaranteed time, and will solicit proposals for observations to be carried out in the period from 3 to 10 months after launch. Due to the large number of observations expected to be proposed for ISO, the proposal-handling system will be automated as much as is feasible. Thus, proposals will be submitted electronically. Observing time on ISO will be allocated on a "per object" basis, as was the case for EXOSAT, rather than on a "per shift" basis as is done with IUE. In order that the best use can be made of ISO's limited lifetime, there will be a review of the implementation of the observing programme about 5 months after launch; if actual instrument performances differ from those predicted, the *Observing Time Allocation Committee* will recommend suitable adjustments to the programme.

5. Operations

The in-orbit operations of the spacecraft and instruments will be carried out by a team of scientists and engineers located at the ISO Control Centre in Villafranca near Madrid, Spain. This site is currently used by the IUE Observatory. During scientific use, the satellite will always be in real-time contact with the ground segment; however, ISO will be operated according to a detailed, pre-planned schedule in order to maximise the overall efficiency of the mission.

Examination of the scientific data will be carried out both on- and off-line. In close to real-time, a "quick-look" output, adequate for an initial estimate of the success or failure of an observation will be available to the Resident Astronomer and Guest Observer (if present). A final product with more detailed data reduction and calibration will be supplied later; the goal is within a few days. This product will be the one with which observers make their astronomical analyses.

6. Present status

ISO results from a mission proposal submitted to ESA in 1979. After a feasibility study (phase A) in 1981-82, it was selected in March 1983 to be the next new start in the ESA Scientific Programme. The definition study (phase B) for the ISO spacecraft was successfully completed early in 1988. The detailed design, development, integration and test phase (C/D) was started on the 15th of March 1988 by an industrial contsortium led by Aerospatiale (F).

The activities are now centered on finalizing the development of the first model of the satellite, the structural-thermal model. The Service Module STM has undergone static load tests and vibration tests; ground lifetime tests are in progress. The thermal behaviour of the Payload Module STM is under investigation, and the telescope is being tested as well. The primary mirror for the flight model has been polished already.

The scientific instruments were selected in mid-1985. Their development is well advanced with the first models, the alignment and mass/thermal dummies, having been delivered to ESA in 1989. The integration and tests of the engineering qualification models, which are very similar to the eventual flight units, is well advanced for a delivery to ESA in the summer of 1990.

The implementation of the ground segment is progressing as well. It is worth noting that ESA has recently signed a contract with Arianespace, for an ISO launch in mid 1993.

7. Conclusions

ISO is a fully-approved and funded mission, which will offer astronomers unique and unprecedented observing opportunities at infrared wavelengths from 2.5–200 μm for a period of at least 18 months. Two thirds of the observatory's time will be available to the european and american astronomical community. Both the spacecraft and its selected complement of instruments are in their main development phase and the scheduled launch date is May 1993.

References

Cesarsky, C., Sibille, F. and Vigroux, L., "ISOCAM, a Camera for the Infrared Space Observatory", New technologies for Astronomy, Proc. SPIE **1130,** 202–213 (1989).

Emery, R.J. *et al,* "The Long Wavelength Spectrometer (LWS) for ISO", Proc. SPIE **589,** 194–200 (1985).

de Graauw, Th. *et al,* "The ISO Short Wavelength Spectrometer", Proc. 22nd ESLAB, Symposium on Infrared Spectroscopy in Astronomy, ESA SP-290 , 549–551 (1989).

Kessler, M.F., " The Infrared Space Observatory (ISO) and its instruments", New technologies for Astronomy, Proc. SPIE **1130,** 194–201 (1989).

Lemke, D., Burgdorf, M., Hajduk, Ch., and Wolf, J., "Detectors and Arrays of ISO's Photopolarimeter", New technologies for Astronomy, Proc. SPIE **1130,** 222–226 (1989).

IRTS: INFRARED TELESCOPE IN SPACE

TOSHIO MATSUMOTO

Department of Astrophysics, Nagoya University, Chikusa-ku, Nagoya, Japan 464-01

Abstract. IRTS is a small cryogenically cooled telescope onboard the small space platform SFU (Space Flyer Unit). SFU will be launched with the new Japanese HII rocket on January 1994 and retrieved by the space shuttle.

The IRTS telescope has an aperture of only 15 cm diameter, but is optimized to observe diffuse extended infrared sources. Four focal plane instruments are being developed under collaboration between Japan and the U.S.A. IRTS covers a wide wavelength range from near-infrared to submillimeter region, and has a capability for the spectroscopic measurement. Due to newly developed detectors, the sky will be surveyed with very high sensitivities. IRTS will provide valuable data on cosmology, galactic structure, cosmic dust, etc.

1. Introduction

Infrared observation in space with cold optics is extremely valuable in astrophysics. After the success of IRAS, new missions, such as ISO and SIRTF, are planned. These two are big observatories in orbit and require a huge amount of man power and money. Since the community of infrared astronomers in Japan is so small, we are planning a small cooled telescope onboard the small space platform called the Space Flyer Unit (SFU). Our telescope, IRTS (Infrared Telescope in Space), has only 15 cm aperture but is specially designed to observe specific objectives, that is, diffuse extended sources.

Small telescopes generally have lower collecting power and spatial resolution than big telescopes. Observation times of the big facilities, however, are shared by many observers and the beam size is fine, resulting in difficulties for the observations which need wide sky coverage. On the other hand, small telescopes are very powerful for this kind of observation if the mission is dedicated and objects are extended compared with the beam size. IRTS is specially designed for this kind of objects and will play a complementary role to the big facilities.

2. Space Flyer Unit: SFU

SFU is a small space platform for experiments in space. Fig. 1 shows IRTS on SFU in space. SFU has an octagonal shape of 4.6 m diameter and weighs 3,500 kg. Outer sections of octagon are used for many kind of experiments, such as material science, life science, etc. IRTS occupies one section as the only astronomical instrument.

SFU will be launched with the Japanese HII rocket on January 1994 and will be retrieved by the space shuttle. Operation for three months in space is scheduled and 3 weeks are allocated for IRTS observation.

SFU will be put in low inclination (31 degrees) and low earth (500 km) orbit. During IRTS observation SFU is rotated around the axis of the octagon once an

Y. Kondo (ed.), Observatories in Earth Orbit and Beyond, 215–222.

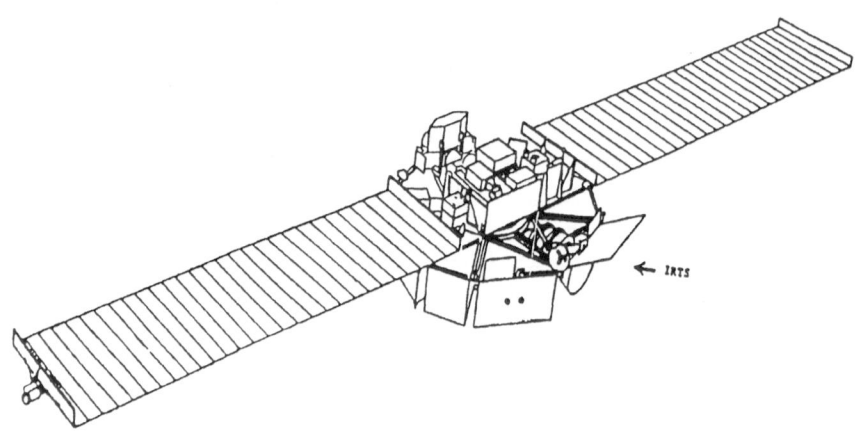

Fig. 1. IRTS on SFU in space.

orbital revolution, so that the direction of IRTS beam is always far away from the sun and the earth's surface. IRTS will survey the sky with constant scan speed and 10% of the sky is observed during the IRTS mission.

3. Cryostat and telescope system

IRTS is a cryogenic system cooled by liquid Helium (Murakami et al. 1989). 100 litres of super fluid Helium is stored in the annular tank which is suspended by the GFRP belts. Phase separation of evaporated gas is made through a porous plug, and the temperature of focal plane instruments is designed to be 1.8 K.

An engeneering model of the IRTS cryostat was fabricated and the required performance of the cryostat was confirmed. The telescope system consists of sunshield, aperture shade, specular cold forebaffle, black cold aftbaffle and telescope. This system is very effective not only to reduce heat load on the cryostat but also to improve baffling performance of the telescope. The telescope itself is 15 cm Ritchy Cretien system made of aluminum alloy and forms a F4 beam.

4. Focal plane instruments

On the focal plane, four instruments are installed; Near-infrared spectrometer (NIRS), Mid-infrared spectrometer (MIRS), Far-infrared line mapper (FILM), and Far-infrared photometer (FIRP). Development of the focal plane instruments are being carried out under the collaboration of the following institutes; Institute of Space and Astronautical Science, Nagoya University, University of Tokyo, NASA Ames

$$\lambda \, I_\lambda = \nu \, I_\nu$$
$$(W \cdot cm^{-2} \cdot sr^{-1})$$

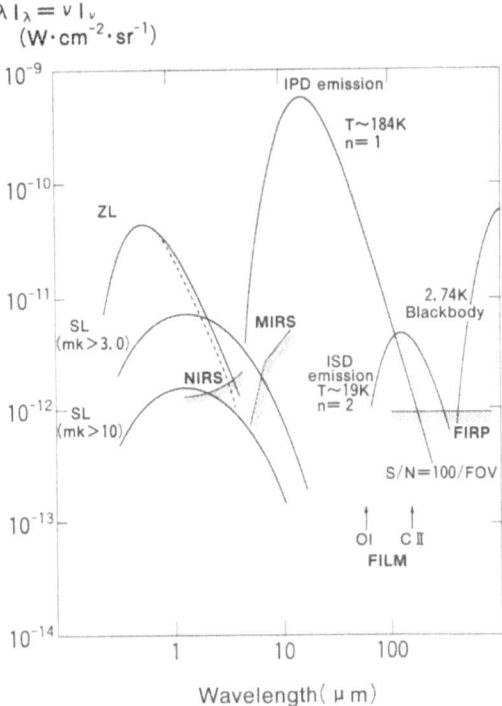

Fig. 2. Infrared surface brightness of the darkest sky and the detection limits of the 4 focal plane instruments.

Research Center, and University of California, Berkeley.

NIRS, MIRS, and FILM use the peripheral area of the beam, while FIRP employs the center of the beam. The star sensor (STS) with an N-slit uses a quarter of the peripheral area of the beam. IRTS is designed to observe diffuse extended objects. Fig. 3 indicates the estimated infrared sky brightness at the darkest region of the sky with the detection limits for IRTS focal plane instruments.

4.1. NEAR-INFRARED SPECTROMETER: NIRS

NIRS is a simple grating spectrometer (see Fig. 3) with a 24 element InSb linear array. Characteristic feature of this optical system is a wide field (0.12 degree square), wide spectral coverage (1.2 μm–4.2 μm), but coarse spectral resolution (0.12 μm band width). A charge integrating amplifier with heated J-FET array is installed (Yamamoto et al. 1989) which results in about 10 times more sensitive detection limit than usual TIA amplifier.

A variety of scientific objectives is expected for NIRS. Interplanetary dust (IPD) scatters sunlight (Zodiacal light: ZL) and emits the thermal radiation in mid-infrared region (IPD emission). Spectroscopic observation will delineate the band

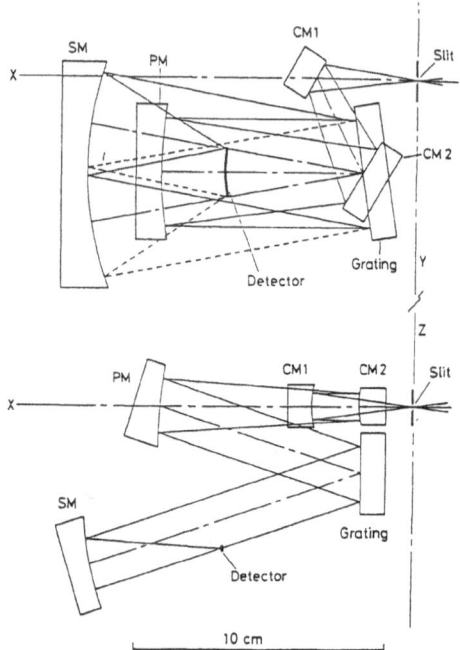

Fig. 3. Optical design of NIRS.

features of the dust, indicating the nature of the dust material.

Star light (SL), that is, an integrated light of the faint stars is also interesting objective. Spectroscopic observation will provide informative data to study the origin of the SL, since important stellar lines of CO and H_2O are included in NIRS spectral range. Observation of SL at high galactic latitude is especially important in studying nature of the galactic halo stars. Diffuse extended emission of interstellar matter is also observable. For example, extended emission of the $3.3\,\mu$m line reported by Giard et al. (1988) will be easily mapped by NIRS.

As is clearly shown in Fig. 2, the sky is very dark at the near-infrared region. This makes the near-infrared region useful as a window to observe extra-galactic diffuse emission. Matsumoto et al. (1988) made a rocket experiment to search for the extragalactic background light (EBL) in the near-infrared region and reported unknown isotropic emission of the order of 10^{-11} W cm^{-2} sr^{-1} which is possibly attributed to EBL. The observation of the near-infrared EBL is particularly important in the study of the early universe, since the redshifted light of the first generation stars could form EBL in the near-infrared region due to their high redshift.

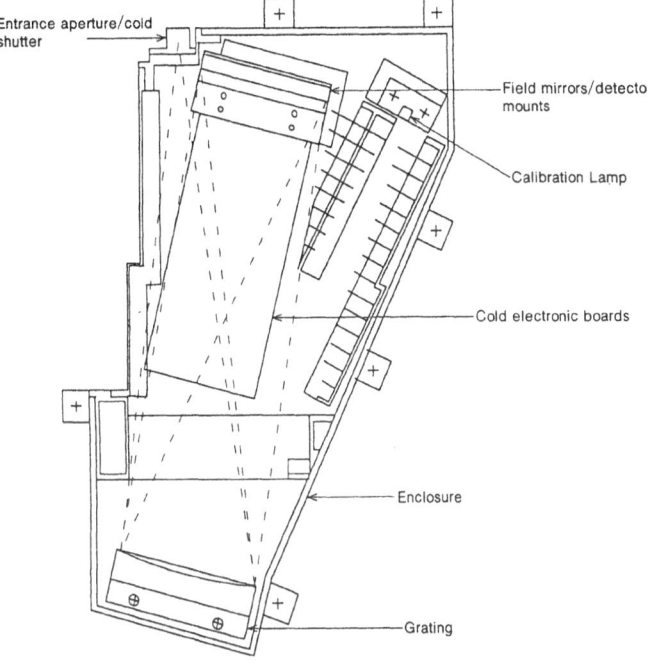

Fig. 4. Optical design of MIRS.

4.2. MID-INFRARED SPECTROMETER: MIRS

MIRS covers wavelength region between 4.3 μm and 11.6 μm with coarse spectral resolution (0.3 μm bandwidth), and has a beam size of 0.12 degree square. The optical system shown in Fig. 4 is very compact due to the special type of the grating, that is, a variable line spacing, concave, variable blaze angle grating. At the focus a 32 element Si:Ga linear array with feed mirrors is placed. A charge integrating amplifier is used, but signals are multiplexed at the cold plate. The detection limit of MIRS is not due to the read out noise but due to the photon noise of the incident radiation.

NIRS and MIRS are aligned along the scan path with same beam size, which makes it possible to obtain spectral features of the sky over a wide wavelength range. The sky at the MIRS wavelength region is very bright due to IPD emission. Spectral features of the dust emission, such as the 10 μm silicate feature, are included in MIRS wavelength range and will be easily identified. Observation of the IPD emission by MIRS will be complementary to the observation of the ZL with NIRS. Combined data will be important in understanding the nature and origin of the IPD.

Dust emission of the interstellar dust (ISD) is also an important target of MIRS. IRAS found the dust emission which has a peak around 10 μm. This hot dust is supposed to be due to fine dust particles (PAH?) heated by a single UV photon and to be closely related with infrared U-lines at 3.3, 6.2, 7.7, 8.6, 11.3 μm. MIRS observations will show how extended hot dust particles are distributed in the interstellar space, and how hot dust is connected with infrared U-lines. MIRS observations will reveal the nature and origin of the hot dust.

Not only extended emission but also point sources will be observed simultaneously. Although aperture of the IRTS telescope is only 15 cm, sensitivity of MIRS for the point source is one order of magnitude higher than LRS of IRAS. Combined data with NIRS will be valuable for studying late type stars and other curious objects.

4.3. FAR-INFRARED LINE MAPPER: FILM

FILM has the highest spectral resolution of the IRTS focal plane instruments. FILM is tuned at the fixed wavelength of the 158 μm and 63 μm, corresponding to [CII] and [OI] lines, respectively. Fig. 5 indicates optical design of FILM. A variable spacing, cylindrical concave grating makes the optical design very simple. The entrance slit sets the beam size to be ($8' \times 20'$). Since the ratio of the [CII] and [OI] wavelengths is 5:2, the optical system is set so that [CII] and [OI] lines correspond to 2nd and 5th order, respectively. Dispersed and focused light is led to the detector system. A beam splitter in the central feed horn reflects the light of the [OI] line and passes the light of the [CII] line. In front of the central feed horn a chopper is installed for modulation. For the [CII] line, three stressed Ge:Ga detectors are employed. One corresponds to the center of the [CII] line with spectral resolution of 430, and other two detectors are used to observe continuum emission outside of the [CII] line with spectral resolution of 130. For the [OI] line, a single non-stressed Ge:Ga detector is used with spectral resolution of 450.

Scientific objective of FILM is the physics of the interstellar matter. Recent balloon and airborne observations have revealed that the [CII] line is a key to understanding the photodissociated region. The [OI] line which is originated in more condensed region plays a complementary role to the [CII] line. The recent theory suggests that the [OI] line is important in cooling the slow shock region. Shibai et al. (1990) found diffuse [CII] emission extending to the galactic plane. Since IRTS is extremely powerful in observing extended object, mapping of the galactic plane with the [CII] line will provide valuable data for understanding interstellar physics. Furthermore, due to the high sensitivity of FILM, detection of the [CII] line at high galactic latitude is possible. Since the physical state outside the galactic plane is not well understood, FILM observations will be exciting.

4.4. FAR-INFRARED PHOTOMETER: FIRP

FIRP is a unique instrument which does not use dispersive optics. The design of FIRP is shown in Fig. 6. The feed horn of FIRP occupies the central part on the focal plane yielding 0.5 degree beam. Incident radiation is divided to 4 channels with beam splitters. 4 composite type bolometers cover the wavelength range between 100 μm and 800 μm with spectral resolution of 40%. Bolometers of low time

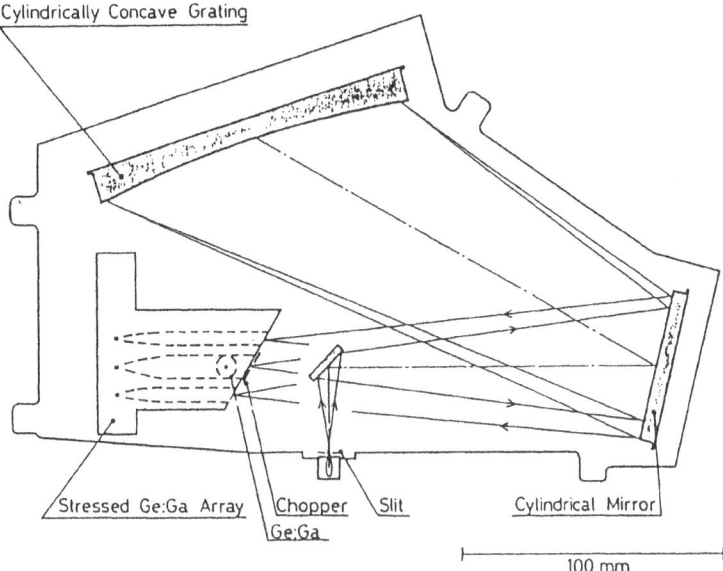

Fig. 5. Optical design of FILM.

constant are used with the AC biased bridge circuit and are cooled down to 0.3 K by a Helium 3 refrigerator, resulting in an NEP of 3×10^{-17} W Hz$^{-1/2}$.

At the short wavelength side of FIRP, ISD emission is dominant. At high galactic latitude, cirrus emission found by IRAS will be observed with significant accuracy. Compared with FILM data, the physical state of the high latitude clouds will be understood more definitely. In the submillimeter region there exists another "window" which enables us to observe EBL. The sky brightness is very low between ISD emission and the 2.7 K cosmic background radiation (Lange et al. 1990). Redshifted radiation of the thermal emission from the dust of the first generation forms the EBL in the submillimeter region. This submillimeter EBL is closely related with the near-infrared EBL.

FIRP and NIRS will play complementary role each other and provide valuable information to understand the evolution of the universe after the recombination era. At the longer wavelength side of FIRP, the 2.7 K cosmic background could be observed. Owing to the high sensitivity, spatial fluctuation, $\Delta T/T$, of the order of 10^{-5} for one beam will be detectable. This detection limit will not only provide strong constraints on the fluctuation of the relict radiation, but also makes it able to observe the Sunyaev-Zeldovitch effect for clusters of galaxies and super clusters.

5. Summary

IRTS is an explorer type mission which is dedicated and optimized to observe diffuse extended objects. In spite of a small aperture, significant scientific results which can not be attained by the big telescopes are expected. IRTS is the first Japanese infrared telescope in orbit and will be an important mile stone for the infrared astronomy in Japan.

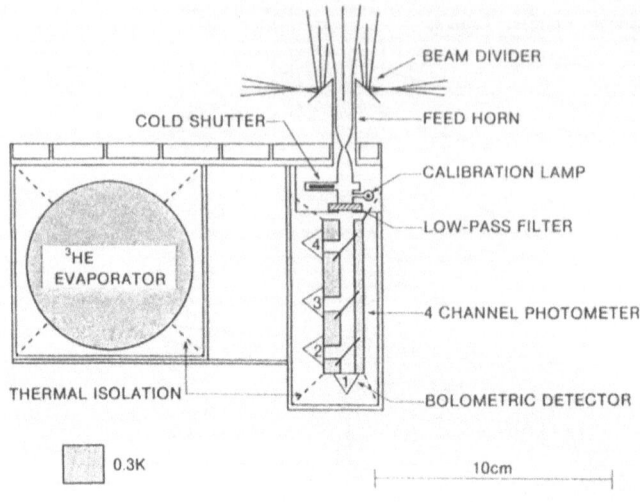

References

Giard,M. *et al.*: 1988, *Astron. and Astroph.* **201**, L1
Lange,A.E. *et al.*: 1990, *Astroph. J.*, in press
Matsumoto.T., Akiba,M., and Murakami,H.: 1988, *Astroph. J.* **332**, 575
Murakami,M.: 1989, *Cryogenics* **29**, 553
Shibai,H., 1990, submitted to *Astroph. J.*
Yamamoto,K.: 1989, *Proc.SPIE* **1157**, 338

FIRST – FAR INFRARED AND SUBMILLIMETRE SPACE TELESCOPE

U. O. FRISK

Astrophysics Division, Space Science Department of ESA ESTEC,
Noordwijk, The Netherlands

Abstract. FIRST is an element of the ESA long term science program. Currently in a detailed study phase it is foreseen as a large (4.5–8 m) diameter passively cooled telescope equipped with a combination of photometer/camera and very high resolution spectrometers. The spectrometers will utilize both direct detection and heterodyne techniques to cover the wavelength band 85–600 micron. The photometers will have both bolometer and photoconductor detector arrays. FIRST is foreseen to be launched shortly after the year 2000 and will be operated as a facility open to the wide scientific community.

1. Introduction

The European Space Agency ESA is studying a large Far-Infrared and Submillimetre Space Telescope (FIRST). The submillimetre range is severely blocked by atmospheric absorption as is clearly demonstrated in Figure 1.

FIRST will open up this virtually unexplored waveband of the electromagnetic spectrum between just below 100 micron and 1 mm for detailed astrophysical measurements.

The planned 4 to 8 m diameter telescope will for the first time allow arcsecond imaging in this wavelength region and, with its high throughput and low thermal background, will result in superb sensitivity for both photometry and spectroscopy. Multi-band high-resolution spectrometers will give unprecedented information on the physics, chemistry and dynamics of interstellar, circumstellar, planetary and

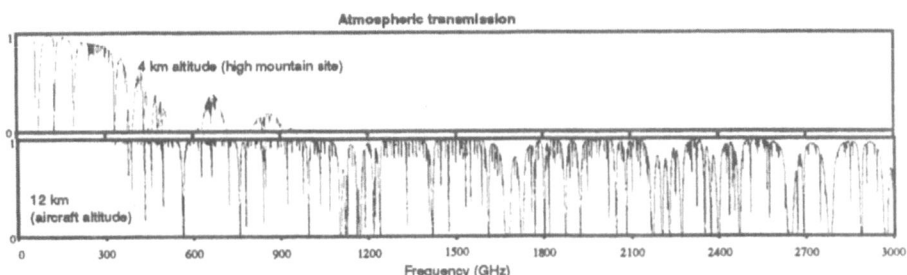

Fig. 1. Transmission of the atmosphere from a 4 km altitude mountain and from the altitude of an airborne telescope such as the Kuiper Airborne Observatory. The blocked regions are caused mostly by molecular line absorption. Many of these lines are the same as of astrophysical interest and cannot even be observed from aircraft altitude.

Y. Kondo (ed.), Observatories in Earth Orbit and Beyond, 223–230.

cometary gas and dust. The ESA Cornerstone FIRST will be a multi-purpose observatory serving the entire astronomical community. It will explore and contribute to a wide range of important problems in modern astronomy and astrophysics.

The concept of FIRST calls for a free-flyer, with a passively cooled telescope which is protected by a thermal shield. The focal plane instruments will be cryogenically cooled by superfluid Helium, and/or by mechanical coolers.

The telescope will have a surface accuracy of about 8 mm r.m.s. in order to efficiently operate at wavelengths as short as about 100 micron, with a possible extension to 50 micron. At a wavelength of 100 micron, the resolution will be 3.5 arc sec for the 8 metre version. As such FIRST will have an angular resolution more than an order of magnitude better than the NASA Kuiper Airborne Observatory (KAO), the Infrared Astronomical Satellite (IRAS), or the ESA Infrared Space Observatory (ISO) at the same wavelengths; objects which IRAS was able to resolve in our Galaxy can be resolved by FIRST in nearby galaxies of the Local Group. FIRST will have an angular resolution comparable to, or approaching that of current ground-based telescopes and will thus allow a direct comparison of molecular material mapped at millimetre or infrared wavelengths from the ground and at far-infrared wavelengths from space. By virtue of its large size, low temperature and emissivity, and location in space FIRST will be 100 to 1000 times more sensitive than the KAO, and at least ten times more sensitive than the Long Wavelength Spectrometer (LWS) on ISO or the largest ground-based submillimetre telescope. FIRST will be able to do detailed 100 micron to 1 mm spectroscopy of the cold and warm components of the Universe out to cosmological distances.

2. Scientific impact of FIRST

FIRST will open up the last major part of of the electromagnetic spectrum still mostly out of reach of astronomers. Each time a new window in the spectrum is opened and an observatory facility is made available, a large variety of scientific objectives appears, and results benefit almost every subject in astrophysics. Furthermore, new and unpredictable scientific problems arise often with the most original outcome. Undoubtedly the same will be true in the submillimetre part of the spectrum. However, with the knowledge we have today we expect the major impact of FIRST to occur in the following areas:

– Physics of the interstellar medium including its chemistry and dynamics both in our and external galaxies.
– Physics of star formation galaxies and in particular its variation from galaxy to galaxy.
– Studies of early evolution of galaxies.
– Properties of primitive solar system material.

As an example, Figure 2 gives an overview of the rich abundance of spectral lines of molecular atoms and ions in the far-infrared and submillimetre waveband.

ATOMIC AND MOLECULAR TRANSITIONS IN THE SUB-MM AND FAR-IR RANGE

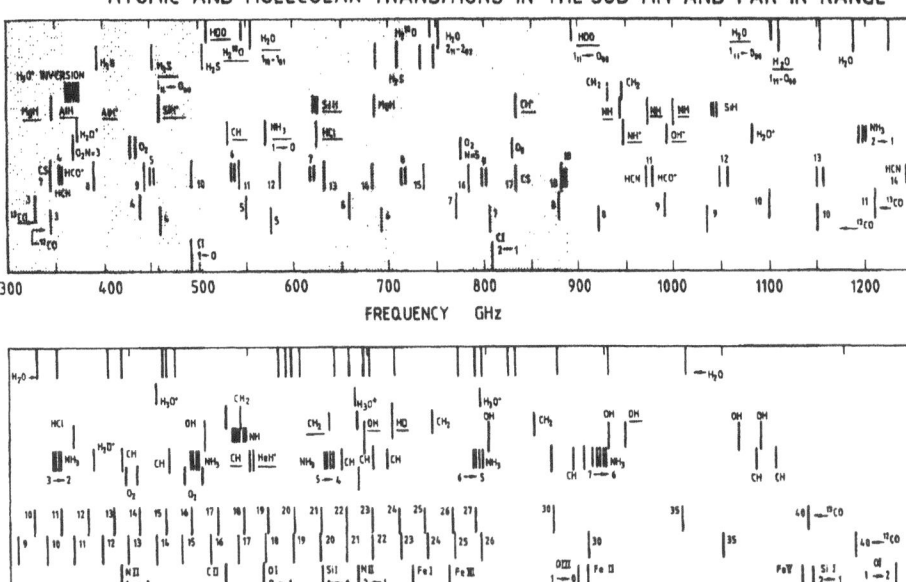

Fig. 2. Some of the important atomic and molecular transitions covered by FIRST. The upper inset gives the frequency range covered by the heterodyne spectrometers, the lower the range (1000 to > 3000 GHz) likely covered by the imaging direct detection spectrometer. Rotational transitions are marked by their upper J–value (e.g. ^{12}CO 10 for $J = 10 \to 9$). Underlining denotes transitions connecting to the ground state of the species. Shaded zones mark frequency ranges that can be observed from the ground.

3. Spacecraft, telescope and orbit

The currently foreseen spacecraft for FIRST is three-axis stabilized by means of momentum wheels. The attitude information is provided by star trackers mounted in close connection to the telescope structure. The calibration of these sensors with respect to the payload viewing direction is performed in orbit by determining the offset of the star trackers from the focal plane sensors at far-infrared wavelengths.

Power comes from fixed body-mounted solar arrays. This limits the instantaneously available fraction of the sky for observation at any given time of the year to less than 50%. Orbital motion of the Earth around the Sun enables different portions of the sky to be viewed at different times during the year very much like from the ground. Like other spacecraft FIRST will have on-board batteries to provide power in periods of eclipse. During all but the very short eclipses scientific operations would be suspended to save power. Over the 6 year lifetime only one period of very long (3.5 hours) eclipses occurs. If cryo-coolers are used then cooling will be suspended during this period and the payload would be allowed to warm up to about 100 K. The choice of a limited range of solar aspect angles also help

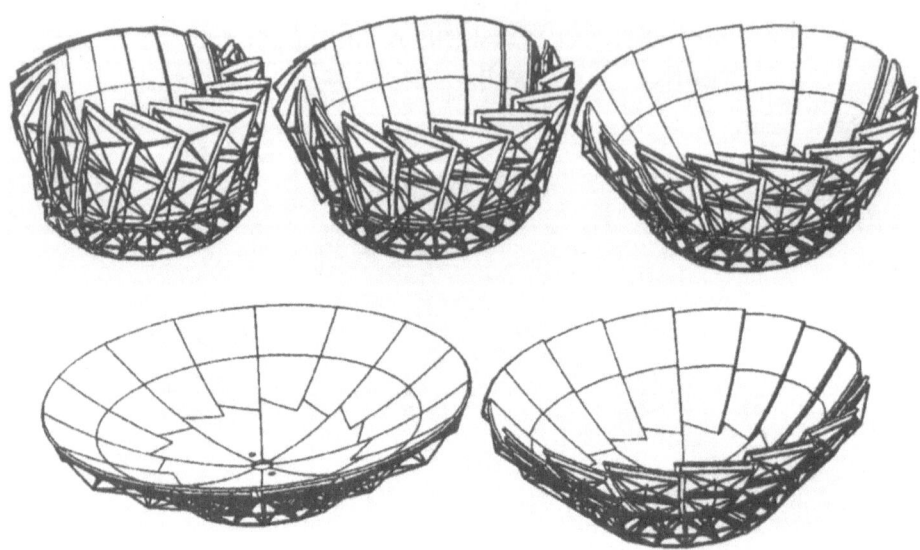

Fig. 3. A sequence of drawings showing how the 8 metre diameter reflector deploys from its stowed to operational state. Each deployable petal is rotating around a single inclined hinge and driven by the same drive system. Synchronism is maintained by the design and requires no additional mechanisms.

to simplify the thermal design. This is important both when coolers and when a cryostat is used.

The telescope for FIRST will be similar in concept to modern ground-based submillimetre antennas. The surface will be made up of panels or sections. Depending on the size of the antenna these sections may be launched in a folded up configuration and then deployed in orbit. The goal for the FIRST antenna is for an overall accuracy of 8 mm rms including allowances for the secondary mirror.

The most promising material for the telescope panels is Carbon-Fibre Reinforced Plastic (CFRP). It is light, strong and thermally stable. CFRP can be "layed-up" over precision moulds to form the exact curved shape required for the reflector and then the epoxy can be "cured" by heating. A spacer that could also be made of CFRP can then be bonded onto the back of the reflector and a further skin added to form a sandwich panel which has very high stiffness but low weight. This technique has been developed for a number of large ground-based telescopes working at millimetre wavelengths and on numerous smaller antennae on spacecraft. Accuracy of better than 2 micron rms has been demonstrated on panels nearly 2 metres in size.

The CFRP panels will be supported by a stiff framework. This frame would also be made primarily from CFRP, this time in the form of tubes. If the diameter is to be as large as 8 metres, the antenna will have to be folded for launch and then unfurled, once in orbit, rather like an opening flower shown in Figure 3.

In this case the structure will incorporate a hinge mechanism to allow for folding. Such a mechanism is currently under development and testing.

The other main component of the antenna is the secondary mirror. It is likely that it will be necessary to provide mechanisms for moving this mirror to the correct position to give the best focussing of the image. A possible further refinement is to make the secondary mirror "wobble" from side to side during observations, this is normally referred to as chopping. The effect of this is to move the image of the astronomical source on and off the detector so that its signals can be separated from the emission from the antenna.

Finally the thermal shield around the whole assembly plays a vital role in protecting the antenna from the solar radiation which would otherwise cause steep temperature gradients across the exposed structures, inevitably introducing severe deformations into the structure despite the use of low-expansion materials. This shield will be constructed of multiple layers of very thin plastic, coated with highly reflecting aluminium and supported by a structure. The shield needs to be large – roughly 10 by 5 metres – and therefore must again be erected in orbit. However very little weight can be assigned to this item. Methods of making such structures have recently been developed. These use inflatable tubes impregnated with a resin that hardens when exposed to the heat from the sun. The thermal shield can therefore be folded for launch, then inflated by a gas when the spacecraft reaches its orbit. The material will harden after a few hours so that the shield will keep its shape indefinitely after the gas is vented. Encased in this protective shield the antenna will cool to about 150 K. This will reduce the background signal emitted from its surface, and differences between the temperatures of its vital parts will be no greater than a few degrees, ensuring that the high accuracy of the antenna will be main tained. The structure for such a thermal shield has been showed in Figure 4 during testing at Contraves.

The whole package will be put in orbit by an Ariane 5 launcher. A highly elliptical 24 hour orbit (apogee altitude 70600 km. perigee altitude 1000 km) appears to best satisfy the combined requirements of low launch cost, large uninterrupted periods and an environment which is both thermally very stable and has a low density of energetic particles.

4. Cooling

There are three classes of thermal control on FIRST: that of the telescope, classical thermal control of space craft systems and the cooling of the payload. The telescope is cooled passively, heat leaking through the thermal shield is radiated into cold space. The equilibrium temperature is around 150 K. It is important that this temperature is stable since large fluctuations would degrade telescope and payload performance. This is achieved by proper design of the thermal shield. However, during the short but close to Earth perigee it is not possible to completely avoid upsetting the thermal equilibrium of the antenna. Therefore (and also because the spacecraft is passing through the radiation belts) observations are suspended during this period. The payload makes use of two different technologies, heterodyne detection and direct detection. In case of the first only the downconverters need low temperatures. The superconducting tunneling junctions used as mixers in the heterodyne receivers about 4–5 K and the associated cold amplifiers about 20–

Fig. 4. The FIRST heat shield as a 1/3 scale model during testing at CONTRAVES. The tubes support Multi-Layer Insulation (MLI) used to reject heat from the Sun and from Earth in order to maintain a stable thermal environment for the FIRST antenna.

30 K. The cold volume is small and we thus refer to this as point cooling. The direct detection instruments use optical methods to achieve their spectral resolution. Since the detectors are performance limited by the thermal background it is important that the thermal radiation from these optical elements is low compared to that from the telescope when spectrally filtered. Hence, large parts of these instruments have to be kept cold (about 7 K). This constitutes a need for volume cooling. The detectors require about 2–4 K operating temperature but have a relatively small volume. Two methods can be used to lift heat at these temperatures: long-life mechanical coolers and the heat capacity of stored cryogens such as superfluid helium. Mechanical cooler developments are on-going within the ESA technology program, see Figure 5.

These are expected to have reached a mature state before the hardware development of FIRST begins. Storing cryogen and using it to cool infrared detectors is a flight proven technique and will also be used for the ISO. Coolers are well suited for point cooling and stored cryogen for volume cooling. The choice for FIRST could be either one or a combination of both.

Fig. 5. This figure shows the demonstration model of the 4 K cooler with one pair of compressors for the stirling cycle and one for the JT stage. The two stage displacer for the stirling cycle with heat exchangers and JT valve on the top is seen to the right.

5. The payload

FIRST will have a full complement of instruments for high and medium resolution spectroscopy, imaging and photometry over the entire range of 80 micron to about 1 mm.

The current plan calls for:
– several sensitive heterodyne receivers covering a significant fraction of the 500 to 1200 GHz range. The heterodyne receivers will provide 0.3 km/s spectral resolution with 2 GHz instantaneous bandwidth. The mixers will be SIS junctions pumped by solid state local oscillators.
– a far-infrared imaging spectrometer covering the 80 to 200 micron range with a Ge:Ga detector array (7 × 7 pixels). Spectral resolution will be selectable in the range of 0.3 km/s to 1000 km/s. Incorporation of a bolometer array may extend the wavelength coverage to longward of 200 micon.
– an imaging photometer for the range 50 to 850 micron, with a 20 × 20 array of Ge:Ga photoconductors for the far-IR near 100 micron and a 2 × 2 array of bolometers for observations near 700 micron.

6. Status of FIRST

Currently a system level study is being carried out by an industrial consortium led by Dornier. The aim of this study is to improve our understanding and definition of the project and to identify areas of technology that need further development. The Science Advisory Group (SAG) consisting of scientists from ESA member countries helps ESA in making decisions where scientific trade-offs are involved. Two further groups: the Payload Working Group (PWG) and the Telescope Working Group (TWG) provide detailed definitions of the model payload and the telescope requirements. This phase was started in 1989 and will end in 1991. During this period major decisions on the design will be made.

Following the completion of this study phase, technology development will be initiated as required lasting up to 1995-96. Assuming that a decision is taken to implement this project as the third cornerstone the project will enter into a detailed design phase followed by hardware development in 1996. The payload that will be built by national institutes will have been selected before this date. The currently foreseen launch with an Ariane vehicle would occur during 2002-3.

A SUBMILLIMETER MISSION FOR THE 1990s: SMMM

T.G. PHILLIPS

Downes Laboratory, California Institute of Technology, Pasadena, CA

1. Introduction

Submillimeter wavelengths hold the key to some of the most important aspects of astronomy. These range from star-forming molecular clouds and proto-planetary disks in our galaxy to infrared emitting galaxies at cosmological distances. Indeed, the essential problems of star-formation and galaxy-formation will be directly probed by the submillimeter spectral lines and continuum radiation emitted by these objects. Other fascinating topics falling into the submillimeter band include the Wien component of the cosmic background radiation, containing information on the nature of the early universe, and nearer to home, the spectroscopy of planetary atmospheres. Since the submillimeter contains fundamental information on the physics and chemistry of so many aspects of our universe, every effort should be made to provide the very best instrumentation for these astronomical studies. We should be capable of detection and analysis of even the most distant objects yet conceived.

Telescopes specifically designed for submillimeter astronomy are now operating on high mountain sites and the field is developing in an exciting and rapid fashion. NASA's airborne program has been in operation for some time and has been of the greatest importance in getting the field started. Both ground and airborne programs will continue to be essential because of their flexibility for implementing new investigations, for instrument development and to support the growth of an active science community, especially students. However, it is now essential to move forward on a space program.

2. The Need to be above the Atmosphere

Space telescopes will be qualitatively and quantitatively superior to any ground-based or suborbital telescope for at least three reasons:

• The Earth's atmosphere blocks the spectral coverage of ground-based telescopes, except for wavelengths longward of 350 μm. Even at the altitudes of airborne telescopes, at least half of the submillimeter spectrum (100 μm to 1 mm) is blocked or confused.

• Emission by the Earth's atmosphere and from the relatively warm telescopes limits the sensitivity of both ground-based and airborne telescopes.

• Observing time is extremely limited with balloon or airborne telescopes and is subject to strict weather requirements for ground-based telescopes operating at the highest frequencies.

Figure 1 shows the transmission of the atmosphere from: a) the 45,000-foot altitude of the KAO or future airborne telescopes; b) a high mountain, such as Mauna Kea, Hawaii. Apart from balloons, low water vapor ground sites and airplanes

Y. Kondo (ed.), Observatories in Earth Orbit and Beyond, 231–249.

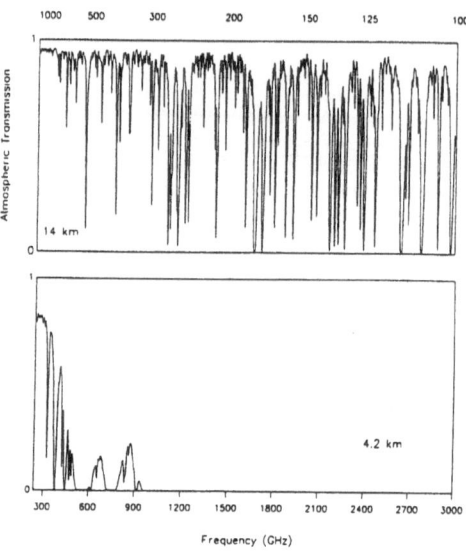

Fig. 1. Transmission of the atmosphere from: a) 45,000-ft KAO altitude and b) a 14,000-ft
mountain site.

represent the only likely non-space platforms for observing in the submillimeter
range. It is clear from Figure 1 that mountain-based telescopes can only be used in
a few selected, longer wavelength windows.

The South Pole site, while an improvement on Hawaii atmospherically, only
allows a restricted range of southern observations and does not approach the transmission quality obtained by the KAO. The atmospheric spectrum from airborne
altitudes is complex and depends upon the resolution of the detection instrumentation. Although certain spectral lines (e.g., those of C II, C I, O I) can be observed
quite clearly through the airborne atmosphere, particularly with high-resolution
Fabry-Perot or heterodyne techniques, it is not possible to make detailed observations of many species, including the critically important molecules H_2O and O_2.

Furthermore, it is not possible to carry out a spectral survey or to obtain a
broad overview of the combined spectral output from dust (which may have as yet
unobserved resonances in the submillimeter band), molecules, and atoms. Finally,
even those specific lines which may be well observed within the Galaxy would not
necessarily be observable for red-shifted galaxies.

3. The Submillimeter Mission

This report describes a 'Moderate Mission' spacecraft concept, designed to make
physical and chemical studies of high red-shift galaxies, molecular clouds, star-
forming regions and planets. SMMM will enormously enhance our knowledge of

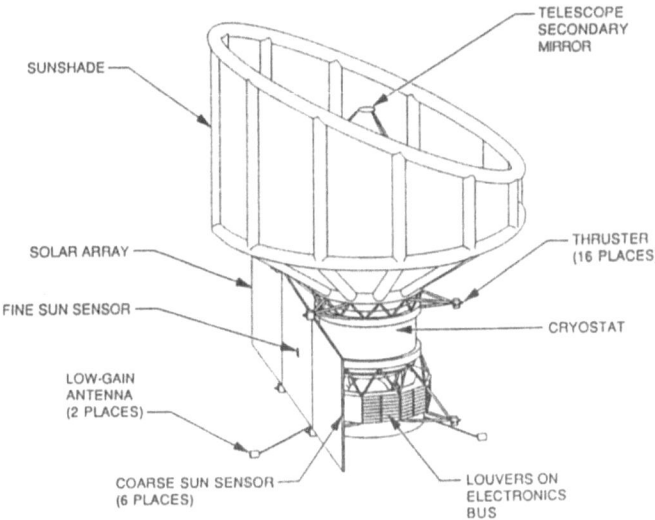

Fig. 2. SMMM spacecraft flight configuration

these objects and provide a pioneering spectroscopic survey of the submillimeter band. It will also have excellent capabilities for cosmological studies in the continuum. At the same time it will generate essential experience for both astronomers and technologists in this new and difficult field and test many of the concepts needed for 'Major Observatory' class missions, such as LDR and the possible Lunar Interferometer, which cannot themselves be achieved in the 1990s time frame.

SMMM (Figure 2) will utilize a roughly 4 m ambient telescope with a liquid-helium cooled focal plane. It will be in a high elliptical orbit (1k × 70k km) which will provide an excellent observing environment. It will achieve a complete submillimeter spectroscopic survey of a large number of sources between 100 and 700 μm. Spectroscopy is to be carried out by heterodyne techniques at the long wavelengths and by direct detection techniques in conjunction with a far-infrared spectrometer at shorter wavelengths. In addition to the spectroscopy, SMMM will have a far-infrared camera with a bolometer array. The expected lifetime is about 2 years.

Very great advances in submillimeter instrumentation have been made in recent years. Heterodyne receivers now use superconducting tunnel junction detectors (SIS) which are extremely sensitive and approach 10% quantum efficiency. Although relatively commonplace in the optical, this is very hard to achieve at radio or submillimeter wavelengths. Very recently major strides have been made in bolometers cooled to 0.1K, so that NEP's of 10^{-17} W/$\sqrt{\text{Hz}}$ can be obtained which will allow SMMM to approach background limited operation, even in the case of the high resolution Fabry-Perot spectrometer. For typical resolutions of 10^4 the line sensitivity will be better than 10^{-18} W m^{-2} in 1000 seconds. In the continuum 0.3K bolometers will achieve point source sensitivities of about 1.5 mJy in 1000 seconds. For the telescope, new carbon epoxy materials are available for panel manufacture and new techniques of construction will provide sufficient accuracy to easily reach diffraction limited performance at 100 μm. The components for submillimeter astronomy in space are approaching readiness.

This mission represents the central 1990's component of a phased NASA plan for the submillimeter region. The overall plan, starting with the Small Scout class Explorer (SWAS), leading to SMMM and finally (although not in the 1990s) to LDR and a possible Lunar Interferometer, is described in a separate "Workshop Report on Submillimeter Planning" submitted separately. SMMM benefits from the initial science investigation and component qualification performed by the .5 m Scout class mission, SWAS, which is already approved.

The current status of SMMM is that it is undergoing pre-phase A study by NASA, jointly with the French Space Agency CNES. Since the programmatic situation is uncertain, it is being studied as either a 'Moderate Mission' with a 3.7 m aperture, to be launched by an Atlas rocket, or as an 'Explorer' with a 2.5 m aperture, to be launched by a Delta. Recently CNES approved the funding for its portion of the phase A study for SMMM.

For the following discussion it is assumed that the 3.7 m version will be constructed.

4. The Submillimeter Astronomy Goals

The submillimeter spectral range contains the clues to processes by which diffuse, dusty gas with temperatures of 10 to 100 K is assembled into stars and planetary systems, both within the Galaxy and in other galaxies. At such temperatures, the peak emission occurs in the submillimeter band, approximately 100 μm to 1 mm. Unlike visible radiation, which scarcely penetrates even diffuse gas, submillimeter radiation can probe across the galaxy and into the cores of even the densest condensations.

Most galaxies and sources in star formation regions, discovered in the Infrared Astronomical Satellite (IRAS) survey, have their greatest flux longward of 100 μm. Very distant and possibly primeval galaxies emit radiation in the mid-infrared, due to dust and atomic fine-structure lines, which will be red-shifted into the submillimeter. The IRAS survey, supplemented by surveys of tracers of molecular gas such as CO, has identified the major nearby sources of far-infrared and submillimeter radiation; however, the resolution of IRAS was limited to 1 to 3 arc minutes in four very broad wavelength bands, shortward of 100 μm. To understand the detailed physics of far-infrared sources, what is needed now is accurate high spatial and spectral resolution measurements of specific objects. A 4 m class telescope will give angular resolutions of between 6 and 18 arc seconds in the 100–300 μm range. Heterodyne and Fabry-Perot techniques will give spectral resolutions up to 3×10^6 (0.1 km/sec) and 10^4 (30 km/sec) respectively. These capabilities are ideal for studies of interstellar gas in the Galaxy and distant galaxies.

Atomic and molecular submillimeter spectra of interstellar gas contain information on a scale rivalling that of optical spectra. They are immensely useful and versatile diagnostics of physical conditions, such as temperature, density, and velocity structure. Because of their rich vibrational-rotational spectra, molecules can probe regions ranging from cold quiescent clouds to hot star-forming regions. In addition, they play an important active role in the energy balance of these objects by providing the major coolants. The amount of energy released by molecular rotational and atomic fine-structure emission at millimeter and submillimeter wavelengths determines, in part, the rates of collapse and, ultimately, the efficiency of star formation in molecular clouds. The amount of energy released is also very sensitive to the exact chemical composition of the gas and dust. Equally, molecular and atomic ions are of great importance, because they determine the coupling of the gas to the magnetic field.

Molecules and molecular ions can also be used as chemical diagnostics of important astrophysical parameters, such as the cosmic-ray ionizing frequency and the overall deuterium abundance and its gradient across the Galaxy. The abundances of some species may be particularly enhanced in star-forming regions, so that these molecules could be used as signposts of the initial stages of the star-formation process. Microwave masers are a traditional example of this. Another application would be the use of molecules as "clocks" to determine ages of molecular clouds and, thus, shed light on the evolution of such clouds. Good progress has been made in understanding aspects of gas-phase chemistry, but the role of grain-surface chemistry, the exchanges between grain-mantles and the gas-phase, and the link between large molecules and very small grains is not understood. This is exactly the area in which SMMM will be of most use, because it would provide us with a complete view of chemical abundances in a large diversity of regions through broadband spectroscopic capabilities, as well as understanding of the dust-grain behavior in the same region, through spectro-photometric and high-angular resolution capabilities.

The major science goals for SMMM fall into two categories: 1) a complete submillimeter line-survey, and 2) galaxies and cosmology. The submillimeter line-survey applies in the main to observations of galactic clouds which require high resolution spectroscopy for analysis of their physical and chemical properties.

4.1. A COMPLETE SUBMILLIMETER LINE-SURVEY

The ability to obtain a complete submillimeter spectrum for approximately 150 sources, in a two year lifetime, will provide a major step forward in fields ranging from star-formation and interstellar medium chemistry and physics to planetary atmospheres. Some important aspects are now discussed.

4.1.1. An unbiased overview of atomic and molecular abundances in gas clouds of various types throughout the galaxy

Figure 3 shows what we anticipate will be the submillimeter spectrum of a typical molecular cloud at a temperature of 30 K. It is a composite of the spectrum of dust, heavy and light molecules and ions and fine structure transitions of atoms and atomic ions. The amazing complexity of heavy molecule rotational spectra at long wavelengths will merge, in the submillimeter, into the fundamental rotation spectra of light molecules, typified by metal hydrides.

Figure 4 indicates the quality of spectrum obtainable with SIS receivers on the ground, through an atmospheric window at 1.3 mm. This quality will be extended through the submillimeter band in the regime of the SIS receivers and, because of

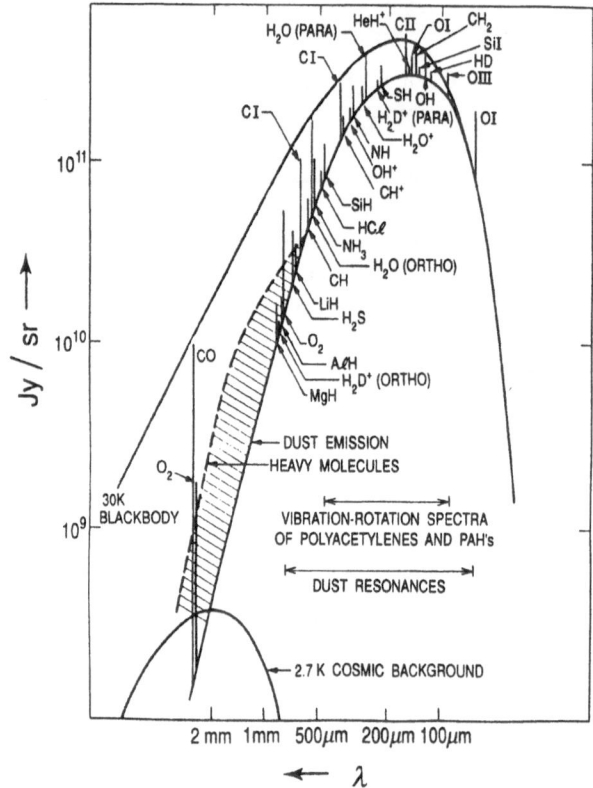

Fig. 3. The anticipated spectrum of a 30K, dense interstellar cloud, showing the dust spectrum, heavy and light molecule rotation spectra and atomic fine-structure lines

the very high sensitivity of the new bolometers, will be approximately matched by the 10^4 resolution Fabry–Perot spectrometer.

The major advantages of a line-survey, aside from the obvious one of completeness are the certainty of identification due to the multiple lines for each species and the ability to solve the radiation transfer and molecule excitation problems, again because of the multiline advantage and also because of the consistent and more accurate calibration. These factors, lead to far better values for abundances than are available from single line studies. Equally important are the possibilities for discoveries!

4.1.2. Thermal Balance and Dense Regions

The heating of the interstellar medium involves both microscopic processes (e.g. absorption of ultraviolet starlight by atoms and molecules) and macroscopic phenomena (e.g. heating by shock waves and cloud-cloud collisions). The heating is bal-

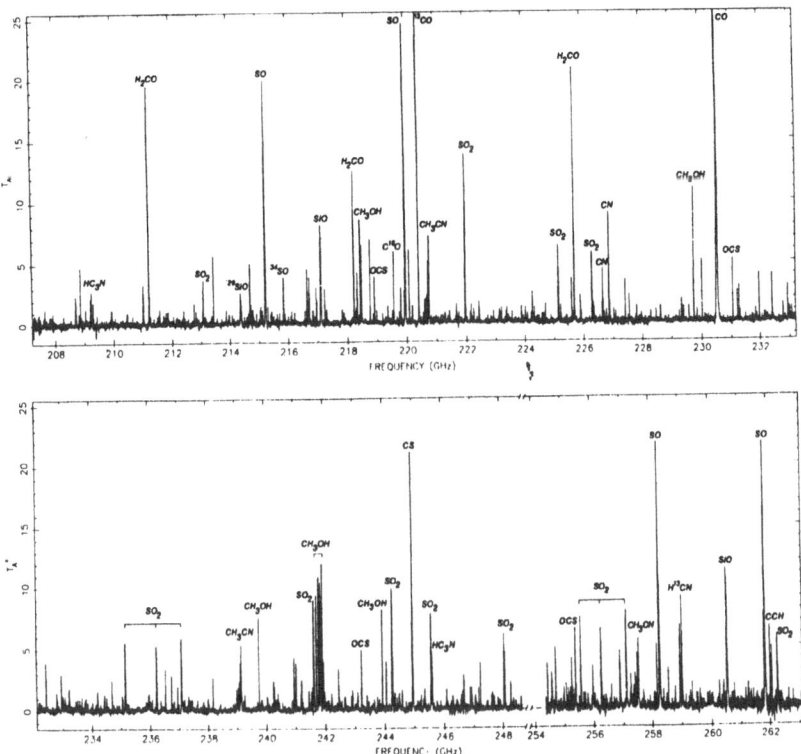

Fig. 4. A spectrum of the core of the Orion Molecular Cloud, taken in the 1.3 mm window using SIS receivers. This shows the richness and quality of spectrum obtainable with sensitive receivers, when the atmospheric absorption is negligible

anced by the radiation emitted by the interstellar gas and dust. The structure of the interstellar medium, i.e., how its pressure responds to gravitational and magnetic forces, is regulated by this thermal balance. Most of the neutral interstellar medium is characterized by temperatures $T \approx 10$ to 100 K, and the bulk of its thermal radiation thus lies in the submillimeter region of the spectrum. Observation of the submillimeter radiation from the gas and dust of interstellar clouds can thus be used to assess their internal and external energy sources and to probe the processes that shape them and turn some of them into stars.

The gas and dust components can behave independently in many interstellar clouds, even though the gas and dust are believed to be well mixed. In the interstellar medium of our Galaxy, the dust exchanges approximately 100 times as much radiant energy with its surroundings as the gas; however, the dust and gas are in poor thermal contact with each other at densities below 10^6 cm^{-3}. The dust radiates continuously with a few broad resonances in the infrared, while the gas radiates through a myriad of discrete, narrow lines. In some warm molecular clouds, notably OMC-1, the lines are known to contribute a significant fraction of the broad band continuum flux at $\lambda \approx 1$ mm and presumably at shorter wavelengths also. A particular advantage of SMMM is that it can survey completely much of the line cooling spectrum of an interstellar cloud simultaneously with the thermal emission of the dust component. A critical point in the collapse of a cloud core to form a star must occur when the dust first becomes opaque at the wavelengths ($\lambda \approx 1$ to 0.1 mm) of maximum emission at $T \approx 10$ to 100 K. At subsequent stages of the collapse, virtually none of the cooling radiation escapes freely from the cloud center and the dust and gas components become strongly coupled through radiation. Submillimeter observations can thus be used to infer the evolutionary status of star-forming clouds.

Owing to its rich submillimeter spectrum and its expected high abundance, H_2O is predicted to be an extremely important coolant, not only in cold clouds, but also in hotter, shock-heated regions. In dense clouds, the rotational excitation of H_2O will be controlled by a complicated combination of collisional processes, radiative transfer, and radiative coupling to the dust. Only with the full spectral coverage of SMMM will it be possible to assess thoroughly the excitation of this important coolant and its relation to cloud structure. Figure 5 displays the cooling rates per H_2 of typical interstellar molecular gas and indicates the dominance of H_2O and other light hydrides at high densities.

The critical density at which a submillimeter transition is observed rapidly increases with frequency. Figure 6 shows the conditions of excitation required for a number of atomic and molecular species. SMMM will be able to study lines throughout the density-temperature space of the figure and will allow the assessment of the thermal balance of clouds over a very wide density range.

4.1.3. Star-Formation regions

The understanding of star-formation is directly related to the determination of the structure of interstellar gas at various scale sizes. The formation and evolution of low-mass stars can be studied by spectral surveys of a few nearby clouds in the Taurus and Ophiuchus regions, by examining density fluctuations in quiescent clouds, and by mapping the distribution of continuum radiation at high spatial resolution. The question of fossil protostellar disks near main sequence stars can also be addressed.

SMMM can make important spectroscopic measurements toward low-mass protostars, such as HL Tau, L1551, and IRAS 16293-2422. Such systems are probably similar to the pre-solar nebula and observations of them will provide insight into the physical and chemical structure of our own primitive solar system. Virtually all theories about the conditions in the pre-solar nebula are based upon very indirect evidence from meteorites, comets, and planetary atmospheres.

The prevailing view is that in formation of the pre-solar nebula, matter from the placental molecular cloud falls onto an accreting disk and passes through a shock front, where the densities increase from 10^4 to 10^6 cm^{-3} in the precursor cloud, to 10^8 to 10^{12} cm^{-3} in the disk. In the inner regions, the temperatures are high (> 1000 K) and all the dust is evaporated. As the disk cools, the atoms and

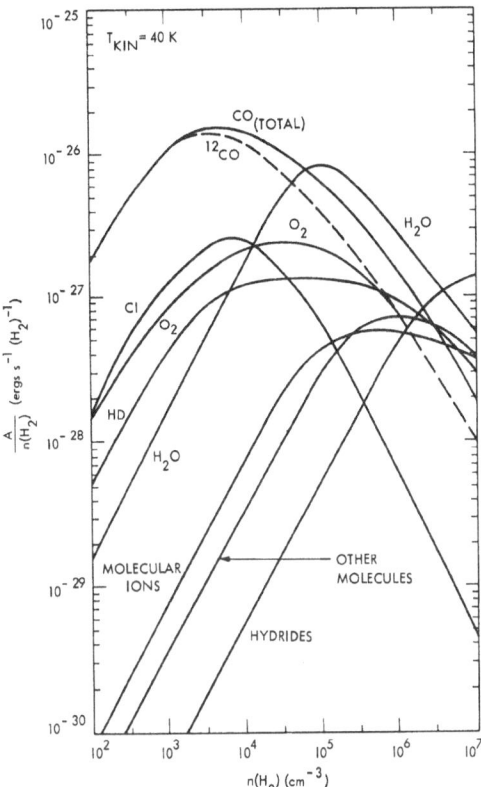

Fig. 5. Cooling rates per H_2 molecule, showing the dominance of H_2O and hydrides at the high densities of star-forming regions

molecules are thought to condense in sequence according to their abundance, local vapor pressure, and ambient temperature, resulting in a very different chemical mix from that of the parent interstellar cloud. SMMM will be able to study the innermost disks surrounding nearby low-mass protostars through the strengths of abundant species and through the velocity distribution as given by emitted spectral line profiles, ev en though these regions are too tiny ($\leq 1''$) to map directly. High-level transitions of molecules such as H_2O, CS, and HCN, which trace high-density regions, will be particularly prominent from the hot, dense inner regions.

A study of the chemical structure of the outer region of such disks is also of interest to cometary studies. Theories of comet formation put their origins anywhere between 30 and 10,000 AU. These objects are, therefore, thought to contain the most primitive solar nebula material, which has undergone the least processing in the inner nebula. However, the inferred abundances for Comet Halley indicate H_2O is by far the most abundant species. The observations of the chemical abundances in circumstellar disks may reveal whether the chemical environment during comet formation is different from that in the dense clouds. If not, this would provide

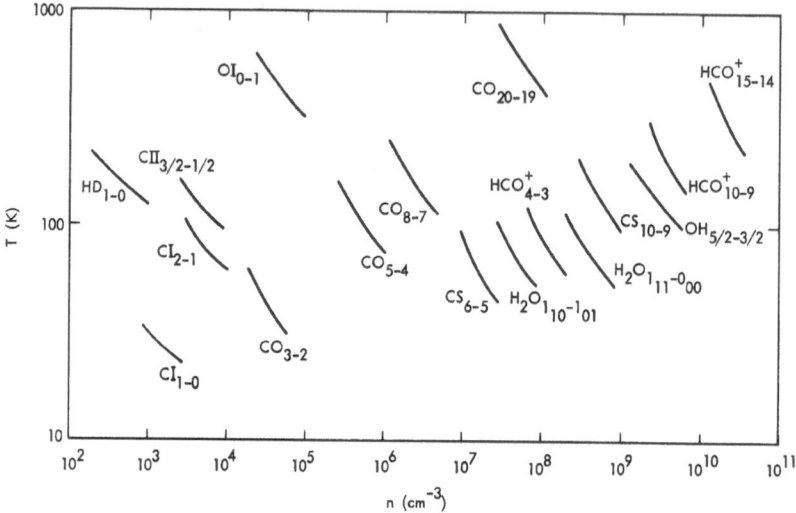

Fig. 6. Indicates the very wide density-temperature range available to SMMM

a strong indication that comets undergo subsequent processing in the inner solar nebula.

The critical molecule for these observations is H_2O, which cannot be observed from the ground. For the strongest lines of H_2O and for an assumed abundance of 10^{-6} relative to H_2, SMMM will provide line profiles for these regions in only minutes. From the structure in the line shapes the mass and temperature distributions may be deduced.

The SMMM continuum camera will be extrememly useful for studies of protoplanetary disks. One of the most exciting results of the IRAS mission was the discovery that approximately one-third of all main sequence stars of types A through K have far-infrared excesses attributable to solid material in stable orbits around the stars. Infrared or visual observations of the brightest of these stars show that this material is located in disks 10 to 500 AU in size, very similar in scale to those seen around younger, but more distant pre-main-sequence stars. If disks are an integral part of the star formation process, then the disks surrounding stars like Beta Pictoris and Alpha Lyrae (Vega) are presumably remnants of the original disks, and their study is of direct relevance to the questions of star and planetary formation.

The sensitivity and spatial resolution of the SMMM camera are well suited to the study of these objects. While SIRTF and ISO have the sensitivity to study these sources, by finding them at distances as large as a kpc or searching for dust spectral features, the spatial resolution of these small cryogenic telescopes is limited, longward of 100 μm, where these sources emit strongly and radiative transfer effects are minimized. In particular, high spatial resolution maps at 100 μm with 6″ resolution (50 to 100 AU at typical distances of 8 to 16 pc) will be critical for understanding the temperature and density structure of the disks around the nearest stars. For example, the presence of gaps or enhancements in the radial density distribution may be signposts to the formation of planets. Further, long wavelength information is critical for determining the size of the emitting grains and the total mass of the disk material. SMMM will have excellent sensitivity for studying the Vega phenomenon.

For example, Vega, which has a disk of average brightness, emits 7 Jy at 100 μm from a region about 25″ (200 AU) in diameter. If this emission were spread out uniformly, the surface brightness would result in 0.4 Jy per 6″ beam of SMMM, detectable in a few seconds of integration. The optical occultation observations of Beta Pictoris reveal emission extending over an arcminute (800 AU). Simple models show that SMMM will be able to follow the infrared emission out at least as far as the presently known optical emission. It will, thus, be possible to make a detailed comparison of the optical and infrared emissions, which is important for determining the distribution and physical properties of dust throughout the disk.

4.1.4. Hydride molecules

The principal rotational transitions of light hydride molecules lie in the submillimeter wavelength region. Many of these transitions coincide with strong atmospheric features, so that observations even from an airplane are not feasible, except in a few specific cases. The hydrides are often the most important molecules both for understanding chemical synthesis and, as already mentioned, for determining physical processes such as cooling. Here we just give a few examples.

One of the most important hydrides that SMMM will detect is H_2D^+. In the now highly favored ion-molecule theory much of the gas-phase chemistry in dense clouds is thought to be driven by reactions of neutrals with the H_3^+ ion. However, since H_3^+ has no strongly allowed rotational transitions, it has so far eluded discovery. Measurements of the H_2D^+ lines would provide direct information on the H_3^+ abundance in dense clouds, which, in turn, would put direct limits on the cosmic-ray ionizing frequencies to which the clouds are exposed. Also, most of the observed remarkable deuterium fractionation in heavy molecules, such as DCN and DCO^+, is thought to derive from reactions with H_2D^+.

The broad frequency coverage of SMMM is especially important for the study of hydrides with identical nuclei, like H_2O, H_2D^+, H_3O^+, and CH_2, for which the fundamental transitions of their semi-independent ortho- and para-species lie at quite different frequencies. Measurements of both species are needed for determining total abundances. Moreover, the relative populations of ortho- and para-species forms may represent a very interesting interplay among processes of formation, destruction, and excitation.

Hydrides whose abundances are of cosmological significance are HD and LiH. The HD $J0 = (1 \rightarrow 0)$ 112-μm line is hard to detect from an airplane due to its low dipole moment, which makes it difficult to distinguish from the dust background and interference from the atmosphere. However, the line will be readily observable with SMMM: Assuming a cosmic deuterium abundance of HD, it can be shown that the intensity relative to the continuum would be 4% at a spectral resolution of 10^4. Observations of HD throughout the Galaxy could provide information on possible gradients with galactic radius. LiH is expected to be significant in the early universe, but primordial LiH may or may not have survived in the interstellar medium. However, in regions exposed to intense cosmic-ray fluxes, such as clouds near the galactic center, the original ^{6}Li/^{7}Li abundance ratio may be significantly enhanced by spallation reactions, which could be probed through observations of the corresponding hydride ratio. LiH should also be detectable in red-giant mass-loss shells where the abundance is higher than usual. Observations across the Galaxy may provide new information on star-formation and nucleosynthesis rates as a function of galactic radius. Such studies can be compared with abundance determinations from visible spectroscopy. Very different information will be obtained this way because submillimeter spectroscopy can probe dense, obscured star-formation regions and galactic nuclei.

In general submillimeter line-survey spectroscopy will tell us the abundances of hydride species over much of the periodic table. It is probably the only way to discover exotic species such as HeH$^+$. However, overall the most important hydride molecule is H_2O, for its role in oxygen chemistry and in cloud cooling.

4.1.5. Dominant Carbon and Oxygen Species

The carbon to oxygen ratio in the interstellar gas is probably the most important quantity in determining the chemical constitution of the molecular and atomic gas in dense regions and the resulting physical properties such as cooling. At the moment it is not even certain whether oxygen or carbon is more abundant.

The principal carbon- and oxygen-bearing species in interstellar clouds are usually predicted to be CO, H_2O, and O_2. Of these, only the column density of CO is reasonably well determined observationally. Little is known about the abundances of H_2O and O_2, especially in cold dark clouds.

KAO observations of atomic C I and C II lines in a limited set of objects have shown that these species are widespread throughout *molecular* clouds, a result which is both unexpected and important in the context of fundamental physical and chemical processes. It is anticipated that future high-angular resolution, ground-based observations of C I from submillimeter telescopes combined with low-angular resolution measurements from SWAS, will provide more insight into the atomic carbon chemistry. Future observations of the C II 158-μm line (and its isotopic variety ^{13}C II) will be possible from space platforms such as ISO and SIRTF, but at a rather coarse angular resolution of only 1.6'. The C II beam size of about 10" obtainable with SMMM is much better matched to the C I beams provided by the ground-based telescopes; especially for the study of star-forming regions, such high angular resolution will be crucial.

Oxygen may be largely in the form of H_2O, the most important and probably the most abundant molecule to be searched for with SMMM. Reliable information about its abundance in interstellar clouds can only be obtained from space-based survey instruments, owing to the severe interference from telluric H_2O features even at 14 km and the need to observe several lines of any species. H_2O is, of course, known to be present in the interstellar medium from its famous maser-transition ($6_{16} - 5_{23}$) in the ortho branch at 22 GHz, but this line involves highly excited states (> 500 K above ground state) and provides little information about the general abundance of water. SWAS will be able to detect a single ortho-H_2O line ($1_{10} - 1_{01}$ at 557 GHz) in a number of warm, dense sources, but that mission will lack the sensitivity and angular resolution to observe H_2O more generally in colder molecular clouds. Since most H_2O-containing regions are likely to be small in extent, the eight times higher angular resolution of SMMM will be crucial. The overall improvement in sensitivity on that one line will be about two orders of magnitude. In fact SMMM would be able to detect the 557 GHz fundamental ortho-line in dark cloud cores, in 1,000 seconds, up to a distance of 2 kpc even if the abundance of H_2O is at the low end of current estimates (10^{-7} relative to H_2). It could easily detect giant cloud cores *throughout* the Galaxy. For the fundamental para-line at 1,100 GHz giant cloud cores could be detected to about 10 kpc, again on the most pessimistic abundance assumption. Moreover, because of its broad wavelength coverage, SMMM will be able to detect many H_2O lines of both ortho- and para-modifications, which will provide accurate abundance determinations. At the same time, these lines will provide interesting information about the H_2O formation and excitation (which is strongly affected by far-infrared pumping) and will, thus, put constraints on the physical conditions in the clouds. Observations of several H_2O lines, simultaneously, within a small beam will also provide insight into the vexing problem of the maser excitation of this molecule in interstellar clouds, and maybe even in external galaxies where "megamasers" are observed.

Molecular oxygen, O_2, is predicted by models to be a major reservoir of oxygen in dense interstellar clouds. However, its observation is also complicated by the presence of numerous terrestrial O_2 features and it has not yet been detected. SWAS will be able to search for the $^{16}O_2$ line at 487 GHz, but may have only marginal detection capability. As for H_2O, SMMM will have the advantage of two orders of magnitude higher sensitivity, higher angular resolution, and multiline capabilities, which will allow detection of many additional O_2 lines. Since the O_2 transitions are only magnetic dipole allowed, the levels are easily populated even in regions of relatively low density, so that, from this point of view, the molecule should be detectable throughout the interstellar medium.

Chemical models predict that a low H_2O abundance is coupled to a high O_2 abundance, or vice versa. This conjecture can be tested directly by SMMM, since it will be able to provide accurate abundances for both species.

4.1.6. Vibration – Rotation Spectroscopy of Small Grains and Heavy Molecules

The spectral range of SMMM is ideally suited for studying the transition region between large, heavy molecules and dust-grains. This poorly understood part of ISM physics and chemistry has recently received considerable attention because of the discovery that small grains consisting of ~ 50 atoms can be transiently heated by the absorption of single photons to radiate strongly throughout the infrared. The possible identification of these small grains with large, planar graphite-like molecules called polycyclic aromatic hydrocarbons suggests that the apparent dearth of interstellar particles between dust-grains 0.1 μm in size and molecules containing about 13 atoms could be due to the difficulty in observing large molecules which rotate too slowly to be found by millimeterwave spectroscopy, rather than a basic

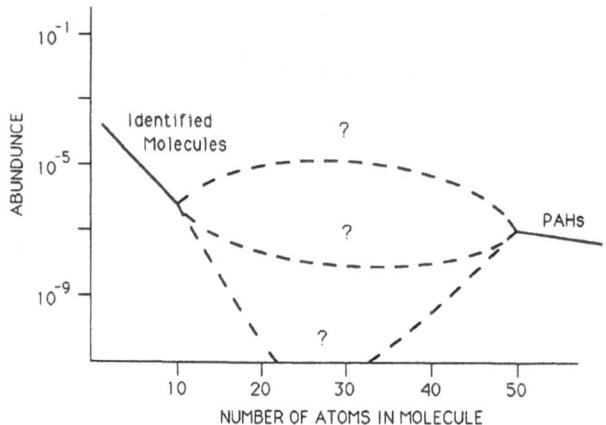

Fig. 7. Showing the gap in observed species in the 13 to 50 atom range.

feature of interstellar chemistry. Figure 7 shows the currently known abundances as a function of the number of atoms in a molecule.

SMMM will address this problem in two ways. First, if there is a continuum of heavy molecules, up to the 50 atom range of the PAHs, *SMMM will be able to observe them by means of their vibration spectra*. For example, the vibration-rotation bands of linear carbon chain molecules such as the well known interstellar cyano-polyacetylenes (HC_3N, $HC_5N...HC_9N$) will fall in the 45–280 μm range. The resolving power of the Fabry-Perot will be well suited to studies of the broad vibration-rotation bands. Second, SMMM will be able to study the PAHs themselves. The PAHs, which were identified by their strong resonances between 3–11 μm, are known from theoretical and laboratory studies to have spectral features out to 100 μm due to low-frequency lattice vibration modes. For example, the lowest frequency vibrational mode of coronene (24 C atoms) is at about 100 μm. Comparable transitions for larger, heavier molecules will appear at longer wavelengths. These features may be quite strong since PAHs are expected to have permanent dipole moments either because they are ionized or because a peripheral hydrogen atom has been replaced by a radical. Finally, it is possible that other forms of dust particles, both with and without mantles, may possess spectral features in the submillimeter, as do many materials in the laboratory at low temperatures. Emission from both polar and non-polar molecules trapped in grain lattice vacancies can lead to observable features in the 70–150 μm range.

4.2. GALAXIES AND COSMOLOGY

The capability of the proposed mission for spectroscopic, high spatial-resolution imaging of molecular and fine-structure transitions and for high angular resolution continuum measurements will make maj or contributions to understanding the structure and evolution of galaxies. Just as for the interstellar medium of the Galaxy, the most important molecular transitions in external galaxies will be those of CO and of hydrides, such as H_2O. Over a broad range of physical conditions, the

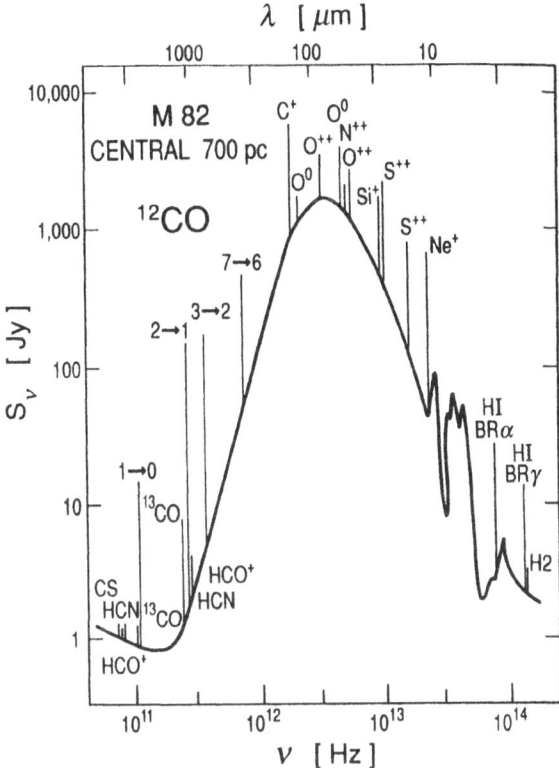

Fig. 8. Observed lines in the nearby starburst galaxy M82.

major cooling transitions for atomic gas are O I and C II. As for galactic clouds dust emission occurs in the submillimeter. These tracers may be observed with virtual immunity to the obscuration which plagues optical studies of galactic structure. Also, for nearby galaxies the opportunity will be present to study the variations of relative atomic abundances by observing the hydrides and comparing these with equivalent variations deduced from optical spectroscopy. Figure 8 shows a wide variety of lines already detected from the nearby galaxy M82. These will, of course, be supplemented by many of the lines from figure 3, particularly H_2O. Because of the freedom from atmospheric absorption these lines can be used, no matter what the red-shift of the galaxy.

4.2.1. Star-formation in nearby galaxies

The positions of atomic, molecular, and ionized gas regions relative to the galactic structures (e.g., the spiral arms and bars) are critical in understanding the large-scale patterns of star-formation in galaxies. Important issues are whether star-forming gas clouds are created within the spiral arms or exist prior to their entry into the arms and how star-formation is initiated in these clouds. In the past, it has been assumed that the molecular clouds are formed within the spiral arms, as low-density atomic gas is compressed by the spiral density-wave. Once formed, the molecular clouds would readily undergo massive star-formation. On the other hand, recent millimeter observations of the molecular gas in nearby external galaxies have clearly shown that molecular clouds exist in the inter-arm regions and, thus, the gas which enters the arms is already in the form of dense molecular clouds. The formation of massive stars selectively in the spiral arm clouds must then be initiated by phenomena unrelated to cloud formation, possibly cloud-cloud collisions, which become more frequent when cloud orbits crowd in the arms. In this picture, the formation of massive stars will be preceded by an enhancement in the number density of clouds and the emission of shock-excited molecular transitions occurring at the interface of colliding clouds. One of the primary signatures of shocked, high density, high temperature regions is H_2O line emission. Mapping selected regions for emission from shocked, high-excitation molecular gas (H_2O and OH) and ionized gas (C II and N II) will permit elucidation of the temporal sequence of cloud evolution and star-formation relative to the spiral arms. To map an entire galaxy of 0.1 square degrees, in a ground-state water line, would take about four hours with SMMM.

4.2.2. IRAS Galaxies.

One of the most spectacular results of the IRAS survey was the discovery of a class of extremely luminous galaxies, typified by Arp 220 and NGC 6240, which emit over 90% of their energy at far-infrared wavelengths. The survey has located about 25,000 galaxies, about half of which were previously uncataloged. Many of these galaxies emit most of their total energy in the submillimeter; many have luminosities in excess of 10^{11} $L_\odot$, and some have luminosities exceeding 10^{12} $L_\odot$. These large luminosities are typical of quasars and BL Lac objects and, thus, the IRAS galaxies form a bridge in luminosity between normal galaxies and the galaxies with active nuclei. Elucidating the origin of the activity in these galaxies will be a vital step towards understanding distant quasars. These galaxies are a significant constituent of the universe. Above 2×10^{11} $L_\odot$, the space density of IRAS galaxies exceeds that of normal galaxies, above $\sim 10^{12}$ $L_\odot$, it exceeds that of the quasars. The IRAS galaxies must be studied at submillimeter wavelengths in order to answer the basic questions relating to their origin and luminosity source. IRAS has vividly demonstrated that interactions can play a dominant role in the evolution of galaxies, possibly leading to the formation of elliptical galaxies, and perhaps even leading to the formation of quasars via the funneling of gas to the nuclei of galaxies and the subsequent evolution of this gas as it falls into a massive black hole. How the interaction of gas-rich galaxies propagates is a problem that must be solved observationally. Because the interactions trigger outbursts of star-formation in dusty regions, the line and continuum mapping capabilities of SMMM are ideal in tracing the locations of gas concentrations, shocked gas, and embedded luminosity sources in galaxies undergoing the full range of interactions from initial contact through to completed merger.

At this time, only a few of the brightest IRAS galaxies have been measured at wavelengths longer than 100 μm. The sensitivity of SMMM (30 mJy, 1 σ in 1 second, in the continuum at 100 μm) is well matched to the faintest galaxies measured by IRAS ($\sim$ 200 mJy at 60 μm), so the bolometers on the proposed SMMM should be able to complete the determination of the full continuum energy distribution of large numbers of galaxies.

4.2.3. Cosmology and the Evolution of High z Galaxies.

Among the central issues in observational cosmology today are the formation of galaxies and the large-scale structure (LSS), and the origins of non-thermal nuclear activity in quasars and other active galaxies. Both issues are inseparable from understanding galaxy evolution at large red-shifts. The former also includes the questions of the initial chemical enrichment, and the nature of the dark matter. Studies at submillimeter wavelengths can provide valuable clues, and perhaps even solutions for some of these important problems.

There is a large gap in our empirical knowledge about the universe between the epoch corresponding to the most distant quasars known, at z $\sim$ 4, and the epoch of decoupling, the cosmic microwave background (CMBR) photosphere, at z $\sim 10^3$. Many cosmologically important processes are likely to have happened in that interval, e.g., the early stages, or perhaps even the bulk of galaxy formation and the initial chemical enrichment, development of the LSS, the appearance of quasars, etc.. Galaxies are now known out to z $\sim$ 3 to 4 and clusters of galaxies out to z $\sim$ 1. The density contrast of the LSS today is about a factor of two, with rich clusters and galaxies representing much higher peaks, whereas the CMBR photosphere is smooth to about one part in 10^5, or even less, on angular scales corresponding to galaxies and the LSS today. This density contrast must have developed, and the accompanying release of binding and other energy must have happened, in the red-shift interval z $\sim$ 1 to 10^3, possibly with the addition of some exotic or highly energetic phenomena. The generic expectation is that the energy released in these epochs would now be observable in the submillimeter region, due to the red-shifting, and possibly also due to reprocessing by dust at large red-shifts.

Hubble Space Telescope will provide much more information on the formative and evolutionary processes of distant galaxies than we have today, but if the initial processes include the generation of dust, the most interesting situations will be hidden to it. *The information from the submillimeter band will then be not just complementary, but critical.*

The extreme sensitivity of SMMM implies the ability to probe galaxies with large red-shifts. As shown in Figure 9, star-burst galaxies of 10^{12} $L_\odot$ and with an energy distribution similar to M82 will be detectable in the continuum (6σ in one hour) up to a red-shift of 2.5. Using the heterodyne receivers, the sensitivity for lines is 2×10^{-18} W/m^2 (6σ in one hour). Assuming that 5×10^{-3} of the total luminosity is in the C II line at 158 μm, such galaxies can be observed in this line up to a red-shift of $z = 2$ for 10^{12} $L_\odot$ galaxies. The high spectral resolution of the heterodyne receivers will provide detailed information on the dynamics of these active objects. There is no other method to obtain this information. The C II line does not enter the heterodyne band until $z \geq 1$, so for lower red-shifts the Fabry-

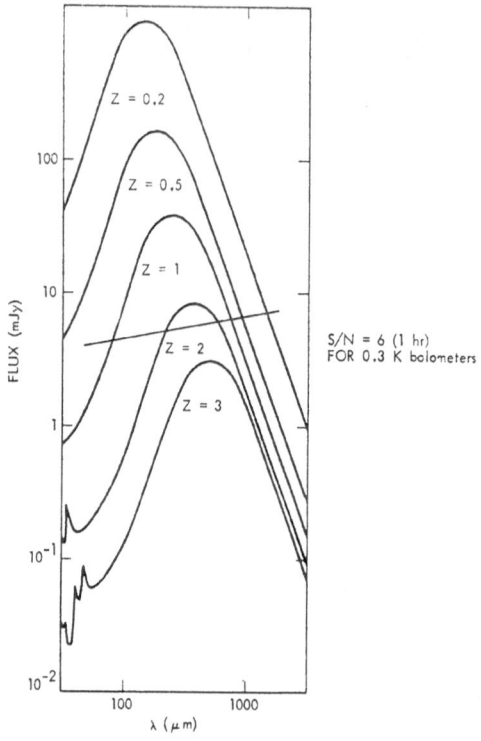

Fig. 9. Plots of the flux from a star-burst galaxy with a spectrum similar to M82, as a function of z. The galaxy is assumed to have an intrinsic luminosity of $10^{12} L_\odot$. Detections will be possible to $z \approx 2.5$, as shown by the line representing the SMMM sensitivity limit for a signal-to-noise ratio of 6 after one hour of integration with 0.3 K bolometers. The line spectra of Figure 8 scale with the continuum shown here.

Perot is used. Again, due to the very high sensitivity of the bolometers, detections of C II will be made in only minutes.

A principal use of the very sensitive continuum measurements will be to search for evolutionary effects in galaxies. The present, best luminosity function of the IRAS galaxies in the local universe is based upon the IRAS bright galaxy sample with $f_\nu(60 \ \mu m) \geq 5$ Jy. The mean red-shift of the galaxies in the bright galaxy sample is $z \sim 0.006$ and red-shifts extend to $z = 0.08$. Although the ultra-deep surveys made with the IRAS satellite go as faint as 100 mJy at 100 μm, these are severely confusion limited, and the real potential of SMMM in studying high red-shift galaxies will be realized only at larger red-shifts than those found by IRAS. With the bolometer array, a survey for very high red-shift galaxies is feasible. In 300 hours of observing time, approximately one square degree of sky could be searched to a sensitivity of about 10 mJy (1 $\sigma = 3$ mJy) in three wavelength bands. If there were no evolution, about 500 galaxies should be detected to this limit within this area, and the sample would have a mean red-shift of $z \sim 0.15$ and would contain some objects with red-shifts ~ 2. The areas chosen for such a survey would have to be carefully selected to avoid galactic "infrared cirrus" and known galaxy clusters. Because of confusion due to multiple sources in the field, SMMM will have very superior sensitivity as compared to the cooled, but smaller telescopes ISO and SIRTF, for wavelengths greater than 200 μm.

An important capability, provided by the long wavelength continuum camera, will be the study of perturbations to the microwave background. Although we are currently waiting for COBE results to define the future areas for study, it is clear that SMMM will be able to make definitive images of any submillimeter spectral distortions such as that detected at 600 μm in the Nagoya–Berkeley experiment. A second distortion which could be measured is that caused by the Sunyaev–Zel'dovich effect. By making measurements at several different wavelengths SMMM would be able to separate the effects of temperature, radial velocity with respect to the CBR and opacity of the cluster gas. The measurement of the peculiar velocities of galaxy clusters could be made with a precision of about 150 km/sec, independent of the absolute magnitude of the red-shift itself.

THE SUBMILLIMETER WAVE ASTRONOMY SATELLITE

G.J. MELNICK

Harvard-Smithsonian Center for Astrophysics

Abstract. The Submillimeter Wave Astronomy Satellite (SWAS) is a NASA Small-Explorer Class experiment whose objective is to study both the chemical composition and the thermal balance in dense ($N_{H_2} > 10^3$ cm^{-3}) molecular clouds and, by observing many clouds throughout our galaxy, relate these conditions to the processes of star formation. To conduct this study SWAS will be capable of carrying out both pointed and scanning observations simultaneously in the lines of four important species: (1) the H_2O ($1_{10}-1_{01}$) 556.963 GHz ground-state ortho transition, (2) the O_2 (3,3–1,2) 487.249 GHz transition, (3) the CI ($^3P_1$ - $^3P_0$) 492.162 GHz ground-state fine structure transition, and (4) the ^{13}CO (J = 5–4) 550.926 GHz rotational transition. These atoms and molecules are predicted to be among the most abundant within molecular clouds and, because they possess low-lying transitions with energy differences ($\Delta E/k$) between 15 and 30K (temperatures typical of many molecular clouds), these species are believed to be dominant coolants of the gas as it collapses to form stars and planets. A large-scale survey in these lines is virtually impossible from any platform within the atmosphere due to telluric absorption.

SWAS will consist of a 55-cm diameter off-axis Cassegrain antenna, two Schottky Barrier diode heterodyne radiometers, and a single broadband (1.4 GHz) acousto-optical spectrometer (AOS) which will give SWAS a velocity resolution of approximately 0.6 km s^{-1}. The spacecraft, supplied by NASA Goddard Space Flight Center, will be a three-axis all-sky pointer utilizing reaction wheels. Use of a star tracker will permit a pointing accuracy of approximately 45 arcseconds. The satellite has a weight of 200 kg, carries no expendables, and, with a projected orbital altitude of 550–600 km, should function for two or more years. Launch is currently scheduled for 1994. SWAS is a collaborative effort between the Smithsonian Astrophysical Observatory, the University of Massachusetts at Amherst, the National Air and Space Museum, NASA Ames Research Center, the University of California at Berkeley, and the University of Cologne (FRG).

Y. Kondo (ed.), Observatories in Earth Orbit and Beyond, 251
© 1990 *Kluwer Academic Publishers.*

(D) RADIO MISSIONS

INTERNATIONAL VLBI SATELLITE (IVS)

R. T. SCHILIZZI

Netherlands Foundation for Research in Astronomy,
Postbus 2, 7990 AA Dwingeloo, The Netherlands

Abstract. IVS is under study in ESA as a second generation space VLBI observatory. The mission concept calls for a 25 m diameter radio telescope in space funded by the principal space agencies. Orbiting the Earth and observing in concert with the established ground-based VLBI arrays in Europe, USA, USSR and Australia, IVS will provide high quality images of galactic and extragalactic radio sources at wavelengths spanning the radio band from decimetres to millimetres with resolution as high as 10 micro arcseconds and sensitivity equal to those of ground-based images. New features of IVS compared to the first generation missions are: a more than order of magnitude increase in sensitivity; an order of magnitude increase in maximum angular resolution; extension of the wavelength range to the millimetre band; and the capability to operate as a stand-alone radio telescope enabling it to explore new frontiers in spectral line and microwave background research, in particular the distribution of galactic molecular oxygen and Compton scattering of the microwave background by foreground cluster gas.

1. Introduction

Radio interferometry plays a central role in high resolution imaging astronomy; instruments like the VLA and MERLIN have, for the last decade, provided images of a wide variety of cosmic objects from stars to quasars with the same detail as can be expected from the Hubble Space Telescope. But it is the Very-Long-Baseline Interferometry (VLBI) networks of telescopes spanning the globe which mark the greatest advance in angular resolution. Current global arrays with up to 18 elements and effective diameters of some 8000 km reach sub-milliarcsecond angular resolution and represent the largest telescopes which have ever probed the depths of the universe. Accompanying this advance has been an increase in the quality of the images, as characterised by their dynamic range, the ratio of the brightest to the faintest reliable features. This increase has come as a result of better aperture plane (or *uv* plane) coverage and improved image construction algorithms.

VLBI observations at high resolution have produced many important discoveries, including apparent superluminal velocities in quasars and radio galaxies, and highly collimated plasma jets in radio galaxies on scales of less than one parsec which are the bases of jets extending to several million parsecs. As the body of discoveries from VLBI has grown, it has become clear that in nearly every compact source observed at cm and mm wavelengths, there remains spatial structure which is unresolved with the best angular resolution achievable with antennas on Earth. To explore these structures in detail, substantially higher resolution, high quality images are essential. This can only be achieved through even longer baselines, which requires one or more interferometer elements in orbit observing in conjunction with Earth-based arrays.

Y. Kondo (ed.), Observatories in Earth Orbit and Beyond, 255–262.

TABLE I

IVS Design Goals

Telescope

diameter: 25 m

operating frequencies: 4.5–90 (220) GHz

symmetric Cassegrain design, focal ratio: 0.4

antenna efficiency (25 m): 0.5 at 43 GHz

surface error (25 m): ≤ 0.3 mm

Science instrument

set of super-heterodyne, dual-polarisation receivers, feeds, and coolers for:

frequency range (GHz)	amplifier	physical temp. (K)
4.5–8.5	HEMT	20
15–23	HEMT	20
42–60	HEMT	20
86–90	HEMT	20
(218–220	SIS	5)

stable total-power detectors

spectrometers (AOS or digital) for 42–60 GHz

2-way Ku-band phase transfer link

Ku-band science data downlink at 512 Mbits/s **Orbits**

It is proposed that IVS occupy three orbits during its lifetime:

Orbit 1: high inclination ($\sim 60°$), apogee altitude ~ 20000 km, perigee altitude ~ 5000 km. This will allow images of unprecedented quality to be made with angular resolution 3 times greater than with Earth baselines alone (eg. 50 μas at 43 GHz).

Orbit 2: the apogee altitude is doubled to 40000 km to increase the angular resolution by a further factor of two (eg 25 μas at 43 GHz) while still providing good quality images. This orbit is the highest for which good quality imaging is possible with only one element of the VLBI array in space.

Orbit 3: the apogee altitude of IVS will be raised to at least 150000 km and perhaps considerably higher (depending on the constraints placed on the spacecraft design) in order to search for ultra-compact radio sources. The angular resolution in this orbit is, for example, $<3\mu$as at 90 GHz. Simple structural information (angular size, orientation, and strength) could be determined with such an orbit.

Lifetime

≥ 3 years operations, ≥ 6 years consumables

A successful proof-of-concept of orbiting VLBI was carried out by a JPL-led international team in the period 1986–1988 using the 4.9 m antenna on the US Tracking and Data Relay Satellite (TDRS) with antennas in Australia and Japan (see R. P. Linfield, these Proceedings). This put the technical feasibility of orbiting VLBI beyond doubt, and demonstated clearly that many radio sources possess structure on interferometer baselines as long as 2 Earth diameters at frequencies of 2.3 and 15.0 GHz.

Two first-generation orbiting VLBI missions have been approved for launch in the 1993–5 period: RADIOASTRON (USSR) and VSOP (Japan), both carrying 10 m diameter antennas. RADIOASTRON (N. S. Kardashev, these Proceedings) will be launched into a highly eccentric orbit with an apogee altitude of at least 75000 km, VSOP (H. Hirabayashi, these Proceedings) into a lower orbit with apogee altitude 20000 km.

A new mission, International VLBI Satellite (IVS), is currently undergoing an Assessment Study in ESA supported by the Soviet Academy of Sciences and NASA/JPL. It is a second-generation mission carrying a 25 m diameter antenna and capable of more than an order of magnitude improvement in sensitivity and angular resolution over the two earlier missions (and more than 200 in sensitivity compared to TDRSS). It is well-matched to the ground VLBI arrays in both sensitivity and wavelength coverage.

2. Mission Description

IVS mission is intended to be a major radio telescope orbiting the Earth. In VLBI mode it will observe together with the ground-based VLBI arrays in Europe, USA, USSR, and Australia (see Figure 1). IVS will also operate as a stand-alone antenna enabling it to explore new frontiers in spectral line and microwave background research. Table I summarises the design goals.

Compared to its precursor VLBI missions, RADIOASTRON and VSOP, IVS will have:

– 20 times greater sensitivity (larger collecting area reflector, wider data downlink bandwidth, decreased receiver temperatures, and ability to increase the integration time by phase referencing),
– 10 times greater maximum angular resolution,
– higher observing frequencies and,
– accurate orbit determination, allowing phase referencing and astrometric observations to be done.

The mission will provide a space-based VLBI element equivalent in sensitivity and frequency range to an element in one of the modern generation of ground-based synthesis arrays (eg the VLBA, the US dedicated VLBI array). At its most sensitive, IVS will reliably detect a 1 mJy radio source in 5^m integration time, and 80 μJy if phase referencing allows a 12^h integration time.

IVS will also be a major advance as a single-dish spectroscopy instrument for O_2 in the 48-60 GHz band. Its sensitivity for extended sources of O_2 will exceed

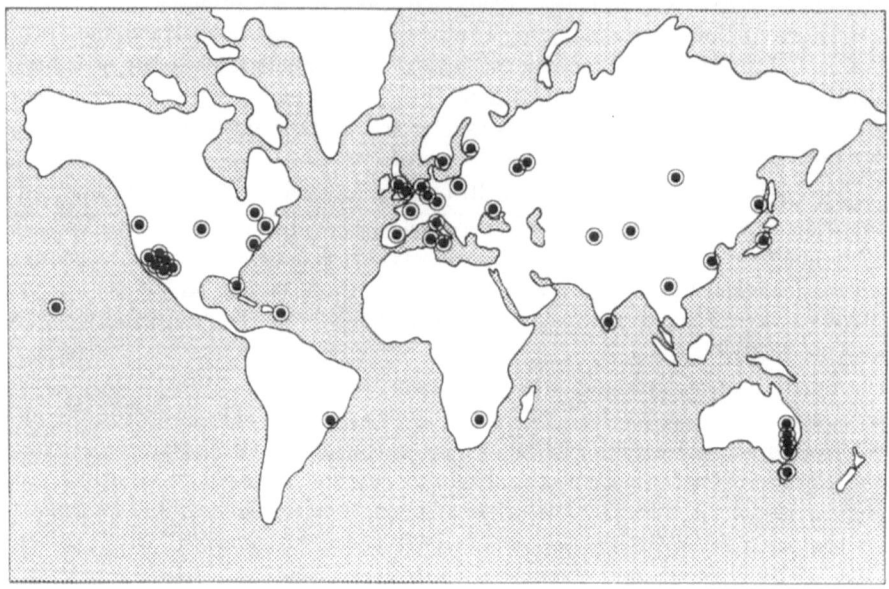

Fig. 1. The global net of VLBI antennas. Forty-four major VLBI stations expected to be operational by 1995. The radio telescopes are located in Australia, Brazil, Canada, China, England, Finland, France, W-Germany, Italy, Japan, the Netherlands, Poland, South Africa, Spain, Sweden, USA and USSR.

that of FIRST, LDR and SWAS (observing at 425 GHz or 487 GHz) by at least an order of magnitude. For point source detection of O_2, IVS will be comparable in sensitivity to the 8 m FIRST.

3. IVS Science

The angular resolution and greatly increased sensitivity of IVS combined with its imaging capability will open up exciting new opportunities in cosmology as well as galactic and extragalactic astronomy, including:

– the mechanisms of relativistic energy generation, transport, collimation, and dissipation in the nuclei of radiogalaxies and quasars (see Figure 2), an independent determination of the cosmic distance scale using H_2O masers (see Figure 3), and the radio emission mechanisms in stellar objects in the galaxy. For most of these studies, an imaging capability is essential.
– On these small scales, changes in radio structure are to be expected in response to surges in the energy release processes, proper motion of the emitting regions at relativistic speeds along jets, and evolution of the emitting regions themselves. Tracing these changes in both galactic and extragalactic sources through repeated multi-frequency imaging in total intensity and polarisation will be one of the pri-

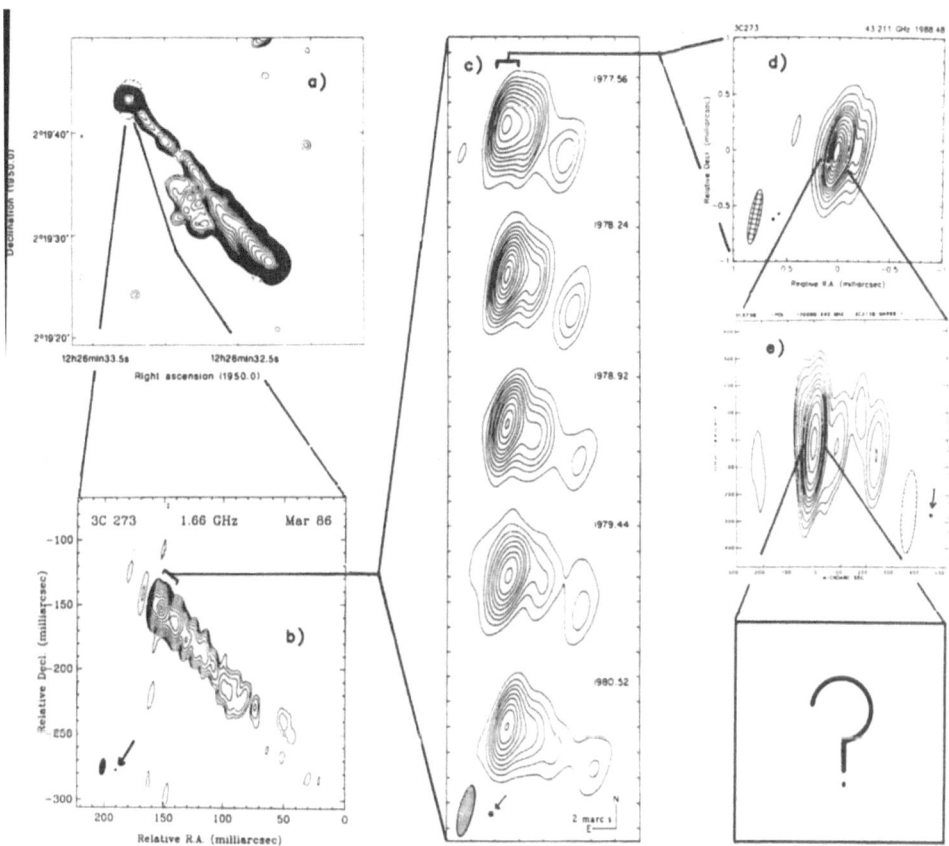

Fig. 2. 3C273, the "classical" quasar. 3C273, imaged at different angular resolutions from 408 MHz to 100 GHz. (a) MERLIN at 408 MHz showing the core (to the NE) and the 21 arcsec long jet which coincides with the optical jet. (b) VLBI at 1.66 GHz on a scale 250 times smaller than panel 'a'. (c) a series of VLBI images at 5 GHz on a scale a further factor of 4 smaller than panel 'b'. The changes in angular position of some of the features over the course of several years imply linear velocities in the plane of the sky well in excess of the speed of light. This "superluminal" motion is usually interpreted in terms of relativistic bulk motion close to the line of sight towards the observer. (d) the first image obtained of 3C273 at 43 GHz on a scale 10 times smaller than panel 'c'. (e) the first image at 100 GHz showing that the "core" at 43 GHz is in fact composed of a number of features. In panels b to e, the synthesised beam of the ground VLBI network used to produce the image is shown together with the IVS beam (indicated by an arrow) at the same frequency. **References**: (a) Davis et al. 1985 Nature, 318,343; (b) Unwin and Davis 1988 Proc. IAU 129, 27; (c) Pearson et al 1981 Nature 290, 365,; (d) Krichbaum et al, submitted; (e) Baath et al, submitted.

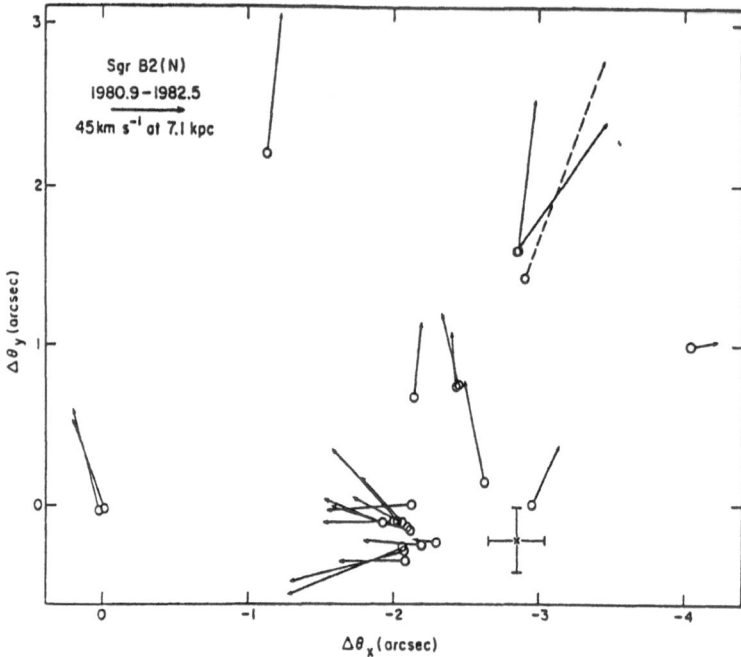

Fig. 3. Sgr B2(N) 1980.9-1982.5. This diagram shows the proper motions of the H₂0 maser spots. The maser spots appear to be expanding outward from the position indicated by the cross. A least squares fit of a uniformly expanding spherical source resulted in an estimate of the distance to the source (within 0.3 kpc to the Galactic Center) of 7.1±1.5 kpc. (Figure from Reid et al.: 1988, Astrophys. J. 330, 809). Such maser complexes will be observable in nearby galaxies. Measurement of their internal motions and/or their galactic rotational motion will allow an independent determination of the cosmic distance scale.

mary tasks of IVS. This should provide unique information on the structure, the kinematics, the magnetic field, and the thermal electron distribution in a wide variety of astrophysically important objects on linear scales of great interest. Many of the changes have timescales of several years, and therefore an operational lifetime of at least three years (with consumables for six or more years) is essential in order for the scientific potential of IVS is to be fully realised.

– One of the most important goals for the 48–60 GHz receiver on IVS is the search for interstellar molecular O_2 which is impossible to carry out from the ground due to atmospheric shielding. Oxygen is more abundant in the universe than all the other heavy elements combined. The evolution of initially diffuse cold interstellar gas to the point of star formation is controlled by the conversion of atomic oxygen into molecular compounds. IVS will be used to measure the distribution and excitation of molecular oxygen (O_2) in interstellar clouds. The transitions which occur near 60 GHz are ideal for this investigation because they sample every N quantum level in the rotational ladder of O_2 (see Figure 4), thus yielding unambiguous information

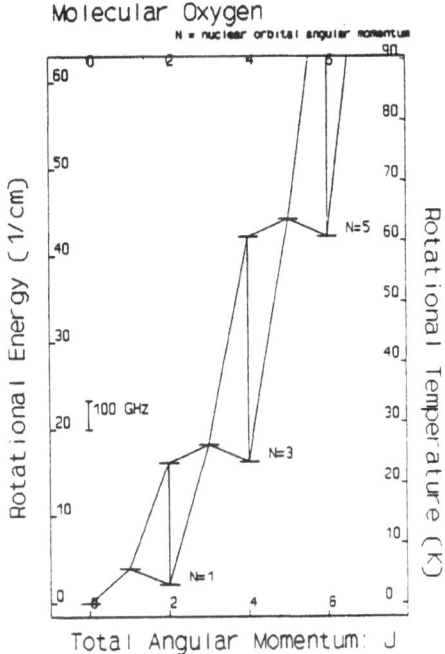

Fig. 4. Rotational energy level diagram of O_2.

on the excitation of the molecule. The lowest transitions, like transitions of CO, are expected to be detectable in most of the molecular clouds of the Galaxy.
– Compton scattering of the Cosmic Background Radiation (CBR) by hot gas associated with galaxy clusters distorts the spectrum of the CBR. This so-called Sunyaev-Zel'dovich effect has been detected at 20 GHz towards three clusters from the ground. However the sensitivity of ground-based observations of this effect is limited primarily by atmospheric noise. IVS with stable sensitive receivers will be a powerful instrument for measurement of the microwave decrement, allowing observations of many clusters in a modest amount of time. Combined with X-ray data, such observations provide an independent measurement of H_0, and the radial velocities of clusters with respect to the CBR.

4. IVS Payload

IVS will be a radio telescope in space that with many of the features and most of the flexibility of a ground-based radio observatory. It will be capable of making almost any observation that is possible from the ground, but its position in space will be exploited to make observations that are not possible from the surface of the Earth. Since IVS functions like a typical ground-based radio observatory, and also has many similarities with a communications satellite, the science payload is conceptually straightforward and the technology is largely in hand.

The payload is to consist of a 25 m diameter microwave antenna, a dual polarisation 4(5)-band receiver at the feed point consisting of low-noise amplifiers and phase stable frequency down converters to produce IF signals, data digitisation equipment for the VLBI observations and the equipment associated with the 48–60 GHz single-dish operation. Stable total power detectors are necessary for observations of the microwave decrement at $\nu \geq 22$ GHz.

The main element of the IVS payload is the 25 m diameter reflector which should be capable of operating fully illuminated with good efficiency at 43 GHz. The Dornier inclined panel concept developed for FIRST is being investigated for IVS. This a solid-panel flower petal arrangement, deployable in space with, for IVS, a 5 m central section and 10 m long petals. The performance at shorter mm wavelengths will also be investigated. Such an antenna will require a large volume fairing such is provided by the payload container attached to the Energia booster (see Section 5).

The receivers are super-heterodynes that select, translate in frequency, digitise, format and relay the astronomical signals to the ground. The signals received on the ground can either be processed immediately in a digital spectrometer, for stand-alone measurements, or recorded on magnetic tape for subsequent processing at a VLBI processing centre. A possibility for stand-alone observations would be to have an on-board spectrometer and memory so that observations could be made with bandwidth wider than that of the link, or when out of view of a ground station.

In its VLBI observational mode, IVS will have the capacity to operate simultaneously any four receivers and relay the received signals via a digital link directly to the telemetry stations on the ground. Typically, simultaneous observations would be conducted in two frequency bands in both hands of circular polarisation.

5. International Collaboration on IVS

The model assumed for the Assessment study is as follows:

– ESA payload module and critical sub-systems including antenna, attitude and orbit control, and link equipment – ESA or NASA DSN data downlink reception and phase link – ESA Science Operations Centre – Soviet service module (standard space platform) – Soviet launch (Energia) – Soviet flight operations control – National agencies for the scientific instrument (e.g. receivers) – National agencies for VLBI data reduction facilities

VSOP, A SPACE VLBI PROGRAMME

H. HIRABAYASHI

The Institute of Space and Astronautical Science
3-1-1 Yoshinodai, Sagamihara, Kanagawa 299, Japan

Abstract. VSOP, VLBI Space Observatory Programme, is an approved space VLBI programme of ISAS for the study of very compact radio sources with the synthesized aperture of 30,000 km diameter, by connecting an orbiting radio observatory with ground radiotelescopes. The VSOP satellite carrying 10 m antenna with 1.6, 5, and 22 GHz band receivers will be launched in early 1995 by M-V rocket of ISAS into an eccentric orbit with 20,000 km in apogee height. The tracking network will be formed for the satellite orbit determination, phase transfer and IF down-link. VSOP aims imaging capability with best resolution of 0.0001 arc second in 22 GHz band. Imaging of active galactic nuclei, star forming regions and stellar objects, and radioastrometry are main scientific targets.

1. Introduction

VLBI observations provide very high angular resolution of the order of 1 milliarcsecond. But serious limits of this outstanding technique is still in angular resolution and picture quality, because the element antennas are fixed on the Earth. Obvious solution to this is having at least one element antenna in space. That this is technically feasible and scientifically valuable has been successfully demonstrated by the TDRSS Orbiting VLBI experiments in 1986 through 1988 (Levy et al., 1986), (Linfield et al., 1990) and (Linfield et al., 1989). In this experiment, the local oscillator onboard the satellite was phase-locked to the ground based hydrogen maser frequency standard and IF signal was down linked to the ground for recording.

The Institute of Space and Astronautical Science (ISAS) has a funded space VLBI programme called VSOP, A VLBI Space Observatory Programme. First, the programme assumed M-3SII rocket for the launching vehicle and the satellite were more modest (Hirabayashi, 1987), but the approval for the new rocket in ISAS resulted in a satellite with reasonable size and sensitivity. The VSOP satellite launch will be in early 1995 and participation from institutes worldwide is assumed. The launch of Soviet Radioastron (Kardashev and Slysh,1987) for space VLBI is reported to occur mid 1990.

The VSOP international symposium was held in ISAS in December 1989, and the proceedings of the symposium is in press with the title "Frontiers of VLBI". So the readers are suggested to consult the book for more detail.

2. General Concepts of VSOP and Science Targets

VSOP will use an orbiting radio observatory with a 10 m antenna in an eccentric orbit to be coherently connected to the radiotelescopes widely spread on the earth. The phase reference signal for local oscillators onboard will be uplinked from the hydrogen masers on the ground and return signal will be compared for two way

Y. Kondo (ed.), Observatories in Earth Orbit and Beyond, 263–269.

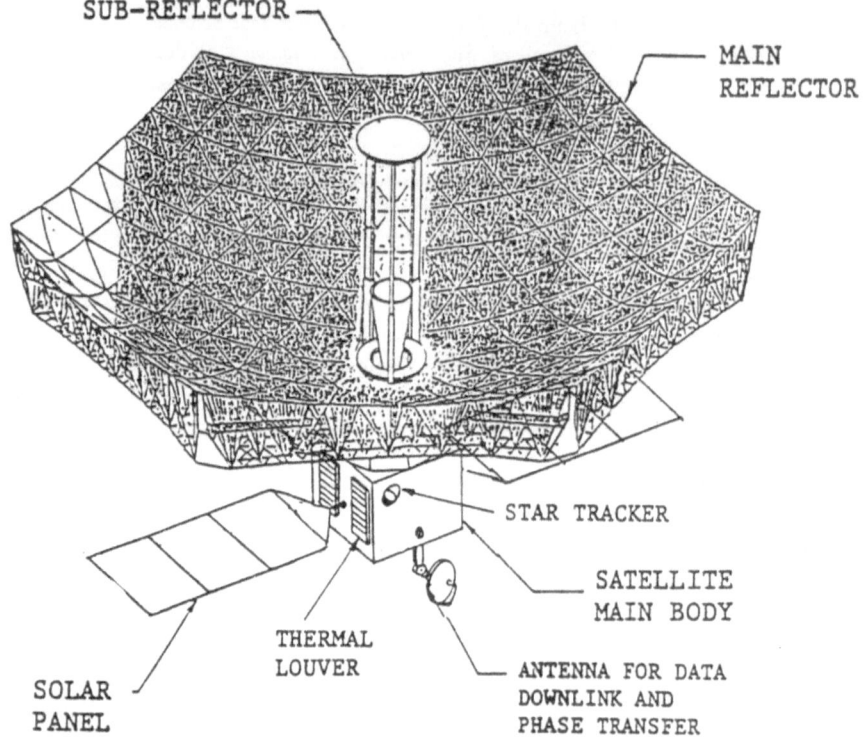

Fig. 1. Present sketch of the Muses-B spacecraft for VSOP

phase monitor and for Doppler tracking. While the IF signal with 64 MHz maximum bandwidth will be downlinked digitally to be recorded to magnetic tapes for later correlation and image processing. The basic scheme of this aperture synthesis is common with the ground based VLBI. The difference is only in the presence of telemetry link for the satellite.

The maximum aperture of 30,000 km will be synthesized to give the best angular resolution of 0.1 milli-arcsecond at 22 GHz. Table 1 is for spatial resolution and scientific rational.

The most important targets of VSOP are Active Galactic Nuclei (AGN). Already by the TDRSS experiment, the sources exceeding the Compton limit of 10E12 K were detected proving the relativistic beaming hypothesis. The projected baseline length of TDRSS experiment and VSOP are the same, but due to the very limited coverage of the UV-coverage, mapping was left undone. The resolution of VSOP is comparable to ground mm-VLBI. VSOP and mm-VLBI will make the accreting, beaming, and radiation mechanisms clear.

Compact stellar objects are accessible by VSOP. Maser sources either in the star forming stage or in late type stage can be imaged and monitored by OH and H2O maser lines. By proper motion measurements of maser spots, we can determine the distance to the region better than by ground VLBI. Pulsars are another group of

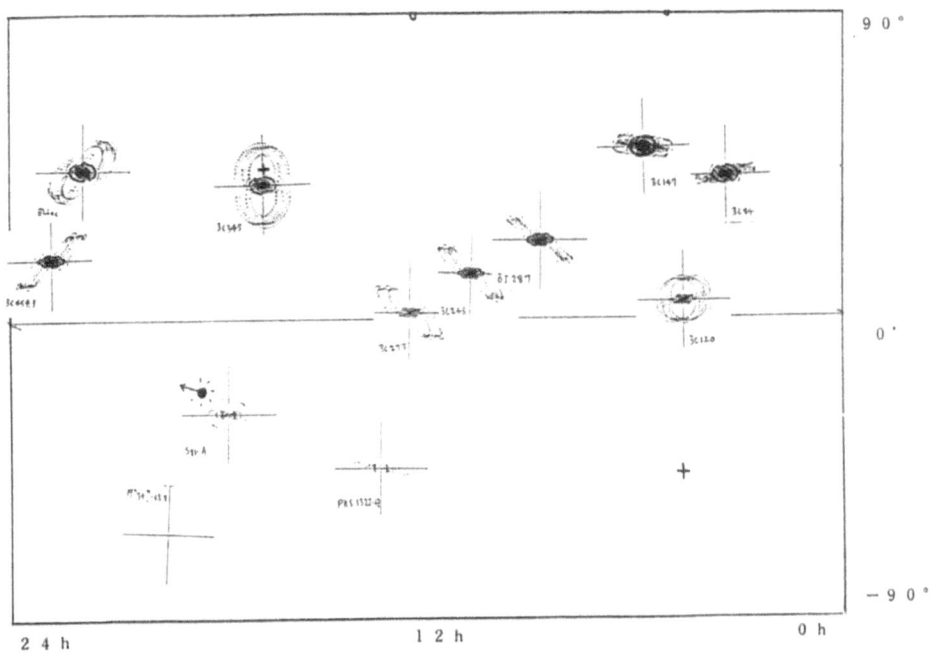

Fig. 2. U-V coverages of typical AGNs in a whole sky map at a fixed epoch

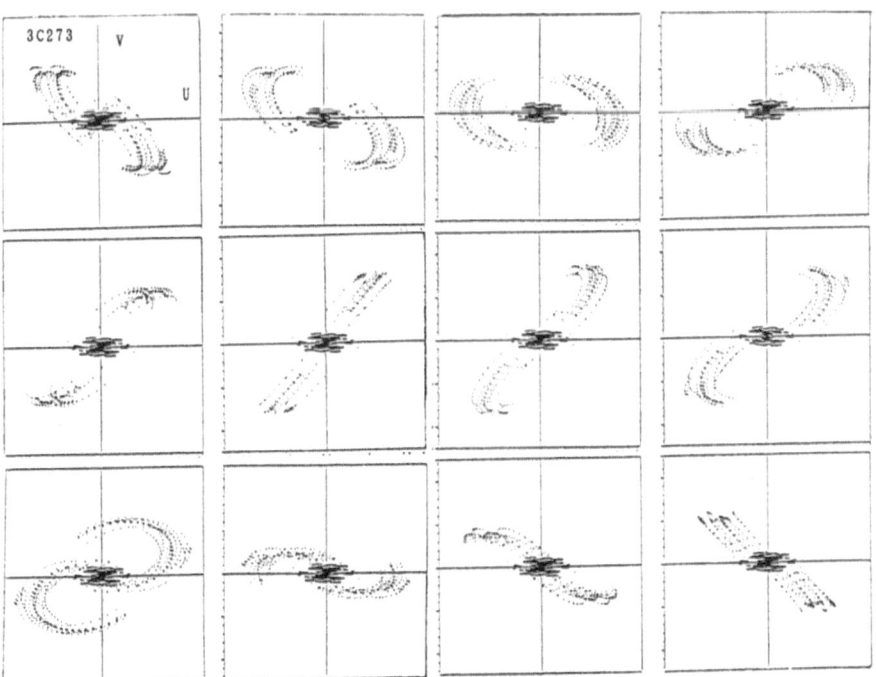

Fig. 3. Time variation of UV-coverage of 3C273 every 2 months.

TABLE I
Observing frequency, maximum angular resolution, and scientific rationale.

Frequency (Wavelength)	Minimum Synthesized Beamwidth	Scientific Objective
22GHz (1.3cm)	0.13 mas*	H$_2$O maser continuum
5GHz (6cm)	0.6 mas	contnuum
1.7GHz (18cm)	1.8 mas	OH maser continuum

- Mapping Extragalactic Radio Sources
- Star Forming Regions
- High Precision Radioastromet
- Distance Scale determination
- Galactic Rotation
- Low Frequency Variables
- Pulsars

*milli arc seconds

stellar sources to be studied.

Radioastrometry can be done by the extension of the baseline lenght but not with the better knowledge of the satellite orbit. Socalled delta-VLBI can be done in some cases.

The initial astronomical requirements are discussed by Hirabayashi (1991a).

3. Muses-B Satellite for VSOP

The fabrication and the launch of the VSOP satellite, currently named as Muses-B, started in 1989. New M-V rocket of ISAS will be used for launching. The M-V rocket will consist of three stage solid propellant boosters and will be able to carry about 2 tons payload to the low earth orbit (LEO). The development of the M-V rocket started from 1990.

Total payload mass will be about 800 kg and the total electric power in the beginning of life (BOL) will be about 550 W. The power will be generated by two solar paddles extended in opposite directions from the satellite main body. Figure 1 show the spacecraft of the present design.

The 10m radio astronomy antenna fixed to the satellite main body will be deployed after launch. The antenna structure is a so-called tension truss design with six extension booms as main back structure, and the reflecting surface is made of wire mesh. The antenna is center fed Cassegrain type, and has 1.6, 5 and 22 GHz feeds. The sub-reflector also will be extended to the nominal position after deployment. 1.6 and 5 GHz frontends will be passively cooled, and 22 GHz frontend will be actively cooled down to about 80K by Stirling cycle refrigerator.

There will be two independent LO and IF channels for radioastronomy signal.

The local oscillators for the mixers will be generated from synthesizers phase-locked to the reference signal sent from the ground telemetry network. The two IF channels are followed by video-converters and A/D converters independently. Again local synthesizers are phase locked to the reference signal from the ground. A/D converters are 1 or 2 bit mode and with 32 MHz or 16 MHz bandwidth. A formatter unit accepts the digital data from the two A/D converters to combine a bit stream with maximum rate of 128 Mbps. This will be formatted with synchronizing, timing and auxiliary bits. The bit streams will be QPSK modulated and down-linked through a power amplifier and a 40 cm high gain antenna in Ku-band. As the wideband down link is digitally coded, the phase fluctuation from atmosphere in downlink path is negligible.

The reference signal from hydrogen masers on the ground will be up-linked in Ku-band for the frequency standards on-board. The Ku-band on-board antenna receives the signal, and the signal will be transmited back to the ground to monitor the phase variation and for precise Doppler tracking. The 40 cm antenna will be equiped on a gimbal at the bottom of the satellite bottom.

The satellite will be 3-axis stabilized, and the pointing accuracy of the radioastronomy antenna is 0.01 degree, and this will be managed by the Reaction Control System (RCS) with star sensors, inertial-gyro and torque motor. The satellite will have thrusters for perigee-up, and angular momentum compensation, etc.

Due to the oblateness of the earth, the argument of perigee and the ascending node of the orbit change. These are about one revolution per 2 years. This means that good direction of the sky for observing changes with time. Observing schedule will be made with this effect in mind.

More details of VSOP satellite and on-board radioastronomy instruments are treated by Hirosawa (1991) and Hirabayashi (1991b).

4. Ground Supports for VSOP and Observing

For VSOP, two kinds of ground networks are necessary; one for the satellite; commanding and housekeeping, orbit determination, and for phase transfer and wideband data link. The other is array of co-observing radio telescopes.

ISAS will provide a tracking station by using 20.9 m antenna at Kagoshima Space Center (KSC). This will be used for the commanding and housekeeping in S-band, and for telemetry in Ku-band. 64 m antenna at Usuda (ISAS), 45 m radiotelescope at Nobeyama (NRO, Nobeyama Radio Observatory), and 34 m antenna at Kashima (CRL, Communication Research Laboratory) are the candidates for co-observing stations in Japan.

Deep Space Network(DSN) of NASA/JPL will support VSOP by building telemetry network with 10 m dedicated antennas at each DSN site. The link power budget for spacecraft is so designed that 10 m antenna can handle the satellite. The Green Bank telemetry station of NRAO also is planned and this, too, helps the VSOP telemetry coverage.

Arraying with VLBI network telescopes like VLBA (Very Long Baseline Interferometer Array, US), EVN (European VLBI Network, Europe), and AT (The Australia Telescope, Australia) and other non-network telescopes are assumed. VLBA is a longer baseline array and will be suitable for high fidelity maps for stronger sources. AT will be essential for southern radio sources with VSOP because VSOP will be the first all-sky mapping instrument.

The tms sensitivities for VSOP-ground interferometer with 64 MHz bandwith, 100 s integration time are about 10 m Jy for 64-70 m class ground antenna and 25 m class ground antenna. Interference fringe several times larger than these level cauld be detected.

For correlating the observed tapes and for sophisticated aperture synthesis imaging, A 10-20 station correlator of FX architecture, with a super computer for imaging capability has been proposed and design study is in progress. We hope the correlator will be partially operational in 1993.

Formation of some kind of worldwide consortium for VSOP is needed in many repects, and discussion is in progress. Inter Agency Consultative Group (IACG) of Space Science Panel-1 meting has been dealing the operational management inviting the representatives from ground telescope arrays.

5. Possible observing scenario

The image quality depends on the UV-coverage, the optical transfer function, and it is a function of source direction and the epoch. The initial orbit and the ground facilities are primary important parameters. The orbit plane changes with $dW/dt = -177$ degrees/year and $dw/dt = 177$ degrees/year where W is right ascension of ascending node and w is argument of perigee. There are two dominant constraint factors in the satellite which limit the UV-coverage. One is that the satellite cannot observe the direction of the Sun with 70 degrees cone angle, because solar paddles cannot be well illuminated due to the shadow of the 10m antenna. The other is the tracking coverage of Ku-band downlink antenna to point telemetry network.

For observing, above constraints must be carefully taken into account. Figure 2 shows an example of UV-coverages of typical AGNs on all sky map at an epoch. Kagoshima and 3 DSN stations are assumed for telemetry network and VLBA.

An example of time variation of UV-coverage is shown in Figure 3 for 3C273. It is seen that the UV-coverage significantly changes with time. So, if we monitor such source, care should be taken for the beam shape and sensitivity changes while interpretation. The coverage recurrs after about 2 years due to the orbit change, and with 2 year span we can compare the radio source structure change very precisely.

6. Discussion

The Soviet Radioastron satellite also will carry a 10 m antenna, and its orbit is 7,400 km in perigee and 77,000 km in apogee. This aims higher angular resolution but less imaging capability if compared to VSOP. The Radioastron is a detection instrument, while VSOP is a mapping instrument, and the two are astrophysically complimentary. If the operational time overlaps between VSOP and Radioastron, we can combine the visibility data, making better UV-coverage for better image of radio sources.

For a limited number of strong sources, there is the possibility of getting fringes between the two satellites. Also, VSOP may be used to calibrate the visibility data of Radioastron-ground system using hte overlaping part of VSOP-Radioastron visibility data. In any case, both space VLBI programmes need ground resources common in nature, and the compatibility of scheme and instruments is an important issue.

Even though VSOP will start new sciences in space, it is still limited in sensitivity. With better sensitivity, we have better image, more sources, and more science. VSOP and Radioastron will answer some astrophysical questions, and discover new questions, and second generation missions have already have firm reason for existence.

7. Conclusion

The concept of VSOP and scientific targets are shown briefly. As a first geneation space-VLBI observatory, VSOP will make first step for the understanding of very compact and bright objects in the universe.

References

Levy, G.S. et al.: 1986, *Science* **234**, 187

Linfield, R. et al.: 1990, *Astrophys. Jour.* **358**, 350

Linfield, R. et al.: 1989, *Astrophys. Jour.* **336**, 1105

Hirabayashi, H.: 1987, in M.R. Reid and J.M. Moran, ed(s)., *Proceedings of IAU Symposium No.129, The Impact of VLBI on Astrophysics and Geophysics*, Reidel, Dordrecht, 441

Kardashev, N.S. and Slysh, V.I.: 1987, in M.R. Reid and J.M. Moran, ed(s)., *Proceedings of IAU Symposium No.129, The Impact of VLBI on Astrophysics and Geophysics*, Reidel, Dordrecht

Hirabayashi, H.: 1991, *Hirabayashi, H., Kobayashi, H. and Inoue, M.*, Frontiers of VLBI, Universal Academy Press, Tokyo

Hirosawa, H.,1991, *ibid*

Hirabayashi, H., 1991,*ibid*

VLBI WITH TDRSS

R. P. LINFIELD

Jet Propulsion Laboratory 4800 Oak Grove Dr., Pasadena, CA 91109, USA

Abstract. VLBI observations using a satellite in earth orbit and ground antennas in Japan and Australia were conducted in 1986, 1987, and 1988. Sources were detected on space-ground baselines at both observing frequencies: 2.3 and 15 GHz. The coherence on space-ground baselines for 340 s was 90% at 2.3 GHz and 76% at 15 GHz. Brightness temperatures in the range 1-4×10^{12} K were measured for 10 sources at 2.3 GHz and 6 sources at 15 GHz.

1. Introduction

The Tracking and Data Relay Satellite System (TDRSS) consists of satellites in geostationary orbit, designed to relay data between a ground station in White Sands, New Mexico, USA (WSGT) and satellites in low earth orbit (*e.g.* the Hubble Space Telescope, the NASA space shuttle). Each satellite has two 4.9 m diameter antennas, equipped with 2.3 and 15 GHz receivers (and transmitters), with bandwidths of 16 and 256 MHz, respectively. A tone from a ground frequency standard is broadcast from WSGT and used to phase-lock all on-board oscillators.

TDRSS satellites appear to be the most suitable existing satellites for space VLBI, due to their local oscillator scheme, high-gain antennas, and large received bandwidths. A TDRSS satellite was used for three space VLBI experiments: July/Aug. 1986 (2.3 GHz only), Jan. 1987 (2.3 GHz only), and Feb./Mar. 1988 (2.3 and 15 GHz). The purpose was to test several technical concepts peculiar to space VLBI, and to perform a survey of the brightest sources to measure their size distribution.

2. Observations

At the time of these experiments, only one TDRSS satellite: TDRSA, was in orbit, above 41° W. longitude. It was constrained to look back towards the earth: within 31° of the nadir in declination, and within 22° in hour angle (Figure 1). The ground radio telescopes used in these experiments therefore needed to be located approximately 180° in longitude away from the sub-earth point of TDRSA. Antennas in Japan (the Usuda 64 m antenna, NRO 45 m, and Kashima 26 m) and Australia (64/70 m and 34 m antennas of the NASA Deep Space Network in Tidbinbilla).

The Mk III recording system was used for all 3 experiments. In the first two, a 14 MHz bandwidth of 2.3 GHz data was recorded. In the third experiment, the Mk III double speed mode was used to record 88 MHz bandwidth at 15 GHz, and 12 MHz bandwidth at 2.3 GHz. Data from TDRSA were broadcast to the ground (WSGT), where they were digitized and recorded. Details of the experimental procedure are given in references 1 and 2. The data correlation was performed at

Y. Kondo (ed.), Observatories in Earth Orbit and Beyond, 271–274.
© 1990 *Kluwer Academic Publishers.*

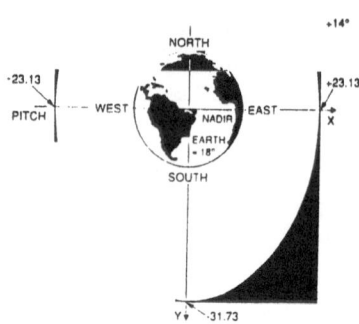

Fig. 1. TDRSS field of view, drawn to scale.

Haystack Observatory, using the Mk IIIA correlator. Software modifications were needed to allow data from an orbiting antenna to be correlated.

3. Coherence

The coherence measured in the third experiment is shown in Figure 2. Both the shape of the coherence curves (substantial loss over short integration times) and the frequency dependence (the coherence degrades quite slowly with increasing frequency) suggest that neither the ground frequency standard nor orbit determination errors are the primary sources of coherence loss. A more likely cause is frequency flicker noise generated in the on-board local oscillator chain (which was not designed to do VLBI).

4. Source Visibilities and Brightness Temperatures

In the second experiment, where the majority of the 2.3 GHz data were obtained, 23 out of 24 sources were detected on TDRSA-ground baselines. The longest projected baseline length (limited by TDRSA pointing constraints) was 2.15 earth diameters ($D_\oplus$). Three sources were detected on baselines longer than $2.0D_\oplus$. The most compact source was 1519−273, with a visibility of 0.66 on a $2.02D_\oplus$ baseline. Sufficient data were obtained on 14 sources to determine brightness temperatures. 10 of those sources had brightness temperatures in the range $1 - 4 \times 10^{12}$ K (Figure 3, reference 3), exceeding the 1×10^{12} K Inverse Compton limit.

At 15 GHz, the detection rate on TDRSA-ground baselines was lower: 11 of 22 sources. However, the sensitivity of the interferometer was much poorer than at 2.3 GHz. The observed brightness distributions of source visibilities and brightness temperatures were similar at the two frequencies.

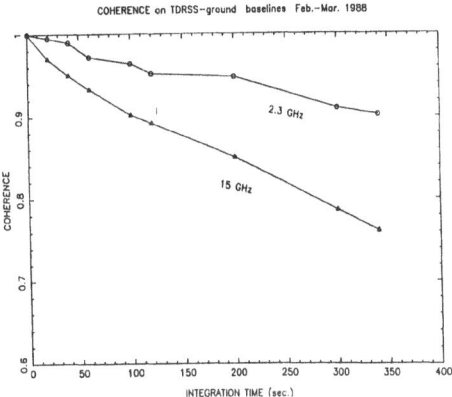

Fig. 2. Coherence values for TDRSS baselines.

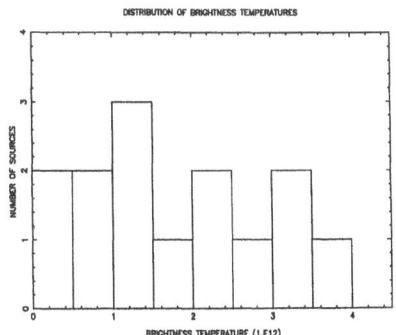

Fig. 3. Histogram of 2.3 GHz brightness temperatures from Jan. 1987 experiment.

5. Discussion

These experiments demonstrated that space VLBI observations can be successfully performed, even with a spacecraft not designed for VLBI. With careful designs, VSOP and Radioastron should be able to achieve excellent coherence.

The measured distribution of sources sizes and brightness temperatures demonstrates that baselines of $1-3D_\oplus$ will be very useful for studying the structure of bright, compact radio sources. The existence of correlated flux on baselines longer than $3D_\oplus$ is still an open question, awaiting Radioastron observations for its answer.

The TDRSS experiments involved a large multinational effort led by G. Levy of JPL, with major roles from groups in Japan and Australia.

References

Levy, G. S. *et al.*: 1989, *Ap. J.* **336**, 1098
Linfield, R. P. *et al.*: 1990, *Ap. J.* , (in press)
Linfield, R. P. *et al.*: 1989, *Ap. J.* **336**, 1105

(E) SOLAR SYSTEM & PLANETARY SYSTEMS

SOHO – AN OBSERVATORY TO STUDY THE SOLAR INTERIOR
AND THE SOLAR ATMOSPHERE

A.I. POLAND

Laboratory for Astronomy and Solar Physics,
NASA Goddard Space Flight Center, Greenbelt, Maryland, USA

and

V. DOMINGO

Space Science Department of ESA, ESTEC, Noordwijk, The Netherlands

Abstract. The Solar and Heliospheric Observatory, Soho, is a joint venture of ESA and NASA. The main objectives of Soho are: a) the study and understanding of solar coronal phenomena; and b) the study of the solar structure and interior dynamics from its core to the photosphere. The primary goals of the coronal and solar wind studies are to understand the coronal heating mechanism and its expansion into the solar wind.

These goal will be achieved both by remote sensing of the solar atmosphere with high resolution spectrometers and telescopes and by "in situ" measurement of the composition and energy of the resulting solar wind and the energetic particles that propagate through it. The structure and interior dynamics will be studied by helio-seismological methods and the measurement of solar irradiance variations. The Soho spacecraft will be three-axis stabilized and located in a halo orbit around the L1 Lagrangian point (approximately 1% of the distance from the Earth to the Sun). It is currently scheduled for launch in July 1995.

1. Introduction

The Solar and Heliospheric Observatory, Soho, is a space mission that forms part of the Solar-Terrestrial Science Program (STSP), developed in a collaborative effort by the european Space Agency, ESA, and by the National Aeronautics and Space Administration, NASA, of the USA.

The STSP, in turn, forms part of ESA's long term science plan known as "Space Science: Horizon 2000" and of NASA's collaborative International Solar-Terrestrial Program (ISTP) with ESA and with ISAS (Institute of Space and Astronautical Science, Japan).

This paper presents an overview of the Soho mission in terms of its overall scientific objectives and its complement of instruments.

2. Scientific Objectives

The Sun is the only star that we can study with high spacial resolution. For example, we can observe images of the Sun and use them to study plasma processes occurring in different parts of its atmosphere. For phenomena occurring in the outer atmosphere (outside the photosphere) we are frequently able to determine many aspects of the interactions between various physical processes such as magnetic field

Y. Kondo (ed.), Observatories in Earth Orbit and Beyond, 277–284.
© 1990 *Kluwer Academic Publishers.*

changes, heating, material flow, conduction, and radiation. An important aspect of these processes is the expansion of the solar atmosphere in the form of a wind (solar wind) that blows past the Earth and most of the solar system. We can study this solar wind "in situ" by means of spacecraft that are outside of the Earth's magnetosphere.

The source region of the solar wind is the layers above the photosphere which cannot be studied from ground observatories for two main reasons. One is that they are so tenuous with respect to the photosphere that any emission that they produce at visible wavelengths cannot be seen against the photospheric background. However, some features in the corona can be observed from the Earth during total solar eclipses when the moon, for a few minutes, blocks out the solar disk. Another reason for not being able to study these regions from the ground is that they are at higher temperatures than the photosphere and therefore produce emission lines due to higher levels of ionization and excitation of the atoms. These lines are generally emitted in the ultra-violet and X-ray regions of the spectrum, to which the Earth's atmosphere is opaque. Thus, the outer layers of the solar atmosphere are not generally observable from the ground.

Previous observations of the Sun from space have been performed by NASA's Orbiting Solar Observatory (OSO) satellites during the 1960's and early 1970's, the Apollo Telescope Mount (ATM) on Skylab during 1972 and 1973, and with the Solar Maximum Mission (SMM) during the 1980's. Many of the observations made by the OSO's and ATM provided the basis for our current understanding of the outer solar atmosphere. The SMM observations further advanced our knowledge, but the experiments were tailored to study solar flares. An important conclusion based on the observations from these previous satellites is that: to understand the processes occurring in the outer solar atmosphere we need simultaneous, high spatial resolution images of the Sun at many wavelengths with a relatively rapid ($\sim$ 1 minute or less) time resolution.

Soho is a solar observatory devoted to furthering our understanding of the outer solar atmosphere and the solar wind. To accomplish this Soho will carry a set of telescopes that will study phenomena that are initiated by processes that commence below the photosphere, and propagate through the photosphere, chromosphere, and the transition region into the corona. These instruments are designed to investigate problems such as how the corona is heated and how it is transformed into the solar wind that blows past the Earth at 400 km/s. To do so it will have spectrometers that will allow the detailed study of the emission and absorption lines produced by the ions present in the different regions of the solar atmosphere. From this information it will be possible to determine densities, temperatures and velocities in the changing structures. These measurements are complemented by the "in situ" study of the composition and energies of the solar wind that results from the coronal structures that have been observed by the telescopes. This is done with the help of particle detectors carried by Soho that sample the solar wind as it passes through it. Soho will thus greatly enhance our knowledge of the solar wind and its source region.

While the solar interior is the region where the kinetic and magnetic energy that drives the outer atmosphere and solar wind is generated, almost no direct information can be obtained about any region below the photosphere. The neutrinos that are generated by the nuclear reactions that take place in the core are the only direct radiation that reaches us from anything that is below the photosphere. But a relatively new technique, helio-seismology, has developed in the last two decades that allows us to study the stratification and certain dynamical aspects of the solar interior. It uses the study of the acoustic and gravity waves that propagate through the interior of the Sun and can be observed as oscillatory movements of the photosphere. An analysis of these oscillations allows us to determine the characteristics of the resonant cavities in which they resonate, much in the same way as the Earth's seismic waves are used to determine the structure of the Earth's interior.

To study the solar interior Soho will carry a complement of instruments whose aim is to study the oscillations at the solar surface by measuring the velocity oscillations via the Doppler effect and by measuring the oscillating changes in intensity that the pressure and gravity waves produce. The study of such oscillations require both high resolution imaging and long uninterrupted time series of observations. In addition, because it is paramount to understand the structure of the Sun in relation to the oscillation measurements, the total solar irradiance, or solar constant, and its variations will be measured.

The Soho satellite will thus enable us to study: the structure, chemical composition, and dynamics of the solar interior, the structure (density, temperature and velocity fields) and dynamics of the outer solar atmosphere, and the solar wind and its relation to the solar atmosphere.

3. Instrumentation

The investigations selected for the Soho satellite are listed in Table I. They can be divided by their areas of research into three main groups: helio-seismology, solar atmospheric remote sensing, and "in situ" solar wind measurements.

The helio-seismology investigations primarily aim at the study of those parts of the solar oscillations spectrum that cannot be obtained from the ground. The required sensitivity for observing the very low modes ($l \lesssim\sim 5$) and the very high modes ($l \gtrsim\sim 200$) is difficult to achieve from the ground because of noise effects introduced by the Earth's diurnal rotation for the low modes, and the transparency and seeing fluctuations of the Earth's atmosphere for the high modes. GOLF and VIRGO aim primarily at the study of the very core of the Sun; for that they have to study oscillations of very low frequencies with very high sensitivity. MDI will observe the whole oscillation spectrum, and will be unique in its study of the upper degree modes that carry information about the composition and dynamics of the outer boundary layer. The three investigations will provide important information about the time varying phenomena in the interior of the Sun.

The Global Oscillations at Low Frequencies (GOLF) investigation will perform uninterrupted velocity oscillation and magnetic field measurements of the full solar disk, spatially unresolved, with extremely high sensitivity ($\lesssim$ 1mm/s and 1 milligauss). To determine the velocity, the Doppler effect is measured by the resonant scattering technique.

Table I. SOHO Investigations

Helioseismology

Investigation	Principal Inv.	Measurements	Technique
Global Oscillations at Low Frequencies (GOLF)	A. Gabriel, IAS, Verrieres-le-Buisson, France	Global Sun Velocity and Magnetic field oscillations l=0-4	Na-vapor resonant cell, Doppler shift & circular polarization
Variability of solar irradiance (VIRGO)	C. Frolich PMOD/WRC, Davos, Switzerland	Low degree (l=0-7) irradiance oscillations and solar constant	Global Sun and low resolution (12 pixels) imaging, active cavity radiometers
Michelson Doppler Imager (MDI/SOI)	P.H. Scherrer, Stanford Univ. USA	Velocity oscillations, high degree modes (up to l=4500)	Doppler shift with Fourier tachometer, 4 & 1.5 arc sec resolution

Solar Wind "In Situ"

Investigation	Principal Inv.	Measurements	Technique
Charge Element and isotope analysis (CELIAS)	D. Hovestadt MPE, Garching, Germany	Energy distribution & composition. 0.1-1000keV/e	Electrostatic deflection, time-of-flight measurements & solid state detectors
Suprathermal and energetic particle analyzer (COSTEP) Energetic particle analy. (ERNE)	H. Kunow Univ. Kiel Germany J. Torsti Univ. Turku, Finland	Energy distribution & composition, ions 1.2-330 MeV/n electrons 0.06-25 MeV	Solid state, and plastic and crystal scintillator detector telescopes.
Solar Ultra-violet emitted radiation (SUMER)	K. Wilhelm MPAE, Lindau Germany	Plasma flow characteristics (T, ρ, velocity) chromosphere - corona	Normal incidence spectrometer, 50-160 nm, spectr. res. 2-40,000 ang. res. 1.2-1.5"
Coronal Diagnostics spectrometer (CDS)	B.E. Patchett RAL Chilton England	Temperature and ρ: transition region & corona	Grazing incidence spectrometer 17-80nm spectr. res. 5000. ang res. 2"

Solar Atmosphere Remote Sensing

Investigation	Principal Inv.	Measurements	Technique
Extreme-ultra-violet imaging telescope (EIT)	J.P. Delaboudin-iere IAS, Verrieres-le-Buisson, France	Evolution of chromospheric & coronal structures	Images in HeII, FeIX, FeXII, & FeXV
Ultra-violet coronagraph spectrometer (UVCS)	J.L. Kohl, SAO Cambridge, Mass. USA	Electron & ion T, ρ, and vel. in corona (1.3-10Ro)	Profiles and/or intensity of several spectral EUV lines
White Light & spectrometric coronagraph (LASCO)	G. Brueckner NRL Washington USA	Evolution, mass, momentum, and energy transport in the corona (1.1-30Ro)	1 internal and 2 externally occulted coronagraphs. Spectrometer.
Solar Wind Anisotropies (SWAN)	J.L. Bertaux, SA Verriers-le-Buisson France	Solar wind mass flux anisotropies	Scanning telescopes with hydrogen abs. cell for Lyα

A resonant cell filled with sodium vapor in a strong magnetic field enables, by the Zeeman effect, the selective absorption of light in the two wings of the solar D lines. A comparison of the intensity on both sides of the line leads to an extremely sensitive determination of the velocity.

The Variability of Irradiance and Gravity Oscillations (VIRGO) Investigation is an experiment to study the solar irradiance variability and oscillations. The instrumentation of VIRGO comprises two active cavity radiometers observing the total irradiance, the sun-photometers measuring the spectral irradiance at three wave-lengths in the near ultraviolet, the visible and near infrared (335, 550 and 865 nm). In addition, narrow band radiance measurements (at 500 nm) are carried out with 12 resolution elements on the solar disc by a high-precision luminosity oscillation imager.

The Michelson Doppler Imager (MDI) will measure velocity oscillations on the surface of the Sun with high angular resolution (4 and 1.5 arcsec). It uses two solid Michelson interferometers as the final elements of a interferometers as the final elements of a tunable narrow-bandpass filter. Prefiltering is accomplished by a blocking filter and a Lyot filter. The Michelson interferometers are tuned by half-waveplates. Doppler shifts are determined by measuring intensities at four points along the line profile. The technique yields line-shift estimates with a linear response over a range of circa 4000 m/s.

The solar atmosphere remote sensing investigations are carried out with a set of telescopes and spectrometers that will produce the data necessary to study the dynamic phenomena above the chromosphere. The plasma will be studied by spectroscopic measurements and high resolution images at different levels of the solar atmosphere. Plasma diagnostics obtained with these instruments will provide temperature, density, and velocity measurements of the material in the outer solar atmosphere.

The SUMER, CDS, and EIT experiments are highly complementary in terms of the scientific objectives they can achieve. CDS obtains data on hotter lines, while SUMER obtains somewhat cooler lines, with an overlap between 50 and 80 nm. They both obtain spectra along a spatial line on the Sun, building up a two dimensional image by moving the Sun's image across a slit in a very short time (on the order of seconds). CDS has a broader spectral range at any one time, but cannot obtain line profiles. SUMER, on the other hand, has a high spectral resolution but a limited simultaneous range. Thus, CDS is better suited for density and temperature diagnostics, while SUMER is better suited for velocity measurements. Finally, EIT provides high resolution images of the whole Sun at several temperatures (HeII 30.4 nm $\sim$ 60,000 K, FeIX 17.1 nm $\sim$ 3 $\times$ 10^{6}K, FeXII 19.5 nm, $\sim$ 1.6 $\times$ 10^{6}K, and FeXV 28.4 nm $\sim$ 3 $\times$ 10^{6}K), thus providing the morphological context of the spectral observations.

While the CDS, SUMER and EIT primarily observe on the solar disk, the UVCS, LASCO, and SWAN experiments make observations in the solar corona. UVCS observes line profiles in the ultra-violet using Doppler dimming and broadening to determine material velocity and temperature. LASCO uses white light measurements to obtain electron density in the corona and spectral observations of emission lines and Fraunhofer lines to determine the hot coronal temperatures and electron temperature respectively.

The SWAN experiment monitors the large scale properties of the solar wind expansion, and in particular the latitude distribution of the solar wind mass flux from equator to pole, and the time variations of this distribution. It does so with Lyman alpha sky maps of interplanetary emission, obtained with an Hydrogen cell and an appropriate optical scanning system.

The instruments to measure "in situ" the composition will determine the elemental and isotopic abundances, the ionic charge states and velocity distributions of ions originating in the solar atmosphere. The energy ranges covered will allow the study of the processes of ion acceleration and fractionation under the various conditions that cause their acceleration from the "slow" solar wind through solar flares. The CELIAS instruments concentrate on "lower" energy process while ERNE and COSTEP put their main emphasis on measurements into the suprathermal energy range.

The CELIAS instrument consists of three sensors, all making use of the time-of-flight technique (TOF). The CTOF determines the elemental composition, the charge state distribution, the temperatures and the velocities of the more abundant solar wind ions from helium to iron. The MTOF determines the elemental and isotopic composition of solar wind ions with a mass resolution $(M/\Delta m)$ better than

100. Finally, the STOF determines mass, charge state and energy of suprathermal and low energy solar energetic particles from hydrogen to the iron group.

The COSTEP and ERNE investigation teams have formed a consortium in which they have defined five sensors that are best suited within the available resources to analyze the composition of the solar suprathermal and energetic particles above the range covered by CELIAS. All the sensors are basically composed of solid state particle detector telescopes. All together they measure, the electron energy spectrum between 60 keV and 5 MeV, the proton flux between 60 keV and 53 MeV, the energy variation of the isotopic composition of ions with energy between 1.4 MeV/n and 53 MeV/n for NI. The charge composition determination extends ups to 540 MeV/n for elements in the Ni region.

In summary, the coronal remote sensing and the "in situ" experiments on Soho will provide a comprehensive data set to study the solar wind heliosphere. The solar imagers and spectrographs will allow the study of the morphology, magnetic structure, and heating and particle acceleration processes occurring at the Sun. At the same time it will be possible to make direct measurements of the particle composition and energy spectrum in the solar wind with the particle experiments.

4. Spacecraft and Orbit

The SOHO spacecraft will be three-axis stabilized and pointing to the sun within an accuracy of 10 arcsec and a pointing stability of 1 arcsec per 15 minutes. The Total mass will be about 1350 kg and 750 W power will be provided by the solar panels. The payload will weigh about 650 kg and consume 350 W in orbit.

Soho is planned to be launched in March 1995 and will be injected in a halo orbit around the L1 Sun-Earth Lagrangian point, about 1.5×10^6 km sunward from the Earth. The halo orbit will have a period of 180 days and has been chosen because, 1) it provides a smooth Sun-spacecraft velocity change throughout the orbit, appropriate for helio-seismology, 2) is permanently outside of the magnetosphere, appropriate for the "in situ" sampling of the solar wind and particles, and 3) allows permanent observation of the Sun, appropriate for all the investigations.

5. Operation, Data and Ground Systems

The Soho telemetry will be received by ground stations of NASA's Deep-Space Network (DSN) during three short (1.3 hr) and one long (8 hr) periods per day. Scientific data acquired outside these periods will be stored on magnetic tape on-board the spacecraft and transmitted to ground during the short ground contact periods. The Soho payload will produce a continuous stream of 40 kilobits/s, however the bitrate will be increased by 160 kilobits/s whenever the solar oscillations imaging instrument is operated in its high bit-rate mode. This will happen either daily during the 8-hr periods or during dedicated campaigns when the satellite will have 24 hour DSN coverage. The campaigns will be organized to provide approximately 2-months un-interrupted observations by the solar oscillation imaging instrument.

An Experiment Operations Facility (EOF), located at NASA's Goddard Space

Flight Center, will be used to coordinate and plan the scientific operations of the payload. Its main task will be to organize in particular the real time operation of the payload and control of the solar remote sensing imaging and spectrometric instruments during the daily 8-hr ground contact interval. ESA also intends to issue an Announcement of Opportunity to invite proposals for a second science operations center in Europe. Moreover, the large amount of solar oscillations imaging data produced by MDI will be the object of a specialized data facility.

6. Coordinated Research

The Soho payload has been conceived as an integrated package that requires coordinated operation and data analysis between the investigations aboard to achieve its scientific aims. The experiment Operation Facility will be the focal point for these coordinated activities. The EOF will also provide a focus for the cooperation with ground observatories. Soho will also collaborate with other space missions. For instance WIND and Cluster, when outside the magnetosphere, will provide solar plasma parameters what will complement the coronal and solar wind measurements of Soho. When Cluster is inside the magnetosphere Soho will be a useful monitor for the conditions of the environment external to the magnetosphere.

In the case of Cluster there is another aspect of cooperation that will occur. Considerable mutual insights and cross-fertilization in the area of plasma physical processes is expected from a coordinated study of the problems investigated individually by each of the two missions. As part of an ESA sponsored workshop, several working groups formulated common scientific objectives and cross-fertilization aspects between Cluster and Soho ESA, 1985. Their recommendations point out that most of the processes studied by the one of the STSP components – Soho or Cluster – have counterparts, – or at least analogues, to be studied by the other component. Joint studies by Cluster and Soho will illustrate the roles of the different parameter regimes and the limits of analogy. This process is particularly important to our attempts to extrapolate and apply knowledge gained in solar-system studies to remote astrophysical objects.

References

Jointly released by ESA (SCI87/1) and NASA (AO-OSSA-1-87): 1987, *Cluster and SOHO, announcement of Opportunity*

ESA SP-235: 1985, *Proceedings of an ESA sponsored workshop on "Future Missions in Solar, Heliospheric and Space Plasma Physics"*, Garmisch-Partenkirchen

AN OVERVIEW OF THE ORBITING SOLAR LABORATORY

D.S. SPICER

Laboratory for Astronomy and Solar Physics, NASA/Goddard Space Flight Center

Abstract. The Orbiting Solar Laboratory (OSL), which is a joint NASA/BMFT/ASI/ AFGL program, will be a free flying, polar orbiting complement of five scientific instruments operating as a facility and designed to observe the surface and upper atmosphere of the sun with high spatial and temporal resolution over a spectral range from the soft X-ray to the near Infrared. OSL's primary scientific objectives are to study the coupling of velocity, magnetic, and radiation fields in the photosphere and chromosphere and to causally relate this coupling to activity in the overlying regions.

To study the photosphere and chromosphere with a spatial resolution approaching 75 km OSL will utilize a 1 meter telescope, optimized for 2000–11,000 Å, to feed a narrow band tunable filter, a set of fixed broad-band filters, and a spectrograph. These three instruments share a common structure, share a unified science data and control system, and use a single-image stabilization system for their common fields of view to form a Coordinated Instrument Package (CIP). To study the transition zone a stigmatic High Resolution Telescope Spectrograph (HRTS) optimized for 1175–1700 Å, will be used capable of providing both images and spectroscopic information. A X-Ray Extreme Ultraviolet Imager and Spectroheliograph (XUVI) optimized for the 40–400 Å regime, will allow study of the corona and of multimillion degree solar transient events such as flares. Both the HRTS and the XUVI instruments will possess image stabilization systems and have a resolving power of 350 km. All OSL instruments will operate overlapping fields of view and be co-aligned. To achieve its goals the OSL will provide a total data stream of 20 Mbits/s, nominally 4 Mbits/s per instrument and has a planned operational lifetime of three years.

Y. Kondo (ed.), Observatories in Earth Orbit and Beyond, 285.

THE PLANETENTELESKOP MISSION

G. NEUKUM

DLR German Aerospace Research Establishment, 8031 Oberpfaffenhofen, FRG

Abstract. The scientific focus of the Planetenteleskop project will be on time-variable solar system phenomena (planet-magnetosphere-satellite interactions, active processes of cometary nuclei, atmospheric circulation and dynamics), on time-invariant solar system phenomena (geochemical provinces on planetary and satellite surfaces, global characteristics of primitive bodies), on planetary environments of other stars, and on general astronomical and astrophysical applications. The proposed Planetenteleskop in elliptical 24 h earth orbit will combine near-simultaneous, high-resolution spectroscopic observations, diffraction-limited imaging quality, long integration times ($<$ 10 h) and excellent target tracking accuracy (nominally 0.05 arc sec/10 h, up to 0.02 arc sec). The excellent tracking accuracy and stability on extended objects and features is provided by a novel real-time on-board image correlation scheme. The Planetenteleskop has been studied in prephase A and phase A by industry and the involved science community in Germany in cooperation with American colleagues.

Y. Kondo (ed.), Observatories in Earth Orbit and Beyond, 286.
© 1990 *Kluwer Academic Publishers.*

THE ASTROMETRIC IMAGING TELESCOPE: NEAR-TERM
DISCOVERY AND STUDY OF OTHER PLANETARY SYSTEMS

EUGENE H. LEVY

Lunar and Planetary Laboratory, University of Arizona, Tucson, AZ, USA

GEORGE D. GATEWOOD

Allegheny Observatory, University of Pittsburgh, Pittsburgh, PA. U.S.A.

and

RICHARD J. TERRILE

Jet Propulsion Laboratory, Pasadena, CA, USA

Abstract. The Astrometric Imaging Telescope (AIT) is a space-based 1.5 to 2 meter diameter telescope designed to discover and study planetary systems around other stars. The measurement objectives and instrument design aim at a definitive search for other planetary systems. The science program includes two planetary-system investigations, and the telescope carries two separate instruments: an astrometric instrument to measure the reflex motions of stars caused by the motions of unseen planets and an imaging corona-graph to directly image planets and other circumstellar matter.

The astrometric techniques is based on the concept of the Multichannel Astrometric Photometer (MAP) which passes a Ronchi ruling over a field of stars to measure the centroid position of the stars in two orthogonal coordinates. Outside of Earth's atmosphere this instrument is designed to be about two orders of magnitude more accurate than the best existing ground-based astrometric instrument, which uses the same measurement technique. The astrometric design accuracy is about 10 microseconds. This will allow detection and study of Uranus-size of larger planets in jovian orbits around several hundred nearby stars. The parallactic study of the parent star of a planetary system will accurately reveal its distance; the star's departure from inertial motion will reveal the masses of individual planets and their orbits. The periods of the orbits are obtained from analysis of the motions of the central star. With sufficient precision and time, analysis of a central star's reflex motion also yields the eccentricities of the planetary orbits. Assuming that the mass of the primary star can be accurately estimated, the measurements will also yield the masses of planets.

The imaging investigation relies on a high efficiency coronagraph to suppress diffraction wings of the bright parent star by a factor of 1000. Laboratory tests on such a coronagraph have demonstrated a high efficiency at concentrating diffracted light into an area where it can be removed by the apodization of an image of the telescope pupil. However, in order to utilize this high efficiency, the scattered light floor of the telescope must also be a factor of 1000 below the diffraction wings. This requirement puts strong constraints on the mid-spatial frequency errors of the primary mirror. Laboratory experiments in mirror fabrication have demonstrated, with sub-scale mirrors, the required flight quality optics. Space-borne, the AIT will directly image planetary systems with the capability of detecting Jupiter-sized planets in hours around the nearby stars. Its sensitivity to faint material near bright stars exceeds that of the Hubble Space Telescope (HST) by 5 stellar magnitudes and makes the instrument ideally suited for the detection and study of the circumstellar material associated with planetary system formation.

Y. Kondo (ed.), Observatories in Earth Orbit and Beyond, 287.

©1990 *Kluwer Academic Publishers.*

(F) SHUTTLE-BORNE ASTRO MISSIONS

ULTRAVIOLET POLARIMETRY

A.D. CODE, K.F. NORDSIECK, C.M. ANDERSON
University of Wisconsin

Abstract. Interstellar polarization and polarization of the sky background is predicted to be significantly smaller in the ultraviolet than in the optical region of the spectrum; while intrinsic polarization is expected to be greater. For these reasons alone ultraviolet polarimetry should provide a powerful diagnostic tool. To date such polarization measurements have not been undertaken. The Wisconsin Ultraviolet Photopolarimeter Experiment (WUPPE) will be the first to carry out a program of UV spectropolarimetry. WUPPE is one of the four ASTRO-1 payloads. It consists of a half meter aperture Cassegrain telescope feeding a spectropolarimeter capable of measuring spectral energy distribution and linear and circular polarization in the spectral region from about 3300 Å to 1400 Å with spectral resolutions from about 4 Å to 40 Å, depending upon the mode of observation and nature of the source. The effective area of WUPPE is approximately 100 cm^2 at 2000 Å and the faint stellar limit for measurement of all four Stokes parameters is about 16 magnitude. The design provides for a large dynamic range and the high precision required for accurate polarimetry. Launch of the first ASTRO Mission is expected in May of 1990.

The science goals for WUPPE include studies of ultraviolet interstellar polarization, intrinsic polarization of a wide variety of stellar types and measurements of polarization of selected extragalactic objects. A description of WUPPE and simulations of expected data will be presented. The paper will also discuss a wide field imaging survey polarimeter (WISP) now undergoing development for a sounding rocket payload at the University of Wisconsin. (This work is supported by NASA contract NAS5-26777).

Y. Kondo (ed.), Observatories in Earth Orbit and Beyond, 291.
©1990 *Kluwer Academic Publishers.*

THE HOPKINS ULTRAVIOLET TELESCOPE

A. DAVIDSEN

Johns Hopkins University

Abstract. The Hopkins Ultraviolet Telescope (HUT) will make pioneering observations in the far ultraviolet (912–1850 Å) and extreme ultraviolet (420–912 Å) bands during its upcoming flight aboard the Astro-1 shuttle mission, currently scheduled for launch on May 9, 1990. HUT employs an iridium-coated 0.9-meter $f/2$ primary mirror, an osmium-coated grating, and a CsI-coated microchannel plate intensifer to achieve a resolution of about 3 A in first order, with a peak effective area of 15 cm^2 at 1100 Å, and time resolution of 2 milliseconds, HUT's EUV response is obtained in second order, with a peak effective area of 10 cm^2 at 600 Å.

HUT is expected to obtain several hundred spectra during its upcoming mission, ranging from Comet Austin to the quasar HS 1700+64 at a redshift of 2.7. The design and operation of the instrument are described, and simulated spectra are used to illustrate a sample of the problems that will be addressed during the Astro-1 mission. In order for HUT to be exploited fully, however, it would be desirable to convert it to a free-flying satellite mode.

Y. Kondo (ed.), Observatories in Earth Orbit and Beyond, 292.
©1990 *Kluwer Academic Publishers.*

THE ULTRAVIOLET IMAGING TELESCOPE FOR ASTRO 1

T.P. STECHER

NASA/Goddard Space Flight Center

Abstract. The Ultraviolet Imaging Telescope (UIT) is a 38-cm Ritchey-Chretien telescope with 2 arc-sec resolution and a 40 arc-min field of view. There are two magnetically-focused image intensifiers which record on film. One has a Cs-I photocathode and 6 filters, the other has a Cs-Te cathode with 5 filters and an objective grating. The film transports contain 1000 frames each of IIa-O film. The point source threshold is m (2200 Å) = 22.0 in a 30 min integration through the widest filter. For extended sources the corresponding mag is 26.0/sq. arc-sec. The UIT complements the instruments on the Hubble Space Telescope in that its field of view is $\sim$ 200 times that of the Wide Field Camera, it has extremely good visible light rejection and is more sensitive to diffuse structures in the UV. The UIT is well-suited to locating interesting UV targets for HST follow-up and the study of large regions such as nearby galaxies and clusters.

Astro 1 will point the co-aligned UV instruments at $\sim$ 200 targets during its 10 day mission. The UIT science team's primary scientific programs include: The search and survey of the UV sky; SN1987A for the UV light echo; The helium burning and collapsing degenerate stars in star clusters; The structure of highly ionized regions in planetary nebulae and supernova remnants; The physics of massive star formation in spiral structure of galaxies; The interstellar medium of other galaxies; A search for active nuclei in 'normal' galaxies; The very faint regions surrounding galaxies that are produced by interactions, mergers, ejection, or accretion flows; The study of stellar populations of galaxies and galactic evolution, ages and chemical content of galaxies, and the study of large samples of galaxies in nearby clusters for comparison with objects at large lookback times.

Y. Kondo (ed.), Observatories in Earth Orbit and Beyond, 293.

THE BROAD BAND X-RAY TELESCOPE (BBXRT) ON ASTRO-1

P.J. SERLEMITSOS

NASA/Goddard Space Flight Center

Abstract. BBXRT is one of four pointed instruments comprising NASA's Astro-1 mission, currently scheduled for a 9–10 day flight in May 1990 on board the Space Shuttle Columbia. X-ray Spectra of over 150 sources will be obtained in the energy range 0.3 to 12 keV, with energy resolution of ~100 eV FWHM and timing resolution of 64 μs. BBXRT's large throughput, broad band mirrors, coupled to segmented cooled Si(Li) detectors, give us the first opportunity to study in detail X-ray spectral features, from oxygen to the iron K band. In developing this instrument, we have made extensive use of detector background reduction techniques in order to be able to observe the many faint extragalactic sources discovered with the Einstein observatory. Sources in our observing plan have been selected to give us in a 2–4 ks observation $< 10^3$ net counts with negligible background. Specific scientific objectives range from refinements of spectra previously obtained at lower resolution, such as a search for relativistic line shifts in compact objects, to observations of faint extragalactic sources for which spectral information is, at best, sketchy, viz. elliptical galaxies and distant quasars. High on our priorities are a search for emission from SN1987A and the first simultaneous observations in the UV and X-ray band of selected sources such as cataclysmic variables and late-type stars.

Y. Kondo (ed.), Observatories in Earth Orbit and Beyond, 294.
©1990 *Kluwer Academic Publishers.*

(G) INTERPLANETARY MISSIONS

FUTURE SOLAR SYSTEM MISSIONS

GEOFFREY A. BRIGGS

Solar System Exploration Division, Office of Space Science and Applications NASA Headquarters, Washington, D.C.

After a decade long hiatus in launches beyond Earth orbit, NASA's planetary exploration program is again moving forward, beginning with the Magellan launch to Venus in May 1989 and the Galileo launch to Jupiter in October 1989. These spacecraft will reach their targets in August 1990 and December 1995, respectively. Both are missions of longstanding priority, Magellan to provide the first global high resolution mapping of the cloud-shrouded Venus surface, Galileo to make comprehensive measurements of the Jovian system in follow-up to the 1979/1980 Voyager flybys.

Beyond these two missions there are other already approved missions: the *Mars Observer* for launch in 1992, the Comet Rendezvous and Asteroid Flyby *CRAF* mission (Fig. 1) for launch in 1995, and the *Cassini* mission (Figs. 2 and 3) to Saturn and its moon Titan for launch in 1996. The very diversity of these five missions and their targets (Venus, the Jovian system, Mars, comet Kopff, asteroids Gaspra and Ida by *Galileo*, Hamburga by *CRAF*, and Maja by *Cassini*, and the Saturnian system, is indicative of the strategy being pursued in the program: one of deliberate breadth that seeks to explore all three main classes of solar system bodies (the terrestrial planets, the outer giants and their moons, and the primitive small bodies).

This broadly-based strategy, as opposed to a focus on, say, Mars and Venus only, was developed in response to recommendations by the National Academy of Sciences and is serving the nation well in that it provides to the science community a comprehensive perspective from which to interpret new results and fit them into our growing body of knowledge: also, the continuing stream of images of the new worlds of the solar system evokes a worldwide public interest, as evidenced by the tremendous impact of the Voyager encounters of the last decade.

The future direction of the program beyond those missions mentioned above, which are already approved, is beginning to take shape as a result of a continuing strategic planning exercise begun a year ago (Fig. 4). This exercise directly involves the science community and the several NASA Centers with an interest in the program; the planning is conducted using as a guide the general principles embodied in the NASA Office of Space Science and Applications Strategic Plan, and it necessarily calls for iteration as the overall NASA program continues to take shape. Specifically, since 20 July 1989, when President Bush announced his commitment to the human exploration of the Moon and Mars, planning for the future robotic exploration of those bodies has taken on a new significance – that of providing appropriate precursor knowledge, both scientific and technological, to lay the ground work for the humans who will follow.

Unlike the Moon, where the Apollo program has clearly established the feasi-

Y. Kondo (ed.), Observatories in Earth Orbit and Beyond, 297–306.
©1990 *Kluwer Academic Publishers.*

bility of human exploration and where there is only one required precursor mission (a polar orbiter to provide a global data base of topographic and geochemistry information; see Figs. 5–7), in the case of Mars much more needs to be done to establish the groundwork – certain essential technologies need proving and a good deal of basic information about Mars (Fig. 8) must be gathered, specifically: surface and atmospheric hazards; the availability of water; knowledge relevant to the issues of planetary protection; and knowledge of the most interesting sites for scientific exploration.

Already a sequence of robotic Mars missions (Fig. 9) has been identified (not much different from the series of science missions long since recommended by the NAS-Space Science Board for the in-depth exploration of Mars) and is being refined in consultation with the Space Exploration Initiative planning team at the Johnson Space Center. These missions include a surface network of small, rugged rough landers to observe the character of a large number of surface sites for future landings (Fig. 10) and to measure the seismic charter of Mars and monitor the atmospheric dynamics. Then surface sample return missions, field geology rovers and high resolution imaging orbiters are required along with a communication satellite infrastructure to support this extended effort. The pace and overall scope of this effort will be defined over the next several years.

As noted earlier, an essential characteristic of the solar system exploration program to date has been breadth; how does one retain that breadth while focussing on the Moon and Mars as national space priorities? First, already-approved missions will continue through about 2005, returning data from the depths of the solar system. Second, other opportunities present themselves: to use ground based and Earth orbital telescopes to search for and characterize planetary systems associated with other stars, and to use newly emerging "lightsat" and microsatellite technologies to inexpensively complete the reconnaissance of our own Solar System (even after Voyager we still have not visited Pluto/Charon or Chiron or the mainbelt and Earth-approaching asteroids).

Planning to begin a serious, deep search for other planetary systems is presently focussed on the application of various approaches to ultra-high precision astrometry (to look for the reflex motion of stars resulting from the orbital motion of associated planets) and the possible use of space-based coronagraphic telescopes (similar in principle to the one used at Cerro Tolollo to discover the dust disk about the star Beta Pictoris). To be effective in observing individual planets near a star of enormously greater brightness, such coronagraphic telescopes would need to have ultra-low light scattering mirrors (super smooth mirror technology is advancing apace because of the needs of the photo-lithography business). In both cases a good deal of technology development lies ahead but one may confidently expect such observatories to be available around the turn of the century; the results from such observing programs may well revolutionize our thinking about our own Solar System and will, assuredly, give us much greater confidence in those theories of planetary origin and evolution that stand the test of being applied to other solar systems.

The prospects of an inexpensive "Explorer-like" program of solar system reconnaissance missions are being assessed. A key ground rule is that launch vehicles must be small – not larger than a Delta – since there is a clear association of space-craft mass (and consequent complexity) and mission cost. Already the feasibility of launching a 300 kg (dry) spacecraft on a delta-VEGA/Jupiter swingby trajectory to Pluto using a Delta launch vehicle has been established; such a spacecraft with a moderately comprehensive science payload of remote sensing and particle and fields instrumentation could reach Pluto within 15 years of a 1998 launch; such a flyby would allow observations of the transient Pluto atmosphere – which has apparently caused Pluto to brighten dramatically over the last twenty years as it passes through the perihelion part of its orbit – before the atmosphere freezes out again around the year 2020. For missions to the Earth-approaching asteroids even smaller spacecraft are contemplated – taking advantage of microspacecraft technology developments that have been made over the last several years. Here launchers like the DARPA's planned Taurus vehicle or equivalent are contemplated, a vehicle significantly smaller than the Delta but larger than the existing Scout and Pegasus vehicles. Mission goals would inevitably have to be quite focused but the range of critical characteristics that can be measured with only a few key instruments is impressive and sensor technologies are creating highly sensitive solid state devices that do not require large telescopic optics as in the past.

Thus there is optimism that an innovative, low cost planetary mission capability – the "Discovery Program" – will emerge over the next few years to maintain scientific balance in the solar system exploration program even as the program focuses on satisfying the requirements of the Space Exploration Initiative. Such a program might well also assume the competitive character of the Explorers where a Principal Investigator proposes the mission and, upon selection, is given the authority and responsibility to carry out the mission in association with an appropriate NASA Center.

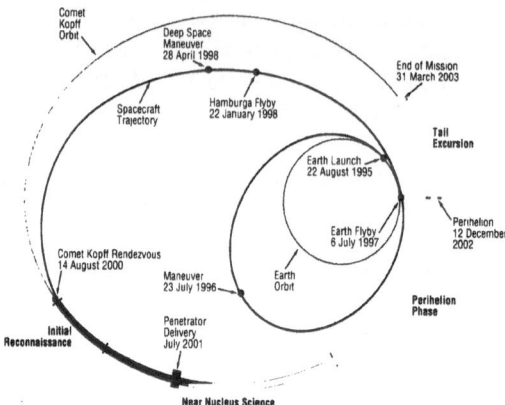

Fig. 1. Trajectory and mission phases of CRAF.

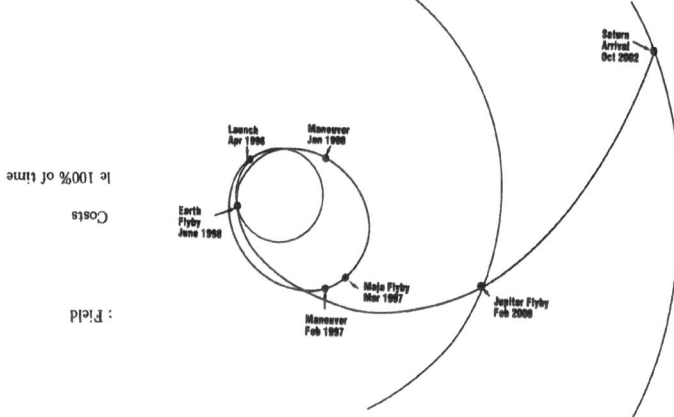

Fig. 2. The interplanetary trajectory of the Cassini spacecraft.

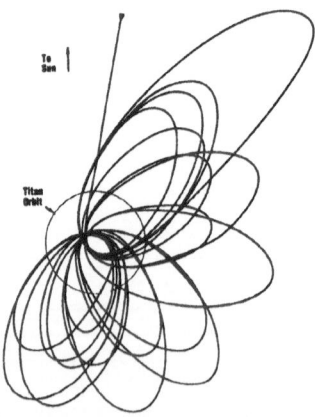

Fig. 3. The Cassini trajectory after arrival at Saturn, passing Titan on each orbit.

Strategy

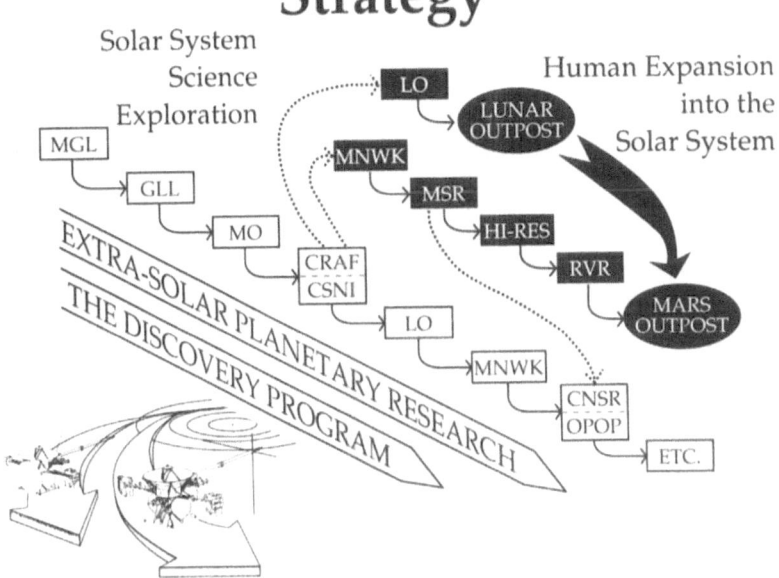

Fig. 4. Strategic Plan for the Exploration of the Solar System.

Lunar Outpost
Precursor Missions

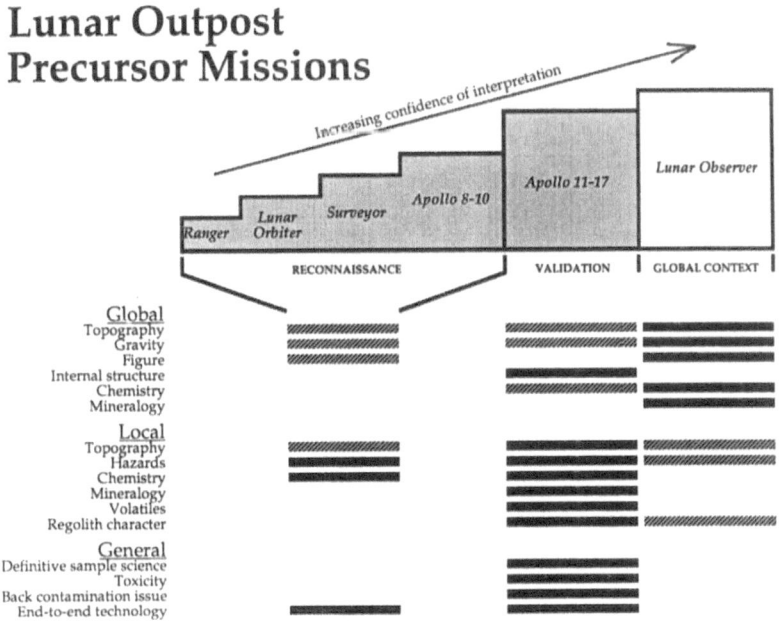

Fig. 5. Lunar Outpost Precursor Missions.

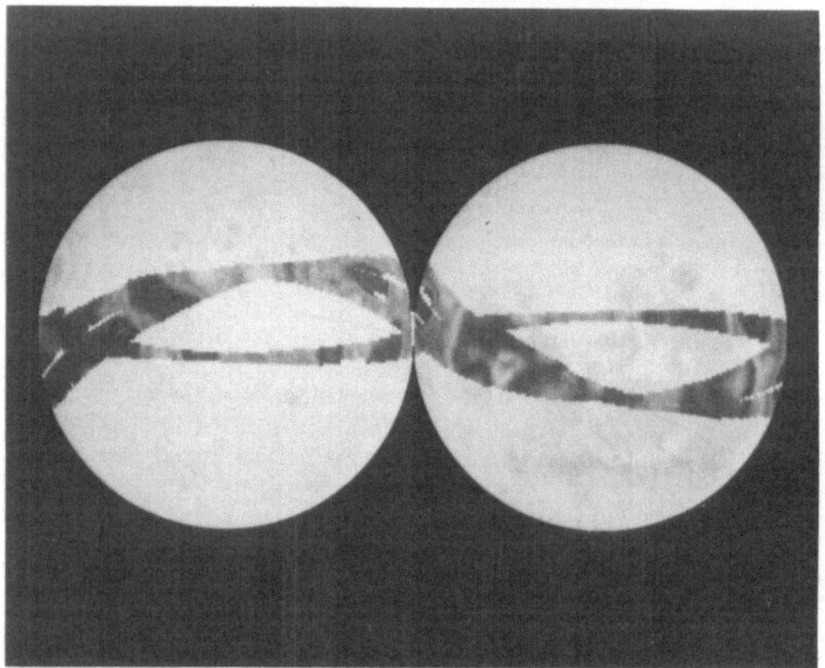

Fig. 6. The bands superimposed on this photograph of the front and back of the Moon represent the only areas of the lunar surface that have been chemically analyzed. The remaining 80 percent is still almost completely unknown.

Fig. 7. An artist's concept of the Lunar Observer mapping the lunar surface. The data the Observer collects will help planners select a site for a manned lunar base and determine what human activities can be accomplished, as well as advancing scientific knowledge of our closest neighbor.

Fig. 8. Mosaic of nearly an entire hemisphere of Mars.

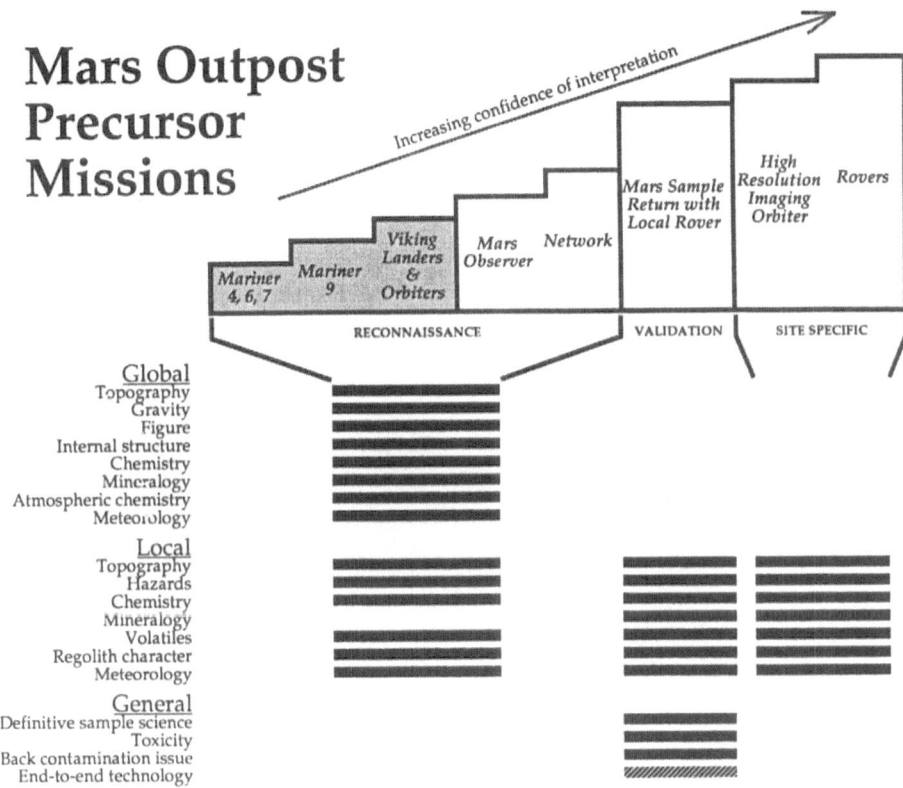

Fig. 9. Mars Outpost Precursor Missions.

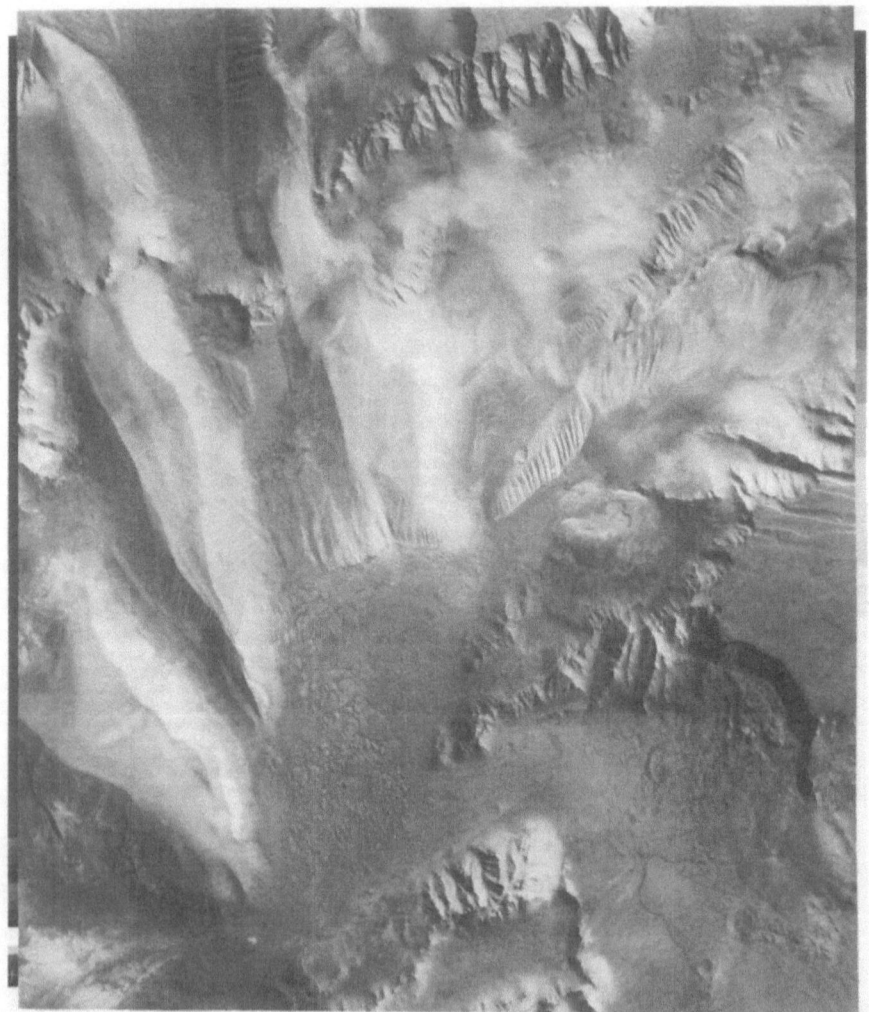

Fig. 10. Detail of Mars' Valles Marineris equatorial rift system, a possible future sample return and landing site.

THE ULYSSES MISSION IN
THE HIGH LATITUDE HELIOSPHERE

D.E. PAGE

ESA Ulysses Office, Jet Propulsion Laboratory, Pasadena, CA 91109, USA

R.G. MARSDEN

Space Science Department, ESTEC, 2200 AG Noordwijk, Netherlands

E.J. SMITH

Jet Propulsion Laboratory, California Institute of Technology, Pasadena, CA 91109, USA

and

K.-P. WENZEL

Space Science Department, ESTEC, 2200 AG Noordwijk, Netherlands

Abstract. Ulysses, a joint ESA/NASA mission launched in October 1990, will be the first to explore the high latitude heliosphere. Launch will be from the Shuttle and a Jupiter gravity assist will be used to send the spacecraft first over the southern solar pole approximately three and one half years after launch and then over the northern solar pole one year later. Instruments will be carried to study the solar wind, the heliospheric magnetic field, energetic solar particles, galactic cosmic rays, solar X- rays, cosmic gamma rays, cosmic dust and interstellar neutral helium. The radio signals used to track and transmit spacecraft data will be used also to sound the corona and to search for gravitational waves.

1. Introduction

All spacecraft to date have remained fairly close to the ecliptic plane. Ulysses, a joint ESA/NASA mission launched in October 1990, will be the first to explore the heliosphere at high solar latitudes. Measurements of the solar wind, the interplanetary magnetic field and high energy solar particles have been carried out since the beginning of the space age but almost certainly the results obtained are applicable only to a narrow belt of helio- latitude close to the ecliptic plane. (Marsden (1986) and references therein, Page (1975)). The solar atmosphere is clearly non-uniform with, for example, flare sites progressing through the latitude range 30 to 5 degrees as the solar cycle progresses. Various observations, including the relationship between geomagnetic activity and the position of activity on the solar disk, lead us to believe that the solar wind blows more or less radially from the sun. To sample what leaves the sun at a particular latitude it is therefore necessary to make *in-situ* measurements of the solar atmosphere at that latitude.

2. Some Science Goals

2.1. THE SOLAR WIND

There is good reason to believe that the solar wind flow pattern shows characteristic latitude differences. The equatorial regions are dominated by the complex stream-stream interactions caused by solar rotation and a characteristic variation of solar

---

Y. Kondo (ed.), Observatories in Earth Orbit and Beyond, 307–313.

wind speed with longitude. At high latitudes, the flow speed is expected to be higher, more uniform and nearly parallel to the magnetic field. Here it may be easier to understand the evolution of the solar wind with distance and to relate its properties to conditions in the solar source region.

The configuration of the Sun and heliosphere evolves throughout the solar cycle. During solar minimum conditions, coronal holes extend equatorwards from the poles to form large, well-defined open magnetic field regions. The corona then exhibits well-developed equatorial streams and departs significantly from spherical symmetry. High speed solar wind streams are observed at low latitudes and unipolar regions form to the north and south of the nearly flat equatorial current sheet. At solar maximum, such long-lasting coronal holes are generally absent and the pattern of solar wind streams and the heliospheric magnetic field becomes less ordered. The polarity of the Sun reverses about a year after solar maximum, leading to a major re-structuring of the heliosphere. There is a possibility that multiple current sheets occur at high latitudes.

2.2. The Heliospheric Magnetic Field

The heliospheric magnetic field is known to vary with latitude. In equatorial regions the field is wound into a spiral by the rotation of the Sun and at large distances becomes azimuthal. In polar regions the influence of solar rotation decreases and presumably the field becomes radial and weaker. (Models indicate that the radial field falls off as R^{-2} whereas the azimuthal component decreases as R^{-1}). Jokipii and Kota (1989) however suggest that large transverse field fluctuations can be expected in polar regions.

2.3. Energetic Charged Particles from the Sun

It is unclear whether energetic particles accelerated in active regions on the sun can propagate to high latitudes. Solar flare particles may originate and propagate primarily in the complex magnetic field structures at low solar latitudes, but their source regions and/or propagation may also extend to considerably higher latitudes, at least during part of the solar cycle. Interplanetary acceleration processes, presumably related to interplanetary shocks, have been shown to occur at least up to the highest heliographic latitude reached by Voyager 1 (38°). Ulysses will extend observations to cover the full range of heliolatitudes.

2.4. The Cosmic Radiation

The solar wind and the frozen-in magnetic field carried by it prevent a part of the cosmic radiation from penetrating to the inner heliosphere. Knowledge of the primary cosmic ray spectrum and composition outside the heliosphere is something which has been sought for many years. If the heliospheric magnetic field at high latitudes is indeed essentially radial, it may be that the cosmic radiation arriving over the solar poles is unmodulated by the solar field and is therefore more representative of the cosmic radiation in the local interstellar medium. There is an almost complete lack of information on how galactic cosmic rays are modulated at high solar latitudes. Measurements of the flux variation with latitude are therefore particularly interesting. Extrapolation to high heliographic latitudes of our present understanding of cosmic ray transport in the heliosphere, derived from observations near the heliospheric equator, suggests that the particle gradient and curvature drifts in the large-scale magnetic field and heliospheric current sheet may play an important role in cosmic ray access to the heliosphere and solar modulation (e.g. McKibben 1986, Smith and Thomas 1986). If drifts are important, then following the next solar field reversal (to positive magnetic field in the northern hemisphere, expected in 1991/92), the net flow will change from outward to inward over the solar poles for positively charged particles. Thus a definite latitude variation of ion intensities should be detected, with the opposite behaviour expected for electrons.

Most of the instrumentation on Ulysses is designed to study features of solar wind, the heliosphere magnetic field and energetic charged particles (Table 1.)

Further instrumentation is designed to study:

(a) Solar x-rays which generally accompany the production of high energy electrons at the sun.

(b) Cosmic gamma-rays. The Ulysses measurements will be combined with measurements from other spacecraft in an attempt to pin-point the origin of cosmic gamma bursts.

(c) Radio and Plasma Waves. These may be produced at the sun, by interactions in the solar wind and as electrons spiral about the interplanetary field.

(d) Cosmic Dust. Investigation of the physical and dynamical properties of the dust as a function of heliocentric distance and ecliptic latitude should help determine the relative significance of comets, asteroids and meteors as dust sources.

(e) Interstellar Neutral Helium. The solar system moves through the interstellar gas at about 20 km/s. Neutral helium entering follows Keplerian orbits in the gravitational field of the sun. Measurements made at the widely separated points which are not co-planar with respect to the symmetry axis of the helium flow direction should permit determination of arrival velocity.

Two other investigations make use of the radio signals which are supplied to transmit data and to track the spacecraft.

(f) Coronal Sounding. When the radio signals from the X and S band transmitters pass through the solar corona en route from spacecraft to earth it should be possible to determine the total electron content through which the radio waves have passed and to follow fluctuations in electron density.

(g) Gravitational Waves. A gravitational wave produced for example by a violent event in a galactic nucleus may perturb the spacecraft orbit sufficiently to be seen in the tracking Doppler data. A measurement accuracy of around 3×10^{-14} is required.

A detailed description of the scientific instruments can be found in Wenzel et al. (1983).

3. Spacecraft Trajectory

Ulysses will be launched in the ecliptic plane toward Jupiter by a combined two-stage IUS/PAM-S upper stage solid rocket motor, carried on board the NASA space shuttle. The Jovian gravity field will deflect the suitably directed spacecraft

TABLE I
Scientific Investigations on ULYSSES

Investigation/ Principal Investigator	Measurement	Collaborating Institutes
Magnetic Fields (HED) A. Balogh, Imperial College, London, UK	Magnetic fields: ± 0.01-± 44000 nT	JPL
Solar Wind Plasma (BAM) S. J. Bame, Los Alamos Nat Lab, USA	Ions: 257 eV/Q - 35 keV/Q e⁻ : 1--903 eV	NASA/ARC JPL HAO, Boulder UCLA NASA/MSFC MPAe, Lindau
Solar Wind Ion Composition (GLG) G. Gloeckler, U. Maryland, USA J. Geiss U. Bern. CH	Ion composition: 145 km/s (H)-- 1352 km/s (Fe^{+8})	NASA/HQ NASA/GSFC TU Braunschweig U. Natal MPAe, Lindau
Low Energy Ions and Electrons (LAN) L. Lanzerotti Bell Laboratories, USA	Ions: 50 keV--5 MeV e⁻: 30--300 keV ion composition: 0.2--15 MeV/n(Fe)	UCB U. Kansas APL Obs. Paris U. Thrace U. Birmingham
Energetic Ion Composition and Interstellar Gas (KEP) E. Keppler, MPAe, Lindau, FRG	Ion Composition: 80 keV--15 MeV/n Neutral He	Imperial College Geophys. Obs. Kiruna MPI, Heidelberg Aerospace Corp. U. Bonn
Cosmic Rays and Solar Particles (SIM) J. A. Simpson, U. Chicago, USA	Nuclei: 0.3 - 600 MeV/n e⁻: 1--2000 MeV	Imperial College ESA/SSD NRC, Canada U. Kiel CEN, Saclay Danish Sp. Res. Inst. NRC, Milan MPI, Heidelberg U. Maryland MPAe, Lindau

TABLE I *(continued)*

Radio and Plasma Waves (STO) R. G. Stone NASA/GSFC, USA	Plasma Waves: 0-60 kHz Radio: 1--940 kHz Magnetic field: 10--500 Hz	Obs. Paris U. Minnesota CNET/CRPE, St. Maur
Solar X-Rays and Cosmic γ-Ray Bursts (HUS) K. Hurley, Space Sciences Lab, Berkeley, USA and CESR, Toulouse, F. M. Sommer, MPI Garching, FRG	X-, γ-rays: 5--150 keV	Obs. Paris SLR, Ultrecht NASA/GSFC CESR, Toulouse
Cosmic Dust (GRU) E. Grün, MPI, Heidelberg, FRG	Dust Particles 10^{-16}-10^{-9}g	U. Canterbury NASA/JSC U. Bochum ESA/SSD
Radio Science/ Coronal Sounding (SCE) H. Volland, U. Bonn, FRG	Spacecraft dual-frequency ranging and Doppler data	U. Bochum
Radio Science/ Gravitational Waves (GWE) B. Bertotti U. Pavia, I	Spacecraft Doppler data	CNR, Frascati U. Uppsala JPL
Interdisciplinary/ Directional Discontinuities J. Lemaire, Inst. d'Aeronomie Spatiale de Belgique, Brussels, B		Aerospace Corp.
Interdisciplinary/Mass Loss and Ion Composition G. Noci, Obs. Arcetri, Florence, I		

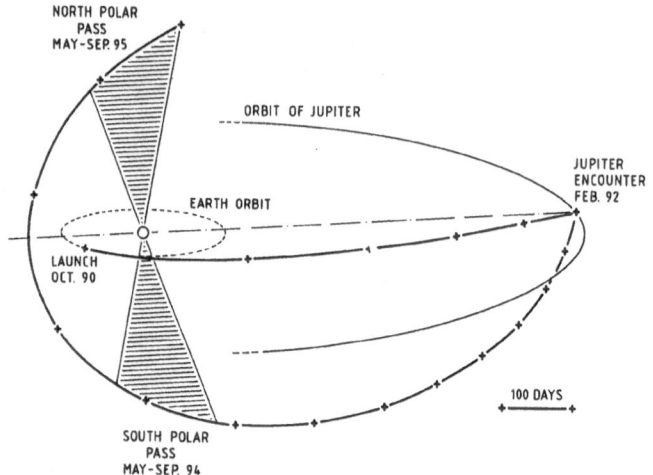

Fig. 1. Typical ULYSSES trajectory viewed from 15 deg above the ecliptic plane. Crosses are shown at 100-day intervals.

so that it travels southward from the ecliptic plane to pass over the south solar pole about 3–1/2 years after launch. At this point Ulysses will be approximately 2.3 AU below the ecliptic. Perihelion at 1.4 AU will be reached about 6 months later still and nominal end of mission will arrive following the north solar polar pass in September 1995. The trajectory is illustrated in Figure 1. Launch in October 1990 (the window is 18 days long) will mean that most of the time spent at high solar latitudes should occur during a fairly quiet part of the solar cycle. This should mean that heliospheric structure will be relatively stable and well ordered and that it will be less difficult to differentiate between genuine latitude variations and random time variations.

4. Operations and Data Recovery

Ulysses will be tracked by NASA's Deep Space Network, operations being directed by a small ESA team resident at the Jet Propulsion Laboratory in Pasadena. Real time data will be transmitted at 1024 bits per second for 8 hours each day. During the remaining 16 hours, data continuity will be assured by using on board tape recorders which will record scientific data at 512 bits per second.

Tapes (Experiment Data Records) of cleaned and sorted data will be provided by JPL to each Principal Investigator, who carries the responsibility to pass on the relevant data to his co- investigators. A Common Data Record will also be prepared to assist data correlation between instruments and to help identify when events of special note occurred. This Record contains selected channels of relatively raw data extracted from each of the main experiments carried on the spacecraft.

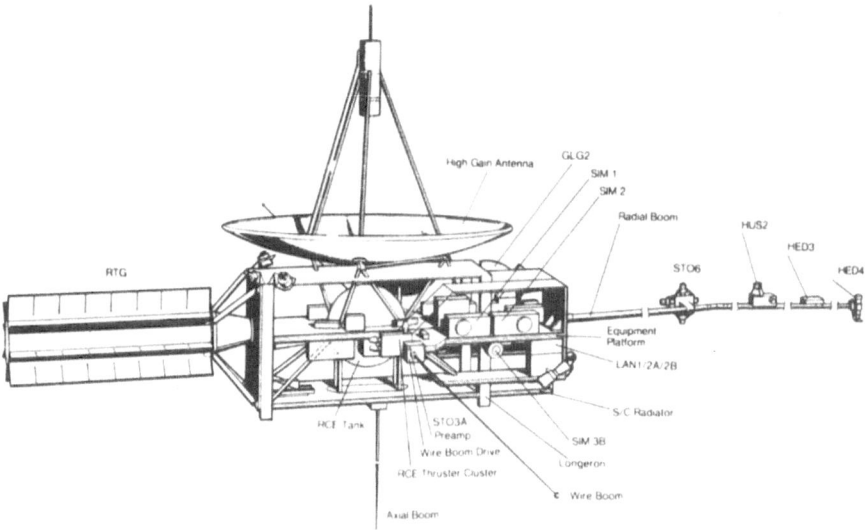

Fig. 2. Schematic of the ULYSSES spacecraft. The experiment codes are listed in Table 1: the number following the code identifies the sub-unit of a specific experiment

5. The Spacecraft

The spacecraft layout is illustrated in Figure 2. It measures about 3 meters across and about 2 meters in height with booms folded. At launch it will weigh approximately 370 kg including the 55 kg of scientific instrumentation. Because Ulysses must operate out to 5.4 AU from the sun, power is obtained from a Radioisotope Thermoelectric Generator (RTG) which will provide close to 290 watts at launch. The spacecraft will be spin stabilized at 5 rpm with information for logic operations coming from a sun sensor. At some parts of the trajectory, it will not be easy to keep the solar aspect angle above the minimum acceptable and at the same time to point the high gain parabolic antenna at the earth. S band telemetry is used for the up-link and both S and X band for the down-link. S band is at 2.3 GHz and X band at 8.4 GHz.

References

Jokipii, J.R. and J. Kota: 1989, 'The polar heliospheric magnetic field', *J. Geophys.Res.*, **16**

Marsden, R.G.: 1986, *The Sun and the Heliosphere in Three Dimensions*, Reidel, Dordrecht

McKibben, R.B.: 1986, in R.G. Marsden, ed(s)., *The Sun and The Heliosphere in Three Dimensions*, Reidel, Dordrecht, 361

Page, D.E.: 1975, *Science* **190**, 845

Smith, E.J. and B. Thomas: 1986, *J. Geoph. Res.* **91**, 2933

Wenzel, K.-P., R.G. Marsden and B. Battrick, 1983, *The International Solar Polar Mission - Its Scientific Objectives*, European Space Agency SP-1050

(H) DATA ANALYSIS ACTIVITIES

SCIENCE OPERATIONS FOR FUTURE
SPACE ASTROPHYSICS MISSIONS

GUENTER R. RIEGLER

Astrophysics Division, NASA, Washington, DC 20546, USA

Abstract. Plans for astrophysics science operations during the decade of the nineties are described from the point of view of a scientist who wishes to make a space-borne astronomical observation or to use archival astronomical data. In the process of preparing a proposal, making an observation, and carrying out data processing, analysis, and dissemination of results, the scientist will be able to use a variety of services and infrastructure, including the "Astrophysics Data System". The current status and plans for these science operations services are described.

1. Introduction to Science Operations

Until recently, the Astrophysics community had acess to data from just a few Astrophysics missions. With a small number of datasets, the use of mission-unique data and analysis tools was considered to be acceptable. With the launch of a large number of Astrophysics missions in the timespan of a few years (see Figure 1), a better approach had to be found. The National Aeronautics and Space Administration (NASA) Astrophysics Science Operations Program was established as an Astrophysics-wide program in order to encourage multi-mission, panchromatic research in Space Astrophysics. By fostering coordination and cooperation among all mission operations and data analysis efforts in Space Astrophysics, NASA expects to maximize the scientific return from operating Astrophysics missions, as well as from existing Space Astrophysics data.

The term *Science Operations* includes four areas:

mission operations, typically carried out at a NASA field center or a mission center,

research programs, consisting of guest observations and archival research by members of the Astrophysics science community,

science operations services, including multi-mission archive centers and science databases, and

Astrophysics Data System (ADS), providing the data-related infrastructure for all of the preceding items.

2. Science Operations: In Transition

The principles and day-to-day execution of Astrophysics science operations are in a state of transition. Although all Astrophysics missions have their unique history and future plans, they tend to evolve towards the same long-term goals.

Y. Kondo (ed.), Observatories in Earth Orbit and Beyond, 317–321.
©1990 *Kluwer Academic Publishers.*

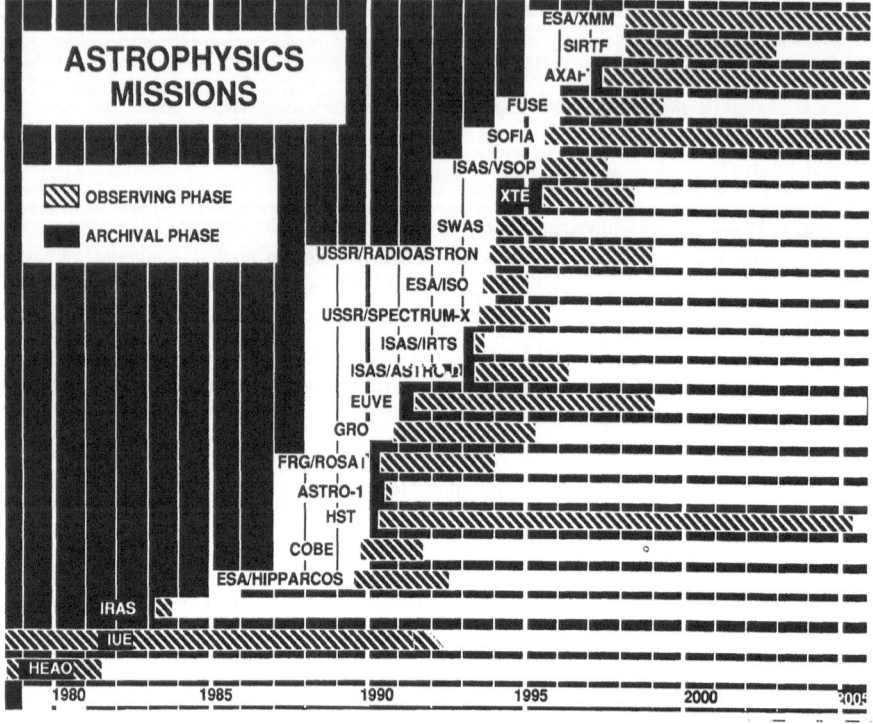

Fig. 1. Timeline of the major Space Astrophysics mission. The active operations/observation phases and subsequent archival research phases are shown as a function of time.

The character of Astrophysics missions is changing from Principal-Investigator (PI) instruments (or even PI-type missions) to facility-class observatories, where the instruments are still built by PIs, but many Guest Observers are expected to use them. Several of these missions are also planned to operate for such long periods of time that we expect a significant turnover in technical and scientific personnel during the active life of the mission. Furthermore, analysis methods and computing hardware will evolve through several generations during the data analysis phase of these missions.

For PI instruments, most of the data analysis was, in the past, carried out by members of the PI team at the PI institution. In the future we expect to see distributed analysis, primarily carried out by Guest Investigators at their home institutions.

The character of scientific research is also expected to evolve, from single-mission or single-wavelength research, to science topic-oriented, panchromatic research. For example, a recent study showed that of all the scientists who used data from the Infrared Astronomy Satellite (IRAS), only 30% considered themselves "IR Astronomers" per se, while 70% came from radio, UV/Optical, high-energy, or theoretical Astrophysics. To enable panchromatic research, we require:

– the ability to execute coordinated (simultaneous or contemporaneous) observa-

ASTROPHYSICS DATA SYSTEM
OPERATIONAL CONFIGURATION - 1991

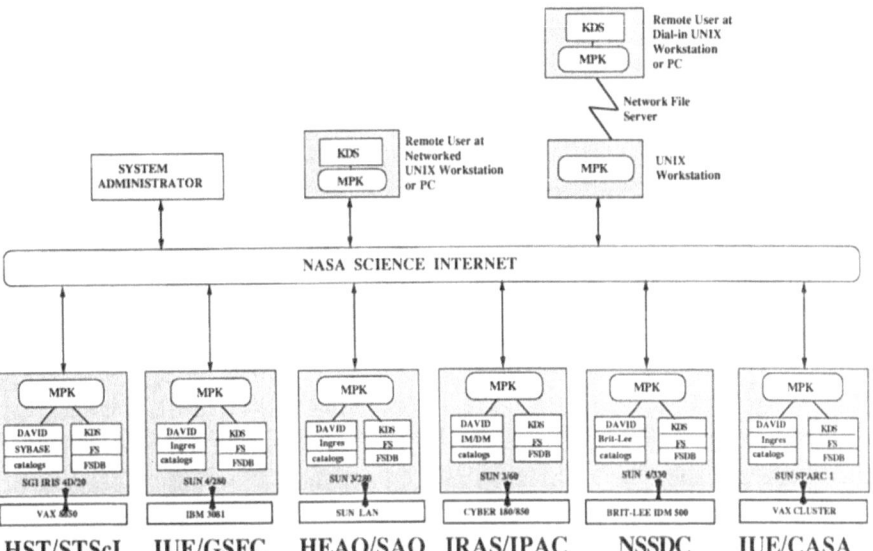

Fig. 2. The Directory Service of the Astrophysics Data System (ADS): Operational configuration of the Directory Service of the Astrophysics Data System, showing the message passing kernel (MPK), the data-independent access method to databases ("DIAM", also called "DAVID"), user interface, documentation manager, factor space (FS), and the factor space database (FSDB) poriton of the Knowledge Data System (KDS).

tions involving space- and ground-based observatories, and
– the ability to carry out multi-mission and multi-wavelength data analysis and interpretation.

This means that Astrophysics missions must supply the necessary expertise, as well as data, data analysis tools, and other services to enable and encourage such topic-oriented research.

3. Science Operations Needs for a Typical Guest Observation

The table below shows steps in a typical Astrophysics investigation, and examples of the types of science operations services required at each step. Some of these services are described in more detail below.

4. Science Databases and Other Science Operations Services

In response to requests from the science community, and after peer review, a number of services are either under development, or are already accessible to remote users. The *Astrophysics Software and Research Aids* Program explicitly solicits proposals for software packages, databases, operational tools, etc., and supports them after competitive science peer review.

TABLE I

INVESTIGATION TASK	SCIENCE OPERATIONS SERVICES REQUIRED BY INVESTIGATOR
Release of the NASA Research Announmcement (NRA)	Supplemental information and proposal preparation available electronically from mission science data center
Proposal writing	Obtain information about existing or approved observations through the Astrophysics Data System (ADS), on-line databases, or science operations services
Proposal submission	Electronic submission, electronic peer review, publication of approved investigations (abstracts, source lists) in databases
Observation planning	Database tools for coordinated observations
Data reduction and analysis	Remote data processing and analysis, portable software, standard software packages, software interchange, ADS, data format standards, databases, discipline-specific archives

The science database and science operations services include:

– convenient and inexpensive access to the SIMBAD database (developed at the Centre de Donnes Stellaires, Strasbourg, France),

– the National Extragalactic Database (NED, developed at the Infrared Processing and Analysis Center, IPAC, Pasadena, California), containing comprehensive data on extragalactic objects, including cross-references, literature citations, and complete abstracts of referenced articles,

– *MultiWaveLink*, an interactive database for the coordination of multiwavelength space and ground-based observing programs (developed at Pennsylvania State University), and

– *Comprehensive Atomic Spectroscopy Database for Astrophysics* (developed at the National Institute of Standards and Technology, Gaithersburg, Maryland).

As a matter of policy, NASA encourages and supports the wide dissemination of data to the astronomical community. Examples of such dissemination are the distribution, on CD-ROMs, of the HST Guide Star Catalog, of the National Space Sciences Data Center's machine-readable versions of frequently used astronomical catalogs, and of the Einstein Observatory's Imaging Proportional Counter (IPC) and High Resolution Imager (HRI) results.

5. Astrophysics Data System (ADS)

The *Astrophysics Data System* provides the infrastructure for locating data, and for the subsequent data analysis. The ADS Project is managed at the Infrared Processing and Analysis Center (IPAC, Dr. John Good, Project Manager) in Pasadena, California. Figure 2 shows the current configuration of the ADS.

It is designed to

– allow remote access by scientists at their home institution through the NASA Science Internet,

– permit scientific inquires (e. g. "where are UV high-resolution spectra of active galactic nuclei?") to be answered simultaneously by all ADS science center nodes ,

– locate data holdings, select data (sensor, correlative, and ancillary data), browse through the data archives, and order data for electronic or mail-order transmission,

– make the exact nature of the operating systems or database management systems at the various data centers (see Figure 2) transparent to the remote user.

After two years of development, the ADS is currently in the test phase, and training of new users and node managers has begun. The full ADS is expected to be operational in mid-1991.

The ADS is very similar in philosophy and design to ESA's European Space Information System (ESIS, developed by ESRIN, Frascati, Italy). Access to the ADS from outside the US will be possible through the NASA Master Directory (developed at the Goddard Space Flight Center, Greenbelt, Maryland) and via direct connection through the NASA Science Internet.

Two very important components of the infrastructure for Astrophysics science operations are communications and data format standards. As a result of the recommendation of the International Astronomical Union for the adoption of the "Flexible Image Transport System (FITS), NASA has adopted a policy for the use of FITS formats for the exchange of data (NOT for data files internal to reduction and analysis programs). In addition to special, mission-specific extensions, the FITS system accomodates basic images, random groups, ASCII tables, IEEE floating-point data, 3-D floating-point data, keyword hierarchies, and single-photon data. In order to assist missions and individual scientists in the use of FITS structure, a FITS Standards Office has been established at the Goddard Space Flight Center, Greenbelt, Maryland.

III. LAUNCH VEHICLES

(A) CURRENT & "NEAR" FUTURE

UNITED STATES LAUNCH VEHICLE SYSTEMS

ROBERT B. KRAUSE

U.S. Civil and International Payloads, Transportation Systems Division,
Office of Space Flight, National Aeronautics and Space Administration,
Washington, DC 20546, U.S.A.

Abstract. United States policy for national space launch capability provides for a balanced mix of launches, utilizing the Space Shuttle and Expendable Launch Vehicles (ELVs). It also directs government agencies to encourage and support the development of a domestic commercial expendable launch vehicle industry. This is to be accomplished by contracting for necessary ELV launch services directly from the private sector and by facilitating access by commercial launch firms to national launch and launch-related property and services they request to support these commercial operations.

The current mixed fleet includes the Space Shuttle and four expendable launch vehicles – Titan, Atlas, Delta and Scout. New small class launch vehicles, including Pegasus, are in development. In addition, studies are underway to assure that the United States has cost-effective, reliable access to space, heavy-lift launch capability, and a new manned spacecraft after the current Space Shuttle reaches the end of its operational life. This paper will highlight the current capabilities of the mixed fleet and summarize the plans for new or modified United States launch vehicles through the first decade of the next century.

1. Introduction

Assured access to space, sufficient to meet defined space goals, is at the heart of United States National Space Policy. Space transportation systems are to provide a balanced, robust, and flexible capability with sufficient resiliency to allow continuous operations even if failures in any one system occur. During the past four years, the focus of United States space launch strategy has been to restore the launch vehicles that were in use prior to 1986 to safe and reliable flight. The Shuttle, Titan, Delta and Atlas/Centaur vehicles all suffered significant failures that resulted in extensive accident investigations for all, as well as extended downtime for both Shuttle and Titan. The Shuttle and Titan programs have since made vehicle and launch processing improvements to increase vehicle reliability. The Space Shuttle has flown 9 successful missions since its return to flight in September 1988. Four of the eight planned flights in Fiscal Year 1990 have been completed. It is scheduled for ten flights in Fiscal Year 1991, eleven in Fiscal Years 1992 and 1993, and will achieve a planned steady flight rate of twelve in Fiscal Year 1994 and beyond. The Delta II launch vehicle has flown seven times, and the Titan III and Atlas I vehicles once each. The first launch of the Titan IV will be in 1990 and the first Atlas II in 1991.

National policy now mandates that the Space Shuttle only be utilized for those missions that require the presence of man, the Shuttle's unique capabilities, or where it is determined that the use of the Shuttle for launch of a payload is dictated by

Y. Kondo (ed.), Observatories in Earth Orbit and Beyond, 325–332.

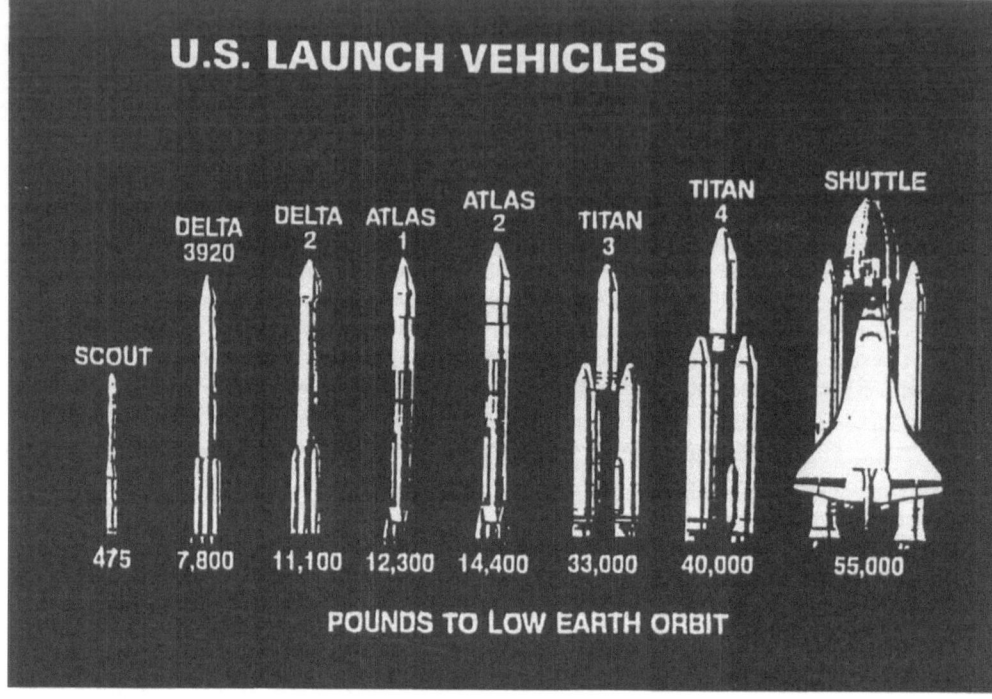

Fig. 1.

national security or foreign policy considerations. Commercial payloads have, there-fore, been off-loaded from the Shuttle and a domestic commercial launch industry has been encouraged by NASA through purchase of needed ELV launch services directly from the private sector.

In the near-term, United States launch strategy is characterized by the introduc-tion of several new production expendable launch vehicles, growth of the commercial launch services industry and continued improvement in the Space Shuttle.

In the long-term, a number of advanced space transportation systems studies and related technology programs are underway. These include the Shuttle-C cargo carrier element, an Advanced Launch System, the next manned spacecraft, and the Aerospace Plane.

2. Current and Near-term Launch Vehicle Studies

The current and near-term launch vehicles in the United States inventory are shown in Figure 1. A summary of Expendable Launch Vehicle performance capabilities is shown in Figure 2. A brief description of the capabilities by launch vehicle follows:

U.S. EXPENDABLE LAUNCH VEHICLES

VEHICLE	AVAILABILITY	PERFORMANCE (I = 28°) LBS			PAYLOAD FAIRING DIAMETER	ESTIMATED VEHICLE LAUNCH SERVICES COSTS 1990 $
		LEO	GTO	GSO		
SCOUT - ETR - WTR	NOW	570 460			2.9 AND 3.5 FT	$10-20M
DELTA II MODEL 6925 MODEL 7925	FEBRUARY 1989 1990	8,780 11,110	3,190 4,010	1,600 2,000	8 AND 9.5 FT 10 FT (FEBRUARY 1990)	$45-50M
ATLAS I	NOW	13,000 12,550	5,150 4,950		11 FT 14 FT	$65-70M
ATLAS II *	EARLY 1991	14,950 14,500	6,100 5,900		11 FT 14 FT	$70-80M
ATLAS IIA	1992	15,700 15,250	6,400 6,200		11 FT 14 FT	$80-90M
ATLAS IIAS	1992	19,000 18,500	8,000 7,700		11 FT 14 FT	$110-120M
TITAN II * (WTR)	SEPTEMBER 1988	4,200			10 FT	$35-40M
TITAN III WITH SRMU	1989 1992	30,500 38,000	11,000		13.1 FT	$145-155M
TITAN III / TOS	1992		13,000		13.1 FT	$190-200M
TITAN III / IUS WITH SRMU	1989 1992			4,200 5,000	10 FT	$245-255M
TITAN IV * / NUS WITH SRMU	1989 MID 1991	39,000 49,000			16.7 FT	$180-240M
TITAN IV * / IUS WITH SRMU	1989 MID 1991	49,000	15,000	5,200 6,600	16.7 FT	$280-340M
TITAN IV * / CENTAUR WITH SRMU	1990 MID 1991			10,200 13,500**	16.7 FT	$260-320M

* NOT COMMERCIALLY AVAILABLE
** CURRENT CENTAUR IS STRUCTURALLY LIMITED TO 11,500 LBS

Fig. 2.

2.1. SPACE SHUTTLE

The Space Shuttle, developed and operated by NASA, provides the nation's only means of manned access to, transportation in, and return from space. The unique capabilities of the Space Shuttle allow support of a wide variety of tasks, including lifting heavy payloads into orbit; retrieving, servicing and repairing satellites; and returning payloads to the earth. The task of satellite servicing and repair will become extremely important in the future as large facility-class assets, such as NASA's Great Observatories and Space Station Freedom, are in orbit for extended periods of time.

With its existing configuration, the Shuttle is capable of delivering up to 55,000 pounds to a 160 nautical mile low earth orbit (LEO) at an inclination of 28.5 degrees. A significant improvement in payload capacity will be achieved with the implementation of the Advanced Solid Rocket Motor (ASRM). The primary need for the ASRM is to improve Space Shuttle flight safety and reliability; however secondary benefits include reduced cost, improved performance, and assured access to space for large payloads. The design goal for the ASRM is to increase payload to orbit capability by 12,000 pounds, which restores the Shuttle to its full design capability.

2.2. TITAN

The Titan II was modified by the Department of Defense (DOD) for launches of smaller payloads to polar orbit. The Titan II vehicles were decommissioned ICBMs that were refurbished, equipped with hardware required for space launch, fully tested, and certified for space flight. The initial launch of the Titan II was in September 1988. It can place up to 4,200 pounds in low earth polar orbit from the United States Air Force (USAF) facility at Vandenberg Air Force Base (VAFB).

The Titan 34D is an operational DOD launch vehicle that is an evolution of the Titan III family of vehicles first launched operationally in 1966. With an Inertial Upper Stage (IUS), it is capable of delivering 4,200 pounds to geosynchronous orbit (GSO) from Cape Canaveral Air Force Station (CCAFS) and 27,600 pounds to polar orbit from VAFB. The commercial Titan III is almost identical to the Titan 34D except for a modified payload fairing.

Titan IV development began as a complement to the Space Shuttle and was expanded following the Challenger accident to launch critical payloads that were expected to be impacted by the Shuttle recovery schedule and cancellation of the Shuttle/Centaur program. The Titan IV with an IUS is scheduled for initial launch capability in 1990. It will be capable of delivering 5,200 pounds to GSO. Initial launch capability of the Titan IV/Centaur is planned for May 1990. It will be capable of delivery of 10,200 pounds to GSO. These Titan IV configurations will be launched from CCAFS. A Titan IV with no upper stage will be used for launches into polar and high inclination orbits from VAFB. It will be capable of delivering 39,000 pounds to a 100 nautical mile polar orbit.

A Titan Solid Rocket Motor Upgrade (SRMU) is under development and will provide improved reliability and producibility as well as 25 percent greater payload capability. It is planned for the eleventh Titan IV launch and can be utilized on the Titan III as well.

2.3. ATLAS

The Atlas family of vehicles includes the Atlas E, which is a decommissioned ICBM, presently used to launch payloads up to 1,800 pounds to low polar orbit from VAFB; the Atlas/Centaur which can launch up to 2,860 pound payloads to GSO from CCAFS; and the Atlas II which will be an upgraded version of the Atlas/Centaur. The Atlas II will have its initial launch capability in early 1991 and will be capable of launching intermediate class payloads (up to 5,900 pounds) into geosynchronous transfer orbit (GTO) from CCAFS.

A commercial Atlas IIA, which will be available in 1992, will have uprated engines. This will increase its capability to GTO to 6,200 pounds. A commercial Atlas IIAS, also available in 1992, will have four strap-on boosters and be capable of placing 7,700 pounds in GTO.

2.4. DELTA

The NASA Delta vehicle was derived from the Thor intermediate range ballistic missle, with improved first and second stages and the addition of several small solid rocket motors. Through a series of upgrades, the capability of this vehicle grew to approximately 1,450 pounds to GTO, 5,500 pounds to polar orbit and 7,600 pounds to LEO. The last NASA Delta vehicle was launched in 1989.

The Delta II launch vehicle was developed by the DOD as a medium class vehicle. The initial version, launched in February 1989 was capable of placing 8,780 pounds into a low earth orbit, 3,190 pounds into GTO, or 1,600 pounds into GSO. In 1990, an upgraded version will be introduced which will increase capability to 11,110 pounds to LEO, 4,010 pounds to GTO, and 2000 pounds to GSO.

2.5. SCOUT

The NASA Scout is a solid propellant, four stage booster with the capability to launch small payloads – approximately 570 pounds to low earth orbit and 460 pounds to polar orbit.

2.6. PEGASUS

The Pegasus is an air launched, solid propellant, three stage booster, which was commercially developed and recently had its initial and successful launch. Its capability is about 600 pounds to a polar low earth orbit or 275 pounds to GTO, using a NASA B-52 aircraft as the launch platform.

3. Long-term Launch Vehicle Systems

For the mid-1990s and beyond, a number of launch systems options are being considered. These include studies of the Shuttle-C cargo carrier element, the Advanced Launch System, the next manned spacecraft, and the National Aerospace Plane.

3.1. SHUTTLE-C

NASA is defining an unmanned cargo concept using the Space Shuttle Vehicle without the Orbiter. In its place is a cargo carrier to which the Space Shuttle main engines are attached. This concept, termed Shuttle-C (for cargo) would build on current Space Shuttle Program experience and infrastructure, using mature elements of the Program's ground and flight systems to provide a limited capability to satisfy potential future heavy-lift launch capability from 1994 through the point where the Advanced Launch System would be available.

Shuttle-C studies now underway will provide further definition of a Shuttle-derived launch vehicle that could be capable of delivering 100,000 to 150,000 pounds due east to low earth orbit or 85,000 to 115,000 pounds into polar orbit. The Shuttle-C concept could support Space Station Freedom assembly, enable planetary and geosynchronous missions, provide an alternate booster for Centaur-class payloads, and serve as a testbed for future launch vehicle elements.

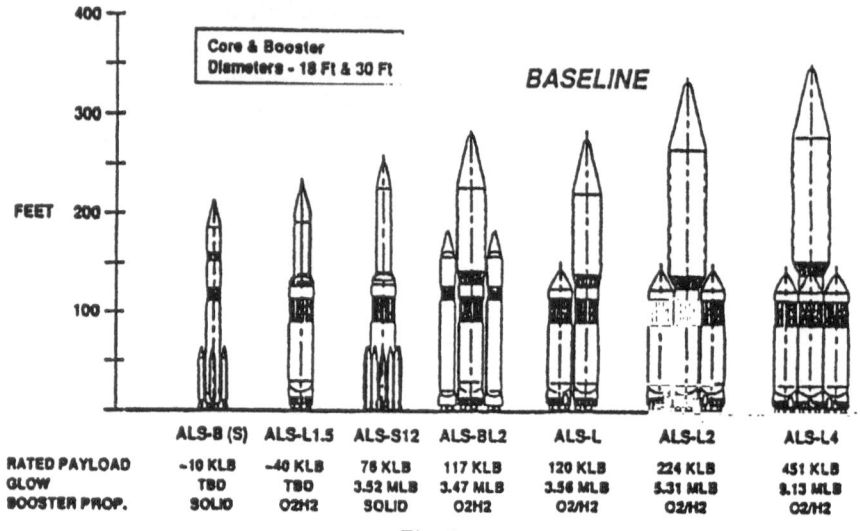

Fig. 3.

3.2. ADVANCED LAUNCH SYSTEM

The Advanced Launch System (ALS) is a joint DOD/NASA program to define concepts and develop technology applicable to a family of unmanned launch vehicles (see Figure 3). They are intended to provide a heavy-lift launch system that meets United States space launch needs and significantly reduces the cost of getting payloads into space. This program is intended to support a broad base of booster and propulsion concepts for both expendable and reusable vehicles with a payload capability range from 85,000 pounds to 150,000 pounds.

3.3. THE NEXT MANNED SPACECRAFT

The present Space Shuttle is expected to reach the end of its operational life in the 2005–2010 timeframe. Since a long lead-time is required to develop space transportation systems, planning for the next manned spacecraft is currently underway. Three concepts are being assessed: evolution of the present Space Shuttle system; a Personnel Launch System (PLS) in combination with man-rated cargo launch and return vehicles; and an Advanced Manned Launch System (AMLS) (see Figure 4). Several key objectives to improve cost effectiveness; increase reliability, maintainability and operability; and improve performance margins will be addressed in examining the three concepts.

The Shuttle evolution concept includes near-term system improvements/evolution of ground and flight operations, fuel cells, the auxiliary propulsion system, and major block upgrades involving the Orbiter, its main engines, and the External Tank.

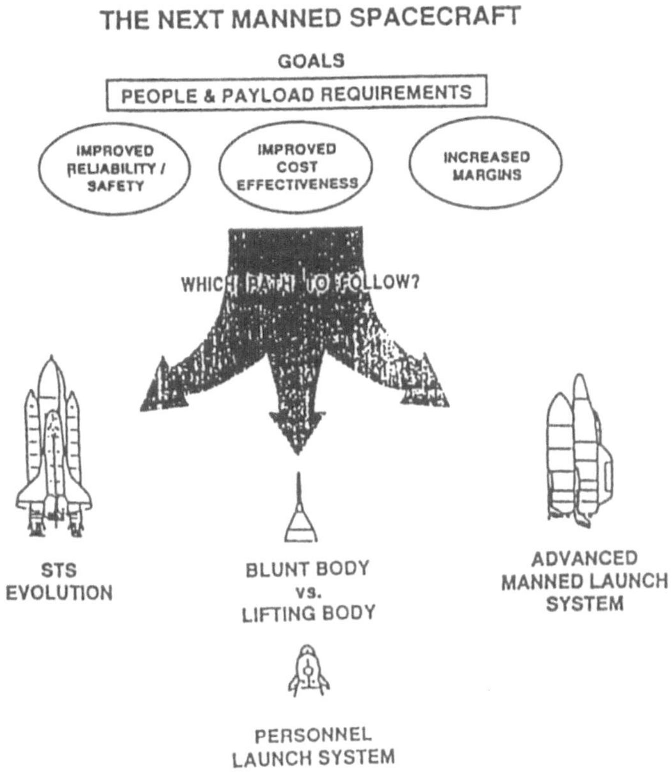

THE NEXT MANNED SPACECRAFT

GOALS

PEOPLE & PAYLOAD REQUIREMENTS

IMPROVED RELIABILITY / SAFETY

IMPROVED COST EFFECTIVENESS

INCREASED MARGINS

WHICH PATH TO FOLLOW?

STS
EVOLUTION

BLUNT BODY
vs.
LIFTING BODY

ADVANCED
MANNED LAUNCH
SYSTEM

PERSONNEL
LAUNCH SYSTEM

Fig. 4.

We will build on the existing engineering data base, while exploiting new technologies. Vehicle mold-line and configuration changes will be minimized.

The PLS concept will focus on possible vehicles dedicated to only carrying people. It includes configurations ranging from blunt body, Apollo-shaped vehicles to high lift-to-drag ratio winged vehicles launched on an expendable launch vehicle. This approach would also require unmanned cargo vehicles to deliver and return cargo and logistics support to the Space Station Freedom, using automated rendezvous and docking systems. The PLS configuration and size are to be studied.

The AMLS concept represents the next generation of a Shuttle system. It would fully exploit new technologies and provide improved design margins. The baseline AMLS concept is a two-stage reusable rocket configuration; however, partially reusable and expendable launch vehicle concepts will also be considered.

3.4. NATIONAL AEROSPACE PLANE (NASP)

The NASP is a program to demonstrate technologies for hypersonic flight and single-stage-to-orbit vehicles with potential space transportation applications. The NASP program is currently a research program intended to design, build and test fly an experimental flight research aircraft in the mid-1990s. If it is successful in demonstrating the technological basis for systems capable of horizontal takeoff from, and landing on, conventional runways; sustained hypersonic cruise and manuever in the atmosphere; and acceleration to orbit and return, the derived technology will be used to develop a follow-on space plane to be operational in the 2000–2005 timeframe.

4. Summary

The United States has established a balanced mix of manned and unmanned launch vehicles. Through government development of the Space Shuttle and Titan IV and commercial launch industry development/improvement of expendable launch vehicles, a capability margin above payload demand can be realized through the mid-1990s. Continued growth of mixed fleet capabilities is required to meet projected demands in the year 2000 and beyond. To this end, NASA is conducting systems definition studies of a Shuttle-derived cargo vehicle (Shuttle-C) and NASA and DOD are jointly conducting Advanced Launch Systems studies designed to introduce a new family of expendable launch vehicles by the turn of the century. Technology demonstrations in support of National Aerospace Plane development are being planned, with success leading to an operational space plane in the early part of the next century.

ESA's SPACE TRANSPORTATION PROGRAMME

J. FEUSTEL-BUECHEL

ESA Headquarters, D/STS, 8-10, rue Mario Nikis, 75738 Paris, France

and

W.WAMSTEKER

ESA IUE Observatory,P.O.Box 50727,28080 Madrid, Spain

1. Introduction

The overall development of ESA's Space Transportation Programme can be perceived in two major phases. The first phase finished with the transfer of the ARIANE 1-4 launch systems to Arianespace for utilisation and commercialization. The second phase was started after the success of the ARIANE launcher programme had created a sound basis for Europe's space industry in the evolving commercial space transportation market. This has lead to the definition of a second phase of the launcher development program which is expected to fulfil the needs for transport into space for many years to come. The wide objectives of the program makes it not only suitable for commercial applications but can also be expected to cover the launching needs for future Space Astrophysics missions.

Market studies of the future commercial space transportation needs show growing satellite masses and an increased worldwide competition. Europe therefore needs a powerful launcher with improved cost-effectiveness and improved reliability for geostationary missions. On the other hand the discussion about the optimum launcher, spacecraft size and orbits for Astronomy missions, as well as for Space Science missions in general, is still a matter under debate, as explained by Dyson elsewhere in these Proceedings. The experience obtained with IUE and EXOSAT -Observatory satellites which, due to their general user nature, do not have a rigidly preplanned observing program - seem to suggest that, at least for operational motives, high orbits appear preferable for such satellites. This is especially true as compared to the operational and planning difficulties foreseen with the recently launched Hubble Space Telescope, which is in low earth orbit (LEO). Of course, for firmly pre-scheduled observatories with a predetermined Science program, such as COBE and HIPPARCOS, this choice may not be obvious.

At this time, the ESA countries also plan for a European manned in-orbit infrastructure which will enable them to use the benefits that can be expected from manned space stations in low earth orbits. This requires a recoverable and reusable transportation system that launches man and equipment and returns them safely to Earth.

ARIANE 5 and HERMES are a unique and forward-looking combination which fulfils both tasks in an effective way. ARIANE 5 allows cost-effective lifting of heavier payloads to geostationary orbit and, at the same time, opens up the possibility of transporting European space station elements into their various low earth orbits.

Y. Kondo (ed.), Observatories in Earth Orbit and Beyond, 333-338.
© 1990 *Kluwer Academic Publishers.*

TABLE I
Ariane 5 Performance Characteristics

	Ariane 44L	Ariane 5
Single launch payload into GTO	4200 kg	6800 kg
Dual Launch into GTO	3800 kg	5900 kg
Payload int polar orbit, unmanned mission	6000 kg	12000 kg
LEO (550 km @ 28.5 degrees), unmanned mission	7000 kg	18000 kg
LEO (500 km @ 28.5 degrees), manned mission	n.a.	21000 kg
Design Reliability, unmanned missions	90%	98%
Design Safety, manned mission	n.a.	99.9%
Recurrent Cost per launch into GTO	100%	90%
Relative launch cost into GTO	100%	55%

The HERMES spaceplane, launched by ARIANE 5, enables Europe to master the technology required for manned space flight. In addition, HERMES is capable of transporting crew, equipment and payload to the European space station elements. This gives a greatly increased in-orbit experimental capacity which enables Europe to participate effectively in the utilisation of the manned space infrastructure.

2. The ARIANE 4 Launcher

The current version of the European launch system is the three stage Ariane, with the modular ARIANE 4 being the workhorse for the 90's. The ARIANE is commercialized by Arianespace, an organization composed all European industries, participating in its developement and based in France. Of the 38 ARIANE launches (August 1990) 33 were successful. The main characteristics of the Ariane 4 are given in Table I. The recent developments in the open launch market and the need for small and cheap launches has generated a development oriented towards additional capabilities in the form of multiple satellite launches in the 400-800 kg range as well as piggyback mode of operation fro up to six satellites of masses no larger than 50 kg each. This mode was first successfully deployed on the V35 launch which brought into orbit not only the SPOT satellite but also deployed six mini-satellites in a Sun-synchronous orbit.

3. The ARIANE 5 Programme

3.1. OBJECTIVES

For ARIANE 5, a number of ambitious goals have been set. These are summarized and compared to the most powerful version of Ariane 4 (AR 44L) in Table I, while Figure 1 shows the overall design together with the main technical data.

The aim is to make the total recurrent cost of using Ariane 5 for a dual launch into GTO at least 10% lower than that of an Ariane 44L vehicle, assuming eight

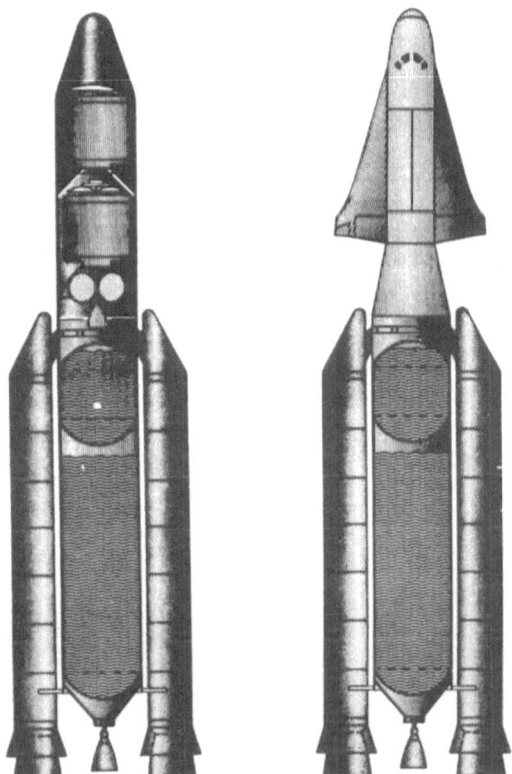

Fig. 1. Ariane-5 and Hermes, the future modular European Launching System

launches a year including four into GTO. On the basis of this target, Ariane 5 should provide a reduction of some 45% in the specific cost (cost per kilogram in orbit) compared with Ariane 44L. The production facilities, the design and development, which are an integral part of Ariane 5 programme, will support the production of 10 launchers a year. The new ELA-3 launch complex in Kourou (French Guyana) will also support this launch rate.

3.2. THE ARIANE 5 LAUNCH VEHICLE

In order to be able to perform such large variety of different missions with only one launcher, ARIANE 5 is made up of a standard lower composite and mission specific upper composites. For unmanned missions in the ARIANE-5 programme, a variety of payload fairings with 5.4 m diameter and lengths up to 21.7 m will be developed (see Figure 2). The launcher's powered flight will have two main phases. The lower-composite flight will be virtually identical for all missions. It begins with ignition of the HM60 Vulcan engine on the ground; once the proper functioning of this has been checked, the main P230 solid boosters will be fired to reach liftoff. At P230 burnout, these are jettisoned and the H150 core stage continues to supply

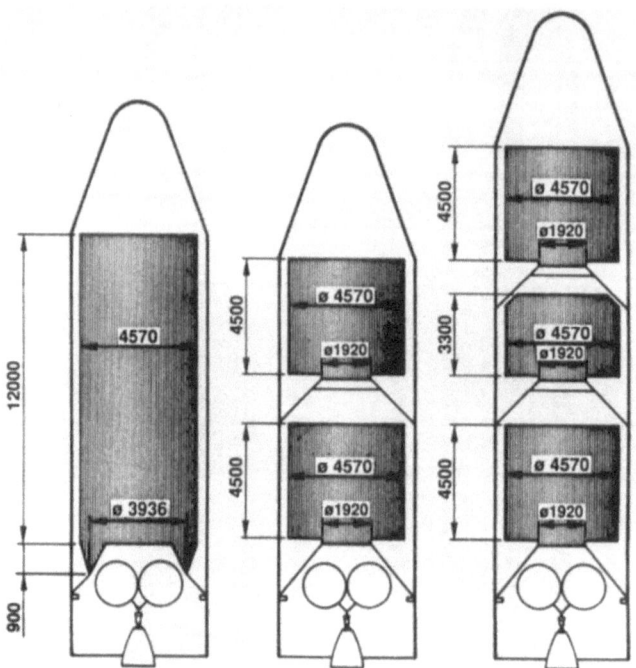

Fig. 2. The various fairings Ariane-5 and their performance characteristics.

the thrust needed for powered flight.

3.3. Schedule of the ARIANE 5 Development Programme

The development of Ariane 5 was started in 1984 with a preparatory phase which primarily concerned the critical cryogenic engine HM60 pre-development, the solid boosters and the overall launcher aspects. This part of the programme was concluded by end 1987. The development programme started early in 1988 and two development launches (501 and 502) for GTO missions are foreseen in early and late 1995. After these qualification launches, the start of the commercial programme is envisaged for 1996. A third development launch (503) in the Ariane 5 programme with an unmanned automatic Hermes is foreseen for mid 1998. These three test flights, intended to validate the in-flight functioning of the entire Ariane 5 vehicle, the functioning of the launch base and associated tracking facilities, are planned as follows:

– Flight A 501: The launch of a dual payload into GTO with the upper section consisting of a short fairing. This corresponds to the dimensioning of a commercial mission in terms of mission duration.
– Flight A 502: The launch of a single payload into a not yet chosen orbit, with the upper section consisting of a long fairing.

– Flight A 503: Launch of the unmanned Hermes spaceplane into a circular 500 km low earth orbit, inclined at 28.5.

Milestones in the Ariane 5 development programme are:

System concept review: late 1987
Preliminary design review of launcher stage and elements: 1987 to 1990
Ground qualification of launcher stage and elements: 1993 to 1994
End of system test in Europe: 1993
ELA 3 available: 1992
First test flight: early 1995
End of development for launcher of automatic payloads: 1995
First operational flight: early 1996
First unmanned Hermes mission: 1998
End of development: 1998.

The cost of the total Ariane 5 development programme is estimated by ESA at 4114 Million AU (1986), which consist of 618 MAU for the nearly completed preparatory programme and 3496 MAU for the actual development phase.

4. The HERMES Programme

4.1. OBJECTIVES

The development of the spaceplane HERMES will take into account the following major requirements:

Primary mission: servicing of the Columbus Free-Flying laboratory
Mission duration: 11 days in the normal servicing mode for the Columbus
Maximum flight rate: 3 mission per year
In-orbit mass: 21000 kg (orbit 500 km @ 28.5 deg. inclination) Crew: 3 astronauts
Cargo capability: 3000 kg upload; 1500 kg download; 18 m^3 total cargo volume
Service life: 15 years or 30 orbital missions.
In order to achieve three flights per year, two spaceplanes will be built.

4.2. SCHEDULE OF THE HERMES DEVELOPMENT PROGRAMME

The HERMES development programme is two-phased. The first phase, which will be concluded by the end of 1990, is to reduce the risks associated with the development of such a novel spaceplane. The reduction of the technological risks is one of the major goals of phase 1. Besides this the consolidation of the definition of the Hermes System and its equipment in order to allow reliable definition of the content, schedule and cost of Phase-2 is another important goal.

The second phase of development programme, culminating in the second flight of a Hermes spaceplane in early 1999, covers the Hermes System's development and qualification for subsequent operational utilisation. The initial qualification will be achieved by means of two orbital qualification flights. The first (H01) will be unmanned, and the second (H02) will include the crew.

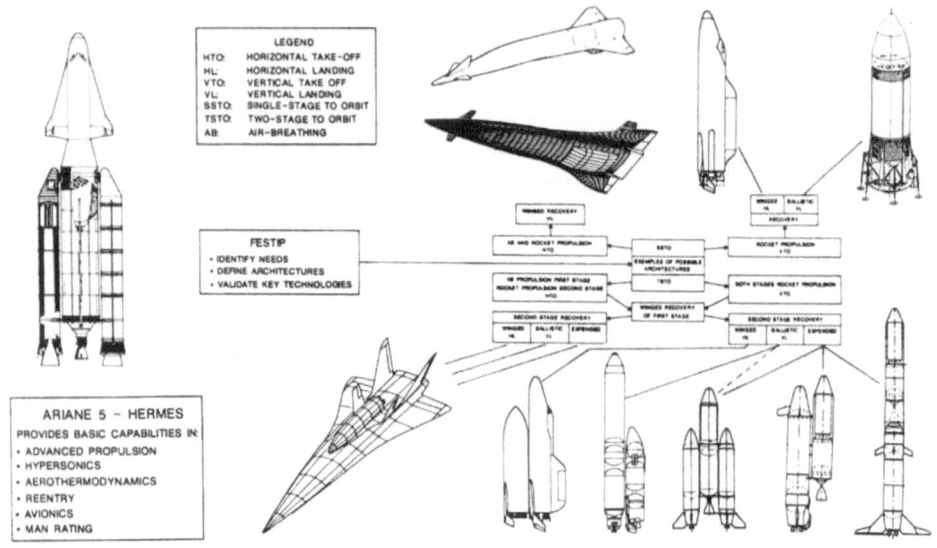

Fig. 3. Possible configuration of future space transportation systems based on the experience gained with ARIANE-5 and HERMES

5. Beyond **ARIANE 5 and HERMES**

Ariane 5 and Hermes will provide Europe with the basic technologies, expertise, industrial structure and management experience required for the subsequent development of future Space Transportation Systems. Many options are discussed today (see Figure 3). Which ones will eventually be chosen can not be predicted at present, but continuous studies are required if Europe is to derive a continuing and expanding benefit from the large effort it is now undertaking with Ariane 5 and Hermes.

6. Conclusion

It is clear that the developments planned for the ESA STS represent an ambitious programme mainly driven by the forward look needs of the communications, industrial and space sciences communities. The good performance in the Ariane 4 programme with a 90% success rate suggests that we might look forward to a launch system which will in the future also be capable to support the various requirements of a vigourous Space Astrophysics programme. One foresees a sufficiently modular launch capability to support both major Space observatories as well as the smaller type missions which are an fundamental part of the overall needs of Astrophysics from space.

LAUNCH VEHICLES OF ISAS

YASUO TANAKA

Institute of Space and Astronautical Science
3-1-1 Yoshinodai, Sagamihara, Kanagawa-ken 229, Japan

Abstract. The Institute of Space and Astronautical Science (ISAS) has developed a series of launch vehicles for delivering modest scale scientific satellites into space. This capability is unique for an inter-university research institute. The present ISAS launch vehicle, M-3S II, carries a 770 kg payload into a low earth orbit (LEO). A new program for the development of a more capable vehicle, M-V, started recently, which will have a LEO capability three times that of M-3S II. The enhanced launch capability will further expand the scope of the ISAS missions to include planetary explorations.

1. Introduction

Before describing the launch vehicles of ISAS, it may be useful to describe the two organizations responsible for all space activities of Japan. One is the ISAS operating under the Ministry of Education, Science and Culture. ISAS is responsible for implementing all aspects of space science research programs. The other organization is National Space Developing Agency (NASDA) under the Science and Technolgy Agency. NASDA is responsible for all applications space programs. The individual programs of both space organizations are reviewed and approved by the Space Activities Commission belonging to the Prime Minister's Office from the point of view of complementarity and coherence.

ISAS is a national inter-university research institute, serving the whole of the space science community in Japan. The role of ISAS is the coordination, planning and implementation of the space research programs for the university scientists throughout the country with their direct involvement.

While ISAS is small in scale (total $\sim$ 300 of which $\sim$ 100 are academic staff), it has some unique features for a research institute. ISAS has developed and possesses its own satellite launch vehicles and launch capability. This feature, in addition to the in-house capabilities of project management, design, tests, tracking and data processing, allows us to conduct all aspects of planning and implementation of scientific missions. ISAS has so far launched nineteen satellites in the past twenty years. Though ISAS missions are all modest in scale, the frequent mission opportunities are enormously important. They ensure continuity of investigations and steady development in each space science discipline. Furthermore, frequent opportunities also allow active engagement of young scientists.

2. Launch Vehicles

In order to avoid duplicate efforts by the two space organizations, the government has accorded ISAS and NASDA complementary roles. ISAS develops launch vehicles consisting of solid propellant motors limited until recently to a booster diameter

Y. Kondo (ed.), Observatories in Earth Orbit and Beyond, 339–342.

TABLE I

Basic parameters of the Japanese launch vehicles

Vehicle		M-3S II	M-V	H-I	H-II
Organization		ISAS	ISAS	NASDA	NASDA
Propellant		all solid	all solid	liquid	LH$_2$/LOX
		3-stage	3-stage	LH$_2$ 2nd stage	2-stage
Total Weight (tons)		62	128	140	260
Payload (tons)	LEO	0.77	~ 2	3.0	10.0
	GEO			0.55	2.2
Nose Fairing (m dia.)		1.6	2.5	2.44	4.0 (or 5.0)
Operational Status		current	1995 ~	current	1993 ~

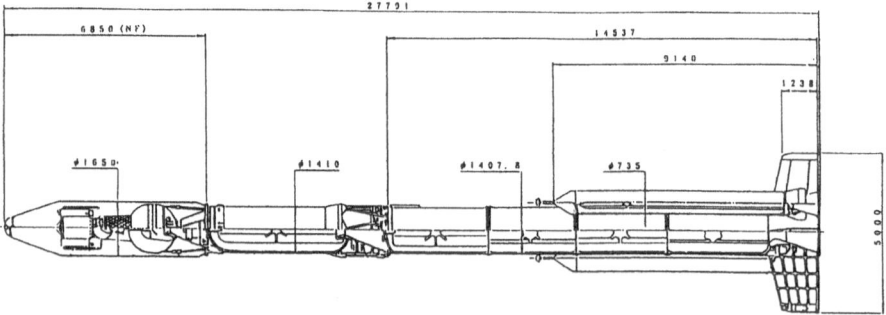

Fig. 1. Schematic diagram of the M-3S II.

of 1.4 m. On the other hand, NASDA is in charge of developing bigger, liquid fuel motors.

ISAS has been steadily improving the capability of its launch vehicle which we call Mu-rocket within the 1.4 m diameter limitation. The current ISAS launch vehicle is named M-3S II and is the latest version of Mu-rocket, which is a three-stage solid propellant motor assisted by two strapped-on boosters. Figure 1 shows a schematic diagram of the M-3S II. This vehicle can launch a 770 kg payload into a low earth orbit (LEO). Also the M-3S II launchers sent two missions, Sakigake and Suisei, to Comet Halley in 1985; these were the first interplanetary missions for ISAS. The X-ray astronomy observatory Ginga, and the auroral mission Akebono were also launched by these vehicles. The M-3S II launchers will bring two more missions, Solar-A for the next solar maximum and Astro-D an X-ray astronomy observatory each weighing about 420 kg, into near earth (500 km circular) orbits in 1991 and 1993, respectively.

Further enhancement of the ISAS launch capability has been strongly desired

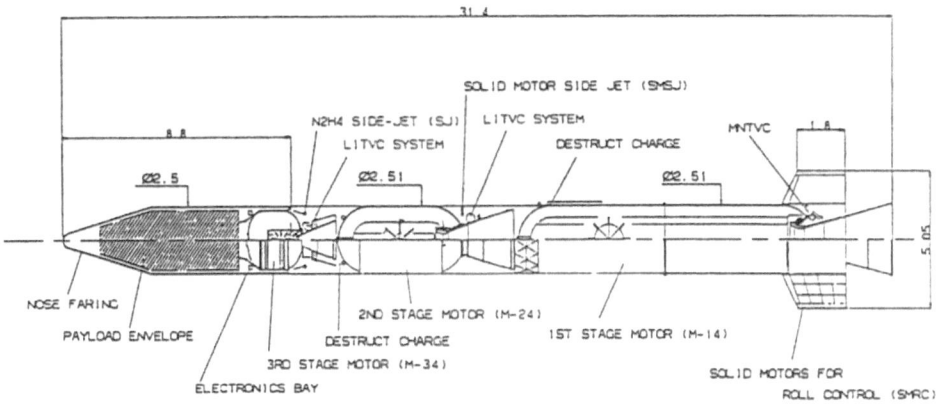

Fig. 2. The configuration of the M-V vehicle.

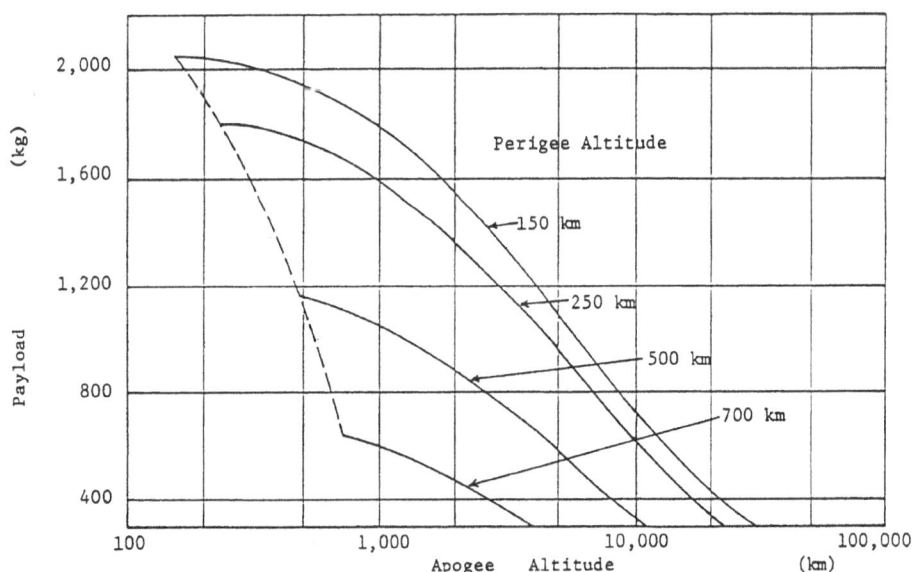

Fig. 3. Payload capability of the M-V vehicle.

by the scientific community in order to fulfill demands in the coming decade for
conducting planetary explorations as well as delivering larger and more sophisticated satellites into near earth orbit. Recently, a new ISAS program for further
development of solid-propellant launch vehicles has been approved.

The new launch vehicle named M-V will have a capability about three times
that of the M-3S II. The present design configuration of the M-V vehicle is shown
schematically in Fig. 2. The launcher consists of three stages of solid propellant
motors with an outer diameter of 2.5 meters. As shown in Fig. 3, the M-V vehicle
will be capable of launching a payload of ~ 2 tons into LEO, or carrying roughly 500
kg to Venus or Mars with an aid of a kick motor. The M-V development program
started this year, and will launch MUSES-B, the first space VLBI mission (VSOP),
in early 1995. The basic parameters of the ISAS launch vehicles, M-3S II and M-V,
are summarized in Table I.

For the above range of payload mass, the solid propellant vehicles are superior to
the liquid-fuel vehicles for modest scale satellites and provide low cost and reliable
delivery to orbit. In the design of the M-V vehicle, advanced technologies are employed to improve performance. For example, the motor cases are fabricated with
a newly developed high-performance steel, and the second-stage and third-stage
motors are equipped with deployable nozzles which give large expansion ratios.

For large payloads, the other space organization, NASDA, has the H-I rocket,
which is currently operational, with a LEO capability of 3 tons. NASDA has also
been developing a next generation launch vehicle, the H-II rocket. The H-II will
be powered by two stages of liquid hydrogen engines and capable of launching 10
tons for LEO or a 2.2-ton geostationary satellite. The main features of H-I and H-II
vehicles are also listed in Table 1. A possibility of occasional use of these vehicles of
NASDA for scientific missions of ISAS is not excluded, whenever the need is fully
justified. Thus, in addition to the regular programs with our own M-V vehicles, this
possibility allows us to consider larger-scale cooperative missions in future.

U.S.S.R. LAUNCH SYSTEMS

The invited speaker from Moscow for this talk was unable to attend the Colloquium due to circumstances beyond his control. The following data concerning the U.S.S.R. Launch Systems have been taken from information published in Soviet journals.

TABLE I

Payload (metric tons).

	L.E.O.	G.S.O.	I.P.T.O.
Soyuz	7	-	1.6
Proton	21	3 (approx.)	5 (approx.)
Energia	105	18	26

Initial inclination = 51.6 degrees

Note:
L.E.O. = Low Earth Orbit
G.S.O. = Geosynchronous Orbit
I.P.T.O. = Interplanetary Transfer Orbit

Y. Kondo (ed.), Observatories in Earth Orbit and Beyond, 343.
©1990 *Kluwer Academic Publishers.*

(B) "LONG-TERM" FUTURE

LAUNCH VEHICLES OF THE FUTURE:
EARTH TO NEAR-EARTH SPACE

G. A. KEYWORTH, II

Director of Research, Hudson Institute, Indianapolis, Indiana, U.S.A.

None of us thought, when this colloquium was scheduled, that the timing would enable it to become a celebration as well. The launch, after years of postponements, of the Hubble Space Telescope, has cast a galactic glow over the proceedings here this week. But at the same time, the frustrating delays caused by the collapse in 1986 and very slow regeneration of the U.S. space launch capabilities since then make this discussion of near-earth access very pointed.

1. Current Situation

As we know, the sheer momentum of the U.S. Space Shuttle Program has dominated our perceptions of space launch for a decade and a half. It reached its peak in the early 1980s when our national policy placed nearly total reliance on the Shuttle as our means of access to space. It was a policy doomed to fail, for obvious and not-so-obvious reasons.

The most obvious is simply the vulnerability of such high reliance on a manned vehicle. Any accident necessitates a review and program of risk-reduction that is as complete as is our respect for the people who fly them.

A second problem with reliance on the Shuttle has been the resulting pressure to come up with a bus-company-like schedule, rather than evolving its flight frequency as experience accrues.

The ability to carry out 24 launches a year looked highly dubious even before the Challenger tragedy. If we could manage even half that number now we'd consider ourselves to be doing well.

A third problem is a strong bias toward large and consequently expensive payloads, a bias that is increasingly anachronistic. Much of the technology developed in the years since the Shuttle was designed, especially in microelectronics, has made smaller payloads more attractive. Smaller, smarter payloads have applications across the board, but they're especially important for space science. We need affordable experiments. We need numerous experiments so that more people can take active roles in them. And we need experiments that can be planned, built, and conducted on much shorter time scales than we've been seeing. That seems to me to be a prerequisite to drawing good young people into space science.

A fourth problem is cost. By any measure current space access is expensive. There are several reasons for that. One is that the kind of rocket technology we use has inherent limitations on performance; those are reflected in the cost. Another is that the sheer complexity of the Shuttle requires an enormous personnel and mechanical infrastructure. Choose your number, but anywhere from 6,000 to 14,000 people are involved, one way or another, in a Shuttle launch. Finally, we don't really

Y. Kondo (ed.), Observatories in Earth Orbit and Beyond, 347–354.
© 1990 *Kluwer Academic Publishers.*

know what the real cost of a Shuttle launch is. When you have a state-run enterprise that dominates its market, accurate financial measures are elusive.

2. Future Systems

All that I've described with regard to current systems is reality, and I don't mean to put value judgment on it. It's simply where we are in the stages of development of U.S. space technology. Over twenty years it's no surprise that the system begins to look dated and inflexible. But it has often been awkward, even painful, to address these realities head-on, simply because so much effort has been invested in the current launch systems and because they must continue to serve our needs for years to come. Yet it is timely, in fact overdue, to look at future technologies and systems for space access. And President Bush's real commitment to and call for us to begin the next chapter in space exploration makes it essential.

Let me set out some criteria that future systems should try to meet.

First, and foremost, is cost. A cost on the order of $5000 per pound to put something into low-earth orbit obviously restricts who or what can get there. From the perspective of space science, even a little experiment becomes a big expenditure, especially when compared to the costs of doing other, non-space research. As I'll discuss, we can set as a target a cost of about two orders of magnitude lower.

My second criterion for a new access system is flexibility of payload size. Not everything we want to put into space is the size of the Space Telescope. In the early years of the space program, bigger was better. Not any more. Clever technology makes it increasingly more effective to use multiple deployments of special purpose satellites.

The small satellite, weighing only a few hundred pounds, is the space-borne equivalent of the personal computer; it can be assembled quickly, using readily available hardware, and it can be launched on short notice with little fanfare. Just as computers are shifting away from emphasis on large mainframes to distributed microcomputers tailored to specific applications, satellite users and builders now look to applications that can be met with smaller and cheaper satellites. The billion-dollar multi-ton satellite is the equivalent of the mainframe computer; it takes a decade to build, is so expensive that only a few can be afforded, and has to line up launch times years in advance. In both cases – satellites and computers – the key technologies are being decentralized and put more and more under the affordable control of individual users. The future looks increasingly like it will be moving to "LightSats" and other PC-equivalents in space. And that, of course, was always supposed to be one of the benefits of a "space age."

This is likely to be particularly important for space science aimed at better understanding of the earth itself, where frequency of sampling has often been the missing element in our ability to create realistic models of natural systems. Just to pick a current example, what would offer a better chance to understand the dynamics of global warming – one 25,000-pound satellite, or 250 100-pound satellites? Yet our existing launch systems are highly biased toward the single large satellite.

My third criterion for future space access is a familiar one: We should try to develop routine access to space. Routine access to space is, of course, what we haven't had for the Space Telescope. Routine access means being able to launch on relatively short notice. And I'll reemphasize that routine access almost certainly goes hand in hand with frequent launches, and that comes back to lower cost as well. Even after forty years of space access with rockets, we're still a long way from that capability, even for unmanned systems. It's eye-opening to compare the way in which manned air flight progressed rapidly to that kind of routine, and safe, operation, compared to space flight.

3. Change

Is there a solution to this set of equations? If there's not, then we may be faced with a demoralizing lid on what we can ever expect to achieve in space, always constrained by high costs that seem as unassaultable as a physical constant. But if that were the case, it would be a rare technology. In fact, in spite of the attention currently focused on extrapolations of Shuttle or current-ELV designs, there are newer, more clever approaches now becoming available. It may be that the greatest barrier we face is not technological, but the inevitable resistance to discontinuities.

While the prospect for rapid improvements in performance would seem to be especially appealing to organizations responsible for space access, it is tempered by the uncertainties – and natural resistence – brought about by rapid change. Machiavelli summed up this universal dilemma centuries ago when he observed that "Change has no constituency." And in those areas where the change promises to be greatest – in changing the cost and operational equations of space access – it creates the most uneasiness among people whose primary mission is to maintain ongoing rocket technology programs and guide its evolution over time.

But if there's anything that's more inevitable than resistance to change, it's change itself. And that's why I'm confident that we're due for some significant breakthroughs. Let me return to those criteria I laid out for future space systems.

My first criterion was cost. And I said we should aim for a two-order-of-magnitude reduction. Call it my Federal Express model for space access. That model cannot be met by most of the follow-on launch systems now being considered. I'm thinking of such options as the Advanced Manned Launch System, essentially a Shuttle follow-on; multi-stage ELVs, which are essentially new faces on old ballistic missile technology; the unmanned heavy lift vehicle, known as Shuttle-C, which would substitute an unmanned payload for the Shuttle's orbiter; or the Advanced Launch System, a next-generation unmanned ELV. Of those, the Advanced Launch System purports to be the most ambitious about taking a big bite out of the cost barrier, with a target rate of about $300 per pound. That's a laudable goal, a one-order-of-magnitude reduction, but few people believe it's realistic – or even close to realistic – based on the kind of technology projected for it.

There is also a class of small expendable launch vehicles, being promoted primarily by small commercial firms. While these don't make significant inroads in cost reduction over larger ELVs, they do address one of the other criteria – flexibility of payload size. In fact, the rapidly intensifying interest in small satellites – in the range from 50 to 1000 pounds – may find that these relatively unsophisticated small launchers fill a role that the larger ones cannot. And as we saw just a few weeks ago, there was a successful launch of Pegasus, which is in effect a multi-stage ELV in which the first stage is a high-flying aircraft. While such an approach lacks some of the elegance of other systems, it illustrates the kind of flexibility of launch and size of payload that is clearly needed.

A more interesting, and potentially promising, class of launch vehicles are single-stage-to-orbit rockets – such as the Phoenix, being developed by Pacific American Launch Systems. These would take advantage of improvements in materials technology to create a lightweight, reusable launch vehicle. They would also take advantage of some novel engine concepts that permit variable fuel/oxidizer ratios matched to ambient conditions as the vehicle climbs through the atmosphere.

In many ways these represent what could be possible in rocket technology when thinking is unconstrained. In the early days of the space program, it was apparent that weight and engine thrust prohibited a single-stage-to-orbit. So we developed staged rockets and got comfortable with them, sometimes forgetting that a design decision based on technology of the 1950s and 1960s might be worth revisiting every so often.

Some people have, and with promising results. So of all the evolutions of rocket technology, these may well be the most promising route for optimizing it. Rather than build a bigger expendable structure to support a heavier payload, the proponents of single-stage-to-orbit may succeed by fresh thinking about the whole launch system. And at least some of the people pushing these concepts have opted at the start to achieve operating simplicity – which translates into lower cost and more flexible operations – by emulating airplane-like operations, rather than rocket operations.

4. Costs

The incentives to do so are strong. About half the cost of a launch goes not to hardware or fuel, but to operations. So the proponents of single-stage-to-orbit, as do many proponents of small ELVs, hope to achieve a substantial cost reduction by reducing that enormous human operating cost.

I continue to focus on cost for two reasons. One, as I'll discuss, it can be cut substantially. The other is that our view of the future of space programs is distorted by assumptions about cost and demand for access. We've all seen estimates of demand for launch capacity that suggest that the Shuttle, plus the existing military and commercial ELVs available worldwide, will be able to handle the demands that are now projected for the next half dozen years. But that's putting the cart before the horse. At $5000, or more, per pound, there are very high barriers to participation. If I'm a potential user, and someone comes to me with an order book in hand, asking me how much launch I want at $5000 a pound, my answer is going to be a lot different than if someone comes to me and says, "If I can provide launch at $50 a pound, how much might you want?" In fact, there are going to be many, many more users – users who aren't even being considered at today's high prices – who will surface if prices drop. There's an assumption at work that presumes demand for space access is essentially inelastic. I think that's nonsense, yet it continues to drive the dominant space access planning processes at work today.

But I'll take it a step farther. I think a look at easily projected space access needs shows that we're already undersupplied with lift capacity, and it may get worse. When I look to space programs over the next several decades, I see the following.

– Expanded need for space-relayed communications, including a new generation of direct broadcast television satellites.
– A proliferation of local and global navigation and position-locating systems, with systems becoming inexpensive enough for everyday use by individuals.
– A next generation of earth-sensing programs, including the ambitious Mission to Planet Earth. The early impact of Spot Image, the French company selling commercial satellite photos, gives a taste of the kind of market that could develop for the even higher quality services that could be made available.
– There's a backlog of planetary exploration options, and manned missions to Mars will require a whole new space infrastructure, for which low-cost, routine access to low earth orbit will be no less than enabling.
– I expect military need for space access to increase, not decrease, as the nature of international relations changes. As nations reduce defenses, there's a correspondingly higher premium on intelligence – nothing more than the old adage to "trust, but verify."
– I also expect military satellites to take increasing advantage of micro-miniaturization of electronics, leading to smaller packages and an increasing mismatch with existing launch systems.

The list of space programs I've just run through is hardly visionary. In fact, the linearity of the extrapolations I've used displays the dilemma we're in, because we'd be hard put to accomplish a majority of them. And, with the appropriate architecture for satellite systems beginning to look more and more like that of distributed computing, with the much heralded space station beginning to look like the proverbial "lead balloon," and with renewed interest in totally different propulsion systems to take years off the round trip to Mars, our current approach to low earth orbit access is about as relevant as worrying about the vitality of U.S. Steel is to U.S. industrial competitiveness.

5. National Aerospace Plane

Happily, there is one research program, well underway, that tackles this dilemma head on. It's a technology whose potential is well recognized in Japan and in Europe, but here it's often overlooked as an oddity – neither airplane nor rocket. Yet it's one that should be able to provide the kind of space access I set the criteria for when I started this talk. That's the National Aerospace Plane, or the NASP. Ironically, but I hope only temporarily so, the NASP hasn't been seriously included in plans for the post-Shuttle space-launch systems. Some see it as lacking a sponsor, although we've already invested nearly two billion dollars in it. Some see it as an airplane, not as access to space, although that's been the driving goal for a decade. And some see it as futuristic, too far away to be part of today's debate. Yet its first flight in 1996 is sooner than the half-width of the policy debate over the Space Station. Lip service is paid to NASP as a means for space access, but the real attention seems focused on things like the Advanced Manned Launch System.

There are a lot of institutional reasons behind this hesitancy about the role of the NASP in space launch. I don't intend to go into those, other than to say that I expect them to fade as the NASP gets closer to reality.

Briefly, for those not familiar with the NASP, it's essentially an airplane that is designed to achieve hypersonic speeds, using a sophisticated air-breathing engine. Little additional oxygen is needed to enter low-earth orbit. And it will return to an airport runway under powered flight.

I said earlier that such novel new approaches as the Phoenix single-stage-to-orbit rocket use engine controls to adjust the mixture for optimal performance during powered flight. That's an important advantage over traditional rockets. The NASP takes that advantage even further. It will alter the engines' internal geometry, as well as fuel and air flows, for optimal combustion in a sequence of speed regimes, from takeoff at airplane speeds from a runway, into supersonic ramjet mode, and eventually into hypersonic scramjet mode.

Let me try to frame how the NASP technology, as it could be embodied in the so-called NASP-derived vehicles, would compare with other space launch technologies.

First, cost and payload. Estimates for a NASP-derived vehicle are a cost of two to five million dollars for a payload of perhaps 20,000 to 30,000 pounds to orbit. That compares to the Shuttle, whose cost for a comparable payload is nominally about $150 million. And here's an interesting comparison between the two. Costs for fuel for a Shuttle flight are only about four percent of the launch costs; about half its costs are for hardware that has to be replaced with each flight. For the NASP, about half its launch costs go for fuel, and virtually none of it is for hardware. Costs for a workhorse ELV like a Titan may be somewhat less than the Shuttle, but still on the order of $100 million for a payload somewhat smaller. Small ELVs, like Pegasus and Space Services' commercial Conestoga rocket, offer a launch sized to small payloads, but, while filling a much-needed gap, their costs per pound are still comparable to larger ELVs. And I simply don't find credible the Advanced Launch System's 300-dollar-per-pound goal.

Finally, I have seen estimates for the new single-stage-to-orbit designs that are substantially lower than existing launch costs. Phoenix, for example, a reusable vertical-takeoff and vertical-landing manned vehicle, projects costs for medium-sized payloads far closer to NASP's projected costs than to existing systems. The Phoenix is an intriguing concept with substantial technical merit. However, it is essentially a paper design at this point, though it might be adaptable to fast-track development that would make it a possible alternative during the transition to air-breathing vehicles.

The other criterion of importance is what we've referred to as routine access to space. For those of you who have been waiting for five years for the Space Telescope Launch, routine access may seem less important than predictable, reliable access. Yet in the long run, space science in particular demands frequent access for small payloads. It shouldn't take years from the time you design an experiment until you can conduct it. Yet it does now. You not only have to wait until it can be incorporated into a larger package, but you're then likely to have to wait until the larger package can win a slot in the launch schedule. One of the prime attributes of the NASP is its fast turnaround and minimal logistics. A single NASP-derived vehicle might be able to make from 40 to 150 flights a year if it can achieve its projected ground turnaround time of 24 hours. And unlike current rockets, and especially the Shuttle with its stringent abort requirements, the NASP shouldn't be affected much more by weather than regular aircraft operations. The NASP also offers the advantage of access to virtually all near-earth orbits. To me, that begins to meet a meaningful definition of routine access.

I am myself highly optimistic about NASP's prospects; other people are understandably skeptical. All of us remember our experience with the projections made for the Shuttle – particularly its mission schedule. But we shouldn't have too long to wait to test many of the projections for the NASP. As it stands now, the NASP program expects to make the first test flight in 1996. Frankly, it could have been sooner, but the development program was stretched out last year, largely for budgetary reasons. Even so, if the test flight program proceeds on schedule, the NASP will be ready to try a single-stage-to-orbit flight in 1998. And we could see a NASP-derived vehicle – that is, an operational version of the NASP – capable of achieving orbit by the year 2000.

But even on a shorter time scale, there will be chances to assess how well the projections are likely to hold up. There have already been several – all encouraging.

6. Prospects of NASP

When NASP was first proposed as a national program in 1985 there were many who were skeptical that materials could be developed to hold up to the extreme operating conditions. The worst of those problems are the nose and leading edge surfaces, whose temperatures would soar at high mach numbers. Yet that potential show stopper looks like it's essentially solved now. And solved in a way worth noting. Early in the NASP program the major contractors formed a unique consortium that enabled them to share data and research on different approaches to the materials problems. So we now have available an array of materials – high-temperature intermetallics, metal-matrix composites, materials with high-thermal-conductivity and high specific creep-strength, and even practical carbon-carbon composites – that seem to be well able to handle the stress expected in the aircraft. And if I were to point to any single area of technology as the most important that will spin off from the NASP, it would be the technology of engineered materials. Ten years from now when manufacturers are routinely designing three-dimensional materials properties into their products, they'll look back on the NASP as the catalyst for that capability.

The other area of uncertainty is the engine technology, which is the fundamental enabler of the NASP concept. The optimistic design objective is to be able to use the engines themselves – without the addition of rocket assist – to propel the aircraft nearly all the way to orbit. So far that optimism is unshaken. Obviously, there are limits to what can be learned by ground testing – limits in the kinds of operating regimes that wind tunnels and shock tunnels can produce. That will only be overcome with flight tests on the actual X-30 research vehicle.

However, the program has gained substantial confidence in the engine designs and variable configurations through advanced numerical simulation techniques in combination with available testing data. There's strong confidence already in the engine performance up to about Mach 10 or 12. It's certainly true that skeptics can, and do, await demonstrated performance. But at this point there's strong reason for confidence that the upper limits of the speed range will be gratifyingly high.

It's also important to remember that, although the NASP's initial, and perhaps most important, role will be for space access, it is being developed as one would develop an airplane. As a manned, reusable vehicle, it will have the advantage of repetitive flight testing in which the operating characteristics will be refined, and in which performance will be explored, step by step, into the unknown flight regimes of altitude and speed.

Also unlike a rocket, the NASP can have varying degrees of success. A rocket that doesn't achieve orbital velocity is a failure. But what if NASP can't achieve more than Mach 15 or Mach 20 in air-breathing flight? It will then inject sufficient oxygen into the same engine to get the final boost. The need to store more oxygen on board will cut into the payload, but most of the performance benefits it set out to capture will be preserved. It still has operating flexibility and fast turnaround because it operates from airport runways. And it will still bring enormous cost reductions to space access by eliminating a vast portion of the most expensive rocket flight regime – from the ground up to perhaps 200,000 feet. In effect, even a NASP that doesn't meet complete design expectations will create a launch platform very nearly in orbit to start with.

NASP, like any development program, will encounter problems as it refines the technology. But two things in particular give me confidence in the ultimate success. One is the state of the engine design. It's derived from a reliable concept, and the work so far to extend and demonstrate its range has been gratifyingly successful. That builds confidence. The other is that there have been several different good solutions developed for most of the critical materials problems, and the new materials have been tested and can be manufactured. So even though the NASP represents a totally new approach to space access, it's a technology that's already has some maturity. Many people are surprised when I talk about the NASP primarily as a means of access to low-earth orbit. There's been a perception that NASP would be primarily a hypersonic airplane, intended for high-speed point-to-point transport, or perhaps as an exotic military reconnaissance or rapid deployment aircraft. In truth, it might be those things as well, and it has as-yet unassessed potential impact – very possibly massive impact – on the aircraft industry. But the nearest term impact will be in space access, and that impact will be profound.

FUTURE DEEP SPACE PROPULSION SYSTEMS

ERNST STUHLINGER

3106 Rowe Drive, Huntsville, AL, 35801, U.S.A.

Abstract. Among several potential future deep space propulsion systems, the two which are closest to realization are selected for closer consideration: solar-electric, and nuclear-electric propulsion. In particular, the paper describes a manned Mars mission using a particle bed reactor and Brayton cycle converter as power source. Technical details of the design and the mission profile of a 4-astronaut expedition to Mars, and a proposed course of action for project implementation are presented.

1. Introduction

Much has been talked and written about future propulsion systems, other than our conventional, time-honored chemical rockets – propulsion systems that can take scientific observatories, unmanned and manned, into orbit, to the Moon, to planets, toward the Sun, and to other places in the solar system.

A number of such propulsion systems have been under consideration for years. They may be grouped into three classes (Table 1): First, 'passive' systems, such as gravity assist or 'planetary flyby' schemes; and the solar sail. Second, 'near-future' systems, among them electric propulsion which includes ion, arc jet, and MPD systems, and also nuclear thermal rockets, of which the famous Nerva project, and the gas core reactor, are examples. The third group contains such far-out concepts as fusion reactor systems, anti-matter propulsion systems, and photon rockets.

2. Selection of Preferred Future Propulsion Systems

Future propulsion systems should have, besides the basic virtues of great reliability, two features in which they surpass existing rocket systems: They should offer a greater propulsive capability, and they should permit higher payloads for the same initial systems mass (Table 2). These features would provide shorter mission times; they would allow more scientific and exploratory activities; and they would lead to a greater overall success of the mission.

Rather than elaborating on all these systems and adding some further speculations, this paper will concentrate on two of the advanced systems which are far closer to realization than the other propulsion schemes, and which will certainly put their imprints on the next phase of the space program. These two systems are the solar-electric and the nuclear-electric propulsion concepts. The two of them have already a long joint history of analytical studies, laboratory development, and even flight testing. They also have features that are very advantageous for many of our space projects of the next 20 to 30 years, particularly the further exploration of the Moon, the Sun, the planets and their moons, comets, and asteroids. They even offer very attractive possibilities for manned missions to the surface of the planet

Y. Kondo (ed.), Observatories in Earth Orbit and Beyond, 355–362.

Mars, a project that received new impetus recently through the Human Exploration Initiative established by President Bush.

3. Electric Space Propulsion

Electric propulsion systems are by no means novel concepts (Table 3). Tsiolkovskii talked about electric rockets almost one hundred years ago; Goddard made some theoretical and experimental studies in 1906; and Oberth wrote about electric propulsion systems in the early 1920s.

Systematic analytical work on electric propulsion began in 1947, and around 1955 the first laboratory experiments with electric thrusters started at the Electro-Optics Company in Pasadena. A few years later, Harold Kaufman at the NASA Lewis Laboratory demonstrated his Kaufman thrusters, followed in 1960 by Horst Loeb in West Germany who used a slightly different ionization scheme. A few years later, flight testing of thrusters onboard satellites began, confirming the validity of the expectations put into electric thrusters. During the following 15 years, about three dozen flight tests were successfully performed in the United States, and also in some other countries.

In spite of all this very promising development work, no real space mission with electric thrusters has materialized as yet, simply because there were always other projects with higher priorities.

Table 4 shows the principles of chemical and electric rocket engines. They are largely identical, except that electric rockets need an external source of power. That source could be the Sun, or a nuclear reactor. Solar power has the advantage of being environmentally clean. However, the density of solar flux decreases with the square of increasing solar distance; also, there is no solar power on the shadow side of the Earth. Solar-electric propulsion is quite attractive for unmanned space missions as far out as the Ateroid Belt, and even for freight transporation to the Moon. Nuclear-power propulsion systems, because of the danger of contamination in case of a catastrophic system failure (such as the impact of a piece of orbital debris), should be operated only at distances greater than about 40,000 km from Earth.

Details of electric (ion) thrusters are shown in Table 5. Table 6 gives some comparative design and performance data of chemical and of electric rocket systems. While the thrust forces and hence the accelerations of electric systems will always be small, obtainable velocity increments are considerable, and so are the payload ratios of electric rockets. An example of an unmanned solar-electric spacecraft for a comet rendezvous and sample return mission with a 50 kWe photo-voltaic power source is shown in Table 7.

Several different systems to convert reactor heat into electricity are available. Table 8 illustrates a particle bed reactor with Brayton-cylce converters; there are two parallel turbo-generator sets, counter-rotating to avoid torques on the space-craft. A powerplant of this type could generate 10 MWe, with a total reactor thermal power of 50 MWth [Ref. 1].

4. Nuclear-Electric Mars Mission

As soon as a reliable, efficient propulsion system of the solar-electric or the nuclear-electric type is available, a broad array of deep space missions will become feasible. One specific mission with a 10MWe nuclear power supply, a manned roundtrip to the planet Mars, will be described in more detail.

The primary objectives of a manned Mars mission will be (1) to assure the safe return of the astronauts, (2) to provide a rich harvest of scientific exploration and technical knowledge, and (3) to become the first step of a continuing program of human exploration of Mars, leading eventually to a permanently occupied base on the planet (Table 9).

These basic objectives immediately lead to a planning philosophy, shown on Table 10. The astronauts should not be left in space longer than about one year. The severe restrictions of "life in the can", much of it even within a space suit, should not be imposed upon them beyond that limit. Besides known physiological effects, there are psychological factors that must not be underestimated. Comfort and potential privacy for the astronauts should be assured. On that first mission, the staytime on Mars should not be longer than 4 to 6 weeks. While traveling through near-Earth space, the transit through the Van Allen Belts should be as short as possible.

5. Artificial Gravity?

Should there be artificial gravity during the long transfer periods between Earth and Mars? To paraphrase Prince Hamlet: "To 'g' or not to 'g', that is the question..." It seems that among those astronauts who have spent weeks or months in orbit, and also among the serious space mission planners, there is definite preference for artificial gravity, perhaps 0.38 g (like on the Martian surface), perhaps even a value that changes gradually from 1.0 to 0.38 g during the transfer. Designing a rotating arm system as part of the transfer spacecraft does not appear to be problematic. A system of this kind could be operated and tested on the Space Station Freedom several years before it would become a part of the Mars vehicle, and much-needed research on artificial gravity could be performed long before the design of the Mars vehicle has to be finalized (Table 11).

6. Supporting Facilities

Irrespective of the details of a manned Mars transfer vehicle, and even of its mode of propulsion, a heavy lift vehicle will be needed to transport the components of the Mars mission into low Earth orbit (LEO) for assembly. A heavy lift vehicle of this type will be needed for other space projects even before the start of the Mars mission, among them the Space Station Freedom, traffic into geosynchronous orbit (GEO), freight and manned flights to the Moon, and as launcher for unmanned large planetary and interplanetary spacecraft. It may be assumed that by the time a manned Mars mission takes shape, there will be a heavy lift launch vehicle (HLV), an orbital maneuvering vehicle (OMV) for transports in the vicinity of the space station, an orbital transfer vehicle (OTV) that can operate between LEO and GEO, a life support system that recycles water, and a booster rocket engine, burning H2 and O2 and developing about 100,000 N (10 tons) of thrust, for use between LEO and GEO (Table 12).

7. Profile of the Proposed Manned Mars Mission

A manned mission to Mars may be structured as follows (Table 13): After assembly in orbit, the complete and manned vehicle will be boosted chemically to nearly escape velocity. During this boost phase, there will be trusses to support the long arms with the habitat capsules (not rotating at that time), as shown in Table 14. The nuclear reactor will be started up at a distance of about 40,000 km from Earth. The propellant, mercury, which doubles as a protective layer around the radiation shelter, will not contaminate LEO, nor will a 'hot' reactor endanger the Earth, or be exposed to orbital debris. The vehicle will proceed to Mars, and will spiral down into a relatively low orbit (500 km?) where it will detach the landing craft (Mars Excursion Vehicle, MEV). Return to Earth will be by nuclear-electric thrust only. After transfer, the vehicle will spiral down into a high (approx. GEO) orbit, from where the astronauts will return to Earth by OTV and Shuttle. No aerobraking will be necessary. The Mars transfer vehicle will not come closer to Earth than GEO. If so desired, the vehicle, including its reactor, could be refurbished in Geo and used again for another mission, or it could propel itself out into space.

There will be two identical transfer vehicles, each of them manned by two astronauts (Table 15), and one similar, but unmanned vehicle for freight (Table 16). Each of the manned vehicles will carry a landing craft (MEV), and the unmanned vehicle will carry two landing crafts. Each landing craft will carry an ascent vehicle, providing substantial redundancy. In an emergency, one ascent vehicle could carry all four astronauts; also, one manned transfer vehicle could accomodate four astronauts (Table 17). Design data for the transfer vehicles are shown in Table 18.

The unmanned freight vehicle will travel first and deploy its MEVs with ascent vehicles, habitat, life support, shelter, and roving vehicle on the surface of Mars. Only when this 'home away from home', including Hertz car and two ascent vehicles, is firmly established on Mars and in operating condition, will the two manned transfer vehicles leave LEO for their voyage to Mars (Table 19).

Details of the transfer trajectories between Earth and Mars, and back to Earth, are shown in Figs. 20 and 21.

It should be noted that the same vehicle designs, and the same mission profile, can be used for further missions to Mars after the first exploratory mission. In particular, both the manned and unmanned versions will represent proper transfer vehicles at the time when a Martial base is to be established, and when regular traffic between Earth and the Mars base is to be maintained (Table 22).

8. A Proposal for Action

Implementation of a manned mission to Mars could proceed along the following course of action (Tables 22 and 23): First, a decision should be made to choose the nuclear-electric propulsion system for the mission. Then, development of the major components – thrusters, nuclear power source, artificial gravity system, life support system, Mars rover – should be started. It should be noted that all these components will be used and even needed on other space projects planned for the near future, even if no manned Mars mission should materialize; for example, electric thrusters will be needed on unmanned solar-electric vehicles for lunar freight and deep space missions; nuclear-electric power sources will be needed for lunar activities, and for unmanned deep space exploration; life support systems sill be needed on the Space Station, and on the Moon; artificial gravity research will be performed on the Space Station; roving vehicles operating under low gravity will be needed on the Moon.

While these developments are underway, a vigorous program of Mars observations should be pursued from the ground, from orbit, from the Moon (once it is revisited), and from unmanned Mars probes. Sample return missions to Mars should be included. This program will certainly be parallelled by similar programs in Europe, the USSR, Japan, perhaps China. There should be a continuous exchange between nations not only of thought, data, observational results, and increasing knowledge, but also of plans for future projects.

Systematic planning for a manned mission to Mars, supposedly taking place around the year 2015, need not be started before about 2000. By that time, all the major components will be far advanced in their development, and even in their testing and applications. Knowledge about environmental details of Mars will have increased decisively. When that systematic planning phase for the Mars mission begins, the time will have come to consider possibilities of sharing project responsibilities with other nations.

Acknowledgements

Present work has been partially supported by General Dynamics Space Systems, San Diego.

References

Powell, J.R., et al.,: "Particle Bed Reactor Multimegawatt Concepts", Brookhaven National Laboratory; Babcock and Wilcox; Grumann Aerospace Corporation; and Garrett Corporation; BNL-39495; March 1987.

FUTURE PROPULSION SYSTEMS	1
o "Passive" Systems:	Gravity assist; Solar sail
o Near Future:	Electric propulsion (1958...) (ion prop.; MPD prop.)
	Nuclear-thermal propulsion (Nerva, 1963-1973)
	Gas core reactor
o Far out:	Fusion rct.; Anti-matter prop.

Future Propulsion Systems	2
Required Features:	They will provide:
More propulsive capability	Shorter mission times
More payload capability	More scientific yield
	More mission success

Tables 1–2

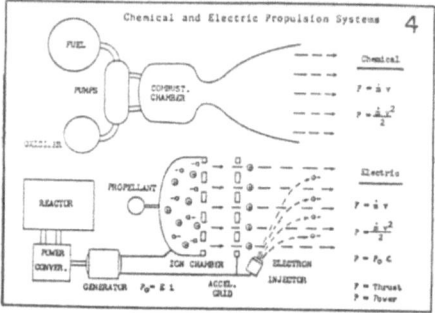

```
        E L E C T R I C   P R O P U L S I O N   3
                    H I S T O R Y

1895      Ziolkovskii; 1906: Goddard;  1920s: Oberth

1947      Comprehensive analytical studies began

1955      Laboratory work began

1957      Kaufman thrusters; 1960 Loeb thrusters

1964      Flight testing started

1964-80   About 35 flights of Elec. thrusters

1968 ...  El. Prop. ready and waiting for missions
```

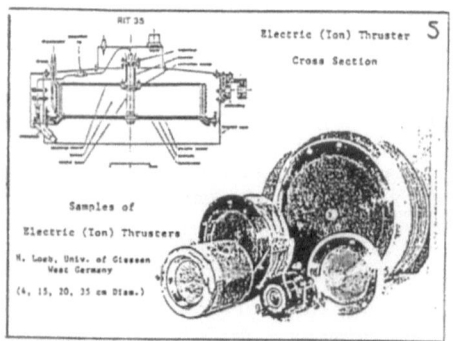

Chemical and Electric Propulsion Systems   4

Electric (Ion) Thruster   5
Cross Section

Samples of
Electric (Ion) Thrusters

H. Loeb, Univ. of Giessen
West Germany

(4, 15, 20, 35 cm Diam.)

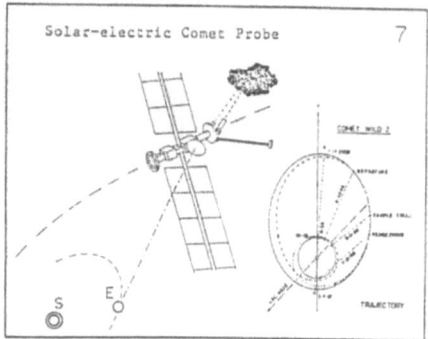

```
      CHEMICAL versus ELECTRIC PROPULSION      6

    Chemical                        Electric

v(ex) = 4.8 km/s              v(ex) = 50...80 km/s

v(ex) optimum:                v(ex) optimum:
as high as possible           √(s.p. p. x t)

                    s.p. = specific power, W/kg
                    t    = total prop. time, s

delta v per stage:            delta v per mission:
    6...8 km/s                    35...45 km/s

payload fraction (2-stg.):    payload fraction:
    2...3 %                       20...30 %
```

Solar-electric Comet Probe   7

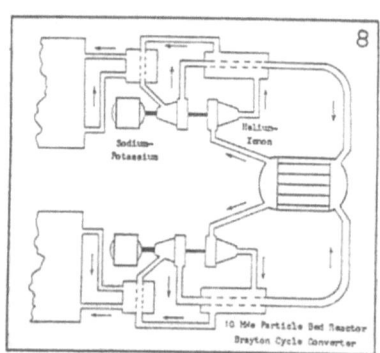

8

10 MWe Particle Bed Reactor
Brayton Cycle Converter

```
M A N N E D   M A R S   M I S S I O N   9

         Primary Objectives:

o  Safe return of the astronauts

o  High yield for science

o  First step of a continuous program
   to explore Mars by man
```

Tables 3-9

```
        M I S S I O N   P L A N N I N G          10
              P H I L O S O P H Y

  o  Total mission time approx. 1 year

  o  Fast passage through Van Allen Belts

  o  Artificial gravity for astronauts (o.4 g)

  o  Comfort and potential privacy for astronauts

  o  Staytime on Mars surface 4 to 6 weeks

  o  First step in a continuing Mars expl. program
```

```
        M A N N E D   M A R S   M I S S I O N    11

   To "g" or not to "g", that is the question...

  o  Astronauts must be in top physical condition
     when they arrive on the Martian surface

  o  "g" is desirable for health, and for comfort

  o  Most astronauts are opting for "g"

     Artificial gravity on the Mars transfer vehicle
     will not be costly in mass or complexity

  o  "g" capsules to be tried out on Space Station
```

```
        M A N N E D   M A R S   M I S S I O N    12

     It is assumed that by 2015, there will be:

  o  A heavy-lift vehicle (180 t from Earth to LEO)
  o  An orbital maneuvering short-range vehicle
  o  An orbital LEO-to-GEO transfer vehicle
  o  A water-recycling life support system
  o  A 100,000 N (10 ton) H2 -O2 rocket engine
```

```
     M A R S   M I S S I O N   P R O F I L E     13

  o  Manned and freight vehicles will be similar.

  o  Freighters (1 or 2?) will stay in space,
     manned veh. (2 or 3?) will return to GEO.

  o  Vehicle assembly in LEO; boost to escape from
     LEO (with astronauts) by H2-O2 booster.

  o  Manned veh. will carry 1, freighter(s) 2 Mars
     Excursion Veh.; each with an ascent vehicle.

  o  At least one freighter MEV will carry a com-
     fortable habitat, a shelter, and a rover.

  o  Astronauts will leave Earth only when MEVs from
     freighter(s) are firmly established on Mars.
```

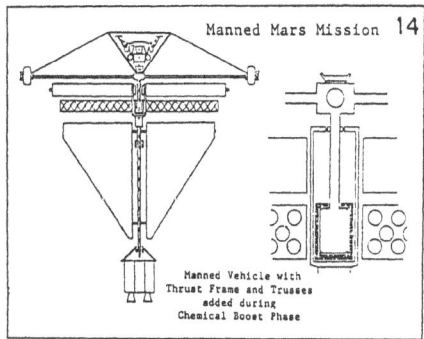

Manned Mars Mission 14

Manned Vehicle with
Thrust Frame and Trusses
added during
Chemical Boost Phase

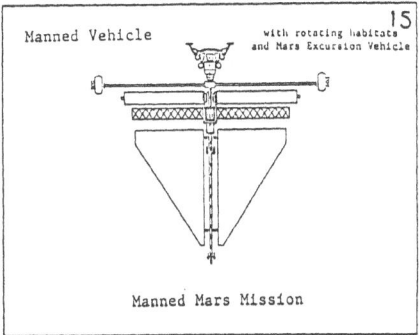

15

Manned Vehicle   with rotating habitats
                             and Mars Excursion Vehicle

Manned Mars Mission

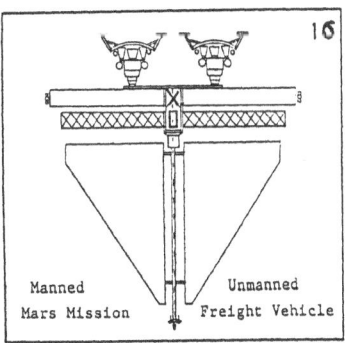

16

Manned   Unmanned
Mars Mission   Freight Vehicle

Tables 10–16

MANNED MARS MISSION 17

Optimum Number
of
Astronauts:

If mission is limited to 2 astronauts
per veh., and 2 manned vehicles per mission,
then in an emergency all 4 astronauts can
ascend from Mars in one ascent veh., and
return to Earth in one manned vehicle.

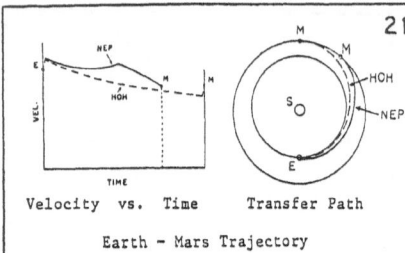

Velocity vs. Time   Transfer Path

Earth – Mars Trajectory

MANNED MARS MISSION 18
Vehicle Design Data:

o Vehicle mass in low Earth orbit: 300...320 tons

o Propellant: mercury; total mass: 140   tons

o Mercury will double as radiation shelter shield

o Total reactor power: 50 MW th., 10 MW electric

o Mass of Mars Excursion Vehicle: 40...50 tons

o One-way travel time: approxim. 180 days

NUCLEAR-ELECTRIC   MISSIONS
(Manned)
22

Mars :   Exploration missions

Establishment of Base

Regular traffic Earth-to-Base

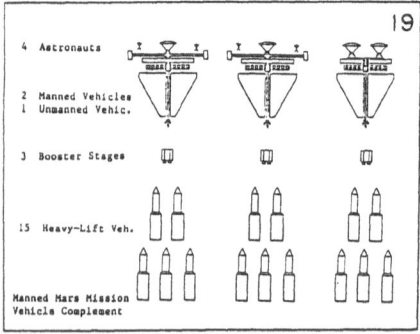

Manned Mars Mission
Vehicle Complement

MANNED MARS MISSION 23
Proposed Actions:

o Decision to use nuclear-electric propulsion

o Immediate development of major components

o Use of all components on unmanned missions

o Continuing unmanned exploration of Mars

o Systematic planning and preparation of
  manned Mars mission, beginning around 2000

o Target date for start of mission: 2015...2017

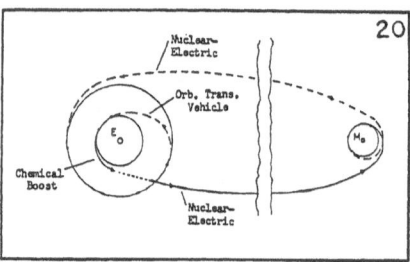

Earth – Mars Trajectory

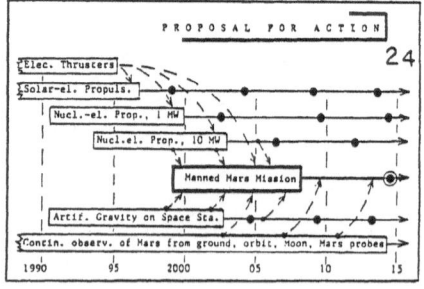

Tables 17–24

IV. RELATIVE MERITS OF VARIOUS OBSERVATORIES

(A) VARIOUS ORBITS AND SITES

Panel: Relative Merits of Various Orbits and Sites in Space

B. G. TAYLOR

MINORU ODA

MICHAEL J. MUMMA

ROBERT W. FARQUHAR and DAVID W. DUNHAM

LUNAR-BASED ASTRONOMY

HARLAN J. SMITH

Astronomy Department, University of Texas, Austin, TX 78712

Abstract. The Moon offers for astronomy a truly impressive array of advantages, many of which are briefly reviewed in this paper. These advantages include especially the vast inertial platform and the expected availability of human and robotic "hands-on" installation, maintenance, and modification. The Earth-orbiting Great Observatories will advance our knowledge to a new plateau, but some of the most fundamental observational questions which we are already asking will require lunar-based instruments, including very large filled-aperture telescopes, interferometers with baselines of tens and ultimately hundreds of kilometers, and the utilization of the radio-quiet backside. It is already time to begin planning the first such installations.

1. Background

The realization is rapidly growing that the Moon is a uniquely desirable place from which to do astronomy, also that it should become accessible again in the relatively near future.

Ever since Apollo-mission days the possibility of lunar-based astronomy has been discussed, though usually as a relatively minor topic. However within the last few years, four workshops have been devoted entirely to the question. The first attempt to interest a wider community of astronomers in the issue was a one-day session held immediately after the American Astronomical Society meeting in Houston in January, 1986 [*NASA Conference Pub. 2489*, 1988], which developed a general overview of some of the areas of astronomy which might profit from lunar siting. It was followed by two specialized lunar-astronomy workshops at Albuquerque, respectively in 1988 on Very Low Frequency Radio Astronomy [*NASA Conference Pub. 3039*, 1989] and in 1989 on Optical/IR Interferometry [*NASA Conference Pub.*, 1990]. Each of these meetings demonstrated the very great value of the Moon as an astronomical site. The most recent workshop, a large one held in Annapolis in February 1990 [*American Inst. of Physics Conf. Proc.*, **207**, in press, June 1990] asked a related question, but with different emphasis – namely whether there is fundamentally important astronomy which appears to demand the Moon for its achievement. Results from all of these workshops have contributed to this review, but most of the references noted below are to the proceedings from Annapolis, and are given in the text as (author, Annapolis 1990).

2. Space Astronomy Alternatives

We are already able to operate relatively modest Earth-orbiting telescopes with good efficiency, albeit with high costs – typical missions now average hundreds of millions of dollars. Over the coming decade we should have demonstrated the art of operating medium-sized telescopes in orbit, at costs probably averaging well over

Y. Kondo (ed.), Observatories in Earth Orbit and Beyond, 365–375.

a billion dollars each. Wonderful as the space environment is for astronomy, in near-Earth orbit these instruments still suffer from deleterious atmospheric effects, including drag with the attendant possibilities of de-orbiting or the need for reboosting, also damage and/or airglow caused by the 8 km/sec encounter with debris and residual atmospheric ions. The zero-g environment requires complicated machinery for setting and tracking, with increasingly inefficient duty-cycle for larger instruments. The 90-minute revolution period enforces short exposures, limited access-time windows for any given object, and rapid thermal cycling of the system. Large ground-based staffs are needed to build and operate these complex instruments, and human access for maintenance and modification ranges from difficult to impossible. Geosynchronous orbit offers a major improvement from the point of view of astronomical efficiency, but in turn is still less accessible and also lacks the opportunity for efficient magnetic dumping of momentum, thereby presenting serious problems for the very large (perhaps 50- to 100-ton) instruments which we visualize needing in the next generation after HST. Likewise, any orbiting system faces extreme difficulties with the ultra-precise station-keeping needed for long-baseline optical/IR interferometry. In spite of such problems, Earth-orbiting systems should and will continue to be used as long they are they are cost-effective.

The Moon is a thousand times more distant than LEO, and still 11 times farther away than even geosynchronous orbit. But in the more important terms of pure energy needed to access the lunar surface, it is only roughly double the cost to LEO and about 25% beyond geosynchronous. Improvements of space transportation technology will continue to bring the actual cost differentials ever closer to the raw energy differences, thus drawing the Moon effectively closer. And, in any event, lunar development and utilization by the human race will not be denied. Accordingly it is both timely and important to look closely at the challenges and opportunities represented by the lunar surface for astronomy.

3. Why is the Moon So Good for Astronomy?

The Moon offers an extraordinary array of features conducive to astronomy. A brief listing of principal ones includes:

Ultra-high vacuum ($\sim 10^5$ particles/cm^3). The lunar surface is effectively in free space. Telescopes can be used over their full frequency range and with their full spatial resolving power. Radio, infrared, and optical interferometers can be established with baselines of ultimately up to tens or even hundreds of kilometers. Exposed optical surfaces should retain their full efficiency for very long times (limited, if unshielded, only by micrometeorite effects). It should even be possible to use "naked" photoelectric surfaces should this be desirable for detector systems of special configurations and optimum efficiency.

Stable solid surface. The Moon offers a solid platform with effectively infinite inertia, allowing ultra-simple low-cost terrestrial-type mounting, pointing, and tracking systems to be used. Its rigidity and minimal seismic activity (10^{-8} that of Earth) ensure extreme stability for elements of interferometers with even the longest spacings.

Cosmic ray protection for humans. Extra-solar-system cosmic rays include high-energy components very difficult to shield against, which build up significant radiation damage in the bodies of astronauts exposed for long times. Solar storms are much more dangerous still; several times a year during periods of high solar activity they supply, with onset warning times which might be only a few minutes, doses which would be lethal to exposed astronauts. The Moon offers essentially full protection, given the simple precaution of having nearby habitations and shelters under some meters of lunar soil, with selenauts exposed only when it is necessary to work on the surface.

Cosmic ray protection for detectors. Especially outside the Earth's magnetic field, the performance of some of the most sensitive detectors for nearly all types of radiation is degraded by noise caused by solar and galactic cosmic rays. Some types of instruments will be able to avoid most of this problem by being located beneath the lunar surface, looking out through a small aperture. Even with optical telescopes on the surface, it may often be possible to have the detector at a coudé focus deep underground. Finally, in those cases where everything must be on the surface, the stable lunar platform and low lunar gravity encourage the provision of truly massive shielding around the detector package.

Dark sky. The Moon should be virtually free of air glow (a kind of permanent aurora) which sets a limit on the darkness of even the darkest terrestrial nights. Absence of atmospheric scattering means that the deepest observations can be made at night even if the observatory is on the near side with the Earth always in the sky, and with the "full Earth" being a hundred times brighter than our full Moon. With proper shading and thermal control of telescopes, it is even likely that relatively deep observations can be made in the daytime. Terrestrial and near-Earth-orbit telescopes have trouble collecting useful data more than about a quarter of the time. Their lunar counterparts should seldom stop observing.

Cold sky. Since thermal infrared wavelengths are generated by every object not at absolute zero temperature, observing at such wavelengths is like trying to do optical astronomy using a telescope every element of which is emitting light, while looking through a glowing daytime sky. The best current solution is to cool every part of the telescope with liquid helium and to raise the system above as much atmosphere as possible using balloons or spacecraft. This is awkward even with relatively small telescopes, and also either suffers from limited lifetime of the helium cryogen supply or demands costly and difficult re-supply. Because the temperature of the sky as seen from space or the Moon's surface is only a few degrees above absolute zero, a telescope on the Moon – if carefully insulated from the ground and shielded from any direct or reflected radiation not coming from the dark sky – would cool to and remain at an exceedingly low temperature. This would greatly reduce and perhaps in some cases eliminate the need for cryogens, as well as allow the telescope to be quite large.

Low gravity. Experiencing only one sixth of the Earth's gravity, lunar structures of any size will be of much lighter, less expensive construction than their terrestrial counterparts. This modest lunar gravity also serves the useful function of causing debris and contaminants to fall to the surface, rather than tagging along in place as they tend to do in space.

Absence of wind. Much of the strength built into terrestrial systems is simply insurance against the threat of extreme wind forces, no matter how unlikely. Lunar structures need attend only to static and thermal loads. For example, telescope "domes" might be simple systems of movable Sun-, Earth-, and dust-shades probably consisting of multiple layers of virtually weightless aluminized foil, stretched over ultra-lightweight but rigid structures. Mechanical designers of telescopes, which on Earth must be carefully strengthened against low-frequency vibration resulting from wind-buffeting, will be able to concentrate solely on thermal passivity and on static forces at various angles.

Rotation. The Moon's roughly month-long rotation period guarantees access to all the sky accessible from the latitude of the observatory site, yet is slow enough to permit very long integrations on the faintest possible objects. Some important modern observations demand unbroken time series with durations of weeks or longer; these are extremely hard to get from the ground or anywhere else except deep space. Lunar rotation also gives free sky-scanning to fixed telescopes and the capability of aperture synthesis to interferometers with fixed elements.

Proximity to Earth. The three-second Earth/Moon round-trip communication time allows Earth-based control of some robotic systems and offers no hindrance at all to operation of telescopes by astronomers on Earth. Massive data streams can be continuously broadcast to the Earth for analysis. Few if any astronomers will need to be on the Moon, even when a great many telescope systems are functioning.

Distance from Earth. Human activities, plus those of the Earth itself, generate noise and interference in nearly all kinds of observations. At 400,000 km, the Moon is far enough away to be relatively well quarantined from most of this pollution – experiencing a hundred-fold reduction below the levels even at geosynchronous orbit. Except at radio frequencies, observations from the Moon will be virtually unaffected by the presence of the Earth in the sky.

The lunar farside. Terrestrial broadcasting and other sources of interference blanket all but a few small slices of the rf spectrum. Yet radio astronomy is a fundamentally important branch of the science, and also includes the part of the electromagnetic spectrum in which detectable emissions from any extraterrestrial civilizations are most likely to lie. For limiting sensitivity, radio astronomers and SETI programs must ultimately go to the only place in the universe which never sees the Earth in its sky – the back side of the Moon. This unique state of radio-quiet needs to be preserved as lunar development proceeds.

Raw material. The Moon offers an inexhaustible supply of many essential materials. In the beginning the raw lunar regolith will form essential shielding against cosmic rays as well as outstanding insulating material. As processing facilities gradually come into operation, various cements and building blocks, ceramics, glasses, fibers, and metals will become available. These will be increasingly important as very large astronomical instruments are eventually undertaken.

Landforms. Perhaps the most cost-effective radio telescope ever built is also the largest – the 300-meter dish lining a hemispherical crater near Arecibo, Puerto Rico. Comparably symmetric lunar craters come in almost every size. Aided by the low lunar gravity, and the lack of wind and weathering action, similar radio telescopes up to several kilometers in diameter may someday be built on the Moon.

Room. The Moon offers almost unlimited area for laying out systems of instruments, which can be added to at any time yet be conveniently located near one or more common bases for supplies of consumables, replacement parts, electric and computer power, etc.

Access. The last, but perhaps most important, lunar advantage with respect to other kinds of astronomy from space will follow from the development of lunar bases, offering the immediate proximity of people and support facilities. For the first time this will allow construction of very large, highly sophisticated space telescopes and instrumentation in extremely simple low-cost mountings and housings, with virtually all components readily accessible for maintenance or change by skilled people in the immediate vicinity. Though nearly all observing will be done by astronomers on Earth, the continuous availability of real-time, hands-on technical support will constitute a revolution in the way cost-effective space astronomy can be done.

For more than a century astronomers have been struggling to develop everbetter sites for their telescopes. Quite apart from the many other factors taking the human race to the Moon, the continued needs of astronomers should soon lead us there – to the best place in the solar system from which to do many if not even most kinds of astronomy.

4. How in Practice Can Some of These Lunar Opportunities be Realized?

Lunar astronomical development can be anticipated from two contrasting viewpoints:

Along for the ride. There are substantial reasons (*e.g.*, Smith, Annapolis 1990) to expect that permanently manned lunar bases will be constructed in the very early years of the coming century, primarily for reasons unrelated to astronomy. To the extent this is true, astronomy can and should be "along for the ride." The cost of establishing and maintaining human operations on the lunar surface will be very great for the first several decades, until a significant degree of self-sufficiency is achieved and until transportation costs come down substantially. However, the *incremental* costs of astronomy done from otherwise-supported lunar bases should be quite modest since, as noted above, even simplified versions of relatively low-cost terrestrial instruments can be installed by the selenauts and maintained by them as intermittently necessary – with nearly all the astronomical operations being directed from, and data sent by radio directly back to, the Earth.

Part of the engine. As astronomers, we can hope that the costs of human presence on the Moon will be accepted as part of the price the human race should expect to pay for the exploration, development of, and expansion into the solar system. However, one can also ask, as at the Annapolis Workshop, whether there are next-generation astronomical questions *requiring* instruments which in turn appear to *require* lunar siting. The answer at Annapolis was emphatically yes. From this point of view astronomy becomes one of the drivers for lunar development and thereby potentially liable for some of the infrastructure costs in addition to those of the astronomical instruments *per se*. To the extent this type of bookkeeping occurs, it should not prevent but will likely have the effect of slowing the development of lunar-based astronomy.

Another conceptual branching arises with the question of whether human attendance is necessary for all lunar-based instruments.

Totally automated systems. This may be the fastest way to get started with lunar-based astronomy. It is surely possible to build softlanders carrying simple astronomical telescopes which could do useful work from the lunar surface. At least initially, these might be relatively small instruments designed in effect primarily for testing the lunar astronomical environment. Also, with some specialized instruments such as elements of radio telescope arrays it may prove cost-effective to design systems to be entirely remotely implanted and operated. However, this approach appears to have problems if adopted as the general rule for lunar telescopes. Compared to free-space operations with their necessarily complicated pointing and tracking systems, it gains the solidity of the lunar surface but pays the price of the extra design and construction cost and complexity associated with totally remote installation and operation, while also giving up continuous accessibility of the entire sky. In addition, it is hard to see how very large and complex systems can be successfully installed and operated without direct human access (or in some cases, perhaps immediately adjacent telerobotic operation).

Human presence. An earlier section has listed a great many opportunities presented by the Moon for astronomy. Many if not most of these come into play when human presence is established. In the long run, and however the bills are paid, significant lunar astronomical developments will, with virtual certainty, involve hands-on human contact. At least in the early stages, this contact might be continual rather than continuous. Telescopes could be installed during relatively short lunar missions, and serviced or modified as necessary by later ones. But the very large systems proposed at the Annapolis workshop will probably require almost constant presence, which fortunately should be available by the time it becomes feasible to build them. It is thus reasonable to expect a variety of kinds and sizes of telescopes to be installed on the Moon. Once the first lunar base is established, its astronomical aspects are likely to develop very much the character of the European Southern Observatory, where groups of nations, individual nations, or even smaller organizations contribute telescopes and equipment while sharing common support and operating facilities. In particular, within the first decade a number of relatively modest instruments are likely to be at the lunar observatory. Each of these – for simplicity, economy, reliability, and stability of calibration – should normally be dedicated to a single function. One such telescope could be a small, even a very small, item in the payload of nearly every lunar re-supply mission. Some candidate instruments are discussed in Sykes *et al.* (Annapolis 1990). We now lump the class of such low-cost but high-yield systems under the acronym DARTs [Dedicated Astronomical Research Telescopes], and visualize that from the beginning of lunar development many of these should be "tossed" at the Moon for unloading, laying out on the lunar surface, plugging in, turning on, and subsequent near-ignoring by the selenauts. A typical list of DARTs might include:

– Fixed transit telescope(s) with clocking CCDs for deep sky studies; – Simple multi-color photometric telescope(s) for studies of stellar variability, stellar seismology, and the possible rare but important detection of planets around other stars; – Simple high-resolution imaging telescope to supplement HST; – Dedicated planetary imaging telescope, especially for detailed studies of the Martian atmosphere in preparation for eventual human landings; – Simple spectroscopic telescope(s) – (sons of IUE); – All-reflecting Schmidt (deep UV and near-IR surveys in many wavelengths); – Passively (later actively) cooled thermal IR telescopes; – Two- (later more-) element optical/IR interferometer.

Similar, non-optical instruments should probably include:

– Prototype very low frequency (VLF) radio interferometer; – Moon-Earth VLBA antenna; – "Suitcase" X-ray telescope; – "Suitcase" gamma-ray telescope.

However, in spite of the great advantages which human-tended lunar operations will offer for the above and similar small to modest instruments, it is unlikely that lunar bases would be established solely for this class of instruments which at least in principle could be operated in Earth orbit.

5. What Astronomical Problems and Instruments Demand the Moon?

The Annapolis workshop did not consider exhaustively all aspects of astronomy – in particular it treated only peripherally a number of challenges in modern physics which impinge on astronomy. Nevertheless, each of the major working groups at the meeting came up with fundamental questions, all of which are sure to challenge astronomy of the 21st century and all of which appear to require lunar basing of the instrumentation needed to tackle them.

5.1. Are there Earth-like Planets Among the Nearby Stars?

Within the next few years any of several techniques should succeed in allowing us to infer the presence of Jupiter-sized planets, assuming they exist, around some of the nearer stars. However, Earth-mass planets approximately an astronomical unit from their primaries will be two to three orders of magnitude harder than Jupiters to discover, and far more difficult still to study in any detail. The problem becomes tractable in principle by going to long wavelengths, where the contrast ratio between such a planet and its central star improves from 10^{-10} (visible light) to 10^{-5} (thermal infrared). To study thoroughly at 10 micrometers wavelength the space as close as 0.2 AU to stars within about ten parsecs of the sun will require an ultra-stable IR interferometer of more than 100-meter baseline. For adequate photon statistics, the total collecting area should be at least several hundred square meters, and the telescopes in the array must be at cryogenic temperatures. To construct and successfully operate such a system in orbit would be a forbidding task, if possible at all, whereas in principle it should be relatively straightforward given human access to the stable lunar surface.

5.2. Do Life and Intelligence Exist Elsewhere?

Life processes drive the Earth's atmosphere far from chemical equilibrium. Detection of both free oxygen and methane in the atmosphere of another planet would be *prima facie* evidence of life at least somewhat as we know it. All candidate planets discovered by techniques such as in the paragraph above, or by any other means, need to be studied by low-resolution spectroscopy, searching especially for O_3, CH_4, NH_3, H_2O, and CO_2 over the wavelength region from about 6 to 15 micrometers. This work requires a preferably single-aperture cryogenic telescope of several hundred square meters aperture with exceptionally precise and stable pointing characteristics. Again it seems necessary to locate such a telescope on the Moon.

Evidence of other intelligence in the universe may be much harder yet to discover. In recommending a modest long-term SETI program, the 1980 Astronomy Survey noted (*Challenges to Astronomy and Astrophysics: Working Documents of the Astronomy Survey Committee*, National Academy Press 1983, pp. 267–272) the probability that, quite apart from the remote possibility of their choosing to establish "beacons," even highly advanced technologies may still use extremely powerful microwave radars and perhaps power beams the spillage from which would be detectable at very great distances. The microwave spectral region is also the quietest in terms of background noise, thus (following the impeccable logic of the drunkard searching for his lost key under the streetlight) we need to study this region minutely, and in effect to search much of the galaxy over an enormous range of microwave wavelengths for evidence of such transmissions. In turn, the back side of the Moon is the only place which is sufficiently radio-quiet for this program to be carried out. Very large antennas – perhaps of the Arecibo type – are needed for the work.

5.3. What's Happening in Our Sun and Other Stars?

Our experience of solar physics and processes is based on a single temporal snapshot in the 4.5-billion-year life of this garden-variety star. We know that the terrestrial climate has changed radically on very long timescales, perhaps in part because of solar changes, and that the sunspot cycle which may affect climate on short (decades to centuries) timescales virtually disappeared during the Maunder Minimum nearly 400 years ago. Even the solar neutrino flux remains an unsolved problem. A fundamentally important way of gaining insight into our sun is to investigate processes occurring in other stars.

Comparative solar-stellar physics requires observing the surfaces of other stars on spatial scales similar to those which appear most relevant for our understanding of the sun (Woolf, Annapolis 1990). Much of the action on the solar surface appears to occur at size scales of 100 km or even smaller, corresponding at nearby stellar distances to micro- or even sub-microarcsecond resolution at optical wavelengths. In turn this mandates optical/near IR interferometers of at least 100-km baselines. The necessary pointing and tracking precision seems unattainable in space but again in principle to be achievable on the lunar surface. So powerful an interferometer will surely require many decades to achieve, but should be an objective toward which initial and intermediate systems can grow – a start can be made with 2-meter class telescopes and 100-meter baselines.

As an important side issue in solar astronomy, very large flares emit particle fluxes which are potentially lethal to unprotected astronauts who are outside the Earth's Van Allen belts. Quite apart from the useful science to be done with a modest several-meter-class diffraction-limited solar telescope on the Moon, its continuous monitoring of the sun should help to warn selenauts of impending danger in time to seek shelter. Full coverage of solar activity will of course require antipodal lunar observatories.

5.4. How Do Stars and Planetary Systems Form?

There are many uncertainties in our understanding of how stars, and in particular solar systems, form. The spatial scales of greatest interest are the zones where protoplanets are presumably forming in the protostellar nebulae – that is, features about 0.1 AU in size. The available tracers are the spectral lines of many gaseous species, particularly CO and H_2O. Preplanetary disks have enormous optical depths in the visible and infrared regions, but become optically thin around 600 micrometers wavelength. In particular the 557 GHz line of water is the best probe of outer low-temperature regions of the disks. Unfortunately, the nearest star-forming regions are about 140 parsecs distant. Even to reach 600-micrometer resolution of 1 AU at this distance requires an interferometer baseline of at least 24 km; ultimately a submillimeter interferometer system will probably utilize a substantial fraction of the lunar diameter. Here again the necessary unhampered space observing conditions along with the required dimensional stability and pointing precision can be achieved on the Moon (Burke, Annapolis 1990).

5.5. What Happened in the Early Universe?

Cosmology appears to be approaching an impasse, wherein the microwave background insists that the early universe was extraordinarily uniform, whereas observations of the most distant quasars imply that the universe had already crystallized out into stars and presumably galaxies by almost comparably early times. Galaxy clumpings and voids are already seen to extend to at least 3.3 billion light-years, but even more provocatively they appear to be periodic at least in some directions (Szalay, Annapolis 1990). The nature of the voids is not understood, nor has "hidden mass" adequate to close the universe been found, though several lines of evidence suggest its presence.

Investigation of the formation and evolution of stars and galaxies in the early epoch $5 < z < 15$ requires photometric and spectroscopic observations of quasars, early galaxies, and perhaps especially the intergalactic medium in the 0.75–10.4 micrometer wavelength range (Weedman, Annapolis 1990). A 16-meter-class cryogenic filled-aperture telescope would meet these needs, but it must be in space to achieve diffraction-limited imagery and full wavelength coverage. Here again the cooling and extreme pointing stability required of this immense and massive telescope speak for its presence on the Moon.

Along with photometry and spectroscopy, ultra-high spatial resolution will be needed for many key investigations. Imaging the structure and kinematics of active galactic nuclei and quasars is basic to understanding their physics and evolution. This core activity is most significant on spatial scales below 0.1 parsec. Microarcsecond resolution is needed to map these structures at distances of several thousand megaparsecs. Fifty-km baseline lunar-based interferometric arrays working at 2 micrometers would satisfy this condition. Large X-ray and gamma-ray telescopes will also be needed to study these questions.

Common themes running through the fundamental problems outlined above are the need for great, usually cryogenic, apertures to collect sufficient photons from the extremely faint and/or distant target objects, the need to see these in great spatial detail reaching ultimately as far as sub-microarcsecond resolution, and the fact that the Moon is almost certainly the only place where these goals can be achieved. Any of the lunar telescopes will doubtless serve many other astronomical objectives in addition to the specific goal or goals justifying its construction – but this has been the history of astronomy.

6. Are We Ready to Return to the Moon?

Thanks to the Apollo missions a great deal is already known about the lunar surface. Nevertheless, it is not clear that we yet know how best to carry out major astronomical developments on the lunar surface. Extensive moving around and actually working on the lunar surface present problems at which the Apollo experience only hints – in particular, how humans can work efficiently in the environment of vacuum and dust, and cope with the threats of micrometeorites, cosmic rays, and especially solar-flare particles. Teleoperated robots probably hold the key to large areas of this problem. Also, there are important practical details such as how to deal with the possibility of major electrostatic effects on sensitive equipment, with extreme temperature changes from day to night, with the influence of cosmic rays on detectors, and with the effects of micrometeorites on optical surfaces. Some experience with operating telescopes of several kinds on the Moon would be very valuable before committing to the final designs of initial large and expensive instruments. A strong case can be made for planning several test instruments – probably first a transit telescope placed remotely by a softlander or at least installed by the very first lunar return party. Likewise, a small fully-steerable telescope could give invaluable experience in the way complex machinery should be built, operated, and shielded from various environmental influences on the Moon.

Political as well as scientific and technical factors will decide how the lunar observatories finally come into existence, and some degree of international cooperation is almost certain to be involved. Most of us are probably tolerant of a rather wide variety of approaches to these questions, but nevertheless hope and expect to see the effort – so important to the future of human understanding of the universe – begun in the very near future. NASA is planning human return to the Moon around the beginning of the new century. In view of the long lead times required for even relatively simple space projects, it is urgent for detailed study and development of lunar telescopes to be undertaken now.

RELATIVE MERITS OF LOW-EARTH, ECCENTRIC, GEOSYNCHRONOUS, AND INTERPLANETARY ORBITS AND SITES IN SPACE

B.G. TAYLOR

Space Science Department, ESTEC, European Space Agency, P.O. Box 299, 2200 AG Noordwijk, The Netherlands

1. Introduction

One criterium for the best orbit (or site) for an observatory has to be "that which is both scientifically and cost-effective". Perhaps the worst orbits are those offered by "infrastructures", which were not designed with astronomy as their primary driver, yet have to be used "because they are there". The ESA astronomy programme currently tends to favour highly eccentric or geostationary orbits (e.g. COS-B, IUE, EXOSAT, HIPPARCOS, ISO), as these appear to satisfy the above criterium. However, depending on the scientific mission to be accomplished and the availability of funds, other orbits are not ruled out priori.

2. Low Earth Orbits (LEO)

The obvious disadvantage is the proximity of the Earth:

– for typical pointed, narrow field-of-view telescopes, the Earth obscures (occults) celestial objects for up to 50% of a typical 90 minute orbital period, hence resulting in poor, on-target, observing efficiencies.
– the high heat input can reduce the longevity of expendable cryogenic systems.
– passage through the South Atlantic anomaly and the horns of the radiation belts can be detrimental to observations due to the particle background.

On the other hand, all-sky surveys can be conducted from low earth orbit, e.g. IRAS, COBE, ROSAT, while a O inclination orbit, below the belts, can give a very low cosmic-ray background as exploited by SAS II, in providing the first measurement of the cosmic gamma-ray diffuse background at about 100 MeV, and intended for SAX.

3. Highly Eccentric Orbits (HEO)

The obvious advantages of these orbits which may range from 24 hour (about 70,000 km apogee, ISO) to 96 hour (about 200,000 km, EXOSAT) or beyond, are that they:

– permit long, uninterrupted observations of celestial objects (as exemplified by EXOSAT, observations of eclipsing, dipping and bursting X-ray binaries);

Y. Kondo (ed.), Observatories in Earth Orbit and Beyond, 377–379.

– provide for the possibility of practically continuous contact through two ground stations (one if the orbital inclination is chosen correctly) allowing real-time, "observatory-like" operations;
– maximise observing time.

On the other hand, the satellite approaches the Earth once per orbit, traversing the radiation belts. At worst the belts are a major source of radiation damage to components or detectors, may require detectors to be switched-off temporarily or at least may raise background levels, reducing instrument sensitivity. When beyond the Earth's geomagnetic cavity, the satellite is, of course, exposed to solar flare radiation, which can, as in the case of EXOSAT, require that observations be interrupted. While the radiation belts and the magnetosphere are relatively well mapped to allow proper design of spacecraft and instruments, nonetheless, it may prove necessary to raise the perigee above the altitudes of peak intensity. For instance in the case of XMM, the perigee will be raised initially to > 1000 km. Nevertheless the yearly dose rate in silicon with 4 mm equivalent of aluminium shielding is of the order of 15 krad.

4. Geostationary/geosynchronous orbits (GSO)

These are a special case of the HEO; the "classical" observatory in this orbit being IUE. It was also the orbit of choice for Hipparcos. The radiation belts by and large are avoided and a single ground station offers full 24 hour, real-time coverage. However, the particle background at 36,000 km is still too high for X-ray astronomy missions and not ideal for IR detectors (the reason ISO's HEO apogee was raised from about 36,000 to about 70,000 km).

5. Interplanetary orbits

With the exception of the occasional solar flare such orbits probably provide the most stable, unchanging environments for an observatory. ESA's SOHO solar observatory will be placed at the L1 Sun-Earth Lagrangian point.

6. Sites

The moon as a location for astronomical instruments has been discussed at this colloquium and indeed will be studied within ESA over the next year. The advantages for radio astronomy (on the far side) and interferometric arrays requiring large, stable structures at UV/optical/IR wavelengths are self-evident. However, a sizeable "infrastructure" is first required for some of the schemes put forward.

7. Launcher size/"delta V"

Generally speaking, the farther from the Earth the observatory has to be placed, the greater the increment in velocity (delta V) from low earth orbit required. Thus a bigger booster or additional motors are required with associated higher costs.

(Perhaps it should be noted here that the costs quoted for ESA's stand-alone scientific missions include launcher costs, which have to be paid from the Agency's Scientific Programme Budget.)

8. Spacecraft complexity and costs

Probably, observatories designed for low earth orbit are more complex, hence costly, than those for the other orbits considered here. For instance:

– on-board data storage is required for later play back via ground stations or data relay satellites.
– the power and thermal subsystems have to be designed (other than for sun-synchronous orbits) to cope with the sunlit and eclipse operations of comparable duration.
– the attitude control system may have to be designed to ensure that the instruments do not view the Earth, or cope with gravity gradient or magnetic torques.
– greater autonomy has to be built-in to maximise fail-safe protection for protracted periods outside real-time control possibilities.

As one moves progressively through the other orbits, the spacecraft complexity and hence cost ought to decrease.

9. Serviceability, man intervention and robotics

Observatories in low-earth orbit can, in principle, be serviced by man from space planes or space stations (or be returned to Earth and re-launched). However, the observatories would have to be designed for that (of course SMM was certainly one notable exception), and man has to be brought into orbit, kept alive and returned at the end of servicing as well as being trained for the servicing mission in the first place. Servicing, of course, perhaps in higher orbits, could also be done robotically.

However, returning to the introductory criterium of "scientific- and cost-effectiveness", taking all cost elements into account (as done in ESA and perhaps we owe that to the tax-payer), the real differences in scientific returns and costs for an observatory launched once and serviced/modified/upgraded n times and the launch of $n + 1$ observatories in an upgradable series should, for once, be determined.

HUMANITY OR ROBOTICS IN SPACE?

MINORU ODA

RIKEN, Hirosawa, Wako, Saitawa, Japan

The diagram (Fig. 1) qualitatively exhibits how several important observations will have to be undertaken. Here, I would like to, at least, raise the question or discuss the problem referring to "Man in Space".

My question is as follows: How the man in space, like the Space Shuttle, and the robotics, like Voyager, precede the man in space "and/or" the robotics in the 21st century? I think, the answer is not obvious and we have to be cautious to make decisions. Once we make the decision, we may not be able to go backwards because the investments will be enormous and the impact to all humans will be very deep.

When we say "Man in Space" (e.g. specifically Man on the Lunar Base) we have to distinguish two cases: one is to use humanity as Scientists or Astronomers to work there or as Engineers to construct, refurbish, repair or assemble facilities. Therefore, we have three choices in the 21st century. They are:

1) Full activities of humanity in space;
2) Engineers in space; and
3) Fully robotics.

I don't want to commit to any directions at the moment, but I think that we must be aware of the existence of choices, and be very cautious as we strive to determine the final goal.

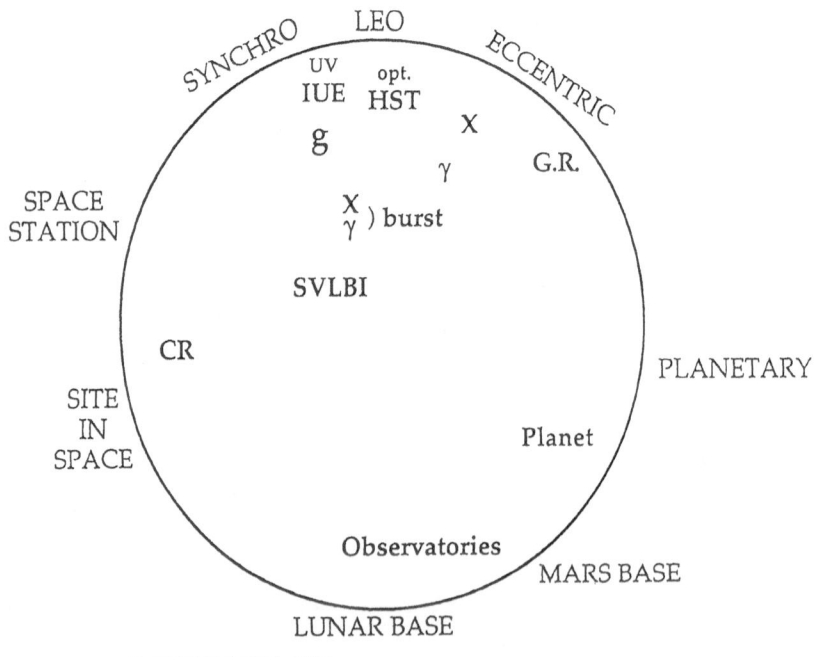

ASTROPHYSICS FROM THE MOON

MICHAEL J. MUMMA

Planetary Systems Branch, Laboratory for Extraterrestrial Physics,
NASA Goddard Space Flight Center, Greenbelt, Maryland 20771

1. Introduction

The moon offers mixed blessings as a site for astronomical observatories, and some of these are presented in Tables I and II (see the paper by Smith for a more complete discussion). The advantages are so strong that they make the moon a unique site for certain kinds of observatories, particularly those which feature very large telescopes or interferometric arrays of large dimensions, and those that require a presence beyond the screening effect of the earth's magnetic field. More than one hundred scientists gathered at Annapolis, Maryland for a Workshop on Astrophysics from the Moon (Feb. 5–7, 1990), and discussed these questions in considerable detail (Mumma and Smith 1990). The objective was to examine the astrophysical frontiers of the 21st century, and to identify those which require the presence of astrophysical observatories on the moon for their successful address. The scope of the Workshop was defined to include all areas of space astronomy. It was assumed that the base program of space astronomy will have been completed prior to the emplacement of lunar-based observatories, thus that the four Great Observatories, the Solar and Heliospheric Observatory, the Orbiting Solar Laboratory, the Coronagraphic Imaging and Astrometric Telescope Facility, and the Earth-Orbital Planetary Telescope will all have been accomplished. The Workshop participants tried to look beyond the scientific objectives to be addressed with those Earth-orbiting facilities to the next generation of scientific frontiers, which are associated with the very small, the very distant, and the very faint. This paper summarizes some of the ideas presented there, but these are presented in much greater detail in the published Proceedings (Mumma and Smith 1990).

2. The Principal Themes

Five great themes of fundamental importance emerged from the discussions. They are:

- *How do Stars and Planetary Systems Form?*
- *Are There Earth-like Planets among the Nearby Stars?*
- *Do Life and Intelligence Exist Elsewhere?*
- *What's Happening in Our Sun and Other Stars?*
- *What Happened in the Early Universe?*

Y. Kondo (ed.), Observatories in Earth Orbit and Beyond, 381–390.
© 1990 *Kluwer Academic Publishers.*

TABLE I
Advantages of the moon as an astronomical site

Ultra-High-Vacuum– Effectively no Atmospheres
        Perfect Transmittance; Diffraction Limited Imaging
        Dark Sky – No Scattered Light – Daytime as well as Night Observing
                Very Small Solar Elongation Possible (Few Solar Radii)
        Cold Sky – Thermal Infrared Background Limited by Zodiacal Emission
                Radiative Cooling of Telescopes for Infrared Observations
        No Wind – Permits Simplest "Sunshade" Domes, and Lightweight Telescopes
High-Lunar-Mass
        Enormous Moment Of Inertia
                Easy Pointing of Massive Telescopes – Completely Smooth Tracking
        Small Gravity
                No Co-Orbiting Dust or Debris in Field-of-View of Telescopes
                Dropped Tools and Parts Easily Retrieved
                Easier Working Environment Than Zero-G
                Biologically Friendlier than Zero-G
        Unlimited Mass Available for Shielding Humans and Sensors From
                Radiation of Solar Storms and Galactic Cosmic Rays
        Vast Area/Dimensionally Stable Real Estate for Interferometers
Small-Sidereal-Rate - 500 Times Smaller than Leo
        14 Day (or More) Availability of Source
                Long Uninterrupted Observations of Variable Phenomena
                Long Integrations on Faint Sources
        Slowly Changing Thermal Environment
                Permits Excellent Dimensional Stability
"Ultimate" Telescopes Probably Possible Only on the Moon
                Giant Filled-Apertures; Large Interferometers; Far-Side Radio Receivers

TABLE II
Disadvantages of the moon as an astronomical site

Lack of Solar Power During Lunar Night
        Requires Enormous Batteries, and Doubled Solar Array Size, or
        Radioisotope Thermal Generators for Small Observatories,
        Nuclear Power Stations for Large Observatories
High Radiation Background - Increased Detector Noise
        Meters of Rock Needed to Match Shielding Effect of Earth's Magnetic Field
        Extractable Resources are Available for Shielding
Expensive to Operate Human-Tended and -Emplaced Observatories
        Soft Landed Automated Observatories are Much Cheaper
        Free-Flyers are Much Cheaper (Even HST!)
        Unmanned Cargo Ships are Needed to Relax Safety-Driven Telescope Costs
Day-Night Cycle Limits Efficiency of Single Observatory to 50%
        Requires two Antipodal Observatories if 100% of Sky must be available 100% of time
        If Equatorial, each Achieves 50% Duty Cycle on the Entire Sky
        If Polar, each Achieves 100% Duty Cycle on Half the Sky

TABLE III
A subset of lunar-specific observatories

Observatories Required by Scientific Considerations
  Large Aperture UV/Optical/IR Telescope
        Detection of Earth-Like Planets, and Search for Exobiology
        Formation And Evolution of Stars, Galaxies, Quasars, and
        Structure in Early Universe
  IR and Sub-MM Interferometers
        Study of Forming and Evolved Planetary Systems
        Structure and Processes in Galactic Nuclei and Quasars
  Optical Interferometer
        Imaging Nearby Stars
        Distances and Distance Scales
  Large Aperture X-Ray and Gamma-Ray Telescopes
        Stellar Accretion Disks
        Physics Of Neutron Stars and Black Holes
  2-Meter Class UVOIR Telescope for Solar System Studies
        Origin of Solar System; Compositions of Comets and Asteroids
        Time Variable Phenomena in Planetary Atmospheres
Observatories Required by Human Safety Considerations
  Solar Flare Network
        Flare Physics and Prediction
        "Early" (Few-Minutes) Warning of Life Threatening Flares for
        Humans on Moon and Mars
  Mars Telescope(s) – Continuous Support of Mars Expeditions
        Two Low Latitude UVOIR Telescopes – 2-Meter or Larger
                Study Mars Atmosphere in Detail Throughout at
                        Least One Solar Cycle for
                        Enhanced Forecasting
                Derive State of Atmospheric Waves and Dust Storms
                        Related to Safe Aerobraking
                        and Accurate Landing

Each theme requires one or more observatories on the moon for its successful address and, in several cases, a given observatory could address more than one theme.

The Workshop of course did not exhaust all possible areas of astronomy and astrophysics. A different set of participants might well have added or stressed other topics (e.g. black holes) out of the rich array of problems waiting to be solved. But we believe any workshop on the subject would agree that the conclusions reached at Annapolis include some of the most important problems, and that these clearly demonstrate the instrumental developments needed to tackle them. In the next sections we examine these themes in slightly greater detail.

3. How do Stars and Planetary Systems Form?

We still do not understand how stars and their systems of planets form, nor do we have a clear understanding of the processes that occurred during the formation of our solar system. Yet there are many young stars in the Taurus-Auriga complex, the nearest star-forming region, and about half of those younger than three million years show clear evidence that they possess nebular-presumably pre-planetary-disks some tens to hundreds of astronomical units in size. The inner regions nearest the star are the sites of mysterious processes which energize stupendous amounts of gaseous material into organized bi-polar outflow along the axis of rotation of the newly forming stellar-planet ary system. We do not yet understand the processes that drive these flows, the mechanism for transporting mass inward toward the star while momentum is transported outward, or the details of formation of planets and comets within the system.

The elemental compositions of the planets preserve the record of fractionation processes in our solar system, while the chemical and mineralogical signatures in cometary material preserve a record of the transition region between unmodified interstellar composition and the chemically equilibrated material of the inner solar nebula. We are beginning to learn how to read this record for comets in their active phase, and will do so using space – as well as Earth-bound observatories in the next decade. However, none of the planned Earth-orbital observatories can observe comets at small solar elongations when they are most active.

A 2-meter class dedicated UVOIR telescope is needed. It requires low temperature optics, a cryogenic focal plane, and a satisfactory solar occulting screen.

It could also provide detailed studies of time variable phenomena such as chemically and temporally variable jetting from individual vents on the cometary nucleus and dynamical and interactive processes in other planetary atmospheres, including wave phenomena and weather on Mars in support of eventual human exploration there.

In order to interpret the fossil record of our solar system, it is essential to identify the physical and chemical processes occurring in pre-planetary disks elsewhere, and to relate this record to current models of formation for our own solar system. For example, it has been suggested that giant gaseous planets form outside the distance where water ice is stable and can be carried in the solid phase into the cores of proto-planets. We also have models for the temperatures, densities, and chemical abundances in the main stellar nebula and in giant planet sub-nebulae for newly forming systems, and these models badly want testing.

The spatial scale of interest is initially 1 astronomical unit (au), later 0.1 au. The available tracers are the spectral signatures of gaseous species, particularly carbon monoxide (CO) and water (and many other species), and of condensed phase matter such as water ice (and other volatile ices), organic grains, and silicious grains. The pre-planetary disks exhibit enormous optical depths in the visible and infrared regions, but become optically thin at sub-millimeter wavelengths. Fortuitously, both CO and water have strong lines in this region, indeed the 557 GHz line of water should be an excellent probe of water in the low temperature regions of the outer disk (beyond $\sim$ 5 au), while the higher excitation lines of water and CO provide important probes of temperature and density in the warmer inner regions. Several low excitation lines of CO can be sensed from the earth's surface, and important preliminary studies will be carried out with interferometric arrays now under construction or in the planning phase, but many important spectral lines are obscured by heavy extinction in the Earth's atmosphere. Ground-based observations of normal water are hopeless, for example.

Observations from space are required, and the aperture diameter must be at least

24 km (at 600 μm wavelength) in order to achieve spatial resolution of 1 au at the 140 parsec distance of the nearest star-forming region. An interferometric array is indicated, and it must be located on the moon to ensure the required dimensional stability and pointing precision. An initial configuration might be remotely deployed, then connected by humans.

4. Are There Earth-like Planets Among The Nearby Stars?

It is now within our ability to directly detect – if they exist – planets similar to those in our solar system, but around the nearby stars. Though giant planets similar to Jupiter will of course be much more conspicuous, the Holy Grail is to detect Earth-like planets. The discovery of even one other planet with size and atmosphere similar to Earth would have profound implications. We are confident that planets from Earth-like to Jovian size can be detected and studied, using instruments of the kinds discussed in this volume. While present and near-term searches may achieve indirect detection of a few Jovian-class planets, direct detection and study of all planets down to Earth-class planets is the real goal, and it is a challenging one. At visible wavelengths, the contrast ratio between a star and an Earth-like planet shining by reflected light is about 10 billion, but in the infrared, where the planet shines in its own thermal radiation, the ratio is about 100,000, a much more favorable value. The diffraction disk of the star is much larger in the infrared, but it is easier to reduce light scattered from the telescope mirror. Thus, on balance, it is probably easier to detect Earth-like planets in the thermal infrared.

The instrumentation to detect Earth-like planets must be based on the moon. Two approaches have been examined: a large filled aperture "conventional" telescope of the 16-meter diameter class, and an interferometric array of comparable total collecting area.

The collecting area is a consequence of the faintness of Earth-like planets and the need to examine a significant number of nearby stars in a reasonable time. When seen at infrared wavelengths from a distance of ten parsecs, the Earth returns only about 50 photons each second to a square meter of collecting area. Thus telescopes several hundred square meters in area are needed to collect sufficient photons for detecting and characterizing the planet. Low temperature optics are required, and they must be located in space, in order to eliminate background emission from both the telescope and the atmosphere. The telescopes are very large, have high mass, and require precise pointing which in turn requires an exceptionally stable platform. This stability is probably not possible to provide in Earth orbit, but is obtained simply for lunar based telescopes by virtue of the moon's enormous moment of inertia and excellent dimensional and seismic stability. Such instruments would have to be assembled on the moon, and maintained from a lunar base. It is certain that both concepts require extensive technical study, and in both cases the critical areas of work are well defined. Detailed studies of these concepts should begin immediately.

5. Do Life and Intelligence Exist elsewhere?

We still know of only one planet on which living things exist, yet several lines of evidence suggest that life may be harbored in many places. The fossil record shows that, at least on Earth, life began almost as soon as conditions permitted. Twenty years of molecular astronomy have demonstrated that many of the pre-biotic molecules thought necessary for the spontaneous origin of living material are found in dense molecular cloud cores, which evolve into new stars, many of which are undoubtedly surrounded by young planetary systems. The discovery of organic grains (CHON), formaldehyde, and complex hydrocarbons in comet Halley demonstrated for the first time that cometary nuclei contain these pre-biotic substances. Cometary planetesimals must be enormously abundant in the cooler regions of proto-planetary nebulae, hence their impacts on young planets must also be numerous. Recent theoretical work argues that a significant portion of the delivered material survives the impact process without significant chemical modification, allowing the surfaces of all young planets to receive pre-biotic material in enormous quantities. The discovery that disks and clouds of debris in fact do orbit many of the nearby stars is a strong indication that all these processes are relatively common. It thus seems plausible that living organisms may exist on at least some other planets.

Life has modified our atmosphere in fundamental ways – indeed the presence of enormous quantities of free oxygen is a consequence of the presence of terrestrial life. Likewise, the presence of abundant methane in an otherwise oxidizing atmosphere is a direct consequence of biological processing of carbon. While the atmospheres of the giant planets in our solar system contain copious methane, and the atmosphere of Mars contains trace oxygen and ozone, only Earth's atmosphere contains both oxygen and methane in significant quantities. Thus, a major objective in exobiology is to characterize the atmospheres of extra-solar planets, and to identify those which have been modified by the presence of life. This can be accomplished by spectroscopy at thermal infrared wavelengths, where the principal biologically-connected gases all have strong characteristic absorption bands. Ozone (a tracer for oxygen), methane, ammonia, water, and carbon dioxide can all be detected at thermal infrared wavelengths, using spectroscopy. *A large aperture telescope having several hundred meters collecting area is required, to identify biological modification of Earth-like planets, and it must be instrumented with low resolution spectrometers (spectral resolving power $\approx$ 100) n the wavelength region 6–15 μm.*

It must feature low temperature optics and cryogenically cooled detectors, be located in space, and have exceptionally precise pointing stability. It seems necessary to locate it on the moon.

Though a successful search for extra-terrestrial biology may require the study of a large number of stars, discovery of life would be profoundly significant. Far more significant yet, but presumably vastly more rare would be the discovery of extra-terrestrial intelligence. A crude feeling for detection probabilities follows from knowing, for example, that the terrestrial atmosphere has testified to biological modification for billions of years, but the evidence of an intelligent presence has only been detectable remotely for less than 100 years. Even supposing human tech-

nical capability to extend for another 10,000 years, the remote detection of biology
through atmospheric evidence would still be favored in the ratio of 100,000 or so.
This greatly over-simplified argument suggests that comparably larger samples of
stars must be examined to provide a comparable probability for the successful detection of intelligence. The Search for Extraterrestrial Intelligence (SETI) is presently
conceived as a search for radio emissions. Its detection sensitivity and completeness
are severely limited by natural atmospheric emissions and opacity, and by artificial
noise sources associated with human activities. The lunar far-side is the quietest
zone in near-Earth space and would provide a unique location for large radio antennae, or arrays. This would provide commensurate improvements in detection
sensitivity and in the range of available frequencies. The potential benefits seem
clear, but there has been no critical and broad evaluation of the potential of the
lunar far-side for SETI.

6. What is Happening in our Sun and in Other Stars?

We have never seen a picture of a star other than our sun. Moreover, essentially all
that we know of solar physics and processes is based on a single temporal snapshot in
the 4.5 billion-year life of this ordinary star. We do n ot know whether predictions
derived from present conditions are representative of future development or not,
but we do know that some aspects of the sun have changed significantly during
recorded history – for example that sunspots virtually disappeared for a period
of time 300 years ago known as the Maunder Minimum. No one knows why this
happened. Our theory of the internal engine that drives the sun predicts a certain
flux of neutrinos which exceeds the detected flux by a major factor. We do not
know whether this represents some basic flaw in our understanding of the interior
processes, of neutrino physics, or of some other missing factor. While many stars
are variable in their energy production, the sun does not seem to vary much, but we
also know that the Earth has experienced major climatic changes which may have
been in part caused by changes in the annual insolation. One way of gaining insight
into our sun is to investigate processes occurring in other stars. Stars of similar age
and mass, or of similar mass but a range of ages, are of obvious and direct interest,
but the sample should also include a range in mass and age beyond the sun's so
that true comparative solar-stellar physics may begin.

We must begin a program of comparative solar-stellar physics, and this requires
observations of the surfaces of other stars on spatial scales similar to that obtained
for the sun. For our sun, the best ground-based observing conditions provide optical
image quality of about 0.4 arc-seconds, or a linear resolution on the surface of about
300 km at visible wavelengths. Correction of atmospheric errors might improve
this to about 150 km, but further improvement requires space observatories. The
nation's plan for solar physics includes the Orbiting Solar Laboratory (OSL), which
features a one-meter diameter telescope instrumented at ultraviolet-optical-infrared
(UVOIR) wavelengths. OSL will achieve about 50 km resolution on the sun's disk
at visual wavelengths.

For stars, we ultimately wish to achieve resolutions comparable with the best
solar images, or about 150 km. Imaging of nearby stars with this spatial resolu-

tion requires angular resolutions in the micro-arc-second range, which at optical wavelengths corresponds to an aperture diameter of 125 km. Interferometric observatories are mandated, and these will need to be based on the moon. They must be based in space in order to escape the limiting atmospheric effects of airglow and turbulence. The observations require low fringe drift rates which imply extreme dimensional control and stability. The needed pointing and tracking precision require a stability and predictability of baseline which cannot be achieved in free space, but which is easily achieved on the moon. This system is the ultimate objective toward which initial and intermediate systems must grow, but a start can be made with 2-meter class telescopes and 100-meter baselines. The dedicated 2-meter class telescopes identified for planetary, stellar, and galactic astronomy might form the elements of an initial simple array for this purpose.

A particularly compelling case is made for a network of dedicated free-flying spacecraft, spaced in solar longitude, to provide predictive and alerting capability for life-threatening solar flares.

These occur relatively often, and the greatly enhanced solar wind they produce could be fatal for unprotected humans. Improved understanding is clearly needed, since at present we are unable to predict when a flare will erupt or whether the particles emitted from it will hit the Earth-moon system. This becomes doubly important for areonauts enroute to Mars, and while they are on the surface of that planet, with its only tenuous atmosphere and tiny magnetic field.

7. What Happened in the Early Universe?

There is a crisis in cosmology, which suggests that our present understanding of processes in the early universe is fundamentally flawed. Measurements of the cosmic background radiation demonstrate the absence of large-scale fluctuations at times corresponding to red shifts of $z < 15$. However, the earliest quasars appear only a bit larger in time, at red shifts of $z \approx 5$ and already exhibit a solar abundance of elements, implying the existence of an earlier generation of stars. Thus, stars and galaxies must have existed near $z \approx 15$ or even earlier, yet we see no evidence of them in the cosmic background radiation. Furthermore, recent measurements of galaxy-pair correlations, based on a deep red-shift survey, suggest that the clumping and voids ("bubbles") seen on the local scale extend to at least 1000 mega-parsecs, or about 3.3 billion light years, and thus that the occurrence and scale-size of these voids appears to be cosmogonic and fundamental. The nature of the voids present in the large scale structure is poorly known at present, but can be addressed by absorption spectroscopy of quasars and galaxies. In addition, the "hidden" mass required to close the universe has not yet been found, even though several lines of evidence suggest that it must exist.

Investigation of the formation and evolution of stars and galaxies in the epoch $5 < z < 15$ requires photometric and spectral observations of quasars and early galaxies in the 0.75–10.4 micrometer wavelength range.

A large filled-aperture telescope of the 16-meter class would meet these needs, but it must be in space to achieve diffraction limited imagery, full wavelength coverage, and sensitivity limited only by the natural background imposed by the zodiacal light. In turn, the low temperature and the extreme pointing stability required of this immense and high-mass telescope speak for its presence on the moon.

Along with photometry and spectroscopy, ultra high spatial resolution will be needed for many key investigations. Imaging the structure and kinematics of active galactic nuclei and quasars is basic to understanding their physics and evolution, but requires far higher angular resolution than can be achieved with filled aperture telescopes. The core activity of galactic nuclei and quasars is most significant on spatial scales below 0.1 parsec. A 16-meter telescope, if diffraction limited at 2 μm wavelength, could achieve angular resolution of 0.03 arc-seconds, but this corresponds to a spatial scale of 0.1 parsec at a distance of only 0.7 mega-parsec. Micro-arcsecond angular resolution is needed to map structure with a resolution of 0.1 parsec at a distance of several thousand mega-parsecs. Lunar-based interferometric arrays with 500 km baseline (at 2 mm wavelength) would achieve this resolution.

8. Environmental and Site Considerations

Two environmental aspects requiring further study include the nature of the surface thermal environment, and the effects of cosmic and solar particulate radiation on instruments, particularly on solid-state detectors and electronic components. Because the moon lacks a strong magnetic field, the particle flux at the lunar surface is much greater than in low-earth orbit and may require burial or heavy shielding of detectors. The high daytime temperatures at low latitudes present a potential problem for unprotected telescopes and instruments, particularly those operating in the thermal infrared and sub-millimeter region, and those requiring active cooling. It appears that simple "sunshade" shielding techniques can be devised for such telescopes and instruments, but the problem requires detailed study.

Candidate sites for initial observatories should be evaluated on the basis of visibility of the celestial sphere, dynamic accessibility for spacecraft landings and rendezvous with lunar transportation nodes or transfer vehicles, trafficable and workable terrain and geological/geophysical interest. A network of 5 sites would be ideal: two at the poles and three at the equator, 120 degrees apart. A minimum of two antipodal sites is needed for support of human exploration of Mars For a single initial site, the NE flank of the Orientale Basin at 80 degrees W longitude on the equator was suggested as a promising candidate. While maintaining line-of-sight communications with the Earth, this site would also permit access by tele-robotic vehicles to the lunar far side for emplacement of radio astronomy instruments, and is of considerable geological interest.

One concept that seems important is the need for site-testing telescopes, to be landed at prospective sites for lunar observatories in order to establish basic feasibility from the perspective of radiation background, thermal factors, achieved operating temperatures, and a host of other practical engineering details.

These should surely precede a final commitment to an individual site. Also, while the Annapolis workshop stressed major instruments and far-reaching programs which require a presence on the moon for successful accomplishment (cf . Table III for a partial listing), several authors have argued for a suite of lesser instruments which may be able to compete scientifically and cost-effectively with their earth-orbital counterparts. The future will determine whether the advantages of the moon will in fact carry over into a much wider range of usage, including a possibly large number of instruments of nearly all sizes and kinds. But whether or not this occurs, the Annapolis Workshop demonstrated that some of the most fundamental goals of astronomy are unlikely to be achieved without lunar basing and the magnificent astronomical instruments which this alone will make possible.

9. Conclusion

In addition to the fundamental themes outlined in this summary, many innovative and useful ideas and concepts were advanced at the Annapolis Workshop, and are described in individual papers and in the Summaries of the Scientific Working Sessions (Mumma and Smith, 1990).

It is clear that the astrophysical frontiers of the 21st century will require very large filled-aperture telescopes and interferometric arrays of many kilometers dimension for their successful address. These must be located in space, and the requirements of accurate and stable pointing alone require that they be placed on the moon. There is yet another important reason for placing them on the moon, namely the need for human interaction with – and tending of – these sophisticated and complex instruments. Compared with free space, the moon presents a familiar environment for humans. Tools stay where they are put, dropped parts fall to the ground for easy retrieval, and the body experiences gravity, thereby avoiding the deleterious health effects of prolonged exposure to zero-g, and there is ready shelter from cosmic rays and solar storms.

Many astrophysical objectives of the 21st century require lunar-based observatories for their full realization, and we should insert these scientific objectives as requirements to be met by a program of renewed human exploration and exploitation of the moon.

Reference

Mumma, M. J., and H. J. Smith: 1990, *Astrophysics from the Moon, AIP Conf Proc.*, **207**, in press

USE OF LIBRATION-POINT ORBITS
FOR SPACE OBSERVATORIES

ROBERT W. FARQUHAR

Goddard Space Flight Center, Greenbelt, Maryland 20770, U.S.A.

and

DAVID W. DUNHAM

Computer Sciences Corp., 10110 Aerospace Road, Lanham, Maryland 20706, U.S.A.

Abstract. The Sun-Earth libration points, L_1 and L_2, are located 1.5 million kilometers from the Earth towards and away from the Sun. Halo orbits about these points have significant advantages for space observatories in terms of viewing geometry, thermal and radiation environment, and delta-V expediture.

1. Introduction

The locations of the seven Lagrangian libration points near the Earth are shown in Figure 1. All five of the Earth-Moon libration points are included, as well as the L_1 and L_2 collinear points of the Sun-Earth system. The latter remain fixed on the Sun-Earth line, which is the horizontal axis of Figure 1. The configuration of the lunar libration points rotates around the Earth once each month.

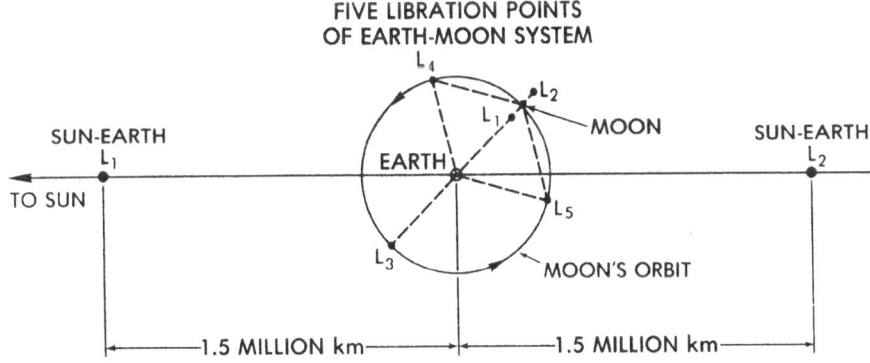

Fig. 1. Libration points in the vicinity of the Earth

The first comprehensive discussion of the use of libration-point orbits by artificial satellites was published by (Farquhar 1969). For space telescopes, the lunar libration points generally offer no special advantages over the Sun–Earth (SE) points. Even the triangular points do not have long-term stability due to strong solar perturbations.

Y. Kondo (ed.), Observatories in Earth Orbit and Beyond, 391–395.

2. Advantages of the Sun-Earth collinear points

The relatively fixed geometry of the Sun, the Earth, and the spacecraft in halo orbits about the L_1 and L_2 points of the SE system give them special advantages for siting space observatories. A halo orbit about the SE L_1 point was used by the third International Sun-Earth Explorer to monitor the solar wind (Farquhar *et al.* 1977; Farquhar *et al.* 1980). The European Space Agency's Solar and Heliospheric Observatory (SOHO) will use a similar halo orbit in 1995.

Orbits around the SE L_2 libration point are best for stellar observatories. The Sun, the Earth, and the Moon can be kept in back of the spacecraft for an unhindered view of over half of the sky at all times. The Soviet Union's Relict-2 spacecraft will use an L_2 orbit in 1993 (Eismont *et al.* 1990).

Spacecraft near either of the SE collinear points seldom need to change their orientation relative to the Earth or the Sun, simplifying thermal control and pointing of high-gain antennas. Large eclipses can be avoided, ensuring a continuous source of solar power.

Orbits around the SE L_1 and L_2 points are unstable, but stationkeeping costs are low, less than 5 m/sec per year. This could be achieved with low-thrust systems such as venting from a helium dewar or solar sails.

The delta-V costs to reach low orbits are shown in Figure 2. The total delta-V is the sum of two delta-V's needed to perform a Hohmann transfer from a low parking orbit (185 km altitude assumed) to a higher circular orbit whose radius is given in Earth radii (Re) of 6378 kilometers on the abscissa. DV1 is the delta-V needed to enter the transfer orbit. DV2 is a circularizing delta-V, performed at the apogee of the transfer orbit to achieve the final orbit. Vertical dashes mark the geosynchronous orbit and the 100,000-km orbit planned for the Space Infra-Red Telescope Facility (SIRTF).

Figure 3 is like Figure 2, but extended beyond the Moon's orbit. The Moon's gravity helps to "circularize" the trajectories of spacecraft sent to the lunar L_1 and L_2 points. These transfers utilize a powered close lunar swingby to decrease the libration-point insertion costs.

The delta-V is largest to achieve an orbit of about 100,000-km. An even bigger saving for SE L_2 orbits is the elimination of DV2; the Sun's gravity can achieve the "circularization". The delta-V cost to reach a halo orbit about the SE L_2 point is a fifth less than that for a 100,000-km circular orbit, meaning that a correspondingly larger payload can be delivered to L_2 than to the currently planned SIRTF orbit using the same launch vehicle.

3. Examples of L_2 orbits

Fig. 4 shows a possible SIRTF trajectory. There are no planned delta-V's following injection on June 29, 1997. Starting a day after launch, trim maneuvers could be performed with helium normally vented from SIRTF's cryogenic cooling system. SIRTF could begin observations just after crossing the lunar orbit. The only eclipses are lunar transits covering less than 20

Relict-2 has plasma instruments to measure the geomagnetic tail. However, the

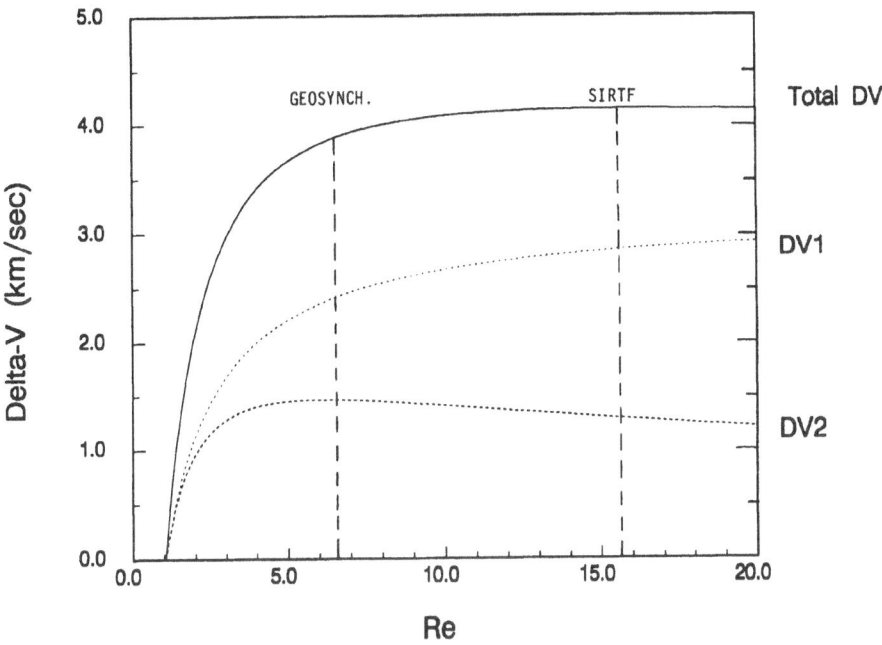

Fig. 2. Delta-V needed to attain circular orbits from 1 to 20 Earth radii

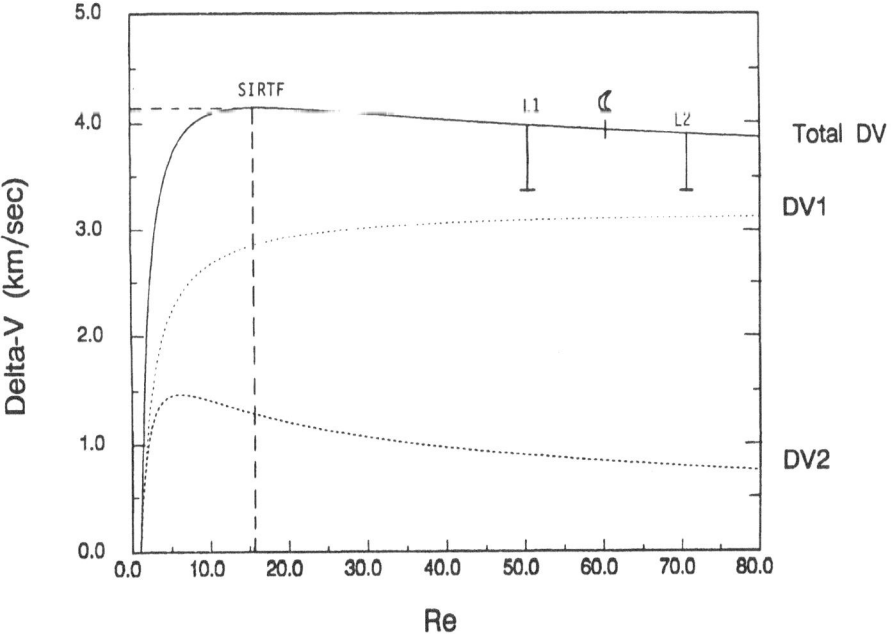

Fig. 3. Delta-V needed to attain circular orbits from 1 to 80 Earth radii

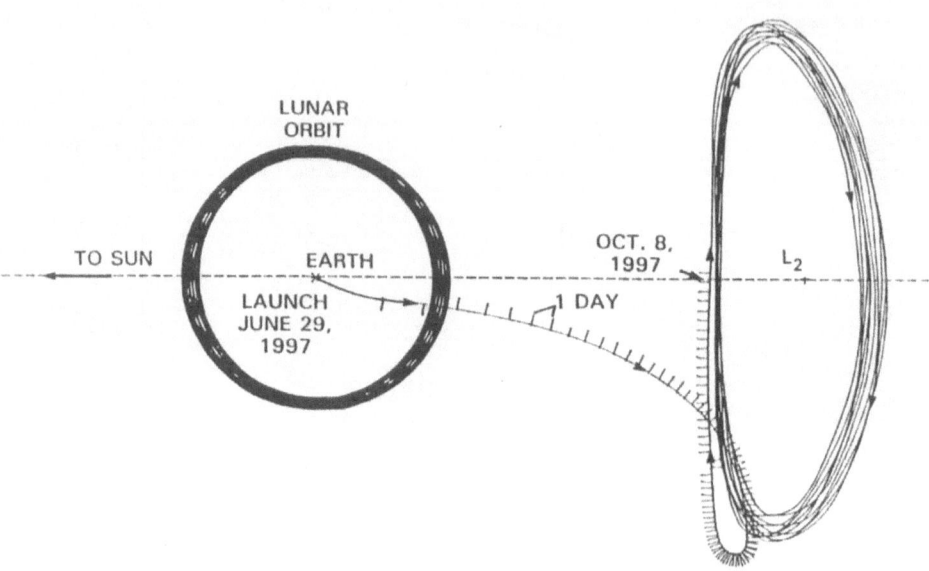

Fig. 4. SIRTF L_2 libration point trajectory

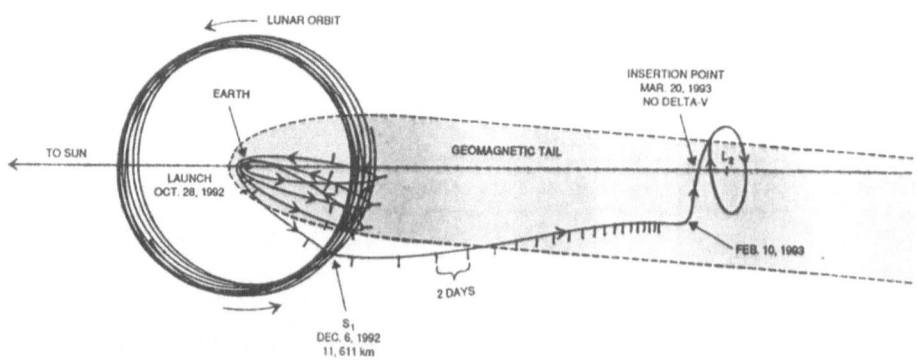

Fig. 5. Relict-2 trajectory using lunar swingby

orbit of Figure 4 quickly crosses this region. Geotail dwell time could be increased considerably with a smaller-amplitude orbit about L_2 achieved with a lunar swingby, as shown in Figure 5. Like the Figure 4 trajectory, there are no planned delta-V maneuvers after injection and no significant eclipses.

4. Conclusions

The advantages of halo-type orbits about the Sun-Earth L_2 libration point have been demonstrated. Future missions already plan to use Sun-Earth L_1 and L_2 orbits, including Relict-2, SOHO, the Advanced Composition Explorer, and one or two spacecraft of the Soviet Regatta series (Farquhar 1990). Until transportation to the lunar surface becomes almost routine, we believe that orbits near the Sun-Earth L_2 point may be the best places for space observatories.

Acknowledgements

Dunham acknowledges support by the National Aeronautics and Space Administration/Goddard Space Flight Center under Contract NAS-5 31500.

References

Eismont, N., Sinitsyn, V., Farquhar, R. W., and Dunham, D. W.: 1990, *Internat. Astronautical Fed. Paper*, 90-302

Farquhar, R.: 1969, *J. of Astronautics and Aeronautics* **7**, 52

Farquhar, R. W.: 1990, *Internat. Astronautical Fed. Paper*, 90-308

Farquhar, R., Muhonen, D., Newman, C., and Heuberger, H.: 1980, *J. of Guidance and Control* **3**, 549

Farquhar, R., Muhonen, D., and Richardson, D.: 1977, *J. of Spacecraft and Rockets* **14**, 170

(B) ALTERNATIVE APPROACHES

MAJOR OBSERVATORIES VERSUS ECONOMY-CLASS
OBSERVATORIES IN SPACE

FREEMAN J. DYSON

Institute for Advanced Study, Princeton, NJ

Abstract. Big space-instruments are essential for observing fine details of selected objects. Small space-instruments are essential for scanning large areas of sky, for discovering and cataloguing new objects, and for monitoring variable or transient sources of radiation. If astronomy is to flourish, both types of instrument must be provided. The problem is not to choose between big and small. The problem is to find the optimum mixture of big and small. Various examples from past history and from future plans illustrate the importance of maintaining a variety of different types of instrument.

1. Baade and Zwicky

The title of this talk was imposed on me by the organizers of the meeting. I did not choose it. I do not like it. The word "versus" implies an antogonism between large and small space-telescopes. No such antagonism ought to exist. Any reasonable program for space-based astronomy must have large and small instruments working in collaboration. Large and small are needed for different jobs. They should complement each other, not comete with each other. Lyman Spitzer wrote in his 1968 article describing the work of the Large Space Telescope: "Quite apart from the engineering desirability of launching smaller space telescopes before building the large instrument, the astronomical requirement for a continuing series of smaller space telescopes should be an overriding consideration in setting the pace of the Large Space Telescope effort." Now it is the job of the IAU to make sure that, in the euphoria surrounding the launch of the Hubble Telescope, these words of Spitzer are not forgotten.

Before looking at future missions I go back to the past. A story from the past may help us to avoid mistakes in the future. I go back fifty years, to the beginning of the Palomar observatory.

The 18-inch Schmidt was the first telescope to be put on Palomar mountain. It was working there for ten years before the 48-inch Schmidt and the 200-inch Hale Telescope arrived. It was there because Fritz Zwicky was interested in supernovae. Zwicky was the first astronomer to observe supernovae systematically. To observe a significant number of supernovae, Zwicky needed a telescope that he could have all to himself, taking pictures of galaxies night after night. He organized the building of the 18-inch Schmidt on Palomar and made sure that he would be in charge. To have unlimited observing time was more important than to have a big mirror. Zwicky made good use of his time. Here is his own account written thirty years later:

"I put this instrument into operation on the night of September 5, 1936, and immediately started a survey of several thousand galaxies. Twenty supernovae were

Y. Kondo (ed.), Observatories in Earth Orbit and Beyond, 399–405.

discovered by my assistant J.J. Johnson and myself in the period from 1936 to 1941. At the same time much of the observational material was gathered for our six-volume catalogue of galaxies and clusters of galaxies. For the construction of the 18-inch Schmidt telescope, its housing, a full-size objective prism, a small remuneration for my assitant, and the operational costs for the whole project during ten years, only about fifty thousand dollars were expended. This probably represents the highest efficiency, as measured in results achieved per dollar invested, of any telescope presently in use, and perhaps of any every built, with the exception of Galilei's little refractor."

Even after making allowances for Zwicky's notorious egotism, I have to agree with his claims for the 18-inch Schmidt. His exploration of the world of the supernovae was of immense importance to the future of astronomy. His sky-survey set the style for many later surveys done with bigger instruments and bigger investments of man-power and money. He was the first astronomer to search deliberately for the most violent events in the universe, and he was the first to understand that if you want to search efficiently for violent events you must cover the whole sky repeatedly. Insofar as the modern style in astronomy is to concentrate attention on short-lived and violent processes, Zwicky was the first modern astronomer. And the 18-inch Schmidt gave him the chance to show what he could do. Among other things, he discovered in his survey of clusters of galaxies the first clear evidence that the universe contains missing mass.

Here is another quote from Zwicky's autobiography:

"Astronomers expect that photoelectronic devices of small aperture and light weight will eventually outperform the giant 200-inch Hale reflector on Palomar mountain. In any case these new instruments will be indispensable if we intend seriously to explore the universe with the aid of rocket-borne instrumentation."

Zwicky was wrong to claim that small telescopes with electronic detectors would put the 200-inch out of business. Zwicky was blinded by his dislike of Walter Baade, the chief observer at the 200-inch in those days. When Zwicky wrote that small electronic telescopes would outperform the 200-inch, what he had in mind was that Zwicky would outperform Baade. Both of them were great astronomers. Neither of them outperformed the other. Zwicky was a cantankerous genius, quick in thinking of usefule schemes and energetic in carrying them out. Baade was an artist, a virtuoso performer who played the 100-inch and 200-inch telescopes like violins. His great contribution to astronomy was the identification of the two populations of stars. The decisive step in his discovery of Population II was to resolve into stars the great red blob at the center of M31. Baade was able to see these stars only by pushing the 100-inch telescope to the limit of its capabilities. For this purpose, any smaller telescope would have been useless.

Once in their lives, when Zwicky and Baade were both young and before they had become enemies, they wrote together a theoretical paper of extraordinary originality. Their paper appeared in 1934, just two years after Chadwick had discovered the neutron. At the end of their paper, Baade and Zwicky said: "With all reserve we advance the view that a supernova represents the transition of an ordinary star into a neutron star, consisting mainly of neutrons. Such a star may possess a very small radius and an extremely high density."

These remarks of Baade and Zwicky were ignored by astronomers for 33 years, until neutron stars were discovered by radio-astronomers. Now we know that almost everything Baade and Zwicky said in their 1934 paper was true. It was a great loss of science that they did not continue their collaboration. Each of them alone was great, but together they would have been greater. If they had remained friends, neutron stars might have been discovered twenty-five years sooner, in 1942 instead of 1967. It happened that in 1942 Baade used the 100-inch to take the classic pictures of the Crab Nebula which you have all seen. He identified the peculiar star near the center of the nebula, which he suspected of being the stellar remnant of the supernova that exploded in 1054 A.D. According to the Baade-Zwicky paper of 1934, the stellar remnant ought to be a neutron star. Baade asked his friend Rudolf Minkowski to take a spectrum. Minkowski, also using the 100-inch, took a spectrum of the star and found it completely featureless with no lines at all. Minkowski calculated the temperature of the star and found it to be 500000 degrees. The spectrum made it certain that this was the supernova remnant, a weird and unique object. But Baade and Minkowski did not go further. They did not look at the star again. They did not in their 1942 publications mention the possibility that it might be a neutron star. Perhaps by 1942, Baade had come to consider a neutron star to be merely one of Zwicky's crazy ideas from which he was glad to dissociate himself. From a human point of view, such a reaction is understandable. But from a scientific point of view, it was a great opportunity missed.

As Cocke, Disney and Taylor discovered, with a one-meter telescope at Steward Observatory in 1969, the star in the Crab Nebula flashes on and off thirty times a second. A few years later, a graduate student observed the flashes and measured their changing period with a one-meter telescope on the Princeton Campus. Whatever a Princeton student did with a one-meter telescope under the polluted sky of New Jersey, Zwicky could have done with his 18-inch under the dark sky of Palomar thirty years earlier.

Zwicky was uniquely qualified to be the discoverer of neutron stars. He was one of the few astronomers in the 1940's who took neutron stars seriously. He was one of the few who saw the potential of electronic photo-detection. And he was one of the few who had unlimited observing time on a good small telescope. If he had put his mind to the problem, he might have seen the star in the Crab Nebula flashing in 1942, and the whole subsequent history of astronomy would have been changed. He missed his chance because he was not talking to Baade.

I told this story of Baade and Zwicky at some length, because it illustrates my general theme. The moral of the story is that you miss making important discoveries if you work with big telescopes alone or with small telescopes alone. You need a big telescope to get good spectra of faint objects. You need a small telescope to search the sky for objects you did not know existed. You need a big telescope to detect objects out to the extreme limits of faintness. You need a small telescope to keep watching a variable object, to see how it changes from night to night or from millisecond to millisecond. But it is not enough to have both big and small telescopes. You need also to have big-telescope astronomers and small-telescope astronomers talking to each other. Zwicky was a classic case of a small-telescope astronomer doing great work with limited means. Baade was a classic big-telescope astronomer, searching the far distances of the universe until, as Hubble said, "At the last dim horizon we search among ghostly errors of observation for landmarks that are hardly more substantial". I won't attempt to apportion the blame between Zwicky and Baade for the fact that they stopped being friends. Both were to blame for missing their chance to put their combined talents to work to solve the mystery that lies at the heart of the Crab Nebula.

2. HST and IUE

I now jump 48 years from 1942 to 1990. We shall have, if all goes well, two ultra-violet telescopes observing from orbit in 1990, HST and IUE. In many respects, the history of the 1940's will be repeating itself. HST is not quite as big as the 200-inch, but it is an enterprise of the same character, massive in scale and cost, plagued by technical difficulties and delays, supported by an effective public-relations campaign, and giving us finally a huge extension of our view of the far reaches of the universe. IUE is not quite as cheap as Zwicky's 18-inch Schmidt, but it is exactly as small, a little half-meter mirror sitting in the sky, unnoticed by the public, pouring our important and unimportant astronomical observations for twelve years while its big sister was still struggling to be born. In one respect history has changed. For better of for worse, the management of telescopes has passed from individuals to committees. No cantankerous genius like Fritz Zwicky is in charge of IUE. IUE is run by an international consortium with input from large numbers of users in many countries. And HST is run by a big organization in Baltimore with input from almost everybody. With luck, IUE may stay alive for another five years. But it is living on borrowed time, its 14-year-old Vidicon detectors have long been obsolete, and it would be ready for retirement if we had a more modern IUE ready to replace it.

The Hubble telescope will not be a replacement for IUE. The two instruments complement each other without much overlapping. HST is needed for deep exploration and high-resolution imaging. IUE is needed for rapid spectroscopy and wide coverage of brighter objects. HST will discover a vast number of things that IUE could not reach. But the greater outreach of HST comes at a high cost in observing time. If you observe with HST at the limit of its capabilities, you need to wait a long time to collect the photons. Since most of the interesting in the sky are faint, HST will never have enough time to observe as many of them as we would wish. The observing time of HST will always be over-subscribed, and no time will be allotted for looking at things which could just as well be observed with IUE. That is the reason why HST cannot be a replacement for IUE.

I jump back again to the past to see what we ought to be doing to replace IUE. Zwicky's 18-inch Schmidt was, like IUE, in need of replacement after 10 years of productive work. Both IUE and the little Schmidt were unnecessarily small for the work they had to do. The replacement for the little Schmidt was not the 200-inch, just as the replacement for IUE is not HST. The replacement for the little Schmidt was the 48-inch Schmidt, which was installed then years after the little Schmidt at Palomar and put to work on the Palomar Sky Survey with Hubble himself in charge. What we need to replace IUE now is the space-based equivalent of the 48-

inch Schmidt. The replacement for IUE does not need a 48-inch mirror. I imagine it as a small ultraviolet telescope with tenth-of-a-second-of-arc resolution and modern CCD detectors, weighing and costing about one-fifth as much as HST. Ted Stecher proposed such a telescope as an Explorer mission a few years ago. NASA turned it down. His telescope will soon fly, under the name UIT, as part of the ASTRO shuttle mission. In a ten-day mission it can do some good observing but it cannot replace IUE. If our space-based astronomy program had been driven by scientific rather than political priorities, if the administrators of the program had followed the policy so clearly enunciated by Lyman Spitzer in his 1968 statement, we should have had the replacement for IUE already in orbit several years ago. We should have had a more capable but still modest-sized IUE, exploring the ultra-violet sky and preparing the way for HST, just as the 48-inch Schmidt explored the sky and prepared the way for the 200-inch fourty years earlier. We could have reconnoitred in depth all the objects which are now candidates for HST observation. We could have picked out and sharpened the crucial questions which only HST can answer. We could have greatly increased the scientific cost-effectiveness of HST by concentrating its precious observing-time upon mysteries that a small telescope could not resolve.

3. Future Projects

After giving ten minutes to the past and five to the present, I have ten minutes left for the future. I will not try to mention all possible future missions and arrange them in order of merit. I take it as given that the program of Great Observatory missions recommended by the Greenstein and Field committees, beginning with HST and continuing with GRO for gamma-ray astronomy, SIRTF for infra-red, AXAF for X-rays and LDR for far infra-red and submillimeter astronomy, will sooner or later be launched and flown. These are all splendid missions and I wish them well. I am especially happy to hear that the plan in now to put SIRTF into a high orbit where it will escape from the worst of the miseries that a low orbit imposes on HST. The trouble with the Great Observatory missions is that they kept us waiting too long before they flew. They have had a bad effect on the progress of astronomy, because they discouraged people from building smaller instruments that cost less and could fly sooner. I shall talk briefly about some modest missions that I would like to see flying while we wait for the big ones. One such mission, COBE, went up recently and is doing spectacularly good science. Fortunately, missions comparable in size and quality with COBE need not wait for NASA funding but can now be undertaken by other countries that are active in space-science. The European X-ray and Infra-red satellites ROSAT and ISO will soon be in orbit, filling two of the serious gaps in the NASA program. We are entering an era in which space-based astronomy will be spread among the nations as widely as ground-based astronomy.

I would like to see a series of modest space-telescopes operating in the same manner as IUE, covering all the various parts of the electromagnetic apectrum and acting as survey instruments to complement the various Great Observatories. We should have had at least six small observatories, one for submillimeter, one for infra-red, one for visible, on for ultra-violet, one for extreme ultraviolet and one for X-rays, already up and running. With luck we could build and fly six small observatories for the price of one great observatory. Together, they would yield an enormous harvest of science.

Another important class of small space-observatories is concerned with development of new optical technology. The great observatories of the next century will undoubtedly include optical interferometers and large thin mirrors with shapes controlled by active optics. Will we try out these new technologies for the first time in huge billion-dollar instruments? I hope not. We ought already to be planning to launch small interferometric arrays and small thin mirrors with active optical control. These missions would not only give us engineering experience essential for learning how to build bigger ones. They would also be important scientific instruments in their own right, allowing us to observe brighter objects with angular resolution surpassing HST. Once the technology of optical arrays is successfully demonstrated on a small scale, big arrays will probably come to dominate optical space-astronomy just as they have dominated radio-astronomy on the ground. But the radio-astronomers did not jump from single dishes to the VLA in one step. The VLA works well because it grew out of long experience in building and using smaller arrays. If we are wise, we will approach the grand optical arrays of the twenty-first century in the same step-by-step fashion.

As a rule, the cost-effectiveness of small telescopes is greatest if they are designed and used as special-purpose rather than general-purpose instruments. IRAS and COBE were designed for the special purpose of sky survey, and they do only that one job. They have been spectacularly successful because they do one job spectacularly well. Zwicky's 18-inch Schmidt, and its successor the 48-inch Schmidt, were likewise designed for sky survey and were likewise cost-effective.

IUE is an apparent exception to the rule, successful in spite of being a general-purpose instrument. But even IUE conforms to the rule in so far as it confines its activities to spectroscopy and does not waste time on imaging. Hipparcos is another example of a special-purpose instrument that promises great scientific return from a modest payload. After Hipparcos completes its mission, as I hope it will, there will be other astrometric satellites, also special-purpose, also modest in size and weight, extending the limits of astrometry both in accuracy and in breadth of coverage.

The majority of the small space-telescopes that I have mentioned would be special-purpose instruments designed for sky survey in a variety of wave-bands. Another instrument of this type is a one-meter diffraction-limited telescope dedicated to a digital sky-survey with tenth-of-a-second-of-arc resolution in visible light. The digital sky-survey would be as important to the future of astronomy as the Palomar sky-survey of 1949–1956. It would not be a small mission. It would be intermediate in size and cost between an Explorer mission like COBE and a Great Observatory like HST. It would require considerably larger CCD detectors than are now available. In round numbers, it requires a CCD array with 10^9 pixels to give a one-degree field of view at 0.1 second resolution and to cover the whole sky with thousand-second exposures within a year or two. The data-stream would come to earth at a rate of one megabyte per second and would require a memory of 100 terabytes for permanent storage. These numbers refer to a survey done in a single color. It would be reasonable to plan the mission to do a survey in several colors simultaneously. Then we could have a 3-color survey completed in five years with the same bit-rate and 300 terabytes of storage. The numbers are formidable, but my computer-expert friends do not find them unreasonable. A major component of the digial sky-survey enterprise will be learning how to digest and distribute data in usable form from the 300 terabyte storage to astronomers all over the world.

But let us not be carried away by grandiose dreams. The main advantage of small special-purpose space-missions is lost if we let them become too ambitious. We have enough Great Observatoties on the drawing-boards already. What we need now is more small missions, using existing hardware, to fill the gaps between the big ones. Let me end as I began with some good advice from Fritz Zwicky. I quote again from his book, "Discovery, Invention, Research".

"I wish to caution all hotheads that it is not advisable to try to do everything at the same time, a mistake which is often committed by individuals and by institutions whose funds are limited. For instance, the construction of multipurpose telescopes is in general not to be recommended. It is better to concentrate one's attention on specific problems and to build instruments best adapted for their solution."

References

Spitzer, Lyman Jr.: 1968, *Science* **161**, 225–229

Zwicky, F.: 1969, *Discovery, Invention, Research*, New York, Macmillan, 91–92, 226

Zwicky, F.: 1948, in Halley Lecture, "Morphological Astronomy", ed(s)., *Observatory*, 68, 121–143

Baade, W. and Zwicky, F.: 1934, *Proc. Nat. Acad. Sci. (USA)* **20**, 259–263

Baade, W.: 1942, *Astrophys. J.* **96**, 188–198

Minkowski, R.: 1942, *Astrophys. J.* **96**, 199–213

Kondo, Y. (Ed.): 1987, *Exploring the Universe with the IUE Satellite*, Dordrecht, Reidel, 347

V. LONG TERM FUTURE ISSUES

Panel: Major Unsolved Problems of Astronomy

HALTON ARP

FREEMAN J. DYSON

FRED HOYLE

M. S. LONGAIR

MINORU ODA

DOES THEORY ADVANCE WITH TECHNOLOGY?

HALTON ARP

Max-Planck-Institut für Astrophysik, 8046 Garching bei München, Germany

Observational technology in astronomy moves ahead. We can see a thousand times fainter and ten to a hundred times more detail than 40 years ago. But does our application to research match the engineering progress? Most of us make the easy assumption that theory is right at the cutting edge, waiting to gobble up each new fact into an even deeper, more detailed insight into the universe. But humans frequently misunderstand the real problems and misapply technology – making everything worse for agonizingly long times.

Is it possible that extragalactic astronomy has serious misconceptions? The key point to appreciate is that its whole structure rests on the belief that we know the distances to objects in the universe. The simple shift to the red of the spectrum of any observed object is assumed to measure its distance. But for 25 years evidence has been increasing that drastically incorrect distances can result. Unfortunately, not only quasar distances but the distances to the vast majority of galaxies also depend on redshifts.

Huge observing facilities can only mislead if the objects observed are assumed to be something they are not! Consider the futility of observing a galaxy which is assumed to be a gigantic monster, able to swallow whole groups of well known galaxies – but is actually only a nearby dwarf. The luminosity of various distances indicators will be derived on this wholly false assumption of great distance. Whatever answer comes out will be treated as a startling discovery and applied to the distance of other objects which are presumed to represent the universe as a whole. Fantastically incorrect conclusions will be drawn about the age, origin and evolution of a completely fictitious universe.

A few examples contradicting paradigms on which current observations are made:

1) *Mark 205/NGC 4319.* Pictures of this famous quasar connected by a luminous filament to a low redshift galaxy have been published extensively[1]. When the Einstein X-ray satellite made a special observation of the pair which was capable of again confirming the connection it found emission between the quasar and the galaxy nucleus. But the observation then lay completely unreported for 11 years! I discovered it in the archives when I was preparing a proposal for the new ROSAT X-ray Telescope. With great difficulty I managed to get the challenge before the appropriate allocation committee to allocate about one thousandth of the available time to observe this connection with five times better signal and the better resolution of ROSAT. Almost everyone admits the existence of this connection would destroy the basis for quasar distances from redshift. Would the allocation committee for U.S. ROSAT time take the opportunity to resolve one of the most important pieces of discordant evidence in extragalactic astronomy? Predictably no. They rejected the observation!

Y. Kondo (ed.), Observatories in Earth Orbit and Beyond, 409–412.

2) *3C 232/NGC 3067.* In 1971 Burbidge, Burbidge, Solomon and Strittmatter showed that the brightest radio quasars fell so close to low redshift galaxies that the chance of being accidental was negligible. One of these quasars, 3C 232, had less than 2 1/2 chances out of 10,000 of not being associated with the bright galaxy NGC 3067. Because the quasar was so bright in apparent magnitude further optical and radio spectra were unavoidably taken of it by other investigators.

The new investigators insisted that the quasar was not associated with the galaxy but in order to account for absorption lines of the low redshift galaxy seen in the spectrum of the quasar they had to postulate an enormous halo around the galaxy through which the background quasar was shining! Moreover they assumed the galaxy was in equilibrium rotation and derived the astonishing result that 16 times the mass of visible matter was unseen ("dark matter"). But the pictures of the galaxy available at that time showed clearly that the galaxy was not in equilibrium rotation, that it was instead ejecting material[2]. Recent measures with the Very Large Array radio telescope in Socorro, New Mexico now reveal a hydrogen filament leading from the galaxy directly to the quasar! If this is not accepted as proof of the ejection of the quasar from the galaxy then what would be considered proof? And why did it take 18 years to make (for another purpose) this crucial observation? Is extragalactic astronomy a science?

In the above case it turns out that X-ray observations again lay unreported in the Einstein Laboratory archives. They show X-rays extending from NGC 3067 northward from the quasar in the direction of the HI filament[2]. In the other 3 quasars reported by Burbidge *et al.* there is also evidence of X-ray disturbances which would associate the quasar with the nearby low redshift galaxy. This evidence also went unreported. Proposals to reobserve these latter quasar/galaxy associations with the more powerful ROSAT have now been accepted by the German side but only on the lowest (C) priority basis.

Several other important morals can be drawn from the case of NGC 3067/3C 323:

• The existence of large halos around galaxies is not supported.
• It is a counter example to the hypothesis that lower redshift absorption lines in quasar spectra originate in intervening galaxies.
• It contradicts claims of proof of relative quasar distance from preferentially low z absorption lines in high z quasars.
• It is a vivid illustration of how missing mass calculations can be completely incorrect.

3) *Numerous Quasars Associated with NGC 1097.* The spectacular straight jets emerging from the active nucleus of NGC 1097 have been pictured in numerous publications. Coveys of X-ray cloudlets and quasars are seen streaming out of the complex central regions which surround the variable, point source nucleus[1]. A team of 7 experienced observers including the astronomers who had originally discovered and published these results submitted a carefully calculated, detailed observing proposal to the Hubble Space Telescope. The proposal called for spectrographic investigation of this central region and its central energy source. Puzzles like formation of young stars, temperature, chemical composition and apparently discontinuous

rotation curves were to be observed. Above all, some of the mysterious "hot spots" in the interior which were *also radio sources* would be revealed as new kinds of objects or confirmed as quasars. The figure illustrates the location and appearance with a low resolution ground based telescope of these objects in the interior of NGC 1097. At last we had a chance to peer into the innermost regions of this quasar factory and explore the creation processes of these objects. The proposal was turned down.

4) *Galaxies as the Framework of the Universe.* Quasars might be peculiar but galaxies define the universe – and we know all about them. Don't we? If we think about the question we realize the following: If the quasars are nearby their redshifts are not a measure of their distance. It is just this redshift which is used to estimate the distance of the overwhelming majority of galaxies. Their luminosities, masses and everything else depend on their assumed distances.

Is there something wrong with our knowledge of galaxies? There certainly is, and it is inescapably, catastrophically wrong. Consider the spiral galaxies with the best defined arms of luminous stars, gas and dust. These are called ScI galaxies. They deviate very strongly from the Hubble relation (redshift proportional to apparent magnitude) for Sc's. But obeying a Hubble relation is the only basis for assigning them distances from their redshifts. When these galaxies are assigned to the groups which they belong they turn out to have excess (non velocity) redshifts of hundreds to thousands of km s^{-1}. Moreover, when the only other independent method of estimating distance (from rotation-mass relations) is applied these same galaxies exhibit distance discrepancies of up to 33 mega parsecs![3] Their redshift distances are up to 70% in excess. The final absurdity is that when these systems are put at their redshift distances they are so large they would completely swallow up any giant systems of which we have accurate knowledge (like M31 and M81, the Sb spirals which dominate our own Local Group and the nearby M81 group.) Such monstrous galaxies would be so voluminous that we should see of the order of one supernova per year in them! Every astronomer knows this is a complete *reductio ad absurdum* nevertheless most continue on in the belief that these galaxies actually are so big and so far away.

Where does this leave the science of extragalactic astronomy at the moment? The technology of astronomical observatories advances very rapidly. Billions of dollars are spent on orbiting ever more sophisticated optical telescopes, X-ray and infrared telescopes. The Hubble Space Telescope by itself costs $ 1.6 billion. Huge new optical telescopes are being built on high mountains in Chile and Hawaii. But it all comes down to nothing if the objects that are observed are assumed to be something that they are not.

We see strong evidence which shows extragalactic objects are not what they are supposed. Until this evidence is thoroughly investigated astronomy risks wasting almost totally its resources and misinforming itself and the public which supports it. My conclusion comes not from what some would claim is *interpretation* of available evidence but is the verdict of the evidence itself – namely, that current theory is violated by numerous straightforward observations at an overwhelming probability level.

Everyone has the responsibility individually to look at this evidence and make up their own mind about it. This becomes necessary because human organizations rarely reform themselves from within. External pressure is needed to effect change. I am no longer naive enough to expect consensus leaders in astronomy to ever say: "This evidence is now too strong, we will change our paradigm". Our only faintly realistic hope is that under enough outside pressure people in control might be reluctantly *forced* to say: "We will apply 90% of our facilities to "respectable" science but allow 10% to be used in innovative observations or testing of apparent contradictions of fundamental assumptions." Everyone who is part of a large telescope facility should insist that time be set aside for this most important end result.

It is clear that maintaining 90% of the present kind of programs will not harm the main thrust of astronomy. Assigning of the order of 10% of available time, however, will be life or death to innovative proposals which are the source of really significant breakthroughs. On the recently published space telescope assignments for example, it is not just that the majority are uninteresting – it's that all the crucially important objects have been deliberately left out.

From what has been said it is clear that I feel the solution must be along the lines of minorities taking some power. A tangible step in the direction would be to insist that all publicly funded research be given minority representation. While not disturbing what is considered main stream research this would enable new directions to be explored, some of which will ultimately become the main stream of the future. The alternative is to have various branches of science wander off into complete delusion and irrelevance. Do you think that science cannot possibly go on for a long time without self correcting? Consider that in 300 B.C. Aristarchus of Samos had quite a true picture of the sun and its family of planets. By 200 A.D. Ptolmy had replaced it with earth centered epicycles. More than 1300 years elapsed before Copernicus, Kepler and Galileo restored the sun as the center. It could have been longer.

References

1. Arp, H.: 1987, *Quasars, Redshifts and Controversies*, Interstellar Media.
2. ESO Workshop on Extranuclear Activity in Galaxies: 1989, p. 89.
3. Arp, H.: 1990, *Astrophysics and Space Science*, **167**, 183

OCCULTATION ASTRONOMY

FREEMAN J. DYSON

Institute for Advanced Study, Princeton, New Jersey

The era of occultation astronomy, studying dark objects by observing occultations of bright ones, has begun. The paper of Duncan, Quinn and Tremaine, "The Origin of Short-Period Comets", *Astrophys. J. Letters*, **328**, 69–73 (1988), greatly improved the prospects for occultation astronomy by demonstrating the existence of a second comet reservoir, the Kuiper Belt, much closer to us than the Oort Cloud and concentrated toward the ecliptic plane. Charles Alcock at Livermore (private communication) has begun work on a practical system of small telescopes to observe occultations of stars by comets. I here propose a system similar in concept of Alcock's but using different hardware. We need to try various systems on a small scale to find out which are most cost-effective. My proposal is based on the Multiple Telescope Robotic Observatory (MTRO) developed by Boyd and Genet at Fairborn Observatory in Arizona. See Russell M. Genet, "Multiple-Telescope Robotic Observatories in Space", submitted to P.A.S.P. (1990).

The following calculations are very rough and should be corrected by more careful estimates. As seen from a star near to the ecliptic, the Kuiper belt projects onto an area of 100 AU East-West by 10 AU North-South. The projected area is

$$10^3 (\text{AU})^2 = 2.10^{19} \, \text{km}^2 \, .$$

Assume that we have an array of telescopes extending 100 km in a North-South direction, packed densely enough so that every comet crossing the array as seen from the star will be detected. The earth's orbital motion causes the array to sweep out the projected area at the rate of

$$3000 \, \text{km}^2/\text{sec} \, .$$

Suppose that there are 10^{12} comets in the belt large enough to be detected. Then the star will be occulted by comets crossing the array at an average rate of one every 2 hours. Suppose further that each telescope in the array is equipped with a 100-channel photometer monitoring the light-intensity from 100 stars in its field of view. Then the array will detect occultations at an average rate of 50 per hour.

It will be advantageous to point the array at regions of the sky where the galactic equator crosses the ecliptic, at the constellation Taurus in winter and at Sagittarius in summer. These are regions where plenty of suitable background stars will be available. We need only find 100 stars, of solar type or bluer, of magnitude 13 or 14, within a half-degree field of view. These stars will be about 1 kiloparsec distant and will have angular diameters of 10^{-10} radians. Their angular diameter is equal to that of a comet in the Kuiper Belt with linear diameter 1 km. We can expect to observe occultations by comets with diameters as small as 1 km. That is why the estimate of 10^{12} for the number of detectable comets is not absurdly optimistic.

Y. Kondo (ed.), Observatories in Earth Orbit and Beyond, 413–415.

The diffraction of light will spread the shadow of a comet over a region of size

$$d \sim (\lambda R)^{1/2} \sim 2\,\mathrm{km}\ ,$$

where λ is the wave-length and R the distance to the comet. By observing the occultation in several colors, we may measure both the distance and the absolute size of the comet. In this way we can obtain a three-dimensional map of the Kuiper Belt.

The scale of the diffraction region also determines the maximum spacing of telescopes in the array. For an array of length 100 km, it would probably be adequate to have 100 telescopes arranged in 3 parellel North-South lines, with a spacing of 3 km between telescopes in each line. A comet-shadow crossing the array will be detected in at least 3 telescopes, and the timing of the successive detections will give good discrimination against birds and bats.

Each photometer should be read out at 10^{-2} second intervals with accurate timing. Using stars of magnitude 14, a telescope aperture of 30 cm will be large enough to give good photon statistics. Fluctuations in light-intensity due to atmospheric seeing will be a more serious problem than fluctuations due to photon statistics. In principle, the effects of atmospheric seeing can be overcome by using larger numbers of smaller telescopes. The optimum size and spacing of the telescopes can only be determined by experience.

Finally, I mention briefly Phase 2 of the occultation-astronomy project. After the comet-detection system, which I call Phase 1, is working satisfactorily, we may proceed to Phase 2, the detection of loose Earth-like planets in the galaxy. Here I am following an idea suggested by Bohdan Paczynski in Ap. J. 304, 1 (1986), to look for planets in the galactic halo. I am interested in testing the hypothesis that planets are made before stars when molecular clouds collapse. Perhaps a collapsing cloud produces a large number of loose planets, only a few of which accrete enough gas to grow into stars. If this hypothesis were true, good places to look for loose planets would be young star-clusters such as the Hyades and Pleiades. By a happy accident, the Hyades and Pleiades lie in a part of the sky which is also good for hunting comets.

An Earth-like planet in the Hyades would have the same angular diameter as a comet in the Kuiper Belt, and the light from a background star would be bent around the planet by gravitational lensing to the same extent as the light is bent around the comet by diffraction. The difference between cometary and planetary occultations is mainly a question of time-scale. The cometary occultation is over in 10^{-1} second, while the planetary occultation takes 1000 seconds or longer, depending on the size and distance of the planet.

Planetary occultations will be very rare and we have only a small chance of seeing one. To maximize the chances, we should aim the 100 photometers on our 100 telescopes at 10^4 different background stars instead of looking at the same 100 stars with each telescope. Since the planetary occultations are slow, we can use fainter stars and read out the photometers less frequently. When working in the Phase 2 mode, the array will not be troubled by effects of atmospheric seeing. The gravitational lensing will give each occultation an easily recognizable signature.

The occultation array lends itself very well to time-sharing. All we have to do is to switch the photometers to Phase 1 mode for 54 seconds and to Phase 2 mode for 6 seconds in each minute. We have then complete coverage of planetary occultations using only ten percent of the time, while losing only ten percent of the comets. The possible discovery of an occasional planet comes as an almost free bonus, while the exploration and mapping of the Kuiper Belt continues.

COMMENTS

102 Admiral's Walk, West Cliff, Bournemouth, England

The word 'origin' is one of the most widely used in science. Yet it seems to me to be always used either improperly or ineffectively. Ineffective uses have a derivative quality about them. As an example, suppose we ask: What was the 'origin' of the magnetic field of the Sun? The best answer I suppose is that the magnetic field of the Sun was formed by the compression of a magnetic field that was present already in the gases of the molecular cloud in which the Sun and Solar System were formed some 4.5×10^9 years ago. But what then was the 'origin' of the field in the molecular cloud? It was present already in the gases from which our galaxy was formed, one might suggest. A further displacement then takes us to the manner of 'origin' of the entire universe, so that no ultimate explanation has really been given. The problem has only been displaced along a chain until it passes into a mental fog through which some claim to see clearly but through which others, including myself, do not see at all.

The simplistic idea of a universe beginning with its laws complete at a particular moment of time will not do at all, in my opinion. Assuming the physical laws to be given begs the question. The price to be paid for escaping from problems of 'origin' through the derivative approach, instanced by the example of magnetic fields, is that the weight of all such problems then falls on the physical laws. Unless we explain why those laws hold and not others we have achieved nothing. The position is no different logically from the religious fundamentalist who claims life to have originated through instant creation. Oddly enough, there are many who are undisturbed by this logical similarity, who pour scorn on the religious fundamentalist and who are yet fundamentalist in their views about the universe itself. Perhaps the explanation is that the many, being fundamentalist at heart, have simply transferred their emotional beliefs from what has become disrespectful to what is still considered respectful. In other words, it is a matter of what those fundamentalists at heart think they can get away with, without incurring the wrath of the peer review system, and thence of being expelled into that outer unfunded darkness from which there is no return.

While the steady-state theory was born more frivolously than all this, as [1] describes, the motivation to persist with the theory in the face of criticisms lay here. An ongoing cosmology such as the steady-state theory could not take refuge in mental fog. It had to face-up to its problems here and now. In retrospect, it is not surprising that the process proved difficult, sometimes even to the point of the problems appearing insurmountable. Sometimes knowledge was missing, as it was for the iron whiskers mentioned in [1], and sometimes objections were artifacts of inaccurate observations.

In the year and a half that has elapsed since [1] was written the situation has moved on in several respects, of which perhaps the most important is the

Y. Kondo (ed.), Observatories in Earth Orbit and Beyond, 417–419.
© *1990 Kluwer Academic Publishers.*

recent emphasis on the observed smoothness of the microwave background. The local smoothness is now down with respect to temperature to about two parts in a hundred thousand, setting severe problems for those who favour theories in which the background was last smoothed at an epoch before galaxies were formed. The observations suggest, almost to the point of compelling, that the background has been smoothed at epochs after galaxy formation, a requirement which can be sustained with an intergalactic density of iron whiskers in the range 10^{-34} to 10^{-35} g cm^{-3}. Iron whiskers have the desirable property of an immensely high absorptivity (and emissivity) at microwave wavelengths of about 1 millimeter, but of much lower absorptivity both in the optical and at radio wavelengths. Expelled by radiation pressure at high speeds from galaxies, the whisker production of many galaxies (say 10^{6}) become averaged together in interstellar space, producing a highly smooth local situation. However, once a radiation field has become thermalized, irregularities in the thermalizing agent are irrelevant. Departures from smoothness then depend only on irregularities in the energy density of the radiation field itself, which because of the high speed of propagation of radiation are likely to be very small, at any rate on scales up to say, a tenth of the Hubble distance (about, 10^{27} cm).

My own endeavours over the past year have been concerned with attempts to calculate particle masses. In units with $c = 1, \hbar = 1$, there is only a single unit, say 1 cm. Then for a Hubble distance of about 10^{28} cm, why is the proton mass 4.75×10^{13} cm^{-1}, and what are the reasons for the variations of mass within the baryon octet and decuplet? In physics, these questions are either passed by simply as the way things are, or are made derivative from other hypotheses. As a beginning of an attempt to derive the physical laws rather than merely assume them, I feel the calculation of masses through a combination of cosmology and quantum mechanics to be a promising point of attack; with what results, I will report in the next year or so. Here I will briefly draw attention to a point which indicates that this might be a fruitful approach.

The closure model of the Friedman cosmologies has

$$(\rho = 3H^2/8\pi G) \tag{1}$$

where ρ is the proper mass density and G the gravitational constant, while the steady-state model derived from the conformally-invariant action of Hoyle and Narlikar has

$$(\rho = 3H^2/2\pi G) \tag{2}$$

Using the observational value of the Hubble constant H, and the empirically-determined value of G, either of these relations tells us that there are from 10^{79} to 10^{80} particles of the mass of the proton (or neutron) within a distance $H^{-1} \sim 10^{28}$ cm, Eddington's famous number. Defining a mass by

$$g^2 \times \text{Eddington Number}/H^{-1} \tag{3}$$

where g^2 is a coupling constant, which for a theory of mass is best set as the strong coupling constant, $g^2 = 15$, we get

$$g^2 = \text{Eddington Number}/H^{-1} \sim 3 \times 10^{52} \text{ cm}^{-1} \tag{4}$$

Interpreting (4) as the inverse of a Compton wavelength, the mass value is enormous. But suppose we multiply (4) by the mass of the proton, 4.75×10^{13} cm^{-1} with $c = 1$, $h/2\pi = 1$. The result is about 10^{66} cm^{-2}, very close to the square of the Planck mass. Hence,

$$g^2/H^{-1} \times \text{Proton Mass} \times \text{Eddington Number} \sim (\text{Planck Mass})^2 \qquad (5)$$

We are accustomed to thinking of (1) or (2) as a purely cosmological result obtained from the gravitational equations. Provided, however, that a cosmologically-generated mass is constructed as in (4), the physically interesting result (5) is obtained. In effect, (5) is an alternative way of writing (1) or (2), with a form that more directly suggests a connection between particle masses and cosmology. While attention has certainly to be given to empirical questions relating cosmology to observation, I believe more sustained progress will be made at the present stage through a theoretical investigation of the relation of cosmology to the laws of physics. The question of course is how to do it!

References

Hoyle, F.: 1989, *Comments on Astrophysics* **13**, 81
Arp, H.C., Burbidge, G., Hoyle, F., Narlikar, J.V., Wickramasinghe, S: 1990, *Nature*, in press

THE ASTROPHYSICS OF THE FUTURE

M.S. LONGAIR

Royal Observatory, Blackford Hill, Edinburgh EH9 3HJ

1. Apologia

It is with some trepidation that I set down these thoughts. The history of physics and astronomy is littered with pontifications about the future, most of which simply end up embarrassing their authors. There are many projects which can be regarded as very safe bets but these might not be the ones which totally transform the nature of the discipline. The situation is analogous to that in the early 1950s when extragalactic astronomy simply meant optical astronomy since there was no other way of carrying out such studies – few would regard that as an adequate position nowadays. Similarly, it is difficult nowadays to imagine cosmology without the Microwave Background Radiation. Thus, the problem for the prognosticator is to tread the narrow line between science fiction and a simple extrapolation of what we do now with our facilities. It is in the spirit of this meeting to concentrate upon space observatories but I believe that it is instructive to look at the whole of astronomy, both from space and from the ground.

2. Theorems for the Established Astronomies

Ask any astronomer from any waveband what they would like and it would be very surprising if they did not all answer in the same vein – they would all say that they need larger telescopes (i.e. more collecting power), higher angular and spectral resolution, larger detector arrays and many more telescopes which can provide all of these. Let me comment on these requirements in a slightly provocative way by enunciating a number of plausible theorems about how these facilities will be promoted and used.

Theorem 1. The scientific goals of all future large facilities are the same.
The reason for this is very simple – in order for any very large facility to be funded nowadays, it must be a facility which can be used by a large fraction of the astronomical user constituency and this means that it must attempt to be all things to all astronomers. For example, the Hubble Space Telescope, the VLBI array and the AXAF Observatory are very different projects but important parts of their scientific rationales are, for example, the determination of Hubble's constant and the deceleration parameter, the origin of quasars, the evolution of extragalactic populations with cosmic epoch, the formation of stars. Even the case for the Gravitational Wave Observatory includes as part of its future programme the determination of Hubble's constant and the deceleration parameter as well as the study of quasars and black holes.

Y. Kondo (ed.), Observatories in Earth Orbit and Beyond, 421–426.
© 1990 *Kluwer Academic Publishers.*

There is a very positive side to this situation. The similarity of the scientific cases means that very different disciplines have much to contribute to each astrophysical area. It is a great tribute to instrumental advances that already sensitivities are such that most classes of object can be observed in all wavebands, with some obvious notable exceptions. The negative side of the situation, which has already been noted by the politicians, is that the astronomers seem to be asking for more and more expensive instruments to tackle the same set of problems for which the last major facility was approved. They ask "When is it all going to stop?", to which the answer is, "It isn't". The scientists will continue to press for larger facilities because that is the very nature of the scientific enterprise. The resource is not, however, unlimited – even in an ideal world, no pure science project is likely to cost more than a small fraction of the gross national product of even the richest nations, although I understand that Tycho Brahe achieved this in building his Observatories at Hven in the 1570s. Perhaps we should learn something from him. In the current climate when pure science may not necessarily have the top priority for funding, I believe we have to be very careful how we sell the large projects. Their apparent similarity to the non-astronomer is not necessarily a good thing.

On a previous occasion, I made provocative remarks to the effect that it might be beneficial if the astronomers concentrated upon the scientific areas which are best matched to the technological capabilities of each waveband. This was not a popular suggestion because everyone is used to designing facilities which cater for very wide communities and we have grown used to the concept that, if you have to put up a large satellite for one purpose, you might as well add on other instruments and broaden the range of the science which can be undertaken at relatively little cost compared to those of the construction and launching of the observatory itself. My concern is that this tendency results in a dilution of the scientific programmes which are best matched to the waveband. This leads to Theorem 2.

Theorem 2. The most important programmes for astrophysics require many years of dedicated effort by groups of astronomers tackling specific major astrophysical programmes. One of the major problems with many of the most important programmes in astronomy is that they require large data sets obtained in a systematic way with well-defined selection criteria. In a survey lecture in 1982, I gave a list of what I considered to have been major observational programmes which had put the subject of astrophysical cosmology on a secure foundation. This included topics such as the definition of the large scale structure of the Universe, systematic surveys of quasars and radio sources, the determination of the distances to nearby galaxies, and so on. It is the characteristic of these programmes that they require an enormous amount of effort to obtain a convincing result. There is no way that these important programmes can be achieved without substantial effort. The great discoveries in astronomy are what one remembers most vividly but what converts a discovery into real astrophysics is the systematic study of the properties of whole classes of object.

A beautiful recent example is the Harvard-Center for Astrophysics galaxy survey. In that project, over 20,000 redshifts for galaxies in a carefully selected portion of the Zwicky Catalogue have already been observed, the complete sample consisting of about 30,000 galaxies. These have been used to produce the most complete three-dimensional map of the distribution of galaxies yet available. This programme was only possible because a telescope was dedicated to this single task. This map is one of the most important probes of the large scale structure of the Universe. In addition, because of the systematic way in which the data have been collected, we are obtaining a completely new view of the properties of the galaxies themselves. In my view, it is projects like this which advance our understanding of the Universe in a substantive way.

Let me give another example in which I believe an opportunity may have been missed. One of the most exciting results concerning active galactic nuclei has been the spectroscopic monitoring of variations in the continuum and broad-line regions of Seyfert-1 galaxies. The discovery of correlated variations between the continuum and line intensity variations with a time-lag of several days is a beautiful example of what was possible with a concerted effort by European astronomers using IUE observations over a period of several years. An enormous amount of information about the properties of the broad-line regions and consequently about the innards of the active nucleus itself can be learned from these studies. It had seemed to me that this was an ideal programme of observations to be undertaken by the Hubble Space Telescope as a key project. One can imagine a systematic campaign of observations of a dozen active galactic nuclei. This programme was not selected as a key project, I believe, because it can at least partially be carried out using optical observations made with ground-based telescopes. I only hope this is the case. This is the type of programme which would have benefitted enormously from a systematic programme of observations under the guaranteed ideal conditions from space.

I believe that as scientists, we should be prepared to make these difficult decisions about which programmes we are going to do properly. The consequence of theorem 1 is that we try to do everything with our telescopes. This is naturally a very exciting way to carry out astronomy and is the way that important discoveries are made. I would argue that it is essential, however, that we make time available in our planning to do a selection of the most important problems properly, which means spending a lot of observing time on them. It also has the consequence that astronomers have to work in larger teams and this certainly runs against one of the major attractions of the field in that, although the facilities for astronomy are large, the science done with them is often small science carried out by an astronomer and a graduate student. This absolutely must be preserved but I would hope that the balance could shift in the direction of the large programmes.

Another problem with the large facilties is exposed in Theorem 3.

Theorem 3. Although the size of the telescope increases, the number of observations remains constant per unit observing time. With some very obvious exceptions, as one increases the power of the telescope, the number of objects observed does not increase. This is because the most interesting objects are always at the very limit of observation. Thus, while we struggle to take the optical spectra of 22nd magnitude galaxies now, we will struggle to measure the spectra of the same number of 24th magnitude galaxies with Very Large Telescopes. This theorem has important consequences for the design of observing programmes. Obviously, if one were to stick to programmes on objects of the same brightness as are observed with smaller telescopes, these could be observed in much larger numbers much more rapidly. However, this is not how the psychology of time allocation panels tend to work. If the programme can be done with a smaller telescope, even if it takes a long time, it is not as attractive as a programme which uses the larger facility to its very limits. My own personal preference would be to ensure that adequate time is allocated to programmes which require large statistics on moderately faint objects as well as the limiting observations which require lots of time per object.

I believe these theorems are important for making the case for future large facilities and for determining the strategies to be adopted in utilising properly the next generation of large space facilities.

3. Future Facilities

I will say very little about the astrophysics because I have already presented a broad survey of the whole of astrophysics (Longair 1989). Those interested in my views on the most important problems of contemporary astrophysics should look there. I will simply highlight some of the major problems and the future space facilities which will be of special importance for these.

Solar and Stellar Seismology. This is a very new field and has already revolutionised the way we study the internal structure of the Sun. We need to be able to undertake the same types of studies for stars as well as the Sun. These studies are complementary to the neutrino astronomy of the Sun and stars.

Brown dwarf astrophysics. The astrophysics of baryonic dark matter is in its infancy. The first tentative identifications of brown dwarfs have been made but we need to convert this into a genuine scientific discipline. Very large space infrared facilities are needed for this task.

Stellar mass black holes. The three reasonably convincing cases for stellar mass black holes in binary X-ray sources need to be consolidated one way or another. The observational understanding of the behaviour of matter about black holes of stellar mass is crucial for fundamental physics. We need to have facilities which enable us to study these types of object in nearby galaxies. Large X-ray astronomy observatories are essential.

The formation of stars and interstellar chemistry. We have yet to catch any star in the process of formation, i.e. actually collapsing to form a main sequence star. The understanding of the processes of star formation in detail is the biggest gap in our understanding of stellar evolution. It is closely tied up with molecular astrophysics and the chemical processes inside the extremely dusty regions in which stars form. These studies have consequences for the whole of the physics of galaxies and astrophysical cosmology. Large infrared, millimetre and sub-millimetre observatories are essential for this problem.

The nature of the dark matter in clusters of galaxies. We need observational tools to enable us to determine the distribution of the dark matter which we now know must be present in large-scale structures such as clusters of galaxies and larger scale systems. In many ways, understanding the spatial distribution of the dark matter is as important as its nature which may end up being the province of the particle physicist. We need large optical facilities to be able to probe the gravitational potential distribution in the Universe, i.e. the ability to measure the three-dimensional distribution of galaxies and their peculiar velocities.

The physics of quasars and active galactic nuclei. Not only are quasars and active galactic nuclei the most powerful sources of energy we know of in the Universe, they display a remarkable range of physical phenomena which have little counterpart in smaller scale phenomena. While the identification of the ultimate energy source as a supermassive black hole is convincing, the details of how we get from that concept to what we see is very far from being firmly established. The needs span the whole of the electromagnetic spectrum from ultra-high resolution radio studies, through optical and ultraviolet spectroscopy to X- and γ-ray studies. In all cases, the prime requirement is for more sensitivity and resolution.

The determination of cosmological parameters. A central goal of cosmology is the determination of a number of basic parameter of the Universe – its rate of expansion (Hubble's constant), its kinematics (the deceleration parameter), its mass density (the density parameter) and the cosmological constant. These are all interrelated through the physics of the large scale dynamics of the Universe. All wavebands have significant contributions to make to these problems.

The origin of the large scale structure of the Universe. There is no more challenging problem than the understanding of how the large scale properties of our Universe came about. Many aspects have become the plaything of the particle theorists and many exciting ideas have been developed. In the end, however, these ideas have to be tested against the real Universe as we observe it. We are only now entering the decades when we have a real chance of studying the astrophysics of galaxies, quasars and other large scale systems at epochs significantly earlier than the present. To convert these studies into the real astrophysics of the early Universe needs sensitivities a order of magnitude and more greater than those which are available from the currently proposed projects.

In addition to the observing facilities necessary for carrying out the astrophysics of the future, complementary facilities are essential for theory and data reduction.

We are now entering the epoch of the **Great Observatories**. Optical, infrared, ultraviolet, X-ray and γ-ray astronomies need the Hubble Space Telescope, AXAF and GRO to make significant advances in the exciting astrophysics in these wavebands. The same can be said of those equally important missions which do not have the accolade "Great" but which are just as important for astrophysics – for example, ROSAT, EUVE, Lyman-FUSE, SOHO-Cluster, XMM, etc. Looking beyond the next decade, it is already evident that we know how to build observatories an order of magnitude larger than these. The availability of Soviet launchers which are capable of launching payloads of 100 tonnes removes many of the constraints which have so far limited the ambitions of the astronomers. Thus, as has been emphasised by Garth Illingworth, it is no longer ridiculous to think about very large optical telescopes in space. Granted the strength of the case for large ground-based telescopes, the case for doing the same in space is very strong.

What I have discussed above is the "dull man's" approach to future space missions – simply put up bigger telescopes to advance all aspects of sensitivity by an order of magnitude or more. In addition, we need the more adventurous missions. I would place the **Hipparcos** mission of ESA in this category as well as the various proposals for VLBI from space. Further studies of the Microwave Background Radiation are essential – we need a "Super-COBE" to increase sensitivities by at least an order of magnitude. Such an experiment must discover fluctuations in the background radiation and start a qualitatively new discipline. I am attracted by simple "probe" experiments, for example, the Solar probe and the interstellar probe, which attempt to sample the material and conditions in these regions directly. Continuing in this vein, I am convinced that optical, infrared and ultraviolet interferometry from space has outstanding potential. The gains in phase stability in the absence of the atmosphere make these programmes orders of magnitude easier from space than from the ground and also orders of magnitude more efficient. This would have the potential of opening up completely new areas of parameter space which are not available now. I have a gut feeling that these types of project have real potential to come up with the qualitatively unexpected as well as a core of outstanding science.

Once one begins this type of extensive thinking, there is literally no end to the possibilities. My own view is that even the most exotic possibilities are now feasible in principle with the availability of heavy launchers and the enormous advances in telescope design and instrumentation. The only problem is that the programmes have to be sold to our funding agencies. Let us make sure that we do full justice to the enormous scientific potential of these facilities by leaving behind major real contributions to our understanding of the Universe and not just a sample of what might have been.

References

Longair, M.S.: 1989, 'The New Astrophysics', in *The New Physics*, (ed. P.C.W. Davies), 94, Cambridge University Press.

EVOLUTION OF THE UNSOLVED PROBLEMS

MINORU ODA

I would like to trace the evolution of the major problems or questions since the 1950's by showing several diagrams (Fig. 1a–e) based on my own recollections. As for the coordinates of the diagrams, the latitude towards North indicates the magnitude of surprise and that towards the South indicated is the deepening of our knowledge. Of course such a visual approach is incomplete and is full of personal prejudices, and many important subjects and considerations may be missing.

I will not go into detail of the interpretation of the diagram. But from the series of the diagram we may see that unification and diversification of the questions alternate as one of our co-panelists, Freeman Dyson, pointed out in one of his inspiring books.

In the 1970's to the 1980's we had good hopes that GUT would develop and in its evidence mono-poles and Proton-decay's would be discovered and that proposed candidates of the dark matter may also be discovered from fairly reasonable particles to much more exotic particles.

Now, as of 1990;

1) COBE showed that the 3K sky is very smooth ,whereas we see also the violent superstructure of the Universe.

2) Neither, the mono-poles nor p-decay's have yet been discovered.

3) The jet is a common phenomenon at a variety of hierarchies of the universe.

4) GINGA, succeeding the EINSTEIN Observatory explores AGN, and 50% or more of CXB (Cosmic X-ray Background) may be interpreted in terms of a collection of AGN, leaving the paradox of the spectrum unresolved.

5) All candidates of the dark matter appear to have disappeared.

6) Superstringis now in the hands of mathematicians and slipping out of the hands of physicists.

7) X-ray and Gamma-ray Astronomy show that the Neutron Star is very much like it was thoroughly expected to be.

8) Theory's of the evolution of the Universe have so developed into two extremes; the "East" theory which assumes the evolution is from the large scale to the small scale and the " West" theory which says the Universe evolves from small scale to large scale. Each has its own serious difficulties.

Maybe, the real question now is "Why are things becoming so normal and, say, dull, while 10–20 years before God brought many surprises, a variety of breaks in symmetry"?

Current experimental and observational programs are also indicated in the diagram (Fig. 1f).

Y. Kondo (ed.), Observatories in Earth Orbit and Beyond, 427–430.
©1990 Kluwer Academic Publishers.

Major Unsolved Problems in Astronomy

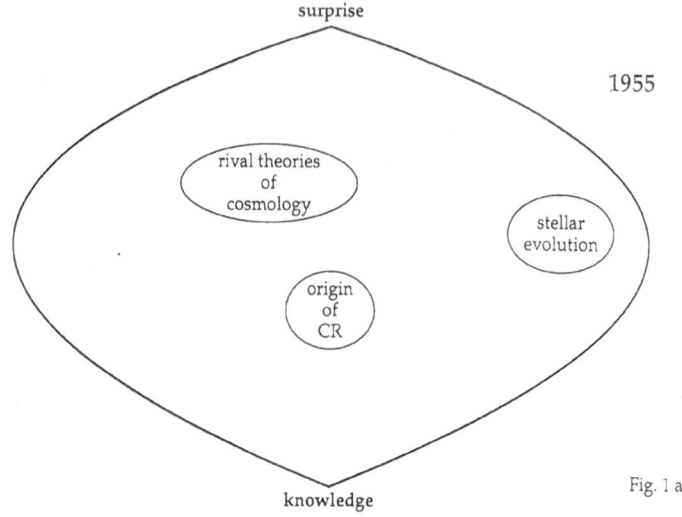

Figure 1a.

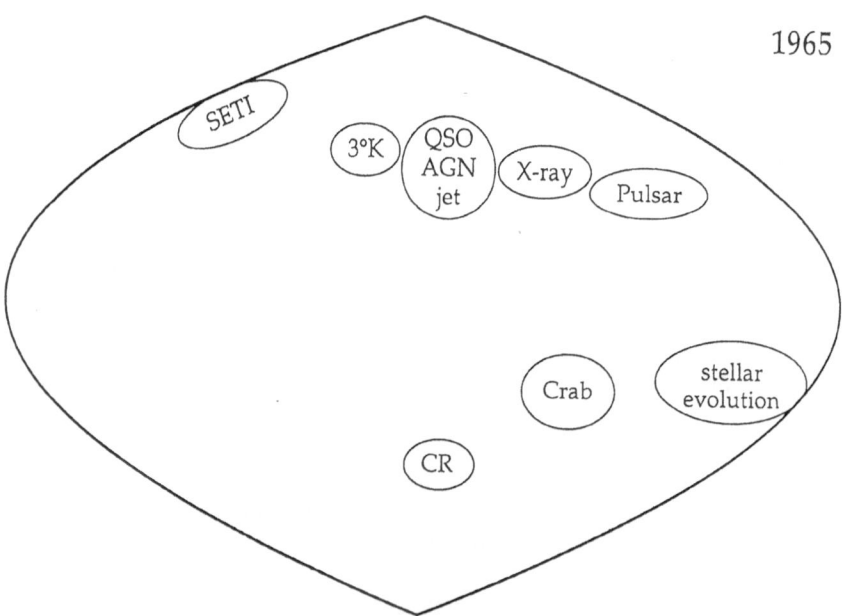

Figure 1b.

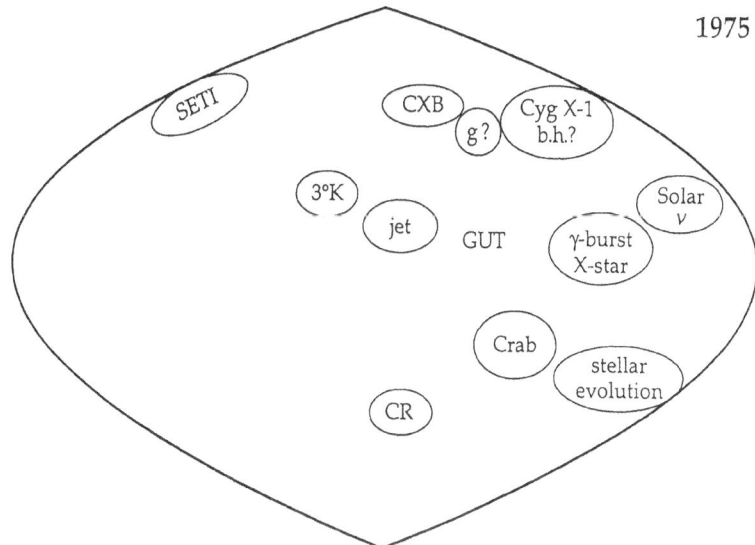

Figure 1c.

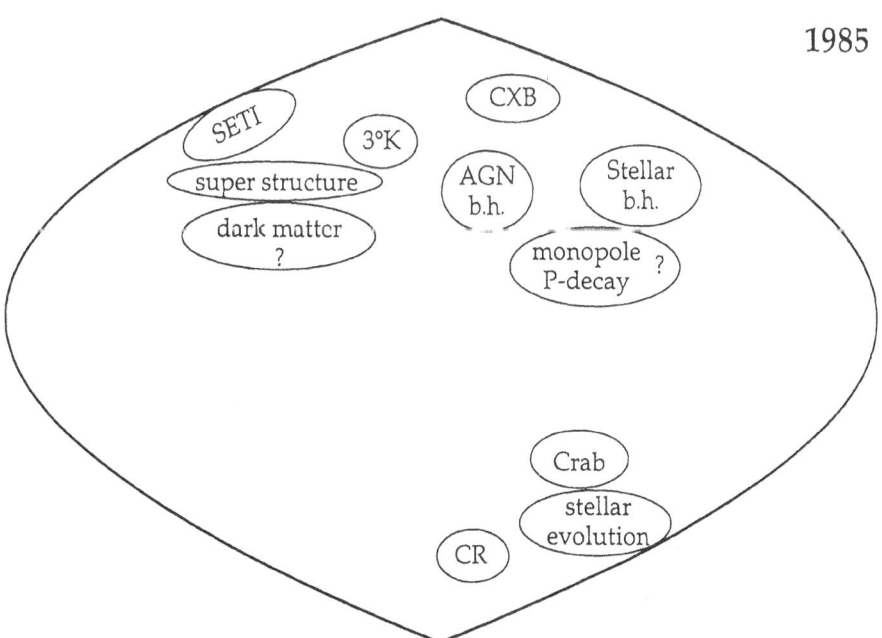

Figure 1d.

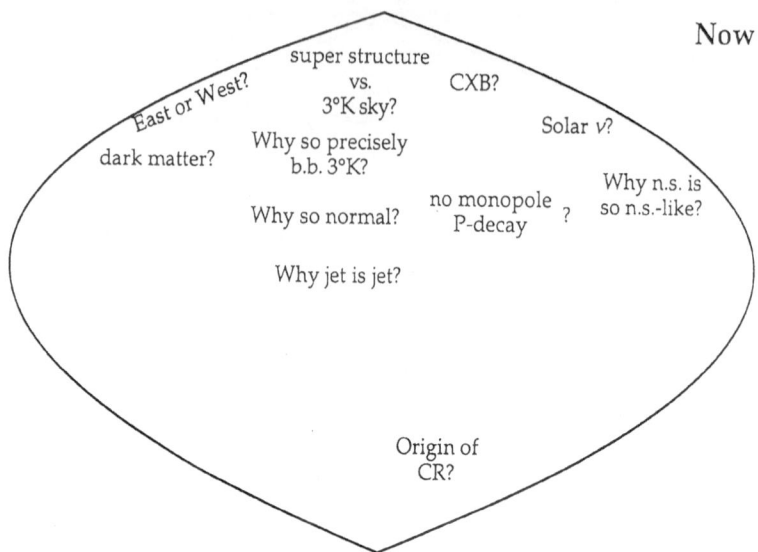

Figure 1e.

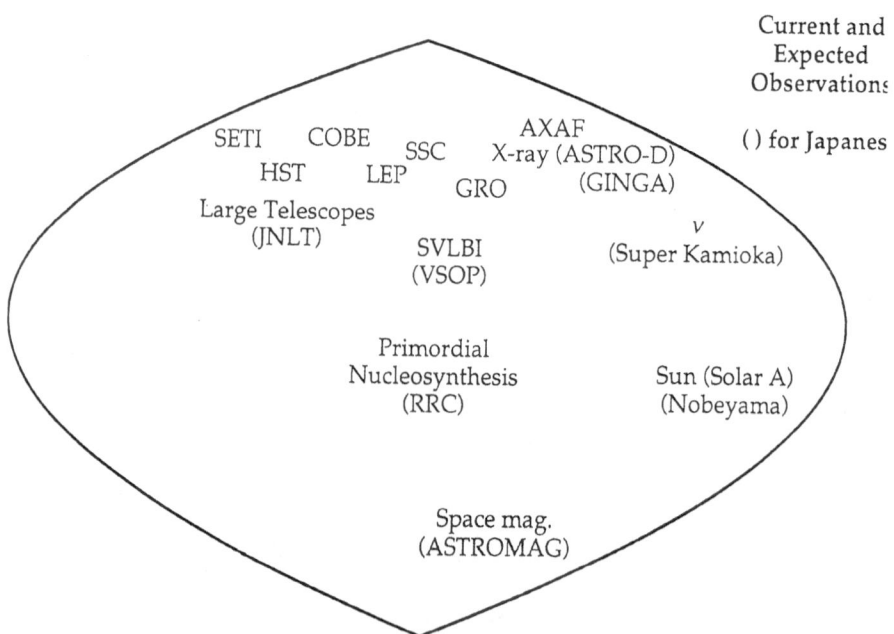

Figure 1f.

CONTRIBUTED PAPERS

SIXA: THE SOLID STATE SPECTROMETER ARRAY
ONBOARD SPECTRUM-X-GAMMA

O. VILHU

Observatory and Astrophysics Laboratory, University of Helsinki,
SF-00130 Helsinki, Finland

H. SIPILÄ

Outokumpu Electronics, Box 85, SF-02201 Espoo, Finland

V.J. KÄMÄRÄINEN

Outokumpu Electronics, Box 85, SF-02201 Espoo, Finland

I. TAYLOR

Princeton Gamma-Tech Inch. PGT, Princeton, NJ, USA

E. LAEGSGAARD

Institute of Physics, Univ. of Aarhus, Denmark

G. LEPPELMEIER

Technical Research Center, Instrument Lab., SF-02150 Espoo, Finland

and

H.W.SCHNOPPER

Danish Space Research Institute, Lundtoftevej 7, DK-2800 Lyngby, Denmark

Abstract. The SPECTRUM – X-GAMMA mission is being developed by the Space Research Institute (IKI), USSR, together with many other countries and is scheduled for launch in 1993 (Sunyaev,1990; Schnopper,1990). Mission objectives include broad and narrow band imaging spectroscopy over a wide range of energies from the EUV through gamma rays, with an emphasis on studying galactic and extragalactic X-ray sources. The Danish Space Research Institute (DSRI) and IKI will provide two thin-foil X-ray telescopes (SODART), each with an aperture of 60 cm and focal length of 8 m. They are designed to have a half-power width of less than 2 arc minutes and will have collecting areas of 1700, 1200 and 100 cm^2 at 1, 8 and 20 keV, respectively. Images and polarization will be recorded by position-sensitive proportional counters. Moderate resolution spectroscopy will be done by the segmented solid state detector SIXA (SIlicon X-ray Array), designed and to be constructed by a consortium in Finland, Denmark and USSR. Finland will have the main responsibility in financing and delivering the detector. The Institute of Electromechanics in Moscow will provide its passive cooling system (110 K). The detector will consist of 19 segments (Si(Li)), each with a diameter of about 8 mm. The spectral resolution of 160 eV (at 6 keV), combined with the large collecting area, provide good opportunities for time-resolved iron line spectroscopy (6–8 keV). The potential observing program includes stellar coronae, cataclysmic variables and X-ray binaries, accretion discs and coronae of neutron stars and black hole candidates, supernova-remnants, active galactic nuclei, clusters of galaxies and the diffuse cosmic X-ray background. We demonstrate the instrument's power through some astrophysical simulations.

Y. Kondo (ed.), Observatories in Earth Orbit and Beyond, 433–438.
© 1990 *Kluwer Academic Publishers.*

TABLE I
The SIXA-detector parameters.

Diameter of active area	4 cm
FOV	15-18 arc min
Number of pixels	19
Pixel diameter	6-8 mm
Number of energy channels	1000
Width of each channel	20 eV
Energy range	0.5–30 keV
Efficiency	over 0.85 (2–20 keV)
Depletion depth	3 mm
Energy resolution (FWHM)	140 eV at 2 keV
	160 eV at 6 keV
	250 eV at 20 keV
Operation temperature	110 K
Power consumption/pixel	13 mW

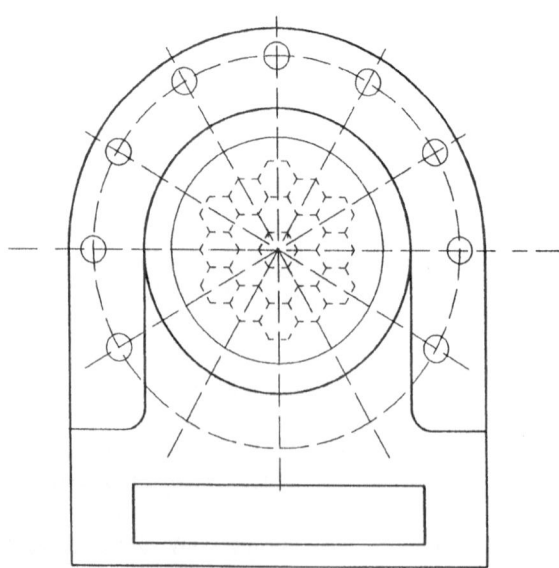

Fig. 1. The SIXA detector unit.

SIXA-BEAM ON G109.1-1.0 (SN-REMNANT)

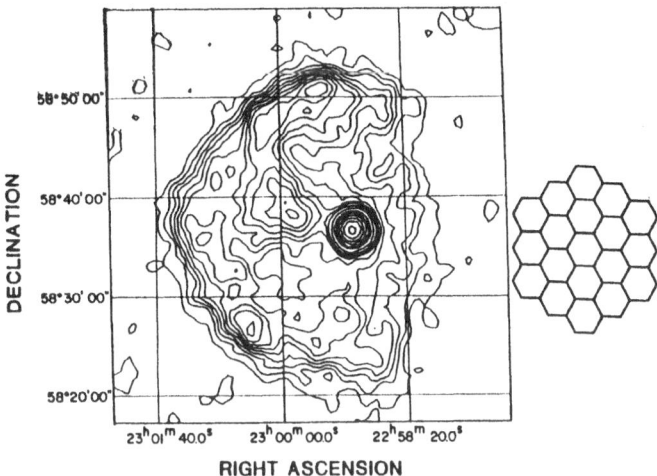

Fig. 2. The SIXA and the Supernova remnant G109.1–1.0.

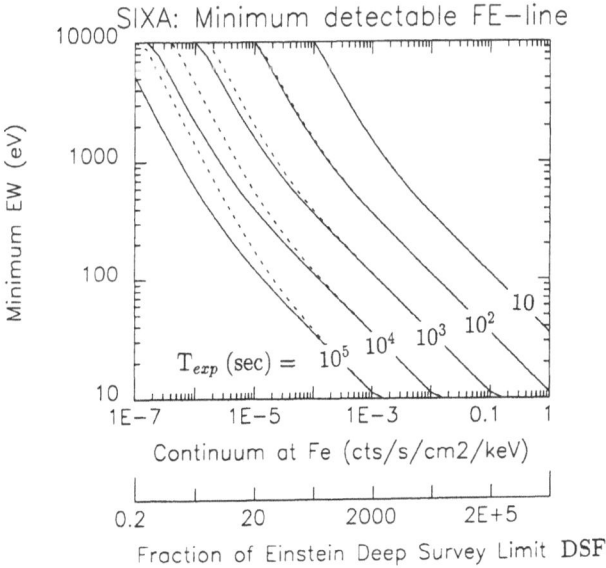

Fig. 3. Exposure times for the iron line detection.

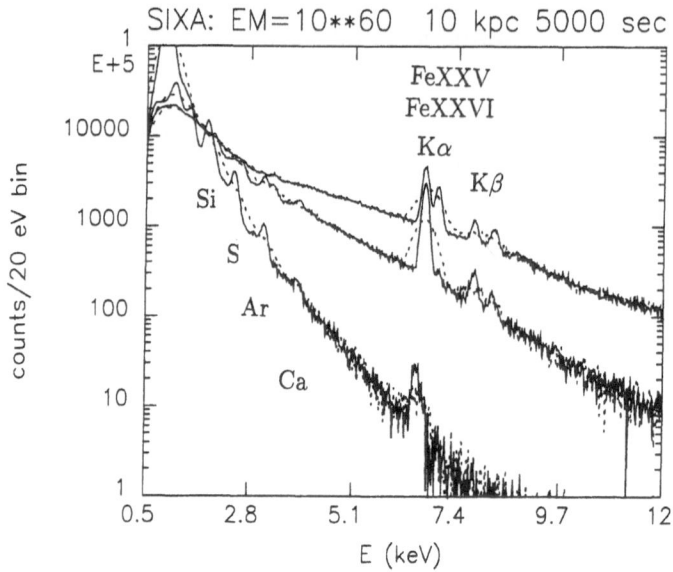

Fig. 4. Spectra of hot thin plasmas as seen by SIXA.

1. The Detector

The significant improvement in energy resolution obtained from solid state devices
(as compared with the gas detectors) more than compensates the requirement that
the device should be cooled. Solid state X-ray detectors are made of lithium-drifted
silicon (Si(Li)). The primary difference from a gas detector is that, although much
less energy is required to liberate an electron within the silicon, there is no equivalent
to gas amplification. The primary charge must be detected by extremely low noise.
Cooled preamplifiers can contribute 100 eV to the FWHM and this is the limiting
resolution for X-rays below 2 keV. At 6 keV, a resolution 150 eV ($R = 40$) can be
reached.

The deep orbit makes it possible to use a radiative cooler to bring the detector
system to the proper operating temperature (110 K). The optimal pixel-size of the
detector matrix is of the order of the point spread function of a point source as
seen by the SODART-telescope, and a pixel size of 6-8 mm was chosen. A larger
pixel size would degrade the energy resolution and also increase the background.
The number of pixels is limited by the minimum power required by the preamplifier
and the heat leaks between the detector assembly and the heat radiator. The SIXA-
instrument will have 19 hexaconal pixels (independent segments), giving a total field
of view of about 15–20 arc minutes (see Figs. 1 and 2). Especially when observing
weak sources and crowded regions, a sufficient number of pixels is important for
background rejection and to retain some imaging capability. Within the expected
accuracy (a few arc minutes) of the satellite pointing system , nineteen segments
serves a safe solution for bright point source observations, as well.

Table I gives the basic parameters of the detector-array and Fig. 1 shows schematically the detector unit. In Fig. 2 the SIXA-beam is overlayed on the supernova remnant G109.1–1.0 (including the central X-ray pulsar 1E2259+586, Fahlman and Gregory, 1981).

2. Spectroscopy and Timing of X-Ray Sources

The high sensitivity, broad energy range and spectral resolution of the SIXA-detector will provide high quality, time-resolved X-ray spectra from a large sample of astronomical objects (for a good review see the XMM science report , ESA SP-1097). The energy range contains the K-shell transitions of oxygen, neon, magnesium, aluminium, silicon, sulphur and calcium, as well as L- and K-shell transitions of iron. Of particular interest is, due to its high abundance, iron at temperatures 10^7–10^8 K. Detailed analysis of these spectral features will permit the determination of the physical characteristics of the emitting region and its surrounding environment. The iron $K\alpha$ line has the energy between 6.4–7.0 keV, depending on the ionization state of the plasma and the excitation mechanism. The equivalent widths of this line in X-ray binaries and active galactic nuclei (AGN) are typically 50–200 eV, as observed by the previous missions (EXOSAT and GINGA).

Fig. 3 shows the minimum detectable iron line with the SIXA-detector (at the 5σ level), computed as a function of the exposure time and continuum level. The strongest galactic X-ray sources have continuum levels 0.1 cts/s/cm^2/keV, and an integration time of 10 seconds can give the detection. The iron line in sources as weak as 20 DSF (DSF=The Einstein Deep Survey Limit) can be detected in one day of observation. An example would be NGC4151 at a distance of 100 Mpc (the actual distance is 10 Mpc).

Compact X-ray binaries are accretion-powered radiation sources where a normal non-degenerate star transfers mass to a compact star (white dwarf, neutron star or black hole). Spectroscopy with sufficient time resolution would give valuable new information about the accretion process and ultimately the compact star itself. The line profiles are affected by the following broadening mechanisms (Kallmann and White,1989):

– blending of different line components (see Fig. 4: T = 10^8K, $10^{7.5}$K and 10^7K (lowest), dashed curves: FWHM = 0.5 keV).
– thermal broadening which is very small even for the highest temperatures , as compared with the SIXA-resolution (160 eV).
– Compton broadening $\delta E = E(4kT-E)/(m_e c^2)$, which is a result of the energy shift due to Compton scattering, and can reach 1 keV for a Thompson depth of 3.
– Doppler broadening due to bulk motions of the plasma in the vicinity of a collapsed object. In the case of a Keplerian motion $\delta E = (r/50r_s)^{-1/2}$ keV, where r_s is the Schwarzschild radius $2GM/c^2$. In addition, general relativistic effects might be present in line profiles (Stella, 1989).

With the SIXA-detector, accretion discs and their coronae can be studied in detail. Accumulating spectra from different X-ray pulse- or burst-phases , the physics of these short-lived phenomena can also be investigated. In AGN's the line pro-

file studies are equally important. The spectral resolution of SIXA at the iron line
($E/\delta E = 40$) is very suitable for these kinds of observations.

3. Observing Modes

Three observing modes will be used: S.E.C.-mode (single event characterization),
E-mode (energy-spectrum) and T-mode (timing in pass-bands). The modes can
be run in parallel or separately. In the S.E.C.-mode every event is stored (i.e. the
photon energy (energy channel), the photon arrival time and the id of the segment).
The arrival time can be determined with accuracy of 1 msec. The detector itself can
separate events within a few microseconds, and this is used (by anticoincidence) for
background rejection.

In the E-mode the integration time is specified by the user. Within this interval
the counts in each of the 1000 channels are summed. The optimal integration time
depends on the source strength. At least several thousand counts have to be recorded
before any useful spectrum can be stored. For point-source observations this mode
is not very memory-economic if spectra from *all* 19 segments are stored. However,
in that case a very good background elimination can be made.

In the T-mode the user selects the integration time (time-bin) and the energy
bands within which the counts are summed. Typically, the energy bands could
be 0.5–2 keV (soft), 2–10 keV (medium) , 10–20 keV (hard) and 6–7.5 keV (the
iron-line).

The memory available will put constraints on the observing modes. If 40
Mbytes of memory will be available, then in the S.E.C.-mode all events from a
100 counts/sec-source can be stored in a 10^5 sec observation. Alternatively, 1000
spectra (from all 19 segments) in the E-mode, or $1.3\ 10^6$ time bins in the T-mode
with 4 windows ($\delta t = 80$ msec), can be stored. In addition to these 3 main observing
modes, specific modes will be developed to study very short timescale (msec size)
and iron line variabilities in strong X-ray binaries.

References

Fahlman G.G. and Gregory P.C.: 1981, *Nature* **293**, 202
Kallman,T. and White N.: 1989, *Astrophys.J.* **341**, 955
Schnopper H.W., 1990, This volume
Stella L.: 1989, *ESA SP* **296**, 19
Sunyaev R., 1990, This volume
XMM The Mission Science Report, 1988, ESA SP, 1097

THE X-RAY LARGE ARRAY

K. S. WOOD, P. HERTZ, AND J. P. NORRIS

*E. O. Hulburt Center for Space Research, Naval Research Laboratory,
Washington, DC 20375-5000, USA*

and

P. F. MICHELSON

Department of Physics, Stanford University, Stanford, CA 94305, USA

Abstract. There is a conspicuous gap in plans for X-ray timing after the *X-ray Timing Explorer* (XTE). Timing science has played a critical role in the development of X-ray astronomy. The need now is to move into a new domain of shorter timescales and weaker modulation, one that can be reached only with very large aperture instruments. XLA is an X-ray facility with an aperture substantially greater than 1 m², nominally 100 m². Most of this area is devoted to a large array of collimated proportional counters. There is also a ~ 1 m² coded aperture. It extends observational parameter space by several orders of magnitude in timing resolution, sensitivity to variability, and angular resolution. This will lead to a qualitatively new kind of X-ray astrophysics that can be applied to the study of a broad range of astrophysical objects. XLA is thus both an advanced timing mission and a general purpose facility whose principal uses are in areas that are not well covered in other aspects of the planned High Energy Astrophysics program.

1. Introduction

In the past two decades some of the most impressive discoveries in any branch of astrophysics have come from X-ray timing, including: X-ray bursters, X-ray binary pulsars, AGN variability, black hole candidates, quasi-periodic oscillators, and transient X-ray sources. Much remains to be learned about these phenomena using better probes at shorter timescales and lower levels of modulation depth – capabilities that often bring out processes taking place close to the central energy release sites, as in QPOs. The key instrumental improvement needed is very large area, since the timescale that can be seen in a given source scales as 1/area.

New phenomena at short timescales will involve new kinds of physics, leading to the involvement of new physics communities in X-ray astronomy. For example, strong-field gravitational phenomena can appear as central regions of compact sources are probed bringing involvement of the relativistic gravitation community. Another community that would benefit from large areas consists of numerical physicists modeling flows and bursts with radiation hydrodynamics codes. Such codes exist now and will develop over the next decade with improved computational power.

XLA is a successor to smaller missions, such as HEAO-1, EXOSAT, and *Ginga*, which discovered many new X-ray phenomena. XLA extends the scope of investigation which was accessible by these earlier missions, making possible studies of all classes of X-ray sources. Its scientific program can be described as "photon rich

Y. Kondo (ed.), Observatories in Earth Orbit and Beyond, 439–442.

TABLE I

XLA SCIENTIFIC CAPABILITIES

FAST PHTOMETRY: 300 TIMES FASTER
1 mCrab in 1 ms, 1 Crab in 1 μs

	flux (Crab)	timescale (s)	modulation (fraction)
Millisecond binary pulsars, CFS instability	~ 1	10^{-3}	10^{-4}
QPOs (high frequency, low modulation depth)	~ 1	10^{-1} to 10^{-2}	10^{-3}
Black hole candidates (shots)	~ 1	10^{-3}	10^{-1}
Period fluctuations, Orbital parameters	~ 1	10^{4}	10^{-12}
Eclipses and dips	$\sim 10^{-2}$	10^{4}	10^{-4}
Radiative shock oscillations in AM Her stars	$\sim 10^{-3}$	1	10^{-2}
Gravitational modulation	~ 1	10^{-2} to 10^{-3}	10^{-4}
Neutron star vibrations	~ 1	10^{-2} to 10^{-4}	10^{-2}

TIME RESOLVED SPECTROSCOPY: 100 TIMES FASTER
30 mCrab Spectrum in 1 ms

Rise time on X-ray bursts	~ 5 spectra per risetime
Photon bubbles	
QPOs, binary pulsars (1 cycle)	5 – 1000 spectra per cycle
Rapid burster fine temporal structure	~ 100 spectra per burst
Pair runaways	

FINE ANGULAR RESOLUTION: 1000 TIMES FINER
1 milliarcsecond on 3 mCrab source

AGN accretion disk diameters	0.3 – 2.5 pc where $R_{core} \simeq 0.5$ pc
Jets (e.g. SS433)	10^{8} cm (10^{-4} jet length)
SNR fine strucure (e.g. Crab)	10^{15} cm
RS CVn systems	$\sim 10^{-1}$ $R_{\odot}$ at 1 pc
Cooling flow galaxy cluster cores	$< 10^{-1}$ core radius

DIFFUSE EXTENDED SOURCES: 10 TIMES FAINTER
2×10^{-9} ct s^{-1} cm^{-2} arcmin^{-2}

Galaxy cluster halos	0.5 degree resolution

X-ray astrophysics", the result of observing ~ 2000 bright X-ray sources with a very large X-ray collector, each source receiving substantial observing time. XLA in two weeks of observation would gather more photons than have been collected in the first 25 years of X-ray astronomy, and this creates a qualitatively new situation. Much of the novelty comes from reaching very short timescales. The large increase in photon throughput brings orders of magnitude of quantitative improvement in four kinds of measurements: (i) fast (microsecond) photometry, (ii) fast (millisecond) time-resolved spectroscopy, (iii) ultrafine (milliarcsecond) angular resolution, and (iv) mapping of highly extended diffuse sources, e.g., outer regions of bright clusters.

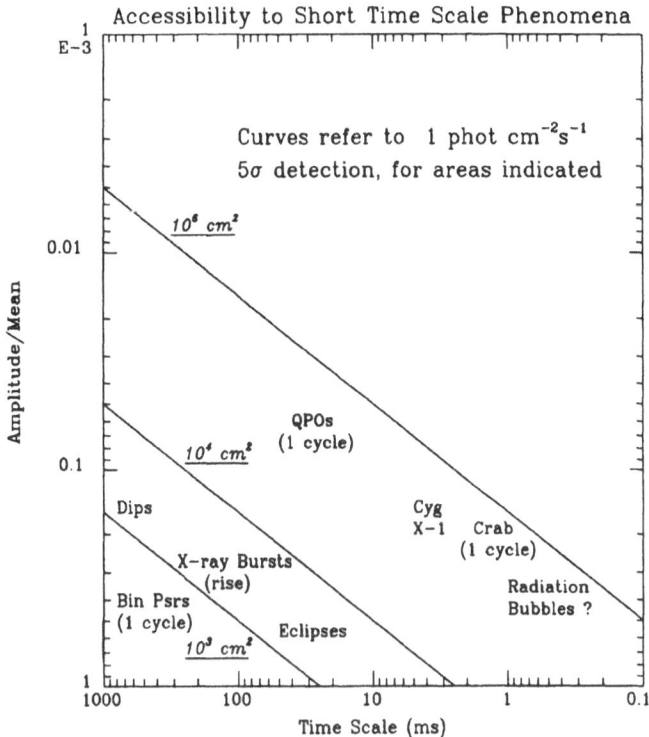

Fig. 1. Low amplitude and short timescale variability require large area instruments, such as XLA, to be detected. Currently planned X-ray timing missions are all smaller than 10^4 cm^2 in area; XLA has an area of 10^6 cm^2.

2. XLA Mission Overview

The scientific applications possible with XLA span the breadth of modern astrophysics (Table I). Fast photometric observations, with time resolution down to microseconds, can be used to search for millisecond X-ray pulsars, single cycle QPOs, shots of noise from black hole candidates, and fluctuations in X-ray pulsar periods.

Fast time resolved spectroscopic observations, with time resolutions down to millisconds, is required to resolve the risetime in X-ray bursts, the spectral variability from photon bubbles, and the energy dependent lags in QPOs. Ultrahigh angular resolution occultation observations, with resolution as fine as milliarcseconds, makes possible the determination of the diameters of X-ray emitting regions in AGN, the lengths of the jets in SS433, and fine structure in SNR. With raster scans, XLA can map extended low surface brightness objects such as galaxy cluster halos.

A pre phase A study of XLA, conducted in 1987 by NASA/MSFC yielded a strawman facility suitable for deployment on the Space Station *Freedom*. XLA contains an effective collecting area of 100 m^2 of proportional counter (PC) arrays and a large co-aligned coded aperture. The PCs have 1 square degree fields of view and are sensitive to 0.25–100 keV X-rays. Pointing stability of ~ 10 arcminutes is required, which is suitable for the Space Station environment. XLA can detect the Crab in 1 μs and 3C273 in 1 ms. The coded aperture not only serves as the monitor for the field being viewed by the large array, but also provides a capability for determining the hard X-ray spectral character of the $\sim 10^4$ brightest sources in the X-ray sky including all the major source classes of X-ray astronomy.

Continuing concept studies by NRL and Stanford University have identified options and tradeoffs for XLA. These include siting options (Space Station, LEO free flyer, lunar surface), assembly options (manned EVA, robotics, self deployment), and instrumentation optimizations (minimize cost and weight per unit area; exploit replication of hardware). Tradeoff studies are currently needed

3. Summary

A large area X-ray array facilitates exploiting bright X-ray sources for a new kind of astrophysics: photon rich X-ray astronomy. Sub-millisecond work requires large areas (Figure 1).

XLA is a general purpose instrument. It has applications in diverse areas of astrophysics, particularly those involving strong gravtitational and magnetic fields and the high densities found in compact objects. Figure 1 shows a few of the phenomena that can only be discovered and studied when very large apertures are used on bright X-ray sources. Facilities with modest area such as AXAF will not fill the need for photon rich X-ray astronomy.

Acknowledgements

The XLA Consortium consists of the Naval Research Laboratory (K. S. Wood (Principal Investigator), H. Friedman, G. G. Fritz, H. Gursky, P. Hertz, J. P. Norris), Stanford University (P. F. Michelson, E. I. Reeves, R. V. Wagoner, M. Yearian), University of Washington (P. E. Boynton, J. E. Deeter), Los Alamos National Laboratory (E. E. Fenimore), NASA Marshall Space Flight Center (J. R. Dabbs), Massachusetts Institute of Technology (W. H. G. Lewin), Pennsylvania State University (E. D. Feigelson), and Sonoma State University (L. R. Cominsky). This work is supported in part by the Office of Naval Research.

THE STELLAR X-RAY POLARIMETER FOR THE
SPECTRUM-X-GAMMA MISSION

P. KAARET, R. NOVICK, C. MARTIN, P. SHAW,
J.R. FLEISCHMAN, T. HAMILTON

Columbia Astrophysics Laboratory, Columbia University, New York, NY

R. SUNYAEV, I. LAPSHOV

Space Research Institute of the USSR Academy of Sciences, USSR

E. SILVER, K. ZIOCK

Lawrence Livermore National Laboratory, USA

M. WEISSKOPF, R. ELSNER, B. RAMSEY

NASA/Marshall Space Flight Center, Huntsville, AL

G. CHANAN

University of California, Irvine, CA

G. MANZO, S. GIARRUSSO, A. SANTANGELO

IFCAI/CNR, Palermo, Italy

E. COSTA, L. PIRO

IAS/CNR, Frascati, Italy

G. FRASER, J.F. PEARSON, J.E. LEES

University of Leicester, UK

and

G.C. PEROLA, E. MASSARO, G. MATT

University of Rome, Italy

Abstract. We describe an X-ray polarimeter which will be flown on the SPECTRUM-X-Gamma mission. The instrument exploits three distinct physical processes to measure polarization: Bragg reflection from a graphite crystal, Thomson scattering from a metallic lithium target, and photoemission from a Cesium Iodide photocathode. These three processes allow polarization measurements over an energy band of 0.3 keV to 12 keV. The polarimeter will make possible sensitive measurements of several hundred known X-ray sources. X-ray polarization measurements will allow us to constrain the geometry of gas flow in X-ray binaries, identify nonthermal emission in supernova remnants, test current models for X-ray emission in radio pulsars, determine the radiation mechanisms in active galactic nuclei, and search for inertial frame dragging (Lense-Thirring effect) around the putative black hole in Cygnus X-1.

1. Introduction

The high resolution instruments available on the next generation of X-ray observatories will allow X-ray observations with angular, energy, and timing resolution comparable to that obtained in optical astronomy. To complete our knowledge of the X-radiation from astrophysical sources it will be necessary to measure the remaining parameter of the radiation, the polarization. The Stellar X-Ray Polarimeter

Y. Kondo (ed.), Observatories in Earth Orbit and Beyond, 443–449.

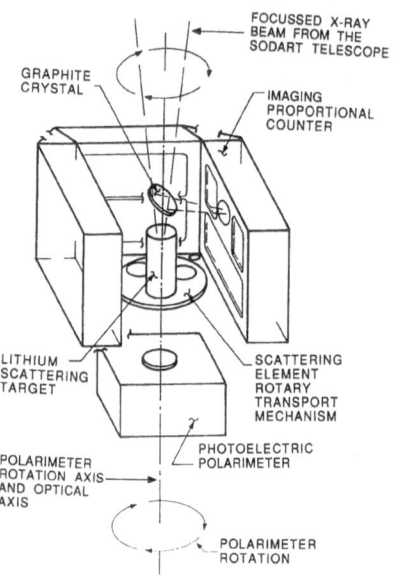

Fig. 1. Elements of the SXRP. When the scattering polarimeter is in use, the graphite crystal and the lithium target are placed on the optical axis of the SODART telescope (shown as a dashed line). The scattered X-rays are detected by four imaging proportional counters, (one of the four is not shown). When the photoelectric polarimeter is in use, the crystal and target are moved off the optical axis so that X-rays may pass through to the photoelectric polarimeter. The entire assembly rotates about the telescope optical axis with a period of approximately ten minutes.

(SXRP) is the only orbiting X-ray polarimeter currently scheduled to flown (Kaaret *et al.* 1989). It will be the only instrument capable of extending X-ray observations to the full parameter space of the radiation. The SXRP, which will be flown at the focus of one of the SODART telescopes on the SPECTRUM-X-Gamma mission (R.A. Sunyaev 1990), will increase the number of sources in which X-ray polarization is detectable from one supernova remnant at present (Novick *et al.* 1972) to several hundred objects including X-ray binaries, isolated radio pulsars, additional supernova remnants, black holes, and active galactic nuclei.

In this paper, we first review the techniques used to make polarization measurements. Then we describe the design of our instrument, highlighting the features necessary to obtain good polarization sensitivity. Finally, we discuss a number polarimetric observations, made possible by the SXRP, which would provide important new information about a variety of astrophysical sources.

2. Instrument Concept

The SXRP exploits the polarization dependence of Bragg reflection from a graphite crystal, of Thomson scattering from a metallic lithium target, and of photoemission from a Cesium Iodide photocathode. The instrument is divided into two parts, as shown in Figure 1. One part, the scattering polarimeter, contains the Bragg crystal, the Thomson scattering target, and four imaging proportional counters (IPC's). The other part, the photoelectric polarimeter (PEP), contains a CsI photocathode and an electron detector. Both parts are mounted on a platform which rotates with a period of approximately 10 minutes.

The graphite crystal and the lithium scattering target are mounted on a mechanism which allows them to be either positioned on the optical axis or removed from the optical path. When these two elements are aligned with the optical axis of the telescope, they scatter the incident X-rays into the surrounding IPC's. When they are off the optical axis, the X-rays from the telescope pass into the photoelectric polarimeter. The movable wheel on which the lithium target is mounted has three positions. One position holds the lithium target. The other positions contain filters for use with the photoelectric polarimeter. Use of two filters gives the PEP two-color energy resolution.

The signature of polarization for any of the three polarization elements is a count rate which varies at twice the rotation frequency of the polarimeter detector assembly. To detect polarization, we must observe the sinusoidal variation in count rate superimposed over the average counting rate. The average counting rate has two components. One component is the background counting rate. The other component arises because the crystal and the scattering target are not perfect polarization analysers. The magnitude of modulation for 100% polarized X-rays and no background is referred to as the modulation factor, μ, and is defined as the ratio of the amplitudes of the modulated and unmodulated components of the signal. We define the minimum detectable polarization (MDP) of a polarimeter element as the minimum level of polarization which it can detect at a 99% statistical confidence level for a given observation. We can relate the MDP to the source counting rate r, the length of the observation T, the modulation factor μ, and background counting rate b using Poisson statistics,

$$\text{MDP} = \frac{4.29}{\mu r} \sqrt{\frac{r+b}{T}}$$

To obtain the best polarization sensitivity we must design the polarimeter with the maximum possible modulation factor and counting rate and the minimum possible background rate. We discuss the design of each polarimeter element below.

The scattering polarimeter contains a graphite crystal mounted above a cylinder of lithium. A graphite crystal oriented at 45° to an incoming x-ray beam will reflect only those X-rays with energies satisfying the Bragg condition and electric vectors lying in the plane of the crystal. If the incident beam is polarized, the intensity of the reflected beam will be modulated at twice the rotation frequency of the polarimeter. The polarization dependence of the Thomson cross-section for scattering of X-rays from electrons gives rise to a sinusoidal dependence in the azimuthal angular distribution. The count rate from a small fixed area of an IPC will vary at twice the rotation frequency of the platform. We have found that a very thin graphite crystal can be used to efficiently reflect X-rays at the first order and second order Bragg peaks, while allowing the X-rays that do not satisfy the Bragg condition to pass through the crystal with only moderate attenuation (Silver *et al.* 1989). Placing the graphite crystal above the lithium target permits us to obtain data from both polarization elements simultaneously. The graphite crystal and lithium target are surrounded by four imaging proportional counters that are used to detect the reflected or scattered X-rays.

The key characteristic in selecting a Bragg crystal is its reflectivity. Higher reflectivities lead to higher counting rates and increased polarization sensitivity. We have chosen graphite crystals because they provide the highest obtainable reflectivity in the energy range of interest. The graphite crystal is sensitive in very narrow energy bands (bandwidth of a few tenths of a keV for the converging X-rays from the SODART telescope) centered on the 45° Bragg reflection peaks at 2.6 and 5.2 keV. The modulation factor for the crystal is above 99%.

The scattering target is a cylinder of metallic lithium 30 mm in diameter and 50 mm long. Lithium is used because it has the highest ratio of Thomsom to photoelectric cross-sections of any workable material. The target is sufficiently long to ensure that more than half the X-rays interact in the target at energies below 15 keV. This part of the instrument provides broad band energy coverage extending from 5 keV, limited by photoelectric absorption in lithium, to about 12 keV, limited by the telescope optics. Because the photon energy changes only slightly during scattering, it is possible to measure polarization versus energy with an energy resolution determined by the proportional counters. The modulation factor for the lithium scattering target, used with an imaging X-ray detector, is 81% (Weisskopf *et al.* 1989).

To detect the scattered X-rays, we use four imaging proportional counters forming a box surrounding the scattering elements. The detectors are sensitive over the scattering polarimeter energy band of 2 to 15 keV. The counters should have an energy resolution of 18% and a position resolution of 1 mm at 6 keV. We have chosen to use imaging detectors because position sensing: increases the modulation factor for lithium scattered X-rays from 28% to 81%, permits us to correct for nonuniformities in the detector response, and allows us to continuously measure the background and its possible spurious polarization signature. The position resolution chosen is a small fraction of the diameter of the SODART telescope blur circle. Because the X-rays from the telescope are either reflected or scattered before they are detected, the flux of X-rays reaching the detector is greatly reduced and low background rates are essential in measuring the polarization of weak sources. We believe we can achieve a background counting rate of less than 10^{-3} counts/sec·cm^2·keV by the use of anticoincidence and pulse shape discrimination background rejection techniques together with careful choice of materials to reduce the inherent and induced radioactivity in the detector body.

The photoelectric polarimeter (PEP) exploits the recently discovered polarization dependence of soft X-ray photoemission. This effect was first reported in the x-ray band by Fraser *et al.* (1988). Further studies of the effect have been performed at Columbia and Leicester (Heckler *et al.* 1989, Fraser *et al.* 1989). A polarization dependence is observed in both the total photocurrent and in the number of photoelectrons produced for each X-ray when an X-ray beam is incident on a photocathode at a very shallow grazing angle (on the order of 10°). When the photocathode is rotated around the axis of the incident X-rays, the photoelectron yield is modulated at twice the rotation frequency. The modulation factor of the polarization signal depends on the energy and angle of incidence of the X-rays and has a maximum magnitude of roughly 30% for 1 keV X-rays at a grazing angle of 8°.

The polarization sensitive element of the PEP will be a microchannel plate (MCP) with square pores. We use a square-pore MCP because the magnitude of the polarization modulated signal depends very sensitively on the grazing angle between the incident X-rays and the photocathode surface. By using a square-pore MCP we can maintain a constant grazing angle over the entire surface of the MCP. Beneath the square-pore MCP there will be a stack of conventional MCP's. The square-pore MCP is operated at low gain to preserve information on the number of photoelectrons produced. The conventional MCP are used to amplify the signal. The electrons will be collected on a wedge and strip anode to give position information. Imaging will improve the performance of the system, allowing us to reduce the background via spatial cuts and permitting us to continually monitor the background. The energy range of the polarimeter is determined at the high end by the quantum efficiency of CsI, the cutoff is about 3 keV, and at the low end of our choice of filters. A thin (0.5 micron) beryllium filter will be permanently mounted over the detector. This gives a low energy cutoff near 0.3 keV. We will have another filter with a higher low energy cutoff mounted on a movable wheel. This will give the photoelectric polarimeter two-color energy resolution.

The photoelectric polarimeter offers a great increase in polarization sensitivity over the scattering polarimeter, due to its vastly superior quantum efficiency. This increase is slightly tempered because the modulation factor of the photoelectric process is lower than for the scattering processes. The net increase in polarization sensitivity, relative to the Bragg crystal, is almost a factor of ten. For observations of sources with fluxes of a few milliCrabs, the PEP can achieve a given level of polarization sensitivity in 1/30th the time required by the Bragg crystal polarimeter.

3. Observations

X-rays are produced with significant polarization when the X-ray emission mechanism is non-thermal or when the X-rays are scattered by electrons in anisotropic geometries. The degree and position angle of the polarization depend on the geometry of the source, thus polarization measurements provide information on the geometry of the X-ray emitting region. In the following, we discuss a polarimetric observations of a few of the classes of objects about which X-ray polarimetry should provide significant new information.

The polarization of X-rays from accreting pulsars is expected to be quite large, near 100% in certain geometries (Gnedin and Sunyaev 1974). Using radiative transfer models it is possible to predict the polarization, and its energy and pulse phase dependence, for pencil and fan beam accretion geometries (Rees 1975, Meszaros et al. 1988). Using the scattering polarimeter of the SXRP we will be able to measure the X-ray polarization of Her X-1 in five pulse phase bins and three energy bins with an average MDP of 1.6% in each bin in a 7×10^5 second observation. This will definitively distinguish between the pencil and fan beam geometries. In addition, we will be able to measure the average polarization of a number of other X-ray binaries (Cen X-3, GX1+4, 4U1626-67, ...) to a level of a few percent in individual 10^5 second observations.

Perhaps the most definitive test for the presence of a black hole is its X-ray polarization signature. The gravitational field near a massive rotating body causes inertial frames near the body to rotate with respect to a distant observer (the Lense-Thirring effect). If a black hole is surrounded by a thin accretion disk, the observed X-rays will be linearly polarized due to scattering in the disk, (Eardley et al. 1978; Lightman and Shapiro 1976; Gnedin and Silat'ev 1978). Because the gravitation field near the black hole is stronger, higher energy X-rays are produced closer to the black hole. The degree of rotation due to the Lense-Thirring effect also increases closer to the hole. Therefore, there is an indirect correlation between the energy of an X-ray and degree of rotation of its polarization vector. Observation of a rotation of X-ray polarization versus energy, would give a clear indication of the presence of a black hole and also probe an untested aspect of general relativity. In a 7×10^5 second observation of Cyg X-1 in its high state, the SXRP, using both the scattering and photoelectric polarimeters, will measure the polarization to better than 0.3% across an energy band from 0.3 to 10 keV and will be able to detect a 10° rotation of the polarization vector across the same energy range. This is sufficient to refute or confirm the theoretically predicted polarization signature (Connors, Piran, and Stark 1980).

The X-ray spectra and total luminosity of active galactic nuclei are not sufficient to uniquely identify the X-ray emission mechanism. X-ray polarization measurements will provide valuable information on the emission process and on the geometry of the central power source. If the emission is due to synchrotron radiation then high X-ray polarization is expected (Ginzburg and Syrovatskii 1965). If the emission is by inverse Compton scattering and the geometry is not spherically symmetric then measurable polarization will be produced (Katz 1976; Shapiro, Lightman, and Eardley 1976; Pozdnyakov, Sobol, and Sunyaev 1976). If the main energy source is thermal, polarization can still arise from electron scattering in the accretion disk around the source (Angel 1969; Basko, Sunyaev, and Titarchuk 1974; Rees 1975; Matt 1989). The scattering polarimeter can provide a measurement of polarization for energies at and above the iron K fluorescence for a few of the brighter AGN. This will allow us to identify reprocessed emission from the accretion disk. However, the main observations of AGN will be done with the PEP. With the high sensitivity of the PEP, the SXRP will be able to measure the polarization, in an energy band from 0.3 to 3 keV, of about a dozen AGN to a level of 3% in individual 10^5 second observations.

Acknowledgements

We wish to thank Richard Spalding, Joseph Chavez, Robert Woods, Patricia Newman, James Daniels, Gary Ahasten, George Peterson, Ronald Akau, and Kate Scurry of the Sandia National Laboratories, and Irwin Rochwarger, Eric Jauch, Alan Goodman, Robert Watkins, and Paul Okun of Columbia University for their numerous contributions to this project. This work was supported by NASA grant

NAG 5-618. This is contribution number 424 of the Columbia Astrophysics Laboratory.

References

Angel, J.R.P.: 1969, *Ap. J.* **158**, 219

Basko, M.M., R.A. Sunyaev, and I.G. Titarchuk: 1974, *Astron. Ap.* **31**, 749

Connors, P.A., T. Piran, and R.F. Stark: 1980, *Ap. J.* **235**, 224–244

Eardley, D.M. *et al.*: 1978, *Comments on Astrophysics* **8**, 151

Fraser, G.W., J.F. Pearson, J.E. Lees and W.B. Feller: 1988, *Proc. SPIE* **982**, 98–107

Fraser, G.W. *et al.*: 1989, *Proc. SPIE* **1160**, 568–579

Ginzburg, V.I., and S.I. Syrovatskii: 1965, *Ann. Rev. Ast. Ap.* **3**, 297

Gnedin, Yu.N., and R.A. Sunyaev: 1974, *Astron. and Astrophys* **36**, 379–394

Gnedin, Yu.N., and N.A. Silat'ev: 1978, *Sov. Astron.* **22**, 325

Heckler, A., A. Blaer, P. Kaaret, and R. Novick: 1989, *Proc. SPIE* **1160**, 580–586

Kaaret, P. *et al.*: 1989, *Proc. SPIE* **1160**, 587–598

Katz, J.I.: 1973, *Nature* **246**, 87

Lightman, A.P., and S.L. Shapiro: 1976, *Ap. J.* **203**, 701

Matt G., E. Costa, G.C. Perola, and L. Piro: 1989, *Proc. 23rd ESLAB Symposium*

Meszaros, P. *et al.*: 1988, *Ap. J.* **324**, 1056–1067

Novick, R., *et al.*: 1972, *Ap. J.* **174**, L1–L8

Pozdynakov, L.A., I.M. Sobol, and R.A. Sunyaev: 1976, *Sov. Astron. Lett* **2**, 55

Rees, M.J.: 1975, *M.N.R.A.S.* **171**, 457

Shapiro, S.L., A.P. Lightman, and D.M. Eardley: 1976, *Ap. J.* **204**, 187

Silver, E.H., *et al.*: 1989, *Proc. SPIE* **1160**, 598–609

Sunyaev, R.A., 1990, these proceedings

Weisskopf, M.C. *et al.*: 1989, *Proc. SPIE* **1159**, 607–616

THE ALL-SKY EXTRAGALACTIC X-RAY FOREGROUND

ELIHU BOLDT

NASA/Goddard Space Flight Center, Greenbelt, MD 20771

1. Introduction

Observations of galaxies in the IR and optical (Lynden-Bell et al. 1989) suggest that the 600 km/s peculiar velocity of the LG (Local Group of galaxies) arises mainly from a foreground of anisotropically distributed mass within $z = 0.013$ (i.e., HR < 4000 km/s). Since the X-ray luminosity of bright extragalactic X-ray sources provides a good mass measure of the radiating objects involved and can be observed relatively free of galactic obscuration effects, such sources are likely candidates for serving as reliable tracers of the total underlying mass (i.e., dark as well as visible) responsible for the acceleration of the LG. In this connection, we note that the local gravitational dipole implied by the fifty X-ray brightest clusters of galaxies at $z > 0.013$ considered by Lahav et al. (1989) is relatively small compared with that inferred from the only three clusters at lower redshifts. Since the local space density of AGN (Active Galactic Nuclei) is about two orders of magnitude greater than rich clusters, however, such compact sources have the potential of providing a vastly improved statistical sample for tracing mass in the low-redshift region of particular interest. Furthermore, recent dipole analysis of the X-ray flux from bright AGN observed with HEAO-1 A2 indicates that they are indeed strong tracers of this matter (Miyaji and Boldt 1990). The implications of this for the very pronounced large-scale foreground anisotropies to be measured via low-redshift AGN resolved in more sensitive all-sky surveys are explored.

For the total extragalactic X-ray sky, i.e. including the relatively large Cosmic X-ray Background (CXB) as well as the contribution of resolved sources, the dipole moment is small compared to the monopole. Because of the large peculiar velocity of the LG, however, it is found that the currently estimated value for this total dipole moment can already be used to set remarkably severe constraints on the volume emissivity arising from all X-ray sources within the present epoch. This provides significant limits on the CXB contributions of all source populations that are well represented in the local universe.

2. Dipole Moments

We know that the extragalactic X-ray sky is dominated by a nearly isotropic, unresolved CXB and suspect that there exists a proper frame in which it shows precise global isotropy. If this frame is anchored to that of the microwave background, then the solar velocity $v_0 = 380$ km/s relative to it yields an apparent dipole anisotropy (Compton and Getting 1935) in the observed surface brightness (B), with an apex

Y. Kondo (ed.), Observatories in Earth Orbit and Beyond, 451–455.

© 1990 *Kluwer Academic Publishers.*

having galactic coordinates $l = 267$ deg, $b = +50$ deg (Lynden-Bell et al. 1988). The dipole moment (D) of this distribution over the whole sky is defined as

$$D \equiv \int_{4\pi} B \cos(\theta) d\Omega \tag{1}$$

where θ is the angle relative to the preferred axis. The monopole (M) is:

$$M = \int_{4\pi} B d\Omega \tag{2}$$

For a power-law energy spectrum of index (a) the dipole/monopole ratio (D/M) is

$$D/M = (v_0/c)[1 + (\alpha/3)]. \tag{3}$$

For $\alpha = 0.4$, the value appropriate to the 2–10 keV CXB, $D/M = 1.4 \times 10^{-3}$ (i.e., a small number). Although this is compatible with the CXB global anisotropy estimated with HEAO-1 A2 (Shafer 1983), the precision of that determination is strongly limited by foreground fluctuations due to unresolved sources and possible extended regions of weakly enhanced surface brightness (Jahoda and Mushotzky 1989).

In sharp contrast to the small value of D/M for the unresolved CXB as a whole, the D/M ratio derived from the all-sky extragalactic foreground of discrete sources (at RH<4000 km/s) is very large. For examining this foreground we define

$$D \equiv \sum_i S_i \cos(\theta_i) \tag{4}$$

and

$$M \equiv \sum_i S_i \tag{5}$$

where S_i is the flux from each individual source and θ_i the angle with respect to a direction ($l = 268$ deg, $b = +27$ deg) defined by the peculiar velocity of the LG (Lahav et al 1989). In particular, for galaxies (mainly spirals) observed with IRAS, $D/M \sim 0.2$ (Lahav et al. 1988). For optically observed galaxies (ellipticals as well as spirals), $D/M \sim 0.4$ (Lynden-Bell et al. 1989). With sources observed in X-rays $D/M \sim 0.5$ for AGN's (Miyaji and Boldt 1990) and $D/M \gtrsim 0.5$ for the statistically weaker sample of clusters (Lahav et al. 1989).

According to linear perturbation theory (Peebles 1980), the peculiar velocity $\mathbf{v}$ of the LG is given by

$$\mathbf{v} \sim \mathbf{g}[2/(3H)(\rho/\rho_c)^{-0.4} \tag{6}$$

where H is the Hubble constant, ρ is the mean total mass density, ρ_c is the critical closure density and $\mathbf{g}$ is the net gravitational acceleration vector at the LG. The magnitude of $\mathbf{g}$ is given by

$$| g | = \sum_i g_i \cos(\theta_i) = 4\pi G(\rho/b)(D/M)R \tag{7}$$

where the sum indicated is over gravity contributions (g_i) arising from all mass elements within a radial distance (R), G is the gravitational constant and $b=$(luminous contrast)/(mass contrast) is a "bias" factor characterizing radiative tracers of mass.

In particular, using the values of D/M observed in various electromagnetic bands we infer from equations (6) and (7) that b (X-ray) $\geq b$ (optical) $\approx 2\times[b(\mathrm{IR})]$, where b (optical) $\approx 2.7 \, (\rho/\rho_c)^{0.6}$; see Lynden-Bell et al.(1989).

3. Local X-ray Volume Emissivity

Although the dipole fit to the global anisotropy of the CXB is compatible with what is expected from the Compton-Getting effect (Shafer 1983), well within statistical errors, the actual "best fit" apex ($l = 313\,\mathrm{deg}, b = +38\,\mathrm{deg}$) is closer to that of the peculiar velocity of the LG than it is to the direction of our own (solar system) motion relative to the proper frame of the microwave background. This suggests that some finite portion of this dipole moment arises from the anisotropic distribution of unresolved X-ray sources which trace the underlying matter responsible for the peculiar velocity of the LG. In any event, the dipole moment resulting from anisotropically distributed unresolved X-ray sources can not exceed that for the overall X-ray sky. As such, the dipole moment of the overall X-ray sky can be used to set a limit on the local volume emissivity corresponding to the *complete* population of present-epoch discrete X-ray sources (including those as yet unresolved) as well as all diffuse emission. In particular, since X-ray emission appears to be a strong tracer of matter, we can use the all-sky dipole moment (D_{All}) and the observed peculiar velocity (v) of the LG to set a significant limit on the local X-ray volume emissivity $(q \equiv M/R)$. To do this we note from equations (6)–(7) that

$$q < (H/v)(D_{\mathrm{All}})/[b(\rho_c/\rho)^{0.6}] \tag{8}$$

where D_{All} is the magnitude of the total dipole moment obtained from the vector addition of the dipole moments associated with the $\underline{C}$ompton-$\underline{G}$etting effect (CG), $\underline{R}$esolved $\underline{A}$GN (RA) and the apparent surface brightness enhancement identified with the $\underline{G}$reat $\underline{A}$ttractor (GA), viz:

$$\mathbf{D}_{\mathrm{All}} = \mathbf{D}_{\mathrm{CG}} + \mathbf{D}_{\mathrm{RA}} + \mathbf{D}_{\mathrm{GA}}. \tag{9}$$

The 4% enhancement in X-ray surface brightness over a 40 deg diameter region centered at a position ($l = 310\,\mathrm{deg}, b = +10\,\mathrm{deg}$) identified with the GA (Jahoda and Mushotzky 1989) corresponds to a dipole moment in that direction of magnitude

$$D_{\mathrm{GA}} = 1.2 \times 10^{-3} M_{\mathrm{CXB}}, \tag{10}$$

where $M_{\mathrm{CXB}} = 6.6 \times 10^{-7}$ erg cm^{-2} s^{-1} is the CXB monopole (2–10 keV). (Such direct effects are discussed in detail by Goicoechea and Martin-Mirones (1990).) For the Compton-Getting effect we obtain via equation (3) that

$$M_{\mathrm{CG}} = 1.4 \times 10^{-3} M_{\mathrm{CXB}}. \tag{11}$$

The dipole moment obtained from resolved AGN (Miyaji and Boldt 1990) is in a direction ($l = 313\,\mathrm{deg}$, $b = +38\,\mathrm{deg}$) 39 deg away from the velocity vector characterizing the peculiar motion of the LG and has a magnitude

$$D_{\mathrm{RA}}(2 - 10\mathrm{keV}) = 5 \times 10^{-10}\mathrm{erg\ cm}^{-2}\ \mathrm{s}^{-1} \tag{12}$$

Using equations (9)–(12) we obtain that D_{All} is in a direction ($l = 297\,\mathrm{deg}$, $b = +35\,\mathrm{deg}$) somewhat closer to that of the peculiar velocity of the LG and has a magnitude

$$D_{\mathrm{All}} = 20 \times 10^{-10}\ \mathrm{erg\ cm}^{-2}\ \mathrm{s}^{-1} \tag{13}$$

From equations (8) and (13) we obtain an upper limit to the local volume X-ray emissivity arising from all sources, viz:

$$q < 5.9 \times 10^{38}\ h\ \mathrm{erg\ s}^{-1}\ \mathrm{Mpc}^{-3} \quad \text{for all sources} \tag{14}$$

where $h = H/(50\ \mathrm{km\ s}^{-1}\ \mathrm{Mpc}^{-1})$ and we have taken $b = 2.7(\rho/\rho_c)^{0.6}$ for all sources of X-ray emission. It is interesting to compare this upper limit on the local composite volume X-ray emissivity with that coming from bright AGN having L(2–10 keV) $> 10^{43}\ h^{-2}\ \mathrm{erg\ s}^{-1}$ (Piccinotti et al.1982), viz :

$$q = 2 \times 10^{38}\ h\ \mathrm{erg\ s}^{-1}\ \mathrm{Mpc}^{-3} \quad \text{for bright AGN.} \tag{15}$$

[The local volume X-ray emissivity (2–10 keV) arising from Abell clusters is $\leq 1 \times 10^{38}\ h\ \mathrm{erg\ s}^{-1}\ \mathrm{Mpc}^{-3}$ (McKee 1980; Piccinotti et al. 1982).] The upper limit given by equation (14) provides a severe constraint on other possible populations of extragalactic X-ray sources in the present epoch, such as the moderate redshift faint radio galaxies proposed by Helfand et al. (1989) to be major CXB contributors. In particular, without evolution, all such other source populations can make no more than a 20% contribution to the CXB (2–10keV).

4. Future Prospects

Assuming a present-epoch density of about $10^{-6}\ h^3\ \mathrm{Mpc}^{-3}$ for rich ($R \geq 1$) clusters (Bahcall 1988) the number expected within the critical foreground region ($HR < 4000$ km/s) corresponds to only a couple of objects or so. This is clearly too few for their effective utilization in tracing the anisotropic distribution of matter responsible for the peculiar motion of the LG. On the other hand, for an AGN space density of about $10^{-4}\ h^3\ \mathrm{Mpc}^{-3}$ for $L_x > 10^{42}\ h^{-2}\ \mathrm{erg\ s}^{-1}$ (Persic et al. 1989) the number expected in this foreground region would be a few hundred. And the number expected over twice the critical depth (i.e., within HR ≈ 8000km s^{-1}) would be more than a thousand, ample for studying the dipole growth with respect to redshift in the interesting region where preliminary results based on a limited sample of AGN (Miyaji and Boldt 1990) indicate the approach of a plateau (i.e., saturation). The spectral homogeneity (2–10keV) of AGN and the good correlation of their X-ray luminosity with the mass of the underlying supermassive blackhole

(Wandel and Mushotzky 1986) renders these sources potentially powerful and interesting tracers of all gravitational mass. A suitable all-sky X-ray foreground survey of AGN could be used to answer the question: "How do supermassive blackholes trace the overall underlying mass distribution responsible for the peculiar velocity of the Local Group?"

The sensitivity (2–10 keV) required for the all-sky survey of foreground AGN described above would be 3.3×10^{-13} erg cm^{-2} s^{-1}. Although the ROSAT all-sky survey is expected to be at this sensitivity level below 2 keV, it does not have sufficient response within the higher energy band needed for an unbiased AGN survey (i.e., one minimizing the effects of galactic absorption and spectral variations). Deepening the all-sky survey to HR$\approx$30,000 km/s (i.e., $z \approx 0.1$) would require a sensitivity of about 2×10^{-14} erg cm^{-2} s^{-1} (2–10 keV), comparable to that of the high sensitivity small-field survey (1–3 keV) of the Einstein Observatory.

Acknowledgements

I thank Ofer Lahav and Takamitsu Miyaji for stimulating discussions.

References

Bahcall, N.: 1988, *Ann. Rev. Astron. Astrophys.* **26**, 631

Compton, A. and Getting, I.: 1935, *Phys. Rev.* **47**, 817

Goicoechea L., and Martin-Mirones, J.: 1990, *MNRAS* , in press

Helfand, D. et al.: 1989, *Bull. AAS* **121**, 1220

Jahoda, K. and Mushotzky, M.: 1989, *Astroph. J.* **346**, 638

Lahav, O., Rowan-Robinson, M. and Lynden-Bell, D.: 1988, *MNRAS* **234**, 677

Lahav, O., Edge, A., Fabian, A. and Putney, A.: 1989, *MNRAS* **238**, 881

Lynden Bell, D. et al.: 1988, *Astroph. J.* **326**, 19

Lynden-Bell, D., Lahav, O. and Burstein, D.: 1989, *MNRAS* **241**, 325

McKee, J. et al.: 1980, *Astroph. J.* **242**, 843

Miyaji, T. and Boldt, E.: 1990, *Astroph. J.(Letters)* , in press

Peebles, P. J. E.: 1980, *The Large Scale Structure of the Universe*, Princeton University Press, Priceton

Persic, M. et al.: 1989, *Astroph. J.* **344**, 125

Shafer, R.: 1983, *PhD thesis*, University of Maryland, NASA TM 85029

Wandel, A. and Mushotzky, R.: 1986, *Astroph. J.* **306**, L61

A LOW ENERGY GAS SCINTILLATION PROPORTIONAL
COUNTER FOR THE SAX-X-RAY ASTRONOMY SATELLITE

A. N. PARMAR, A. SMITH, AND M. BAVDAZ

Space Science Department, European Space Agency, ESTEC,
Noordwijk, The Netherlands

Abstract. The payload of the italian/Dutch satellite SAX will include a set of four concentrators each with a geometric area of 90 cm^2. Imaging GSPCs will be located at the focal planes of the concectators. The Space Science Department of ESA will provide one of these GSPCs which will be sensitive to X-rays with energies between 0.1–10 keV. In order to achieve such a low-energy energy response, a driftless configuration and a thin plastic window have been adopted. At 6 keV the collecting area will be 50 cm^2 and the energy and angular resolutions 8% and 1.6' FWHM, respectively.

1. Introduction

The Italian/Dutch satellite Sax is scheduled for launch into a 600 km equatorial orbit in December 1993. This orbit ensures a relatively constant, low particle background. During the 2 year mission the instruments will make imaging, spectral and timing studies of a wide range of X-ray sources in the energy range 0.1–200 keV (Scarsi 1983; Perola 1983). The payload of scientific instruments includes a high pressure GSPC, a phoswich detector system, two wide field cameras (WFCs) and four imaging GSPCs (Spada 1983). All the instruments are coaligned except for the WFCs which point in opposite directions along an axis perpendicular to the other instruments. The imaging GSPCs are located at the focal planes of four X-ray concentrators each with a geometric area of 90 cm^2. In order to extend the sensitive energy range of the concentrator spectrometers below 1 keV, one of these GSPCs will be a new design of low-energy gas scintillation proportional counter (LEGSPC). This instrument is being provided by the Space Science Department of ESA. It will have energy resolution comparable to prototype CCDs at low energies and a factor of ~ 2 better than proportional counters at iron K (6.4 keV). Its high sensitivity, imaging capability and low background will allow detailed spectral and timing studies of both point and extended sources to be made.

2. The SAX LEGSPC

The general performance characteristics of the LEGSPC are given in Table I. The low energy response requires the use of a thin plastic window since the low-energy cut-off of a conventional 50μ beryllium window is $\gtrsim 1$ keV. The current solution for this window is 1μ polypropelene with a thin Lexan coating, although an alternative approach involving a diamond coated polyimide foil is presently under development at Outokumpu OY, Finland. Impurities due to outgassing from the window and detector (the detector cannot be baked at high temperature because of the presence of the polypropelene) are removed by a passive getter.

Y. Kondo (ed.), Observatories in Earth Orbit and Beyond, 457–461.

TABLE I
The LEGSPC Performance

Parameter	Value
Energy Range	0.1–10.0 keV
Energy Resolution at 6 keV	8% FWHM
Angular Resolution at 6 keV	1.6' FWHM
Effective Area at 6 keV	50 cm^2
Field of view	40'
Time Resolution	16μs
Maximum Throughput	2000 events s^{-1}
Background Rejection	98%

Xenon was chosen as the filling gas because its scintillation spectrum closely matches available photomultiplier entrance windows. Since the penetration depth of low energy X-rays in xenon is low, a conventional drift region would unacceptably degrade the energy resolution of the detector because of the loss of electrons to the front window (Inoue et al. 1979). This problem is overcome in the LEGSPC since the X-rays are absorbed directly into the high field scintillation region. The absence of a drift region and the associated drift/scintillation grid leads to an intrinsically better performance (Simons et al. 1985), but does cause a dependance of pulse height on penetration depth that has to be considered in the signal processing.

The first prototype SAX LEGSPC detector is described in Smith, Peacock, and Kowalski (1987). In later designs a cylindrical geometry was chosen in order to overcome field non-uniformities and the electric field near the window was more carefully controlled. Two detectors of this design were built by SIRA Ltd, UK under contract from ESA. These early detectors were found to be sufficiently sensitive to allow the study of various aspects of GSPC gas physics including the light emission process, electron drift velocity and longitudinal diffusion, and the Fano factor (Favata et al. 1990; Smith, Favata, and Kowalsky 1989; Kowalsky, Smith, and Peacock 1989). After improvements and increased confidence in the reliability of the entrance window a larger (2 cm) diameter was chosen which lead to the final design as shown in Figure 1. Two Fe^{55} radioactive sources constantly illuminate regions of the detector that do not see the sky allowing the position and energy gains to be constantly monitored. The first SAX engineering model detector is presently being built to this design by AEG Inc. Ulm, FRG.

The readout photomultiplier is a Hamamatsu 3 × 3-anode tube R2488. The anodes are arranged in a square configuration providing 9 separate signals for each observed X-ray. These are combined to produce pulse height, burst length, X_{pos}, Y_{pos} and veto signals (see Favata and Smith 1989). Each signal is then converted to a digital value via an ADC and then handed to the digital electronics and microprocessor system. This is based around an 80C86 microprocessor with 32 kbytes of ROM and 104 kbytes of RAM memory and is presently under development at MBB

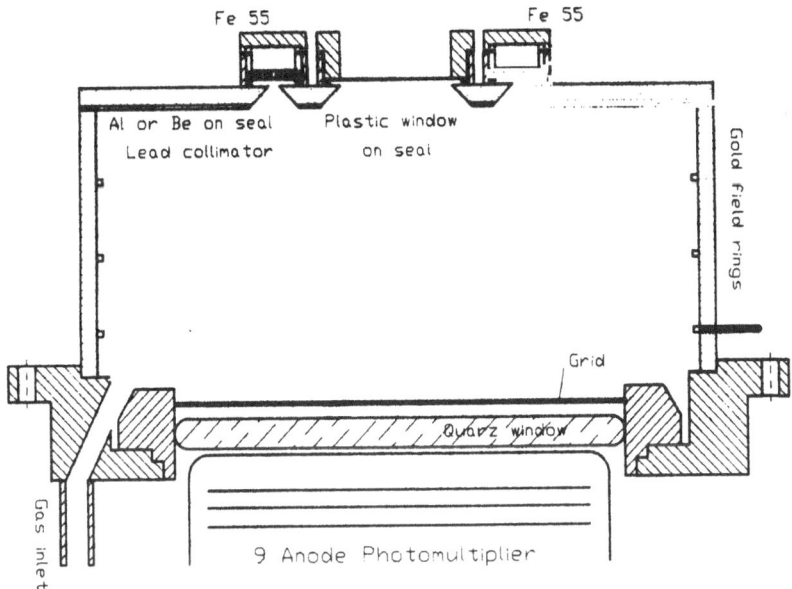

Fig. 1. Schematic of the SAX LEGSPC gas cell.

Ottobrunn, FRG. There are a number of commandable modes. In Direct mode, the energy, time, position, burst length and veto state of each event is passed to the on-board data handling system. Various Indirect modes exist in which either spectra and/or images of selected regions of the field of view may be accumulated on-board prior to transmission to the ground in order to reduce the telemetry load. In both modes the correction to pulse height due to varying penetration depth of the X-rays can be performed in flight and/or during ground analysis.

3. Scientific Objectives

The mid-1990's will be an exciting time for X-ray astronomy with the launches of SAX, Astro-D, XTE and Spectrum-X-Γ expected within a short interval. Astro-D and Spectrum-X-Γ will use CCD detectors to provide a resolution of 100–150 eV over a ~ 0.5–10 keV energy range while XTE will concentrate on timing studies in the 2–200 keV energy range. While CCD development is still proceeding at a rapid pace it is worth comparing their properties with those of GSPCs:

– The resolution of the LEGSPC at iron K of 500 eV is about a factor 4 worse than a CCD. However, since it scales as $E^{1/2}$ it becomes comparable to that of a CCD at low ($\lesssim 1$ keV) energies.
– The LEGSPC provides extremely good time resolution of up to 16μs.

This should be compared with a typical time of 2s needed to read out a typical TV format CCD.

There are, however various ways of improving this last figure. One of these is

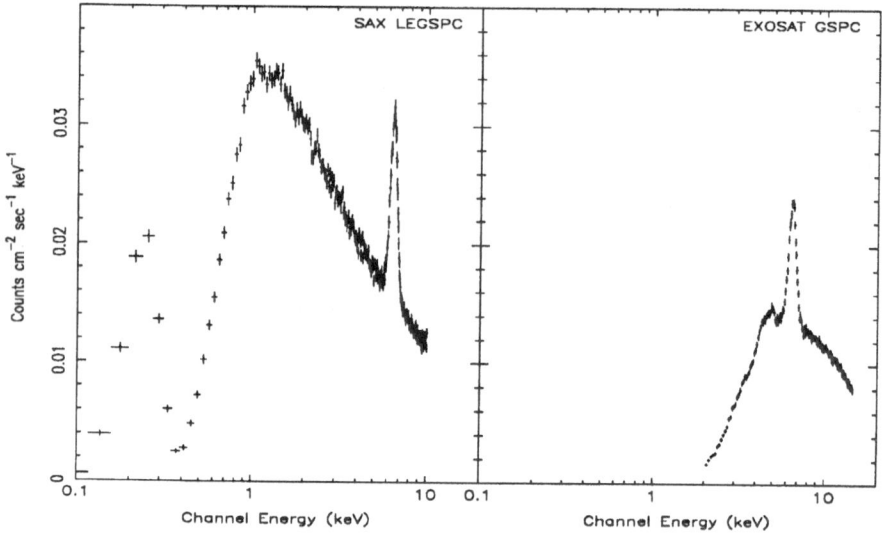

Fig. 2. The same source observed by the LEGSPC and EXOSAT GSPC instruments

to read out the CCD continuously, the spatial distribution of the observed X-rays then represents timing information.

– In X-ray astronomy CCD detectors are used in photon counting mode. This means that the mean arrival rate should be < 0.1 photons/pixel/frame which can place severe constraints on the maximum source strengths allowed. There are no such problems with the LEGSPC which has a maximum throughput of 2000 counts s^{-1}.

– The energy resolution of GSPCs is known to be extremely stable, whereas there are still uncertainties as to the effects of large radiation doses on CCDs (e.g., Holland, Lumb, and Castelli 1989).

The instrumentation on SAX is ideally suited to the study of the brighter time-varying sources such as X-ray transients. These sources were first discovered as bright sources that appeared in previously source free locations and then slowly decayed. The brightest known to date, Nova Mon 1975, reached a flux 50 times that of the Crab Nebula. Transient outbursts above 50 mCrab are roughly estimated to occur at a rate of 5×10^{-4} deg^{-2} yr^{-2}. X-ray transients can be divided into three classes; pulsing sources with hard spectra, many of which are identified with Be star companions; bursting sources with soft spectra; and the ultra-soft transients. SAX with its WFCs monitoring large regions of sky will detect and study some 10–30 bright transient events per year. Optical identifications of many of these sources will lead to important further information. The combination of the wide energy range of the SAX instruments and the large field of view of the WFCs makes SAX ideally suited for detecting these sources and especially for carrying out detailed follow-up studies over a wide luminosity range. These studies will provide important new insights into the nature of the accretion process onto both neutron stars and black holes.

Some idea of the quality of the spectra that will be obtained using the LEG-PSC is shown in Figure 2. A spectrum with an equivalent hydrogen column density of 10^{20} atoms cm^{-2}, a power-law photon index of 0.61, and a 6.4 keV iron line of 0.2 keV FWHM width and 0.85 keV equivalent width (the flux in a line divided by the strength of the continuum under the line) was assumed. The assumed 2–10 keV flux of 1.7×10^{-9} erg cm^{-2} s^{-1} is typical of bright transient sources. The significantly better resolution of the LEGSPC compared to the EXOSAT GSPC (Peacock *et al.* 1981) results in a narrower line with higher peak counts. The loss in efficiency of the SAX mirrors above 10 keV means that the EXOSAT GSPC is more sensitive at $\gtrsim$ 10 keV. At low energies, the different window construction and improved design of the detector allows the LEGSPC to observe the hypothetical source down to 0.1 keV.

Acknowledgements

The assistance of E. Dutruel, F. Favata and N. Stricker is gratefully acknowledged.

References

Favata, F., and Smith, A.: 1989, *SPIE Proceedings of EUV, X-ray and Gamma-Ray Instrumentation for Astronomy and Atomic Physics*, SPIE, Bellingham, 1159, 488

Favata, F., Smith, A., Bavdaz, M. and Kowalski, T. Z.: 1990, Submitted to *Nucl. Instr. Meth.*

Holland, A. D., Lumb, D. H., and Castelli, C. M.: 1989, *SPIE Proceedings of EUV, X-ray, and Gamma-Ray Instrumentation for Astronomy and Atomic Physics*, SPIE, Bellingham, 1159, 113

Inoue, H., Koyama, K., Matsuoka, M., Ohashi, T. and Tanaka, Y.: 1979, *Nucl. Instr. Meth.* **157**, 295

Kowalsky, T. Z., Smith, A. and Peacock, A.: 1989, *Nucl. Instr. Meth.* **A279**, 567

Peacock, A., etal.: 1981, *Space Sci. Rev.* **30**, 479

Perola G. C.: 1983, *Proceedings of the Workshop on Non-thermal and Very High Temperature Phenomena in X-ray Astronomy, Rome*, 175

Scarsi, L.: 1983, *Proceedings of the Workshop on Non-thermal and Very High Temperature Phenomena in X-ray Astronomy, Rome*, 171

Simons, D.G., de Korte, P. A. J., Peacock, A., and Bleeker, J. A. M.: 1985, *SPIE Proceedings of X-ray Instrumentation in Astronomy*, SPIE, Bellingham, 597, 190

Smith, A., Favata, F. and Kowalsky, T. Z.: 1989, *Nucl. Instr. Meth.* **A284**, 375

Smith, A, Peacock, A., and Kowalski, T. Z.: 1987, *IEEE Trans. Nucl. Sci.* **NS-34**, 57

Spada, G. F. L.: 1983, *Proceedings of the Workshop on Non-thermal and Very High Temperature Phenomena in X-ray Astronomy, Rome*, 217

THE EXOSAT RESULTS DATABASE

A.N. PARMAR AND N.E. WHITE

Space Science Department of ESA, ESTEC, The Netherlands

Abstract. The EXOSAT database provides on-line access to the results and data products (spectra, images and lightcurves) from the EXOSAT mission as well as EINSTEIN SSS and MPC spectra, EINSTEIN HRI images and IUE spectra of X-ray sources. In addition, a number of well-known optical, infrared, and X-ray catalogs, including the HST guide star catalog are available. The complete database is located at the EXOSAT observatory at ESTEC in the Netherlands and is accessible remotely via SPAn by typing 'set host 28703' and logging in to account 'XRAY'. Alternatively, a sub-set of the database is available at GSFC and may be accessed by typing 'set host 6467'. Access is also possible via Arpanet/Internet, GTE TELENET and by direct mail.

The database management system has been specifically developed to efficiently access the database and to allow the user to perform statistical studies on large samples of astronomical objects as well as to retrieve information on single sources. The system has been written to be mission independent and includes timing, image processing and spectral analysis package as well as software to allow the easy transfer of analysis results and products to the user's own institute.

Further information about the system including user's guides is available from A.N. Parmar and N.E. White at ESTEC and from their colleagues J. Behnke/Code 634 and K. Arnaud/Code 666 at GSFC.

Y. Kondo (ed.), Observatories in Earth Orbit and Beyond, 462.
©1990 *Kluwer Academic Publishers.*

AN X-RAY ALL SKY MONITOR FOR A JAPANESE EXPERIMENTAL MODULE ON THE SPACE STATION

M. MATSUOKA, N. KAWAI, T. IMAI, M. YAMAUCHI, A. YOSHIDA,
T. KOHNO, A. YONEDA

The Institute of Physical and Chemical Research,
2-1 Hirosawa, Wako, Saitama 351-01, Japan

and

H.TSUNEMI

Department of Physics, Faculty of Science, Osaka University,
1-1 Machikaneyama-cho, Toyonaka, Osaka 560, Japan

Abstract. We propose an X-ray all sky monitor for Japanese Experimental Module (JEM) on the space station. Considering practical circumstances, we show as a case study that the all sky monitor with slit hole cameras is most promising for monitoring the short-term and long-term X-ray transients. We call this all sky monitor as MAXI (Monitor of All-sky X-ray Image). Position determination of gamma-ray bursts could be achieved with accuracy less than one degree observing the X-ray component of the burst. Weak X-ray sources such as active galactic nuclei could be also monitored with time resolution less than one day. The X-ray all sky monitor will work to discover X-ray novae and transient phenomena and give us the alarm for further detailed observations. The obtained data will be also used for archival study.

1. Introduction

Most celestial objects have potential to emit X-rays. X-ray luminosities from various objects are variable on timescales of wide range. Limited space and observation time on satellites disable us to monitor these variable X-ray sources continuously.

Although X-ray all sky monitors have been performed by several satellites, duty cycles for observations are not enough for monitoring all the sky and also they have not yet monitored weak X-ray sources such as active galactic nuclei (Priedhorsky and Holt, 1987). It is, however, very promising to achieve a comprehensive X-ray all sky monitor on the space station (Matsuoka et al. 1988).

Space station program being promoted by NASA has just entered into its detailed design and development phases. As the first step of scientific payload to be attached to Japanese Experimental Module (JEM) we propose the MAXI (Monitor of All-sky X-ray Image) which is able to detect rapid variabilities from strong X-ray sources as well as long-term variabilities from weak X-ray sources. This paper gives an outline of the MAXI on the space station.

2. The all sky monitor suitable for the space station

We discuss on a practical X-ray all sky monitor suitable for the NASA space station as a scientific payload in the first phase. It is suggested that the MAXI (Monitor

Y. Kondo (ed.), Observatories in Earth Orbit and Beyond, 463–468.

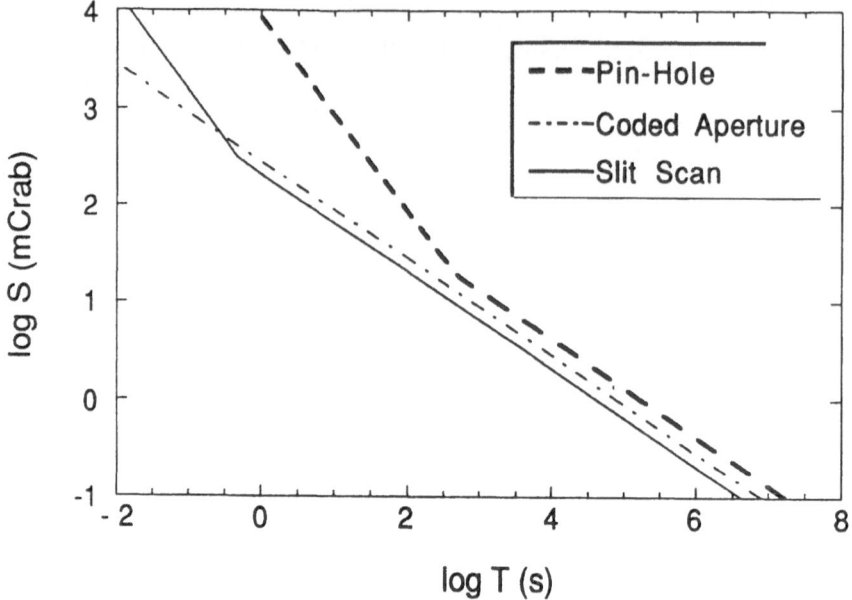

Fig. 1. The time vs. intensities to be required to detect a source with counts more than 5 photons as well as with a significant level of 5σ.

of All-sky X-ray Image) is attached to Japanese Experimental Module (JEM) on the Space Station. We will take into account the following problems to design the practical MAXI:

(1) Low cost and simple system for manufacturing, operation and maintenance.
(2) Moderate requirements for weight, electric power and data transmitting rate.
(3) Capability of partial achievement of scientific objectives by a single module.
(4) No requirement of an attitude control system.
(5) Providing of useful data for the community in astrophysics.
(6) Promotion of international collaboration for this payload.
(7) Desirable joint observations with wide band burst monitors such as optical/UV/II and gamma-rays.
(8) Capability of optional instruments to grade up the data.

Here we supplement a few points for items (3) and (8). The item (3) means that the payload consists of several modules, but a single module is capable of partial achievement of scientific objectives independently. In item (8) astronauts can sometimes operate the optional instruments. This is a merit of the manned space station, but this requirement is not indispensible.

On the other hand the following requirements for the MAXI are considered in addition to the above problems:

(a) Detection limit of the order of 1 mCrab enables us to detect X-rays from active galactic nuclei.

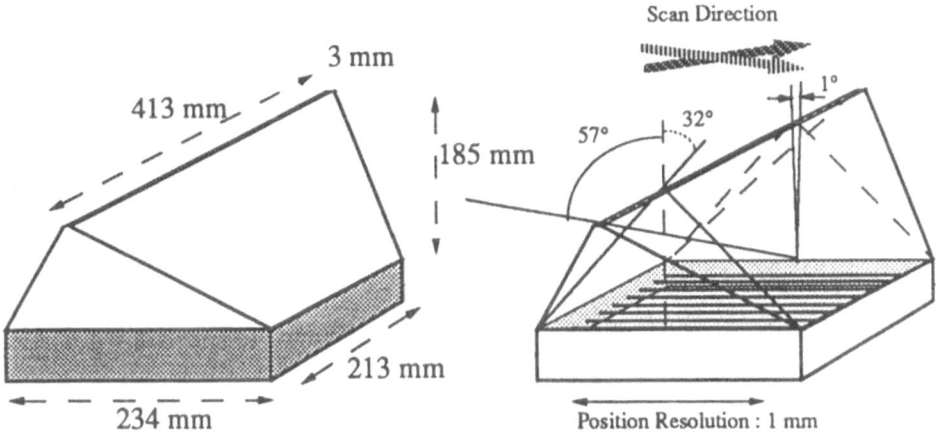

Fig. 2. A minimum unit of a pair of slit-hole cameras.

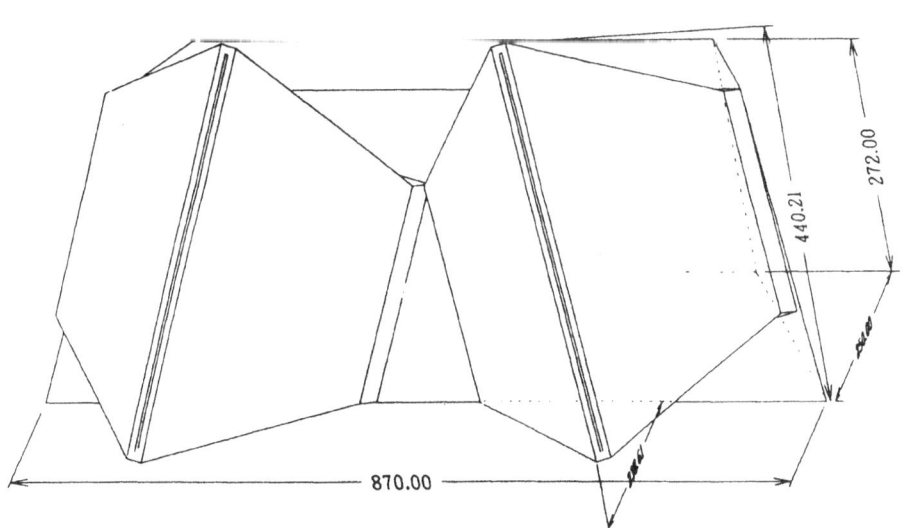

Fig. 3. An overview of a pair of cameras.

(b) Spatial resolution is less than one degree to avoid confusion of sources with intensities stronger than 1 mCrab.

(c) A large duty cycle of observation time requires the X-ray image in a wide field of view.

Several kinds of X-ray all sky monitor systems have been proposed and most of them have been performed with satellites for each objective (Holt and Priedhorsky 1987; Caroli et al. 1987). We summarize these systems in the followings:

(1) Scanning survey with slit-like fields of view using rotation of the satellite or instrument.

(2) Scanning or rotating instruments with modulation collimators.

(3) A pinhole camera.

(4) A slit-hole camera.

(5) A coded mask camera.

For the space station we should employ the all sky monitor which has a duty cycle of 100% for intended field in the sky; that is, four systems above, (2), (3), (4) and (5).

The burst monitor of rotating modulation collimators on Hakucho have achieved fruitful discovery of new X-ray burst sources (Kondo et al. 1981). A modulation collimator on balloon gondola could determine a gamma-ray burst source in the hard X-ray band for the first time (Nishimura et al. 1978). The hard X-ray or soft gamma-ray all sky monitor using rotating modulation collimators is still being performed (Lund 1985). However, it has been pointed out mathematically that for a particular correlation map (or the corresponding data) X-ray intensity pattern is not uniquely determined even in the absence of statistical noise (Doi 1988). Therefore, the rotating modulation collimator is not suitable for imaging the all sky. Thus, in the following we will examine three systems of (3), (4) and (5) in detail.

We will compare three systems on the assumption that each system consists of 6 units of each effective area of 1000 cm^2 and each covering field of one steradian; i.e., this system has total effective area of 6000 cm^2, covering 2π str. of sky. Non X-ray background rate is assumed to be $6 \times 10^{-3}s^{-1}$ cm^{-2} for 2–10 keV. Position resolution is assumed to be $1° \times 1°$; i.e., the area of the pinhole camera and one pixel for the coded camera correspond to 0.29 cm^2, while the area of the slit-hole camera is 17 cm^2.

Fig.1 shows the time which is necessary to detect a source with detection counts more than 5 photons as well as with a significant level of 5 σ. This figure indicates that the slit-hole camera is most promising in respect of S/N ratios over wide intensity range. A photon limit of this camera constrains a detection of bursts weaker than 1 Crab with a duration less than about 0.1 sec.

Although sources more than two would be often in the field of a slit pixel, each location could be determined using criss-cross method. Here, we can also use scanning information, because platforms on the JEM make one revolution every satellite orbit.

3. An outline of the MAXI

The MAXI consists of 6 units, covering 2π str. in total. One unit has a pair of cameras with two slit-holes, covering the field of one str. A pair of slit-holes are slanted each other by about 30 degrees. Data from a pair of cameras are similar to those in scanning mode from a pair of counters with slanting slat collimators. A large duty cycle of the sky by a slit-hole camera is an essential difference from a small duty cycle by a slat collimator system. A conceptional outline of a minimum unit of a pair of cameras is shown in Fig. 2, in which this detector is a one-dimentional proportional counter, having an effective area of 500 cm^2. Fig. 3 shows an overview of a pair of counters.

Since locations of most bright X-ray sources are known precisely, we could determine their coordinates on the image map of the MAXI from the data themselves. Therefore, we do not need a star sensor for data analysis of the MAXI. On the other hand a lot of data must be processed. Total data points with resolution of one degree are estimated to be about 3×10^4 for all the sky except sky region near the sun. We estimate telemetry requirement to send the data. Telemetry data will be arranged with four modes; i.e., low, medium, and high bit rate modes and a burst mode. In the following we have considered that platforms on the JEM rotate one degree every about 15 sec.

Low bit rate mode: 1.15 k bits/s for one energy band, time resolution of 7.5 sec and spatial data of 120 pixels. Medium bit rate mode: 7 k bits/s for 4 energy bands, time resolution of 1 sec and spatial data of 120 pixels. High bit rate mode: 32 k bit/s for 64 energy channels, time resolution of 1 m sec and spatial data of 240 pixels. Burst mode: total memory of about 100 k bits for one burst data with duration of 10 sec, 64 energy channels and time resolution of about 1 msec. It would be possible to set a demountable mass storage memory such as optical disks for further information. In addition to the standard telemetry, more data can be taken if an optional mass storage system (MSS) is available. This MSS has to be operated by astronauts for change and retrieve of this memory once ten days or a few months. Optical disks or magnetic tapes could be available for the MSS.

The weight, size and power consumption of MAXI can be made to conform to the JEM specification. The one dimensional proportional counters should be manufactured carefully to keep the long-term operation. Detailed design can be modified flexibly if necessary. For example, we can transport each unit of the MAXI separately with space shuttles or Japanese HII rockets and build up one by one on the space station.

References

Caroli, E. *et al.*: 1987, *Space Sci. Rev.* **45**, 349

Doi, K.: 1988, LxR Report "Intrinsic ambiguity in the X-ray image data obtained with the rotating modulation collimator system", No.8801

Holt, S.S. and Priedhorsky, W.: 1987, *Space Sci. Rev.* **45**, 269

Kondo, I. *et al.*: 1981, *Space Sci. Instr.* **5**, 211

Lund, N.: 1985, *SPIE Proc. of X-ray Instrumentation in Astronomy*, 597, 95

Matsuoka, M., Kawai, N., and Tsunemi, H.: 1988, IPCR CR-13 prepared for a reference
    mission to be attached to the JEM on the space station,
Nishimura, J., et al.: 1978, *Nature* **272**, 337
Priedhorsky, W.C. and Holt, S.S.: 1987, *Space Sci. Rev.* **45**, 291

THE ASTRO MISSION

T.R. GULL

NASA/Goddard Space Flight Center

Abstract. The Astro Mission, an attached payload dedicated to astronomy, is now scheduled to fly May 9, 1990 in the shuttle Columbia. Four instruments make up the complement. Broad Band X-Ray Telescope (BBXRT, Peter Serlemitsos, Principal Investigator
GSFC) utilizes the Two Axis Pointing System (TAPS). The three ultraviolet instruments
are Hopkins Ultraviolet Telescope (HUT, Arthur Davidson, JHU), Ultraviolet Imaging
Telescope (UIT, Ted Steccher, GSFC) and Wisconsin Ultraviolet Photopolarimeter Experiment (WUPPE, Art Code, UWI). The instruments are co-aligned in the Spacelabs Instrument Pointing System (IPS). Five of the seven astronauts are professional astronomers.
Mission Specialists Mike Lounge, Jeff Hoffman, and Robert Parker, and Payload Specialists Samuel Durrance and Ron Parise. During the planned ten-day mission, over 290
shuttle manoeuvres will enable approximately 1000 individual instrument observations of
230 separate astronomical objects. The Astro mission is designed for multiple flights.

Y. Kondo (ed.), Observatories in Earth Orbit and Beyond, 469.
©1990 *Kluwer Academic Publishers.*

THE UV IMAGER FOR THE ISRAELI SCIENTIFIC SATELLITE

N. BROSCH

The Wise Observatory, Tel Aviv University, Israel

Abstract. Israel will orbit a satellite dedicated to scientific research. One of the two experiments studied for deployment on this platform is a three-channel imager in the ultraviolet proposed by Tel Aviv University and designed jointly by staff of the Wise Observatory and of El-Op (Electro-Optical Industries, Ltd.). The design provides very significant scientific returns in a small payload and for a moderate cost.

1. Rationale

Israel orbited two technology demonstrator satellites, in September 1988 and in April 1990. In preparation for its entry in the "space nations" club, the Israeli Space Agency (ISA) called for proposals of instruments to be orbited on a dedicated scientific satellite platform (the National Scientific Satellite-NSS). One of the two proposals selected for Phase A study is the UV imager payload (TAUVEX).

The constraints of the scientific payload imposed by ISA are mass less than 20 kg, power consumption less than 25 W and size to fit inside a cylindrical envelope of 50 cm diameter and 150 cm length. The launch constraints are similar to those of the Scout and the design lifetime is two years.

The UV imager mission seeks to exploit the dark skies in the UV region between Lyman α and the atmospheric cutoff with the constraint of a small sized payload. We proposed an imaging experiment of reasonably wide field and imaging quality, that can be accomodated in the payload envelope.

Any new mission competes with the Hubble Space Telescope (HST) and with the UIT on the ASTRO platform. The advantage of TAUVEX relative to the HST will be in the size of the imaged field, because the large collecting area and superb imaging qualities of HST are unique. The advantages relative to UIT will be in the duration of the mission, *i.e.* in the portion of the sky imaged by the payload, and possibly in the dynamic range of the experiment.

2. Design

2.1. OPTICS

The size constraint implies that at most a single telescope of about 40 cm aperture could be accomodated. An imaging experiment to provide interesting science must do so in a number of spectral bands. This is normally done with a filter-changing mechanism. We felt that, in the interest of reliability, we should design away from moving parts, possibly at the expense of some efficiency.

The baseline design is of three 20 cm diameter bore-sighted Ritchey-Cretien telescopes mounted on the same bezel. The telescopes have an effective f/number of 5 and the image quality is 5″ over a 1.7° diameter field. This is achieved with a field

Y. Kondo (ed.), Observatories in Earth Orbit and Beyond, 471–474.

corrector composed of two doublet groups with almost zero power. The corrector lenses also focus the optics on-orbit, should this be necessary. The physical length of each telescope is about 36 cm. The obscuration by the secondary mirror and the baffles is ~50% of the area presented to the sky.

The entire mechanical construction is of graphite-epoxy, with mirrors made of lightened Zerodur and lenses of LiF or CaF_2. A number of light shields are designed into the structure. They reduce the amount of scattered light as well as decouple the optics and structure from the cooling load of space. Thus only 11.5 W are required to keep the telescopes at $+20°C$.

2.2. DETECTORS

We requested for TAUVEX a space-qualified detector, that is available more or less off-the-shelf. We selected for the baseline design the wedge and strip (W+S) detectors developed by the Berkeley group (Siegmund et al., 1986) with 30 mm cathodes, 50 μm pixels and 500x500 pixel format. The W+S detectors are photon-counting devices whose sensitivity depends on the type of cathode and trasmission of glasses used in the window construction.

For the baseline design we configured the detectors with CsI or $CsTe$ cathodes and with windows made of suitable materials to provide cutoff wavelengths and suitable bandpasses. The payload will image in three spectral bands, of ~300Å width each, centered at 1600Å, 2200Å and 2600Å.

2.3. DATA STORAGE

A severe constraint is the communication schedule with the ground. Israel provides at present a single ground station (GS) located within its boundaries. The NSS will be seen by the GS for at most one hour per day, in a number of passes. The collected data must be stored on-board and dumped whenever the GS is in view. Thus the need for fairly large storage that conflicts with the mass and power constraints and for reasonably fast data links.

The storage medium we selected is a solid-state memory (OBM) of 60 MB, that should suffice to store one image per telescope per orbit, for an average of three orbits before downloading. In order to comply with the memory size, we have to relinquish the photon-counting option. We shall therefore store intermediate images, of some four second integration, that will be compressed by run -length coding, by a factor of ~100. Only the compressed frames will be stored in the OBM and will be telemetered to the GS, with the reconstruction of the image to be done on the ground.

3. Performance

The payload described above provides a reasonable performance in imaging faint details as well as bright objects. As an example, the event rate from the sky background, assuming a sky brightness of 25 mag/$\square''$ at the short and long bandpasses, and 26 mag/$\square''$ at the intermediate bandpass, is such that we expect to measure it with S/N~5 in a 2000 second exposure and with 10'' pixels.

Stellar objects of UV monochromatic magnitude of 17 will be measured with a S/N~10 in a similar exposure. The brightest objects for which we expect to obtain photometric information are of UV monochromatic magnitude ~8. Diffuse objects will be detected with a S/N~5 when they are 10% or brighter than the sky background and when their size is 1□′ or larger.

The *nominal* performance of UIT is better in detecting stellar objects. However, UIT is limited by its method of recording images and by the time in orbit. TAUVEX will allow more than 5000 separate pointings during one year, that will cover about 1/3 of the entire sky. Moreover, TAUVEX will collect *simultaneous* data in the three spectral channels; this is not possible with UIT.

4. Science

A few projects were studied in some detail for the baseline Phase A mission. It appears that TAUVEX is a very good QSO detector, because of the simultaneous imaging at three spectral bands. It is possible, using the UV and visual color indices, to separate QSOs from stars. Thus we plan a survey phase early in the mission, where 1500 □° will be mapped around each galactic pole, with 10% overlap between the fields. This requires a three month period on orbit and will yield an unbiased catalog containing 6000-10000 QSOs.

The same is true for white dwarf (WD) detection; it is possible to separate WDs from normal stars using the three UV bands. We expect to compile a catalog of about 10000 WDs at the completion of the Polar Caps'survey.

Another project has to do with the star formation processes in galaxies. Here we plan to image all the Local Group galaxies, to map the entire Virgo cluster and to search for supernovae by revisiting the entire cluster after six months, and for a third time after another six months. We shall also image a sample of about 40 late-type galaxies and another of some 50 early-type galaxies.

Finally, one very interesting study has to do with fast variations in the UV emission of cataclysmic systems. TAUVEX offers the option of downloading individual photons as fast as msec rates, for objects brighter than ~12-13 UV monochromatic magnitudes and when the platform is in direct view of the GS. Thus WDs and physics of accretion disks can be studied at wavelengths close to maximal emission.

5. Status

The Phase A reports of the two competing groups were submitted to ISA at the beginning of 1990. The two proposals are now in the process of being evaluated and we expect the evaluation reports by the end of June 1990. At this point ISA shall decide which of the two experiments will proceed to Phase B (detailed design leading to the Final Design Review) and eventual construction and launch. ISA shall also decide whether to advance both experiments and fit both on the same launch.

Summary description of observatory

– The payload is the three-band UV imager proposed by Tel Aviv University and accepted for Phase A study by the Israeli Space Agency.

– The design lifetime is two years in orbit.

– The payload is 150 cm long and 50 cm in diameter. It masses 24.7 kg and consumes about 25 W. The telescope section is about 50 cm long; a 100 cm long shield reduces cold loading.

– Three 20 cm aperture R-C telescopes are co-aligned on the same platform.

– Each telescope has a FOV of 1.7° in diameter.

– Each telescope is "tuned" to a different bandpass in the UV: centers of bands are at 1600, 2200 and 2600Å.

– Detectors are of wedge-and-strip type, with CsI or $CsTe$ cathodes, and filters.

– Images are stored in buffer memories for a few sec, then are compressed and stored in solid-state memory for later downloading.

– The sensitivity is such that a 17th mag star will be detected after 2000 sec with a S/N~10. A diffuse object that is only 10% above the sky will be detected with a S/N~5 after 2000 sec, by averaging over $10^4 \square$".

Acknowledgements

The TAUVEX team of Tel Aviv University consists of E. M. Leibowitz, H. Netzer, S. C. Beck, E. Ribak and N. Brosch. We benefited from discussions with the Berkeley group (S. Bowyer, M. Lampton, O. Siegmund, and others). We are grateful to C. Martin for allowing us to read the PAX proposal. The Phase A of this study is supported by a grant from the Ministry of Science and Technology.

References

O'Connell, R.W. 1987, A.J. **94**, 876-882.

Siegmund, O.H.W., Lampton, M., Bixler, J., Chakrabati, S., Valerga, J., Bowyer, S. and Malina, R.F. 1986, J. Opt. Soc. America A **3**, 2139-2145.

ORFEUS-SPAS:

THE BERKELEY EUV SPECTROMETER

STUART BOWYER

*Space Astrophysics Group, Center for EUV Astrophysics,
University of California, Berkeley, California 94720*

and

MARK HURWITZ

*Space Astrophysics Group, Center for EUV Astrophysics,
University of California, Berkeley, California 94720*

1. Introduction

ORFEUS-SPAS, the *Orbiting Retrievable Far and Extreme Ultraviolet Spectrometers*, on the Astro-SPAS space platform, is a joint project of NASA and the BMFT in the Federal Republic of Germany (see Kraemer et al. 1990, this volume). The Berkeley spectrograph for this mission will obtain high-resolution ($\lambda/\Delta\lambda > 7000$) spectra of point sources between 390 and 1200 Å. The Berkeley instrument incorporates a set of four novel spherically figured, varied line-space (SVLS) gratings used in a geometry that is similar to that of the classic Rowland mount to span this large wavelength interval. Two spectral detector units containing curved MCPs and delay-line anodes encode the arriving photons in digital format for telemetry. An additional optic directs the image of the source in the entrance aperture onto a sealed FUV detector which is used to track the source as it drifts during an observation, enabling a post-flight reconstruction of the spacecraft pointing vector. This in turn will allow us to define the wavelength of each recorded photon with precision.

2. The Berkeley Spectrometer

In Figure 1 we show a schematic layout of the optical components of the spectrograph. The instrument contains four diffraction gratings, arranged so that each is illuminated by a "wedge" on the annular aperture of the telescope primary mirror. Each wedge subtends about 20% of the available geometrical area of the 1 m diameter primary and therefore has a focal ratio of about $f/5$ rather than the full $f/2.4$ from the primary mirror. Each grating is used to study only one-fourth of the total 390–1200 Å bandpass, and operates over a relatively small range of diffraction angle β, allowing high diffraction efficiency to be maintained. Two spectral detector units each accept the spectra from two diffraction gratings.

In our design, each diffraction grating in the instrument is spherical, with a radius of curvature of 1 m. Unlike the conventional Rowland spectrograph, the groove spacing of each grating varies as a fourth order function of position on the optic. Furthermore the distances from the telescope image to the grating center, and from

Y. Kondo (ed.), Observatories in Earth Orbit and Beyond, 475–480.

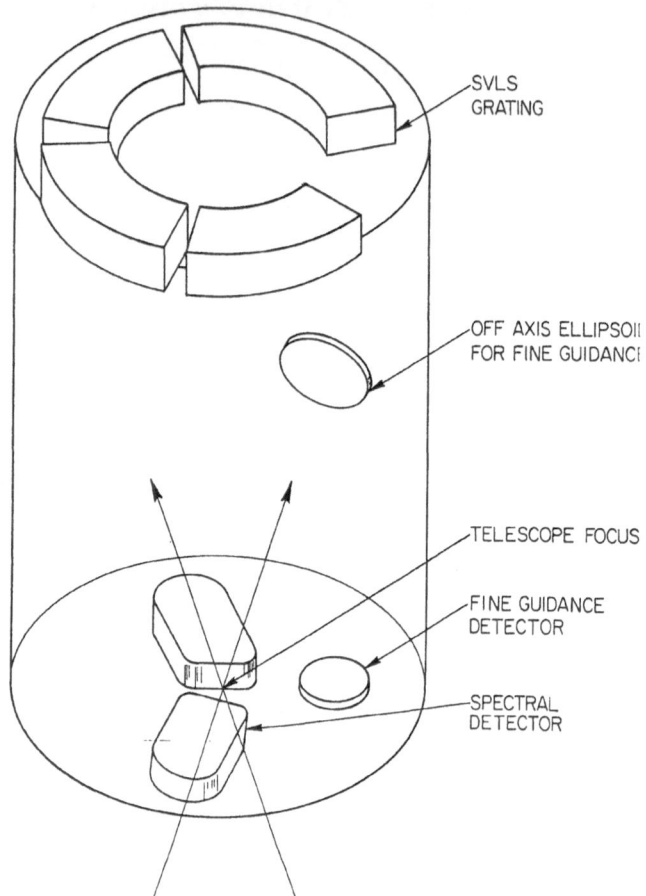

SVLS
GRATING

OFF AXIS ELLIPSOID
FOR FINE GUIDANCE

TELESCOPE FOCUS

FINE GUIDANCE
DETECTOR

SPECTRAL
DETECTOR

Fig. 1. Schematic layout of the Berkeley spectrograph.

the grating center to the detector surface, are not equal to the values on the corresponding Rowland circle. The astigmatism and aberrations of this spectrograph are substantially lower than what would be achievable with a conventional Rowland spectrograph and toroidal, uniform line-space diffraction gratings. A more detailed description of SVLS spectrometers can be found in Harada and Kita (1980).

Two detectors are used, each accepting the dispersed spectra from two diffraction gratings. The MCPs are spherically curved, with a radius chosen to provide a close match to the tangential focal surface of the diffraction gratings. A flat detector would have limited the resolution to ∼1500. Each detector contains a stack of two MCPs, the electron charge clouds from which strike a delay-line anode readout system (Lampton, Siegmund and Raffanti 1987).

The fine guidance system consists of a small off-axis ellipsoid, which intercepts a fraction of the beam from the primary mirror and images the target onto a sealed-tube MCP detector equipped with a wedge-and-strip readout system (Martin et al. 1981). The fine guidance system is sensitive to photons in the 1400–1800 Å

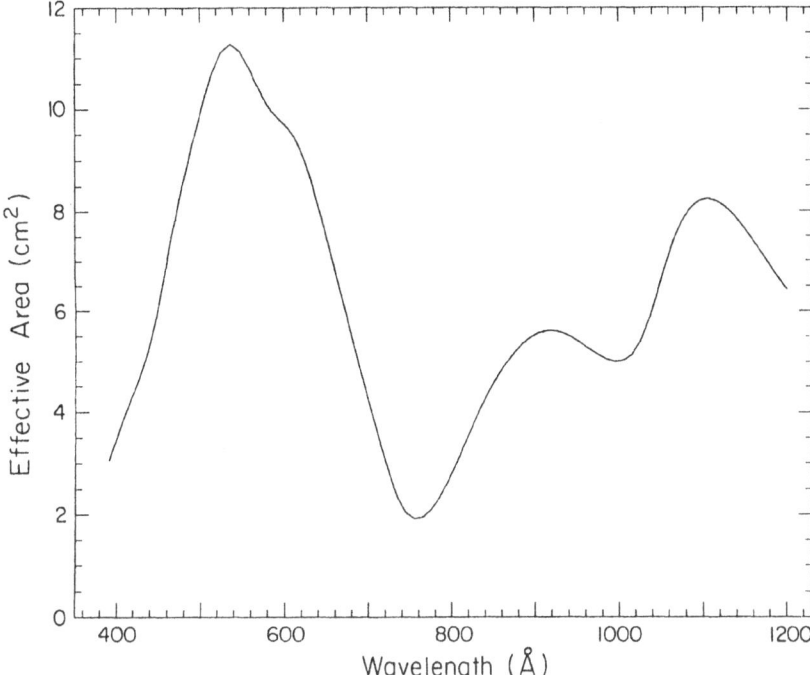

Fig. 2. Predicted effective area of the instrument vs. wavelength. Fabricated instrument will display small discontinuities at crossover wavelengths between gratings due to variations in diffraction efficiency and coating reflectivity.

band. Each pixel on the detector represents about 3″ on the sky. To determine the location of the source to a precision of about 1″, we will place the detector so that the image is defocused to cover ~3 pixels and calculate a centroid of the photons recorded during a given integration period. The integration period is determined by the expected drift rate of the target in the aperture; during one second the target should move no more than about 1″. Even a relatively faint target such as the hot WD HZ 43 should produce over 100 events s^{-1} on the detector, enabling a centroid to be calculated to an accuracy better than 1″.

3. Instrument Performance

The telescope primary mirror will be coated with iridium, a stable coating which provides good normal-incidence reflectivity across the instrument bandpass. The diffraction gratings will be coated with the same material. We expect that the gratings will achieve a groove efficiency of ~50%, and that the quantum detection efficiency will be equal to that measured for the stable photocathode KBr by Siegmund et al. (1988). The resulting curve of effective-area-versus-wavelength is shown in Figure 2.

Prior to the launch of *ORFEUS* in 1992, NASA will have flown two other major instruments with spectroscopic capabilities in the EUV. These are *HUT* (Davidsen

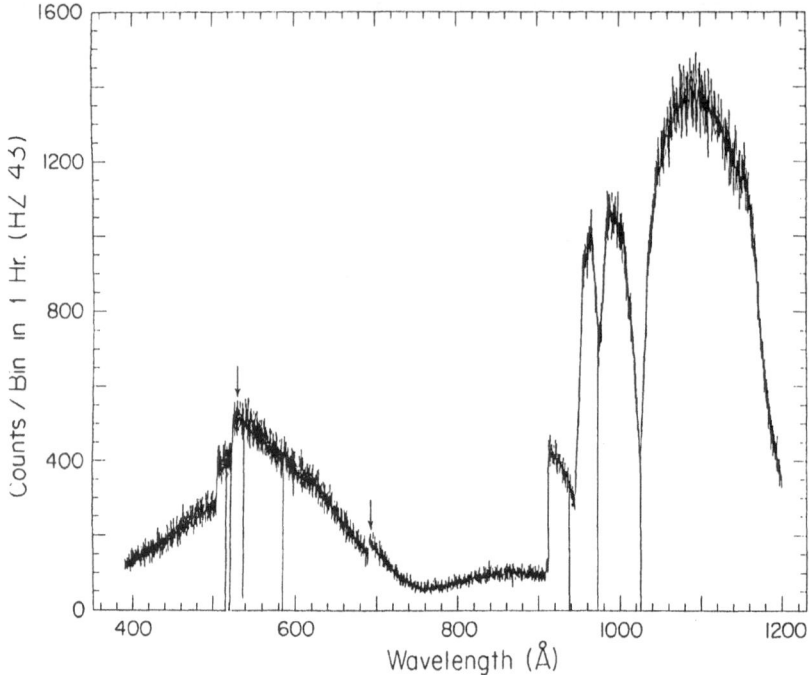

Fig. 3. Sample spectrum of hot WD HZ 43. Counts per spectral bin for a 3600 s integration are plotted vs. wavelength. Column density of H I assumed to be 2×10^{17} cm^{-2}; column of He I assumed to be 1/10 that of H I. Absorption edges of interstellar H I and He I are shown, as are strong resonance lines of these species. Because the spectral resolution changes discontinuously at grating crossover wavelengths, the simulated counts per spectral bin display those discontinuities (marked with an *arrow*).

et al. 1989) and *EUVE* (Malina and Bowyer 1989). Although the relative *sensitivity* of these instruments depends on quantities that may be difficult to estimate (e.g., the detector background rate), for relatively bright sources it is the effective area that limits the achievable signal-to-noise, so it is the effective area – and spectral resolution – that we compare here.

HUT contains a normal-incidence mirror which is only slightly smaller than that of *ORFEUS*. The spectrograph design is a classical Rowland mount. At wavelengths between 840 and 1580 Å, the *HUT* grating is used in first order and should achieve a net effective area that is ~1.5 times that of *ORFEUS*. Between 420 and 925 Å, the *HUT* grating is used in second order and can achieve an effective area roughly equal to that of *ORFEUS*. But because many targets will be substantially fainter in that band than at first-order wavelengths, an aluminum filter will certainly be needed for continuum measurements to attenuate the first-order radiation and may well be needed for studies of line radiation. This filter reduces the *HUT* effective area at the short wavelengths to 1/6 to 1/10 that of *ORFEUS*, depending on wavelength. The spectral resolution of *HUT* is ~400, approximately 1/20 that of *ORFEUS*. The

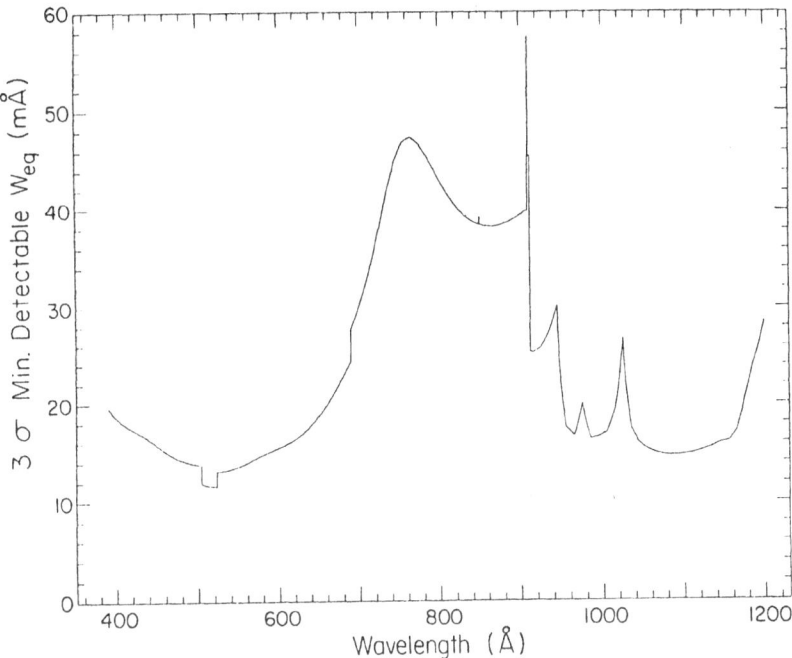

Fig. 4. Minimum equivalent width (mÅ) for 3 σ detection of an absorption line in the spectrum shown in Figure 3 vs. wavelength.

EUVE spectrometer overlaps the *ORFEUS* bandpass below ∼760 Å. Its effective area is 1/4 to 1/20 that of *ORFEUS* at overlap wavelengths. The resolution of the *EUVE* spectrograph is about 275, roughly 1/30 that of *ORFEUS*.

In order to demonstrate the power of this instrument, we have simulated the spectra of a source that is a likely candidate for observation during the *ORFEUS* mission: a hot WD. Because *ORFEUS* will provide unique data about the ISM as well as the point sources themselves, we have also estimated the sensitivity of the instrument to detection of absorption line features. This is by no means an exhaustive set of sources we expect to observe, which includes stars from many regions of the H–R diagram, interacting binaries, and other celestial objects.

In Figure 3 we show a simulated spectrum of the hot WD HZ 43 as would be observed with this instrument. We display the background-subtracted photon counts per spectral resolution element versus wavelength for an integration time of 3600 s. Discontinuities in the spectrum occur at wavelength breaks between the diffraction gratings and also at the absorption edges of neutral hydrogen and neutral helium. Absorption lines due to the strongest resonance transitions of interstellar neutral hydrogen and neutral helium are shown. These strong absorption features are detected at a high level of significance; it is of interest to evaluate what equivalent width an absorption feature would have that would be detected in this spectrum at some threshold, e.g., 3 σ significance. We show this minimum detectable equivalent width in Figure 4. Given the measured column density of H I towards HZ 43, we expect to detect line features of many elements in a variety of ionization states, greatly improving our understanding of the LISM. Although only a few sources are expected to be as EUV-bright as HZ 43, with longer integration times it will be possible to perform similar studies on a variety of WD targets.

Acknowledgements

This research was supported in part by NASA grant NGR–003–450.

References

Davidsen, A.F., Kimble, R.A., Durrance, S.T., Bowers, C.W., and Long, K.S.: 1989, R. F. Malina and S. Bowyer (eds.), *Extreme Ultraviolet Astronomy*, New York: Pergamon Press

Harada, T. and Kita, T.: 1980, *Appl. Opt* **19**, (23), 3987

Kraemer *et al.*: 1990, This volume, pp. 177–184

Lampton, M., Siegmund, O.H.W., and Raffanti, R.: 1987, *Rev. Sci. Instrum.* **58**, (12), 2298

Malina, R.F. and Bowyer, S.: 1989, R. F. Malina and S. Bowyer (eds.), *Extreme Ultraviolet Astronomy*, New York: Pergamon Press

Martin, C., Jelinsky, P., Lampton, M., Malina, R.F., and Anger, H.O.: 1981, *Rev. Sci. Instrum.* **52**, (7), 1067

Siegmund, O.H.W., Everman, E., Vallerga, J.V., and Lampton, M.: 1988, *Appl. Opt.* **27**, (8), 1568

AN OBSERVATORY FOR MAPPING THE FAR UV DIFFUSE
GALACTIC EMISSION LINE BACKGROUND

F. L. ROESLER, J. HARLANDER and R. J. REYNOLDS

Dept. of Physics, University of Wisconsin, Madison, WI 53706

Abstract. A new instrumental concept for interference spectroscopy called Spatial Heterodyne Spectroscopy (SHS) is described. This instrument as currently demonstrated could provide important information on the structure, excitation, and dynamics of the $\simeq 10^5$ K component of the interstellar medium by providing velocity-resolved (20 km s^{-1}) maps of the faint FUV emission line background over a hemisphere of the sky within a 5–6 year observation period. We are currently studying concepts expected to reduce this time by at least an order of magnitude.

In the SHS technique, an all-reflection dispersive interferometer produces a Fourier transform of the spectrum as two-dimensional spatial frequencies on an imaging detector. The system does not require scanning, and measures its own internal alignment state. Although the system suffers the conventional Fourier transform multiplex disadvantage associated with the photon noise in the background FUV continuum, we estimate that for a broad-band survey Spatial Heterodyne Spectroscopy as currently demonstrated can provide 4–5 fold gains over practical grating spectrometers of similar dimensions and spatial and spectral resolution. Field widened methods currently being studied promise additional gains of two orders of magnitude.

1. Scientific Motivation

The discovery of ultraviolet absorption lines (Jenkins 1978) and diffuse x-ray emission (Williamson et al. 1974) from highly ionized trace elements within the interstellar medium established the existence of a hot (10^5–10^6 K) component of the gas. This gas is generally believed to be produced by high velocity shocks from supernova remnants, expanding into the ambient ISM creating hot, low density bubbles, 50 to 200 pc in radius. If long lived and frequent, these bubbles will profoundly influence the ISM structure by interconnecting and occupying most of the interstellar volume (Cox and Smith 1974; McKee and Ostriker 1977). Buoyant forces may carry the hot gas far above the Galactic midplane, producing a hot Galactic corona (Spitzer 1956) and large scale vertical circulations (Bregman 1980). While the existence of the hot component is firmly established by observations, its pervasiveness and influence on the structure of the interstellar medium has not yet been established.

The recent discovery of far-UV emission lines from hot (10^5 K) interstellar gas by Martin and Bowyer (1990) using a Rowland spectrograph on the Berkeley Extreme Ultraviolet/Far-Ultraviolet Shuttle Telescope has opened an important new window on the interstellar medium. The observations clearly show CIV $\lambda1550$ and OIII $\lambda1663$ emission lines from the diffuse interstellar medium with intensities typically 4500 photons cm^{-2} s^{-1} sr^{-1} ($0.06R$) and 2400 photons cm^{-2} s^{-1} sr^{-1} ($0.03R$) respectively, at high Galactic latitudes. Also reported were detections at

Y. Kondo (ed.), Observatories in Earth Orbit and Beyond, 481–486.

greater than 90% confidence of NIII $\lambda1750$ and OIV plus SiIV $\lambda1400$ with intensities of 1000–2000 photons cm^{-2} s^{-1} sr^{-1} ($\simeq 0.02R$). A map over the sky of the intensities, radial velocities, and line profiles of these far-UV emission lines would provide important insights into the distribution and nature of hot interstellar gas. For example by comparison with 21 cm maps one could see whether the FUV occurred mainly inside the boundries of hot bubbles or on the outer surface of clouds immersed in a more widespread hot medium.

We are now developing a new instrumental concept for interference spectroscopy called Spatial Heterodyne Spectroscopy (SHS) (Harlander and Roesler 1990) that will provide the large throughputs needed to carry out such a survey. The instrument could be deployed as a dedicated compact observatory or as a part of a larger facility, preferably in far earth orbit or beyond to avoid the geocoronal photon haze. In the remainder of this paper we will describe the new concept and compare its projected performance with that of a conventional grating spectrometer of similar size and resolving power.

2. The SHS

We believe that SHS provides the first practical approach to interference spectroscopy in the far ultraviolet. The basic concept is most easily understood using the transmitting beamsplitter arrangement in Fig. 1 in which a conventional Michelson interferometer is shown with the return mirrors replaced by diffraction gratings G_1 and G_2. Light enters through aperture A_1 and is collimated by lens L_1. At the exit, lenses L_2 and L_3 relay the superposed coherent images of gratings G_1 and G_2 onto the image plane I where a position sensitive detector records the Fizeau fringe pattern produced by the interferometer. The generation of Fizeau fringes of wavelength-dependent spatial frequency follows from the grating equation

$$\sigma(\sin\theta + \sin(\theta - \gamma)) = m/d \qquad (1)$$

where σ is the wavenumber of light, m is the order of diffraction, θ is the Littrow angle and $1/d$ is the grating groove density. For an input point source of wavenumber σ, two coherent plane wavefronts are produced at the output (one from each arm) whose normals are inclined to the optical axis by angles γ and $-\gamma$ respectively. Ignoring magnification by L_2 and L_3 these crossed wavefronts produce Fizeau fringes of spatial frequency

$$f_x = 2\sigma\sin\gamma \simeq 4(\sigma - \sigma_0)\tan\theta \qquad (2)$$

where σ_0 is the Littrow ($\gamma = 0$) wavenumber. For input spectrum $B(\sigma)$, the intensity on the detector as a function of position x is given by

$$I(x) = \int_0^\infty B(\sigma)(1 + \cos(2\pi(4(\sigma - \sigma_0)x\tan\theta)))d\sigma \qquad (3)$$

The Fourier transform of $I(x)$ will recover the input spectrum. No element has been mechanically scanned in this process. Zero spatial frequency does not correspond to zero wavenumber as in conventional FTS but to $\sigma = \sigma_0$. Effectively,

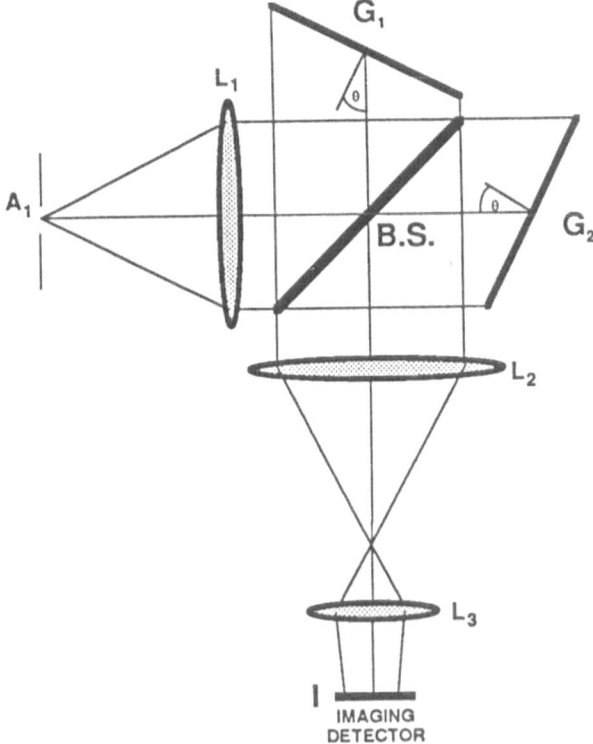

Fig. 1. Schematic diagram of the basic SHS configuration. Wavelength dependent Fizeau fringes produces by diffraction gratings G_1 and G_2 are recorded by a position sensitive detector I. The Fourier transform of the fringe pattern recovers the spectrum as described in the text.

SHS instruments record the entire path difference scanned by FTS simultaneously on a position sensitive detector without scanning, and heterodyne the interferogram with a frequency corresponding to the Littrow wavenumber of the gratings. For the geometry in Fig. 1, the limiting resolving power ($\sigma/\delta\sigma$) is $R_0 = 4W\sigma\sin\theta$ where W is the width of the gratings. By including off axis input angles in the grating equation (Equation (1)), it can be shown that an SHS instrument achieves the same etendue (or throughput) as conventional FTS or Fabry-Perot interferometers. Optical tolerances are relaxed compared to conventional FTS because the Fizeau pattern from a monochromatic line can be used to correct phase errors in software.

An all-reflection version of SHS suitable for FUV operation can be achieved by replacing the transmitting beam splitter by the symmetrically blazed (or alternatively Holographic) grating G_0 as shown in Fig. 2. We have calculated and demonstrated in laboratory tests that instruments of this general design perform as predicted and achieve the high etendue characteristic of conventional interference spectrometers (Harlander and Roesler 1990). The resolving power of the system in Fig. 2 is $R_0 = 4W\sigma\sin\theta(1 - \tan\theta'\cot\theta)$.

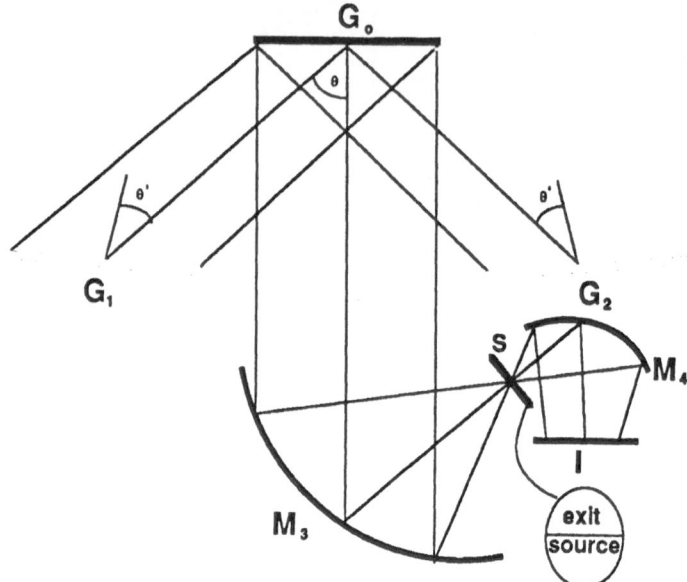

Fig. 2. Schematic diagram of the basic all-reflection SHS configuration. Light enters the system through the upper half of split aperture S and exits through the lower half.

3. Performance Prediction for a FUV Mapping Interferometer

Figure 3 shows a concept suitable for comparing predicted SHS performance with that of a conventional grating spectrometer optimized for mapping intensities, velocities and line profiles of far UV emission lines. It is analogous to the configuration in Fig. 2, but has the advantage of separated input and output. Although is has an extra grating reflection its efficiency is comparable to the instrument in Fig. 2, which requires a reduced entrance aperture.

We have not had the means to test the configuration of Fig. 3, but ray tracing techniques verified on the other configurations indicate that its performance can be predicted with reasonable confidence. The comparison is complicated because SHS, like conventional FTS, is a true multiplexing system, and thus achieves a multiplex advantage when significant detector dark current is present, or a multiplex disadvantage when photon noise dominates.

For the comparison we have chosen echelle grating configurations (Harlander and Roesler, 1990, briefly discuss the SHS with echelle gratings) covering a 400 Å range centered on 1500 Å. The conventional spectrometer uses a 30 cm × 60 cm 63° echelle grating and a $f/2.5$ camera while the SHS system uses 30 cm × 30 cm gratings symmetrically blazed at 30°, approximately. Both detectors are assumed to be 1000 × 1000 arrays of 20 μm pixels. We have used mirror efficiencies of 0.8, grating efficiencies of 0.4 and detector QE of 0.2. Lyman alpha has been excluded by a CaF_2 because of the large photon noise that it could introduce in the multiplexing SHS system especially in near-earth orbit. Full multiplex disadvantage is

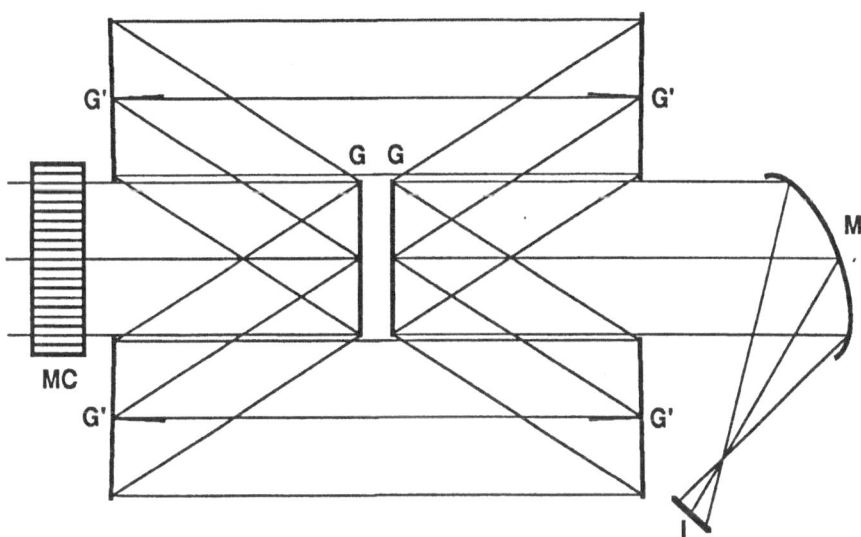

Fig. 3. SHS system used for comparison with an echelle spectrograph. Light enters the system through a mechanical collimator MC. Gratings G and G' have slightly different groove densities so as to produce the required dispersion while maintaining a relatively compact size.

assumed although non-negligible detector noise might be present in the required long exposure times. A single measurement reaching $SN = 10$ for a .025R line in a continuous background of 400 photon cm^{-2} s^{-1} sr^{-1} $Å^{-1}$ is estimated to take about 90 min with the SHS system and about 400 min with the grating system. Thus SHS has an advantage between 4 and 5.

A survey with 1° angular resolution and 20 kms^{-1} radial velocity resolution, providing 5–10σ detections of the observed CIV and OIII emission is estimated to take 5–6 years with a modest size ($\simeq$ 30 cm diameter optics) SHS instrument. By binning into larger angular resolution elements, fainter lines such as NIII λ1750, OIV and SiIV λ1400, and NIV λ1490 might be measured.

We are also studying field-widened methods for SHS systems. We have calculated gains ¡beyond¿ that of simple SHS of $\simeq$ 4 for an all-reflection system not yet optimized for minimum aberrations, and are confident that further refinement will push the predicted gains well over an order of magnitude. For a system using a thin transmitting wedge of MgF_2 or CaF_2, additional gains of more than two orders of magnitude appear practical.

We believe that SHS can have an important impact on planned and future NASA programs, and should be vigorously developed to allow timely decisions on its place in the mainstream of space observatory instrumentation for the next decade and the next century of space exploration.

References

Bergman, J.N.: 1980, *Ap.J.* **236**, 577

Cox, D.P. and B.W. Smith: 1974, *Ap. J. (Letters)* **189**, L105

Harlander, J. and F. L. Roesler: 1990, *SPIE* **1235**, in Press

Jenkins, E.B.: 1978, *Ap. J.* **219**, 845

Martin, C. and S. Bowger: 1990, *Ap. J.* **350**, 242

McKee, C.F. and J.P. Ostriker: 1977, *Ap. J.* **218**, 148

Spitzer, L., Jr.: 1956, *Ap. J.* **124**, 20

Williamson, F.O., W.T. Sanders, W.L. Kraushaar, D. McCammon, R. Borken, and A.N. Bunner: 1974, *Ap. J. (Letters)* **193**, L133

PROJECT OF A THREE REFLECTION TELESCOPE
FOR WIDE FIELD ULTRAVIOLET OBSERVATIONS

A. AMORETTI, M. BADIALI, A. PREITE-MARTINEZ and R. VIOTTI

Istituto Astrofisica Spaziale, CNR, Via Fermi, 21, 00044 Frascati RM, Italy

Abstract. We present a project of a three-reflection 1.5 m $f/3$ aplanatic wide-field telescope, in view of a small-satellite mission for the all-sky survey in the Ultraviolet.

1. Wide Field Astronomy

Of fundamental importance for our knowledge of the Universe is the exploration
of the full sky in the whole electromagnetic spectrum. Complete sky surveys have
been made in radio, infrared, optical and X-ray frequencies which have unveiled
many unexpected phenomena, and have disclosed new categories of astrophysical
objects, which have been later studied in more details with "pointed" experiments.
But, as far as the Ultraviolet is concerned, no complete sky survey has been so far
made, in spite of the important results obtained by the many UV space experiments
in the past two decades, which have clearly indicated that a very large amount of
astrophysical phenomena still remain to be discovered in the UV.

In order to perform the UV all-sky survey, a well corrected wide field telescope
is required. The investigation should be done in different wavelength bands, typically three, in order to determine the gross energy distribution of each source, and
the interstellar extinction. The FOV should be large enough (one to a few square
degrees) to complete the survey in a reasonable time. With square fields of $90' \times 90'$,
and with $10'$ overlapping of adjacent fields, the whole sky can be imaged with some
23000 pointings, which should require a 2–3 year mission. The experiment does not
ask for a very high angular resolution, because this would bring the information
to be stored, on-board analyzed, and transmitted to Earth to a level which can
hardly be handled with the current techniques. A reasonable value for the resolution should be 1 arcsec. The main requirements of the telescope should be a FOV
diameter of at least 2 deg, unvignetted, protected from straylights, with an image
size of less than $1''$ in the whole FOV. The linear scale should be around 10 to
30 micron/arcsec to be compatible with the currently used detectors. Clearly, very
large detectors with a large number of elements (pixels) are needed.

Other important requirements should be the flatness of the surface of better image definition, the absence of refractive media, and the compactness and structural
simplicity of the system. Ideal instruments capable to fullfill all these requirements
are the three reflection telescopes described by Amoretti et al. (1989). In view of
a small-satellite mission for the all-sky UV survey, we are considering telescopes of
this kind with 1 m $f/2$ to 1.5 m $f/3$ apertures. Here we describe the 1.5 m $f/3$
solution.

Y. Kondo (ed.), Observatories in Earth Orbit and Beyond, 487–491.

© 1990 *Kluwer Academic Publishers.*

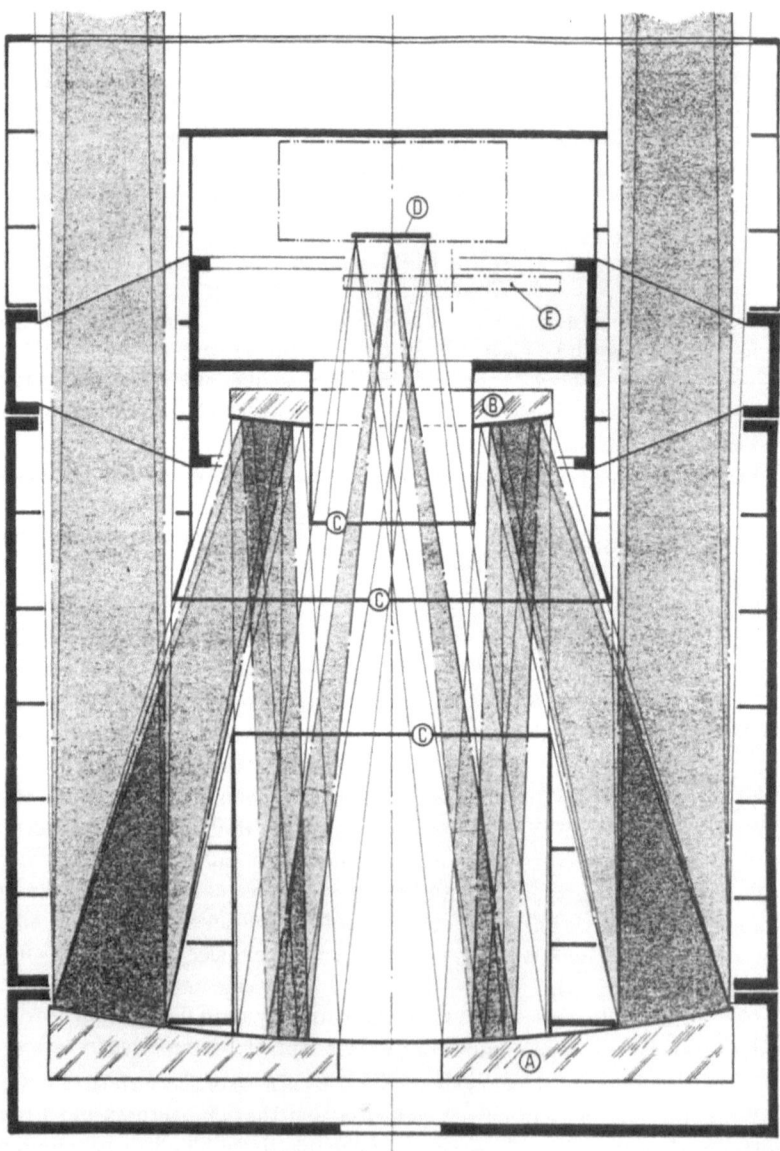

Fig. 1. The schematic layout of the TRUST telescope. A is the primary mirror which also acts as third reflecting surface. B is the secondary mirror. C are the baffles. D is the focal plane, and E is the filter wheel.

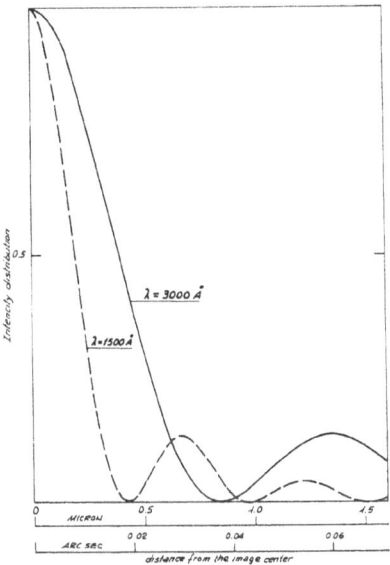

Fig. 2. The point spread function on the axis at the focal plane.

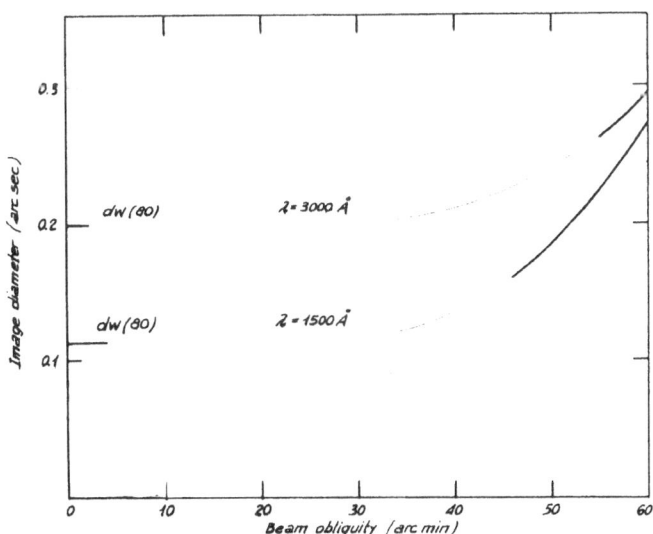

Fig. 3. The diameter of the circumference encircling 80% of the beam energy at 150 and 300 nm as a function of the beam obliquity.

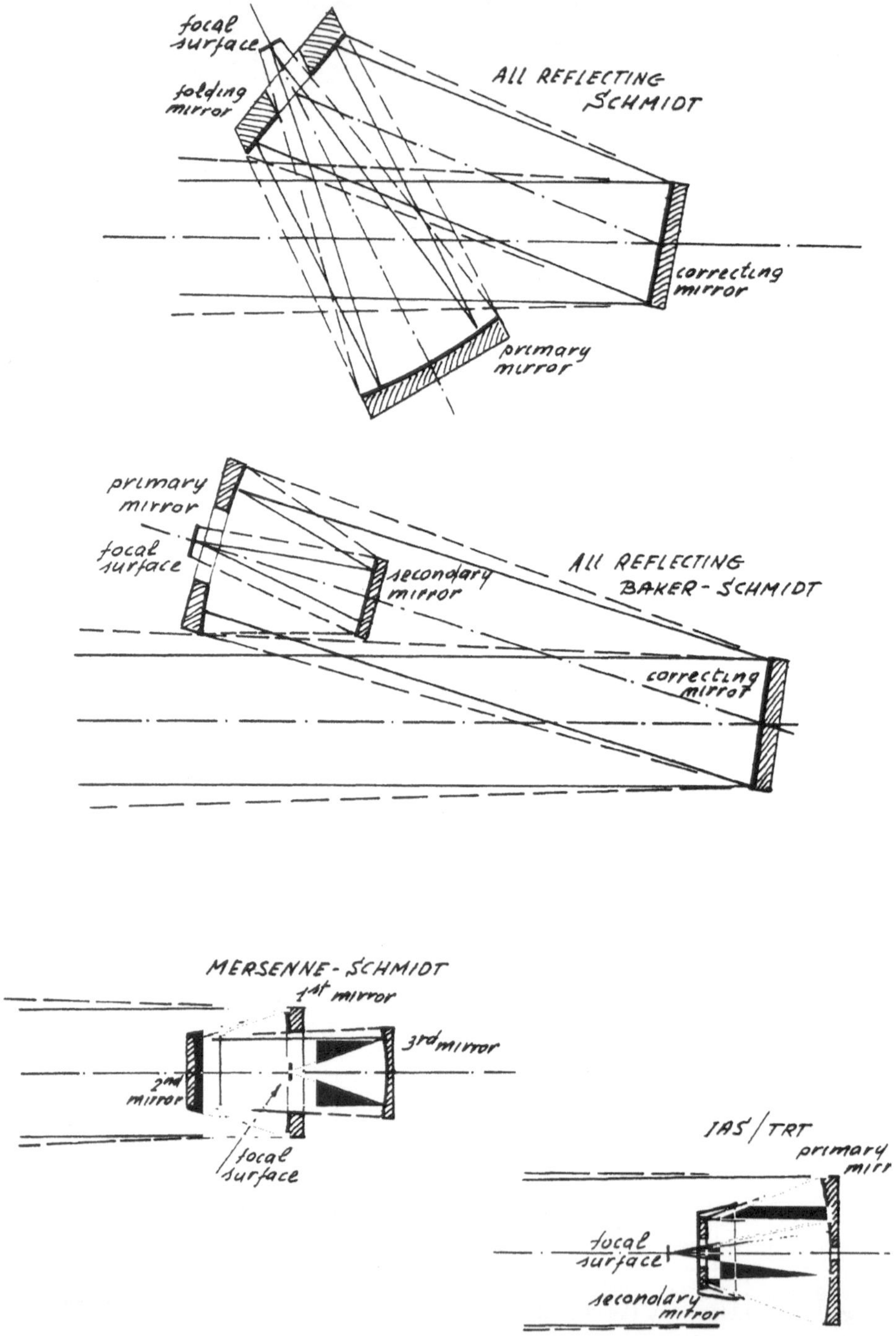

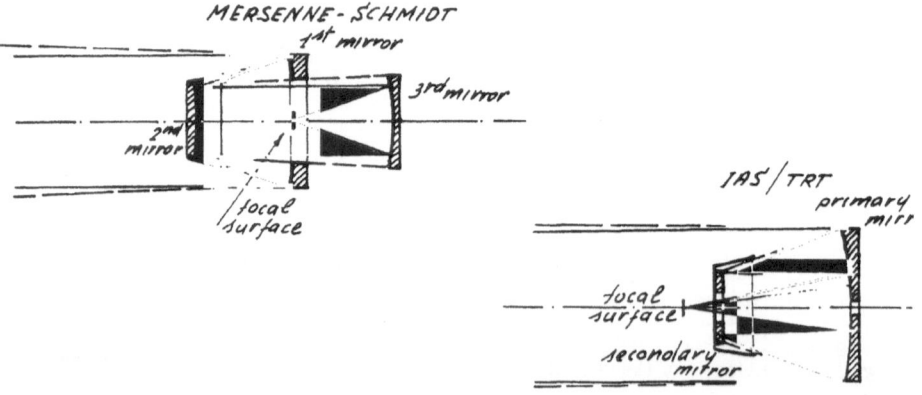

focal surface

folding mirror

ALL REFLECTING SCHMIDT

correcting mirror

primary mirror

primary mirror

focal surface

secondary mirror

ALL REFLECTING BAKER-SCHMIDT

correcting mirror

MERSENNE-SCHMIDT

1st mirror

3rd mirror

2nd mirror

focal surface

IAS/TRT

primary mirr

focal surface

secondary mirror

Fig. 4. Comparison of TRUST with other wide-field telescopes.

2. The Trust Project

The TRUST (Three Reflection Ultraviolet Space Telescope) project is based on a 1.5 m $f/3$ three reflection telescope which is shown in Figure 1. The main characteristics of the instrument is the $f/1.5$ primary mirror (A) which is acting both as first and as a third reflecting surface. This configuration greatly simplifies the mechanical structure of the telescope, ensures an easier optical alignement and a high compactness of the telescope (see Amoretti et al. 1989). There is also large room for instrumentation near the focal plane (D). High order spherical aberration and coma, as well as third order astigmatism, are corrected by a suitable choise of the figuring parameters, which results in a 2 deg-diameter corrected FOV. Figure 2 shows the on-axis PSF at 150 and 300 nm. The figure clearly shows that, in spite of the large obscuration of the secondary mirror (B), the resolution (FWHM of the central peak) is 0.02–0.04 arcsec. At 60' from the axis the resolution is still smaller than 0.2". Figure 3 gives the dependence of the image size (diameter of the circumference encircling 80% of the beam energy) on the beam obliquity.

In Figure 4 we finally compare TRUST with other proposed wide-field telescopes. The all-reflecting Schmidt and Baker-Schmidt telescopes (SWAT 1979) are much less compact than TRUST, and the resolution is much worser. Moreover, in the first telescope the focal plane is not flat. Both have a FOV of 4 deg, but the field cannot be explored with a good resolution. The Marsenne-Schmidt (Willstrop 1984) has about the same compactness and resolution of TRUST, and a larger FOV, but its configuration is not convenient since the first and third reflecting surfaces are separated, the position of the focal plane is not satisfying, and there is little room for instrumentation. The larger FOV is not a real advantage as it would be difficult to find a detector which would exploit the whole field at full resolution. In conclusion, the TRUST project, for its compactness (hence less weight and smaller on-board encumbrance), wide corrected and unvignetted FOV, high angular resolution, and flat focal surface, appears an ideal economy-class telescope for a space-borne all-sky survey experiment.

References

Amoretti, M., Badiali, M., and Preite-Martinez, A.: 1989, *Astron. Astrophys.* **211**, 250

SWAT: 1979, *Spacelab Wide Angle Telescope, Working Group Report*, NASA, Goddard Space Flight Center

Willstrop, R.V.: 1984, *Mon. Not. R. Astr. Soc.* **210**, 597

SANTA MARIA:

AN ORBITING MULTISPECTRAL OBSERVATORY

C. MORALES, L. SABAU and A. GIMENEZ

INTA, Madrid

A.L. BROADFOOT and B.R. SANDEL

LPL, University of Arizona, Tucson

R. STALIO

Department of Astronomy, University of Trieste

A. TALAVERA

VILSPA, Madrid

and

A. BUCCONI

CARSO, Trieste

1. Experiment concept

SANTA MARIA is a program of international scope, being a collaboration between scientific groups in Italy, Spain and the United States. The experiment consists of a cooperative space project with the objective of designing, developing, launching, data processing and scientific exploitation of an astronomical satellite under the name of SANTA MARIA.

The three research organizations will contribute to the project with tasks shared in the most practical way to guarantee the success of SANTA MARIA. The distribution of responsibilities is as follows:

– The spacecraft will be provided by Spain (financing, construction and integration).
– USA will provide the launcher under the Small Explorer Program.
– Italy will provide the launching and mission operations from the San Marco base in Kenya.
– The payload will be shared between the three countries.

SANTA MARIA is a small, low cost satellite which will perform imaging spectroscopy in the range from 200 to 7000 Å of astrophysical and solar system sources. It will also perform continuous measurements of the earth plasmasphere and the solar flux.

To achieve this task, SANTA MARIA will be equipped with several spectrographs operating simultaneously to provide coverage of the whole wavelength range. The detectors will be ICCD's. The scientific payload will include a particle analyser for plasmasphere studies and a solar spectrometer.

SANTA MARIA will be capable of obtaining simultaneous multiwavelength observations over long time scales of a large variety of astrophysical and planetary

Y. Kondo (ed.), Observatories in Earth Orbit and Beyond, 493–496.

objects. The understanding of many such objects is, in fact, often limited by a fragmentary picture of their behaviour in the different wavelength regions and by lack of extended temporal coverage. Much of modern astrophysics is now concerned with situations where significant variations are observed to occur over a wide range of time scales in many spectral regions.

2. Scientific instrumentation

The scientific instrumentation for SANTA MARIA is divided into three sections :

1) The primary experiment package consisting of a set of imaging spectrographs covering the spectral region from 200 Å to the visible. They are designed for simultaneous pointed observations of solar system and astrophysical objects. Table I gives the instrumental parameters.

2) The main spectrographs are complemented for earth atmospheric studies by an earth plasmasphere experiment.

3) The solar system observations are further complemented by a solar experiment.

Table I: Instrument Parameters					
Channel	Wavelength range (Å)	Resolution (Å)	Aperture (cm^2)	Sensitivity (*)	Configuration
EUV2	200–600	10	4×12	6	Grazing;ICCD; windowless intensifier
EUV1	550–850	5	4×12	7	Grazing;ICCD; windowless intensifier
FUV	800–1300	1-2	25×36	8–4	Telesc+Rowland grat.; ICCD; windowless int.
UV	1150–3200	5	10.5×15	0.4	Telesc+Rowland grat. ICCD; sealed int.
Vis	3000–7000	7.5	10.5×15	0.1	Telesc+Rowland grat.; ICCD; S20 cathode
(*) Point source continuum, S/N=10, in units 10^{-13} erg s^{-1} cm^{-2}Å^{-1}					

The largest fraction of the experimental payload is dedicated to the primary instrument. The earth plasmasphere and solar experiments are small instruments integrated into the spacecraft structure and will use only a small portion of the spacecraft resources. In the primary instrument, the spectrum between 200 and 7000 Å will be recorded simultaneously in 5 spectral channels with intensified charge coupled devices (ICCD's). In all these channels spatial resolution is maintained in the cross dispersion direction thereby providing a monochromatic imaging capability for targets with strong emission lines. An angular resolution of 4.5 arc seconds is foreseen. The largest aperture is that of the 800–1300 Å FUV spectrometer. It consists of a telescope and a Rowland grating spectrograph with two ICCD's which will provide resolution of 1 A/pixel. The apertures of the remaining channels are scaled to yield sensitivities to be compatible with the FUV channel.

The longer wavelength channels, UV (1150–3200), and visual (3000–7000 Å), will have the same optical configuration as the FUV channel, but only one ICCD will be needed to obtain the required resolution. The preliminary configuration of the two EUV channels EUV1 (400–860) and EUV2 (200–600) consists of small grazing incidence telescopes and spectrographs. Other configurations have been proposed by Naletto and Tondello (1988) and will be considered in a further study.

The 5 spectral channels can also function as broad band photometers simply by summing contiguous spectral elements on the detector, sensitivities in this mode are correspondingly higher.

SANTA MARIA will contain several small instruments to provide continuous spatial photometry of the Earth's plasmasphere and atmosphere which will be complemented by absolute measurements of the solar ionizing flux. Miniature optical imagers will measure the distribution of $He^+(304$ Å$)$, $He(584$ Å$)$, $O^+(834$ Å$)$, $H(1216$ Å$)$ and $O(1304$ Å$)$ from a swathe 30 degrees wide through the orbit. An Imaging spectrometer for Energetic Neutral Particles (ISENA) will detect and image the ENA particle flux from the same look directions as the optical imagers. The ENA particles originate from charge exchange reactions of neutral hydrogen with the ions of the plasmasphere. The ISENA will preserve information on the arrival direction of the particles and will be able to distinguish H, He, and O particles while measuring their energy.

The solar experiment consists of pinhole spectrographs (200–3200 Å) giving spectral half width of about 4 Å. An absolute measure of the solar ionizing flux will be made periodically using a gas ionization cell.

The very ambitious scientific goals of SANTA MARIA are possible due to the detector technology. The use of intensified charge coupled devices provides us with an imaging detector system that simplifies the mechanical design of the instrumentation and constrains the optical system to be small.

The instrumentation and technique for SANTA MARIA are now state-of-the-art, i.e. no instrument development is proposed. All elements of the various instruments and detectors will be or have been flight proven in presently and previous funded programs.

3. Scientific program

Much of modern solar system physics and astrophysics are now concerned with situations where significant variations are observed to occur in the UV and optical bands over a wide range of time scales. Obtaining a reasonably complete and coherent picture of how activity in different wavelength regions is interrelated is fundamental to our basic understanding of many type of objects. Such variable phenomena occur at all time scales ranging from very short, of the order of minutes, to very long, of the order of years.

Studying active phenomena frequently involves attempts to coordinate simultaneous observations involving various astronomy satellites and ground based-observers In spite of the large logistical problems which lead to discouragingly low success rates, the enormous scientific value of coordinated observations has produced an increasing amount of this type of observation. Several successful observing programs involving coordinated spacecraft and groundbased observations can attest to the value of simultaneous and long term coverage (Barstow et al. 1986 and Pringle et al. 1987).

SANTA MARIA instrumental sensitivities are determined by the primary goal of achieving time-resolved, multispectral observations of at least several key objects in a variety of important astrophysical classes. An important feature of SANTA MARIA will be the ability of observers to trade off increased signal-to-noise ratio on fainter targets against time or spectral resolution. As a dedicated multispectral observatory it is anticipated that SANTA MARIA will focus on those problems which, by their nature, will require a commitment of large segments of observing time.

SANTA MARIA provides an ideal complement to many of the actual and future facilities which will operate during the next decade, such as IUE, HST, ISO, etc., and the ground based telescopes of new generation.

SANTA MARIA will include the following research fields:

- Solar, Planetary and Terrestrial observations.
- Hot and cool Stars.
- Cataclysmic Variables.
- Active Galactic Nuclei.

References

Barstow, M.A., Holberg, J.B., Grauer, A.D., and Winget, D.E.: 1986, *Astrophys. J.* **306**, L25

Naletto, G., and Tondello, G.: 1988, *Internal Repport, Dipartim. Elettronica e Informatica*, Univ. di Padova

Pringle, J.E. *et al*: 1987, *Mon. Not. R. Astr. Soc.* **225**, 73

CRYOGENIC TESTING OF OPTICS FOR ISOCAM

JOHN K. DAVIES

Royal Observatory, Edinburgh, EH9 3HJ, Scotland

Abstract. The ROE ISOCAM optical test facility is described. The results of testing of lenses shows good agreement with theory and remarkable consistency.

1. Introduction

ISOCAM is an infrared camera operating in the 2.5 to 16 micron range which forms part of the payload of the ESA Infrared Space Observatory. The camera has two selectable channels, each with its own optics and detector system. Each channel includes 4 lenses, giving image scales from 1.5 to 12 arcseconds per pixel, interference filters and CVF sets. The lenses, which are of plano-aspheric form, were diamond machined from silicon (short wavelength channel) and germanium (long wavelength channel). After the application of anti-reflection coatings by OCLI and mounting of the lenses by ISOCAM prime contracter Aerospatiale, the individual lenses were tested by the Royal Observatory Edinburgh at 4 K to verify their performance. Filters were procured from Spectrogon via the Stockholm Observatory and, after mounting by Aerospatiale, tested by ROE to confirm that the required passbands and transmissions were achieved at 4 K.

2. The ROE Test Facility

The test facility is designed to provide an environment optically similar to that encountered in ISOCAM. The test cryostat is representative of one channel of the camera and lenses and filters are mounted in their holders, although not in the actual ISOC AM lens/filter wheels. All tests are carried out with the cryostat cooled to 4 K by a combination of a closed cycle cooler and liquid helium boiloff.

A schematic diagram of the test facility is shown in Figure 1, it consists of the following elements.

a. A blackbody source and chopper.

b. A monochromator using an in-line, off axis, Ebert configuration and interchangable gratings covering the range 2.5–16 microns. The exit of the monochromator is a 100 micron diameter pinhole.

c. A flat, gold coated, mirror tilted at 44 degrees, to reflect radiation from the pinhole to the f/15 scanning mirror.

d. An f/15 mirror, radius of curvature 1314 mm, tilted at 1 degree. This mimics the ISO telescope secondary mirror and can tipped independently in two axes to scan the image of the pinhole across the Fabry mirror and hence, via the lens under test, across the detector array.

e. The cryostat which includes an ISOCAM Fabry mirror, a filter wheel, a lens mechanism and a detector. The lens mechanism holds a single lens in its ISOCAM

Y. Kondo (ed.), Observatories in Earth Orbit and Beyond, 497–499.

INFRARED SPACE OBSERVATORY OPTICAL TEST FACILITY

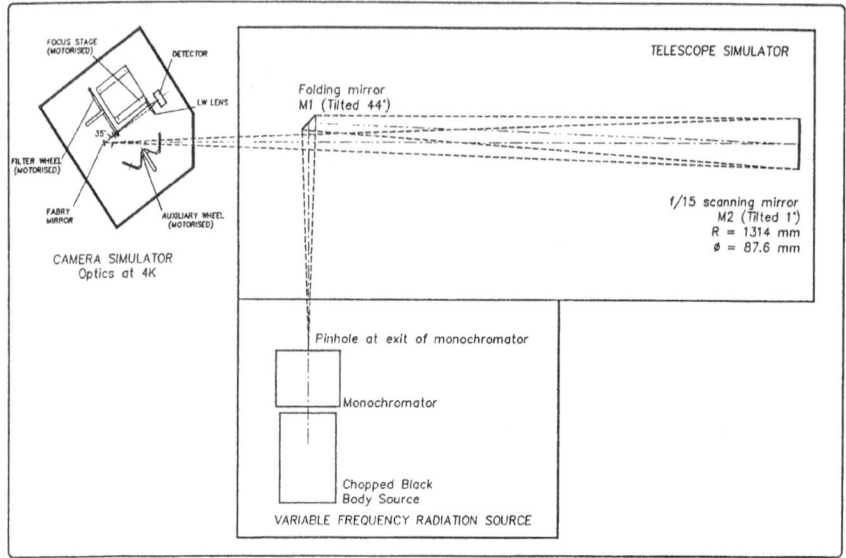

Fig. 1. Schematic Drawing of the ROE ISOCAM test facility.

mount and can be moved axially with a precision of 1 micron using a stepper motor located outsid e the cryostat.

In place of the detector array planned for ISOCAM, the test facility uses five individual SiGa elements mounted at the centre and edges of a simulated array. Each detector is fitted with a mask 60 microns in diameter.

The entire system is operated via an HP9816 mini computer which controls the stepper motors that move the lens, turn the filter wheel and scan the f/15 mirror. The HP9816 is also used to interrogate the detectors and store test data. An automated test se quence consisting of measurements over a range of different focus positions and filters can be run unattended overnight.

3. Test Procedures

By scanning the image of the monochromator exit pinhole across a detector and recording the signal at a number of mirror positions it is possible to reconstruct a beam profile, i.e. scanning mirror position versus signal received. This is a convolution of the object size, the point spread function of the lens and detector mask diameter. Since the size of the exit pinhole and the detector mask are known it is possible to extract the Point Spread Function (PSF) from the beam profile. By repeating the process at a number of different lens positions it is possible to determine the position of best focus for the lens, ie the Back Focal Length (BFL). The use of different detectors allows tests to be made both on and off axis. The derived BFL and PSF, at selected wavelengths, can then be compared with the theoretical values from the Code V package. The experimental results are used in the assembly

TABLE I

Comparison of ISOCAM 6 arc sec/pixel silicon lenses.

	QM	FM	FS	Theory
PSF (μm)	43 ±6	45 ±5	46 ±5	45 (at 3.5 μm)
PSF (μm)	59 ±6	57 ±5	62 ±5	60 (at 4.5 μm)
BFL (mm)	32.90 ±0.3	32.86 ±0.4	32.84 ±0.4	32.6 ±0.2 (at 3.5 μm)
BFL (mm)	32.98 ±0.35	32.93 ±0.35	32.91 ±0.35	32.8 ±0.2 (at 4.5 μm)

of the camera; correct positioning of the lens is critical since the camera has no focussing capability once assembled.

Filter testing is done by placing an appropriate filter in the beam and scanning the monocromator over a range of wavelengths which include the nominal filter passband. Comparison with a similar scan made without the filter in the beam enables the filter transmission to be found. Although this method is subject to complicating factors such as baseline drift between scans and the defocussing caused by inserting a filter into a converging optical beam it has proved satisfactory in practice.

4. Results

At this writing ROE have tested a total of eighteen ISOCAM lenses, comprising 8 for the Qualification Model, 4 for the Flight Model and 6 Flight Spares. It has been found that agreement between theory and experiment is good and that the BFL and PSF derived for nominally identical lenses, ie those of a given pixel size for the QM,FM and FS, are in very close agreement with each other. This is notable because for contractural reasons the QM optics were machined by a different company to the FM and FS optics. An example is given in Table I.

A complete set of ISOCAM interference filters (comprising 21 filters) has also been tested. The transmissions and passbands for the entire set were found to be satisfactory and in agreement with manufacturer's data taken at 77 K and extrapolated to 4 K.

5. Conclusions

The ROE test facility has been used successfully to demonstrate that optics procured for ISOCAM are within specification and to supply values of BFL required in the assembly of the camera. The test facility is also available to provide experimental data in support of testing required during integration of the QM and FM cameras.

EDISON:

A SECOND GENERATION INFRARED SPACE OBSERVATORY

H. A. THRONSON, JR.

Royal Observatory, Edinburgh and Wyoming Infrared Observatory,
University of Wyoming

T. G. HAWARDEN

Royal Observatory, Edinburgh and Joint Astronomy Centre, Hilo

and

C. M. MOUNTAIN, J. K. DAVIES, T. J. LEE, AND M. LONGAIR

Royal Observatory, Edinburgh

Abstract. EDISON is a large-aperture telescope under study for the second generation of infrared space observatories, whether in orbit or on the moon. The optics equilibrate via radiative cooling to temperatures between 40 and 80 K, depending upon, for example, telescope structure and location. At these temperatures, telescope emission is below that of the astronomical background at all wavelengths shortward of 20–40 μm. The detector components can be cooled via mechanical refrigerators now in an advanced stage of development. A mixture of radiative and mechanical cooling means that there is no natural limit to EDISON's lifetime. In addition, the upper stage rocket fairing can be almost filled with light-collecting optics and alternative low-emissivity optical designs, such as off-axis systems, can be easily engineered. We are presently evaluating a design for a 2.5 m observatory to be launched in collaboration with European astronomers as part of the NASA Explorer program. In this presentation, we describe possible spectroscopic and spectrophotometric studies of very faint infrared sources that will require large-aperture space telescopes working at the celestial background limit.

1. Program Summary

A class of **P**assively-cooled **O**rbiting **I**nfra**R**ed **O**bservatory **T**elescopes (POIROTs) are being studied for second-generation infrared space observatories, both in orbit and on the moon. These are long-lived, large-aperture facilities, using a mixture of radiative and mechanical cooling of optics and detector elements. In our current models, the telescope optics equilibrate at between 40 and 80 K, depending upon radiator design, orbit selection, role of mechanical cooler, and heat load from the electronics. Although POIROTs are much warmer than cryogenic telescopes, radiation from the telescope is less than that from the natural (astronomical) background at all wavelengths shortward of 20–40 μm. Unencumbered by a toroidal cryogen tank, an upper stage fairing can be nearly filled with the light-collecting optics of a radiatively-cooled observatory.

Our current proposal is for EDISON,[1] a 2.5 m POIROT-type telescope, designed to match the capacity of the presently-available large Atlas shroud. Freeing the

---

[1] Thomas A. Edison was probably the premier infrared astronomer of the 19th century. From a henhouse near Rawlins, Wyoming in July, 1878, he detected the near-infrared emission from both α Boo and the solar corona.

Y. Kondo (ed.), Observatories in Earth Orbit and Beyond, 501–505.
© 1990 *Kluwer Academic Publishers.*

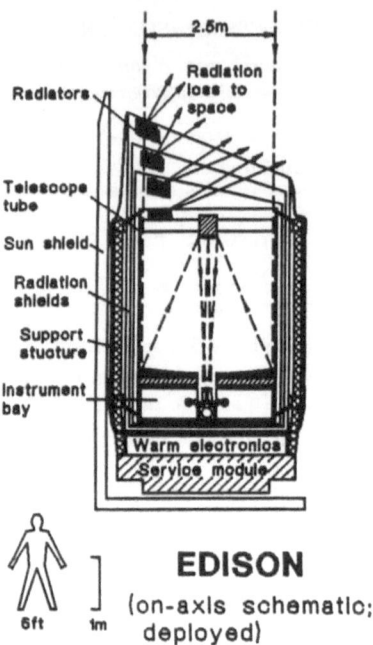

Fig. 1. Schematic design of EDISON as deployed in an on-axis configuration. Significant improvement in performance should be realized with an off-axis design.

telescope from cryogen tanks allows us to explore alternative designs, such as very-low-emissivity and off-axis optics, to further reduce contaminating emission while maintaining a large aperture. We are also studying the feasibility of even larger monolithic telescopes, such as a ~ 3.3 m EDISON launched by a Titan IV rocket.

As described in the following pages, a large aperture, in addition to a very long lifetime, will allow EDISON to, for example, [1] deeply survey for brown dwarfs, protogalaxies, protostars, and circumstellar matter; [2] study solid state and gaseous features in objects as diverse as Jovian planets and high-z galaxies; [3] image such objects as circumstellar disks and extragalactic circumnuclear material at arcsecond resolution or better in the near- and mid-infrared; and [4] easily search for nearby sub-stellar and planet-like objects, and undertake a program of classification of their atmospheres.

EDISON is being proposed as a joint American-European venture, launched as part of the NASA Explorer program.

2. EDISON Design

2.1. RADIATIVE COOLING

Unheated instruments in space naturally cool to temperatures well below that of the same structure on the surface of the earth. For example, the *IRAS* satellite telescope,

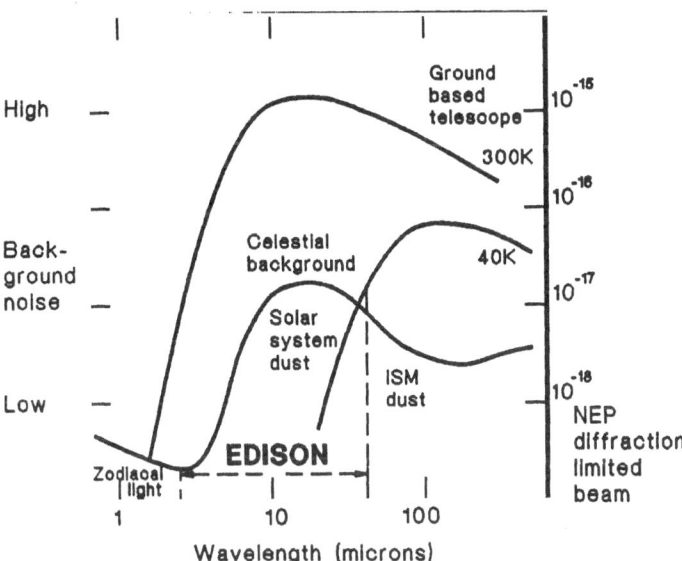

Fig. 2. The limiting sensitivity of an infrared telescope in space is determined in a large measure by the radiation on the detectors. Here we show the thermal emission from our 40 K EDISON model and a warm groundbased telescope, compared to the celestial background in space.

which was not optimized for radiative cooling, equilibrated at about 100 K after cryogen boiloff. Large-aperture infrared telescopes which are specifically designed to be cooled via radiation should, therefore, reach background-limited performance throughout the near- and much of the mid-infrared. A schematic drawing of one design for EDISON is shown in Figure 1. In this particular model, a deployable or inflatable sun shield provides significant shading and the set of three nested shields and radiators dominates the radiative cooling.

One preliminary thermal model for EDISON equilibrates at a telescope temperature of roughly 40 K. As shown in Figure 2, radiation from optics at this temperature is below that of the celestial background at wavelengths shortward of about 30 μm. As the figure demonstrates, in addition to atmospheric extinction and radiation throughout the infrared, emission from the telescope is a major limitation to observations of faint infrared sources from the earth's surface.

2.2. CLOSED-CYCLE REFRIGERATORS

Although some possible detector components operate well at the temperatures achievable by radiative cooling alone, long-lived closed-cycle coolers will be necessary to achieve maximum performance. Temperatures as low as $\sim$ 4 K are the

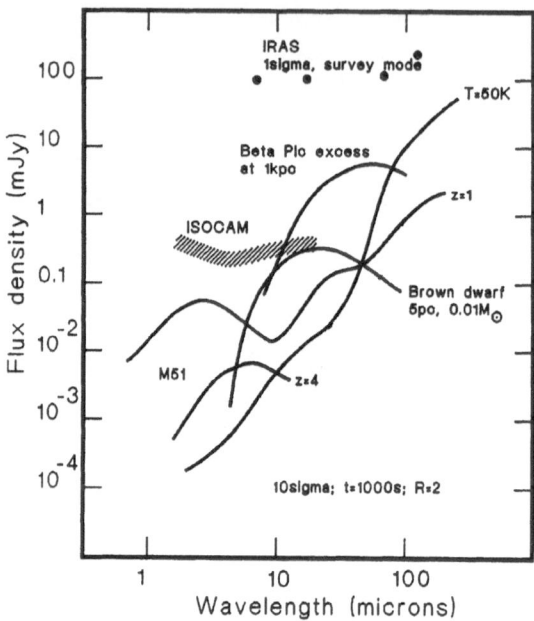

Fig. 3. High-sensitivity photometry with EDISON based on a telescope equilibrium temperature of 50 K.

goal of the three-stage Joule-Thomson/Stirling cycle refrigerators now in an advanced stage of development by Rutherford Appleton Laboratory under contract to the European Space Agency.

3. EDISON Scientific Capability

3.1. PHOTOMETRIC PROGRAMS

Figure 3 demonstrates the sensitivity of a proposed broadband camera using infrared arrays with detector capabilities expected for the early- to mid-1990's. Also shown are a variety of astronomically interesting objects, as well as the performance of *IRAS* in survey mode and that expected for ISOCAM. Note that nearly-normal galaxies such as M51 can be investigated at high z, allowing EDISON to study, for example, evolution of stellar populations using broadband photometry. The high angular resolution of EDISON will be necessary in the confusion-limited environment of galaxies at large redshift. This sub-arcsecond resolution at wavelengths shortward of about 10 μm will also be useful in studying "Vega"-like circumstellar material around stars at great distances within the Milky Way.

3.2. SPECTROPHOTOMETRIC AND SPECTROSCOPIC PROGRAMS

A major justification for infrared space observatories is their ability to obtain spectra of galaxies at high redshift. EDISON will be able to obtain spectra of infrared

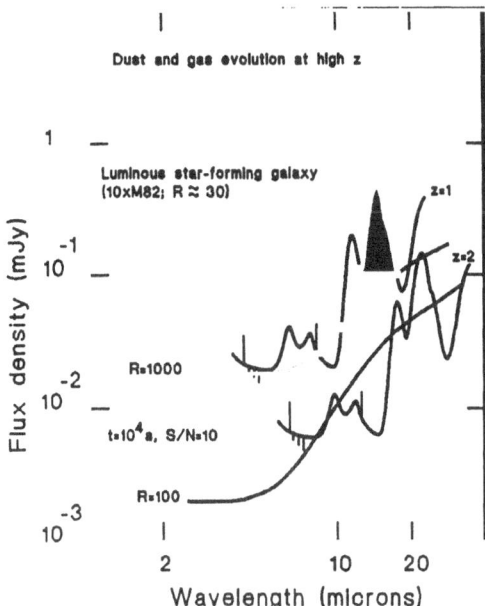

Fig. 4. Spectrophotometry of the near-infrared gas and dust features in high-z galaxies can be obtained by EDISON to investigate the evolution of the ISM. The model spectra show the near- and mid-infrared dust features for an object 10 times more luminous than M82. Also shown are the positions of the hydrogen Brα and Brγ lines, plus the first overtone CO absorption band.

gas and dust features after modest integration times, again taking advantage of its high angular resolution to lower the confusion limit for distant objects. One program of particular importance, as illustrated in Figure 4, will be study of the evolution of interstellar gas and dust features via intermediate-resolution spectra of very distant galaxies as a function of redshift.

A decade from now, many candidate brown dwarfs and other sub-stellar planet-like objects should have been identified and the essential task of classification will be carried on with high-sensitivity infrared spectroscopy. Our current proposal for a 2.5 m EDISON will be sufficiently sensitive to not only be used for study of the atmospheres of brown dwarfs, but hypothetical Jovian and sub-Jovian objects will also be investigated. As an example of performance, the earth could be clearly identified as a non-Jovian planet at a distance of about 5 pc after an integration time of only 2 days. The major limitation to a campaign of planetary classification using EDISON is not sensitivity, but the extreme difficulty of discrimination between the target and a neighboring star. On the other hand, it is reasonable to suppose that many sub-stellar, planet-like objects exist either at great distances from a central star or entirely isolated. Furthermore, EDISON should be considered as the prototype for one element of an interferometer or spatial array that would have the capability to discriminate between a star and surrounding planets.

MERITS OF SPACE VLBI MISSIONS FOR GEODYNAMICS

JOZSEF ADAM * and IVAN I. MUELLER

*Department of Geodetic Science and Surveying, Ohio State University,
Columbus, Ohio, U.S.A. 43210-1247*

Abstract. Two dedicated space VLBI projects are currently in preparation to launch one or more VLBI radio telescopes in orbit between 1993–1996. One in the Soviet Union called RADIOASTRON is already an approved and funded mission. The second one is a Japanese orbiting VLBI mission called VSOP. There is a Western European mission with NASA participation in Phase A Study at ESA called QUASAT. Although this project was not approved by ESA at the end of 1988, it might still be taken into consideration during the next decade. In the meantime in 1986 a successful demonstration of space VLBI was made at the first attempt using the NASA TDRSS satellite.

It is expected that space VLBI will be a reality in the next decade. Orbiting radio telescopes will be used to make interferometric observations of extragalactic radio sources in conjunction with the major ground-based VLBI arrays in Europe, USA, Australia, Japan and the USSR. It is planned to determine the orbits of these radio telescopes in space with high accuracy by the missions themselves and possibly using additional tracking systems (e.g., GPS, PRARE). The main goals of all space VLBI projects are to carry out astrophysical investigations. However, astrophysical space VLBI data may be used for geodesy and geodynamics as well. Therefore we started to investigate potential possibilities of space VLBI missions for these areas. The space VLBI observables may be useful to improve Earth's gravity field and in the unification and connection of reference frames inherent in the space VLBI technique. Our work is to explore the feasibility of these potential applications and to provide sufficient background for the inclusion of space VLBI observables in geodetic data processing programs (e.g., GEODYN). In this presentation, the mathematical models for space VLBI observables suitable for covariance analysis are described. A summary of estimable parameters is included. Singularity problems arising from coordinate system definition, observability conditions and critical configurations will be studied. Finally, the scope for space VLBI in geodesy and geodynamics is outlined.
Acknowledgement. This research is supported by NASA under Research Grant No.

NSG 5265.

* J. Adams is on leave from Satellite Geodetic Observatory, Hungary.

Y. Kondo (ed.), Observatories in Earth Orbit and Beyond, 507.
©1990 *Kluwer Academic Publishers*.

LOW FREQUENCY RADIO ASTRONOMY FROM EARTH ORBIT

KURT W. WEILER AND NAMIR E. KASSIM

NRL, Code 4030, Washington, DC 20375-5000

Abstract. Low frequency radio astronomy for the purpose of this discussion is defined as frequencies $\lesssim 100$ MHz. Since the technology is fairly simple at these frequencies and even Jansky's original observations were made at 20.5 MHz, there have been many years of research at these wavelengths. However, though radio astronomers have been working at low frequencies since the first days of science, the observing limitations and the move of much of the effort to ever shorter wavelengths has meant that most areas still remain to be fully exploited with modern techniques and instruments. In particular, the possibilities for pursuing the very lowest frequencies by interferometry of ground to space, in Earth orbit, or from the Moon promises a rebirth of work in this wavelength range.

We present concepts for space-ground VLBI and a fully space-based array in high Earth orbit to pursue the astrophysics which can only be probed at these frequencies. An Orbiting Low Frequency Radio Astronomy Satellite (OLFRAS) and a Low Frequency Space Array (LFSA) are two concepts which will open this last, poorly explored area of astronomy at relatively low cost and well within the limits of current technology.

Y. Kondo (ed.), Observatories in Earth Orbit and Beyond, 508.

HIGH-RESOLUTION IMAGING SPECTROSCOPY AT
TERAHERTZ FREQUENCIES

ROBERT L. BROWN, ANTHONY R. KERR,
A. RICHARD THOMPSON, FREDERIC R. SCHWAB

National Radio Astronomy Observatory, * *Charlottesville, VA*

Abstract. HISAT, a multi-element heterodyne interferometer attached to Space Station Freedom, will provide spectroscopic images with unprecedented detail of those submillimeter lines of C, O and C^+ which are critical diagnostics of UV excitation in the Galaxy. With the arcsecond angular resolution achievable from the space station, HISAT will reveal:

– The distribution of sources of ultraviolet radiation in the Galaxy;
– The effective temperature of the UV radiation as a function of galactocentric radius;
– The chemical and isotopic enhancement of atomic carbon and oxygen with galactic radius;
– The propagation of UV radiation in molecular clouds and its stimulative, or inhibitive, effect on star formation;
– The density structure, clumpiness or fragmentation, of molecular clouds throughout the Galaxy. HISAT has been selected by NASA for a concept-phase study.

1. Overview

The goal of this mission is to determine the origin and propagation of ultraviolet (UV) radiation in the Galaxy by studying the excitation of atomic carbon and oxygen at submillimeter wavelengths. Obscured by the Galactic disk, UV radiation from most of the Galaxy is invisible at the earth. However, we can infer its presence from its effect on interstellar material. We intend to obtain spectroscopic images at high angular resolution of those submillimeter lines of C, O and C^+ to which the Galactic disk is transparent and which are critical diagnostics of UV excitation.

Near the Sun, 8.5 kpc from the center of the Galaxy, the energy density of UV radiation, at wavelengths $91 < \lambda < 300$ nm, is nearly equal to that of visible light or infrared (IR) radiation. But the influence of the UV radiation on the interstellar medium is disproportionately large: the UV ionizes all elements with ionization potentials less than one Rydberg, and it heats the gas via collisions between the resultant photoelectrons and the atomic and ionic gas. Further, because the UV radiation has its origin in early-type stars, it serves to delineate regions of active star formation. We can study such regions in the solar neighborhood with UV detectors stationed above the atmosphere. However, the UV opacity of the Galactic disk is large, and therefore we cannot directly observe UV light that arises at distances greater than a few hundred parsecs from the Sun: we have little knowledge of the UV energy density, its origin, propagation and effect throughout most of the Galaxy.

* The National Radio Astronomy Observatory is operated by Associated Universities, Inc., under cooperative agreement with the National Science Foundation.

Y. Kondo (ed.), Observatories in Earth Orbit and Beyond, 509–515.

HISAT

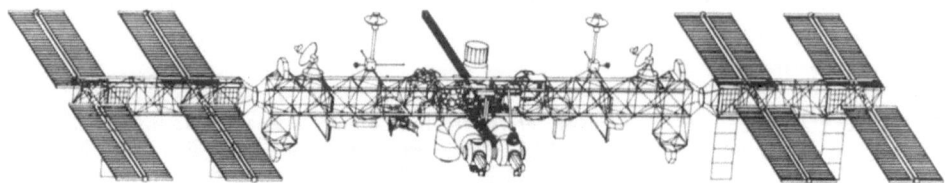

HIGH RESOLUTION IMAGING SPECTROSCOPY
AT TERAHERTZ FREQUENCIES

Fig. 1. HISAT, shown in its simplest form as a two-element interferometer, mounted on the upper truss of Space Station Freedom.

We intend to bridge these gaps in our understanding by use of spectroscopic observations from an imaging interferometer attached to Space Station Freedom. We will investigate secondary effects of the UV radiation in enough detail to permit us to infer the nature of the UV emission and its propagation throughout the Galaxy. The unique capabilities of the space station provide an ideal environment for operation of an interferometer that will provide images with arcsecond angular resolution of the crucial submillimeter lines of C, O and C^+.

Below we describe a Space Station Attached Payload that allows us to achieve this objective. The plan is evolutionary: as technology matures and is incorporated into the instrument the scientific yield increases. At all phases the instrument yields unique scientific insight complementary to that provided by the premier orbiting instruments, HST and SIRTF, with which it will be contemporary.

2. Galactic UV Radiation in Perspective

Understanding of the origin and propagation of UV radiation in the Galaxy must be based on observations of the effects of that radiation on its surroundings. The observed effects, in turn, provide us with insight into the extent of nucleosynthesis in the Galaxy, the galactic abundance gradient, interstellar chemistry, and the structure of molecular clouds which presage star formation.

2.1. The Distribution of Galactic UV Radiation

The single unmistakable manifestation of a stellar source of UV radiation is a region of atomic hydrogen ionized by the radiation $\lambda < 91$ nm of the hot star. From observations of their radio and IR emission such HII regions are found throughout the Galaxy, preferentially in the Galactic plane and preferentially interior to the solar circle.

Each of these HII regions is a copious source of UV emission longward of 91 nm, with energies of 10^{36} to 10^{38} ergs s^{-1}, which freely escapes the region of ionized hydrogen and is absorbed by gas and dust in its vicinity. We can study its effects and infer its properties by observing spectral lines of the most abundant elements, O, C and C^+. As noted below, these observations need to be made from a space platform.

2.2. TEMPERATURE OF THE UV SOURCES

The effective blackbody temperature of stellar sources of UV radiation can be computed accurately using numerical models of stellar atmospheres. The emergent stellar intensity, at all wavelengths, depends in these models on the chemical abundance adopted for the star. As the abundance of heavy elements (O, C, N, Ne, Si, S, Fe, Mg) increases relative to the solar values, the effective temperature of the exciting star decreases, as does the average energy of the nebular photoelectrons, and hence the heating rate in the HII region is lowered. The temperature of the HII region excited by such a star is lower than it would be if the excitation were by a similar star having solar abundances.

Present indications are that nebulae near the Galactic center are systematically cooler. Whether this temperature gradient represents an increase in heavy element abundance in the UV source that excites the HII region or an increase in the abundances in the nebular gas – or both – awaits the [CI], [OI] and [CII] observations noted below.

2.3. CHEMICAL AND ISOTOPIC ENRICHMENT IN INTERSTELLAR CLOUDS

Galactic abundance gradients play an important role in discriminating between nucleosynthesis models involving primary and secondary species. Primary elements, C and O, have H or He as their direct progenitors, whereas secondary species, such as ^{13}C and ^{17}O, are products of subsequent nucleosynthesis processes. The ratio of the abundances of any two primary elements, for example the ratio C/O, should thus be a constant, whereas the abundance of a secondary element should increase as the abundance of its primary progenitor, e.g., $^{13}C/C$. Thus these two ratios together provide a complete picture of the efficacy of nucleosynthesis as a function of position in the Galaxy. The space station interferometric images of [OI], [CI] and [^{13}CI] obtained in this mission will directly address these questions.

2.4. PROPAGATION OF ULTRAVIOLET RADIATION IN MOLECULAR CLOUDS AND ITS EFFECT ON STAR FORMATION

The stellar UV radiation longward of the Lyman limit, $\lambda > 91$ nm, which freely escapes the HII region and is absorbed by the gas and dust in molecular material found near early-type stars, provides an ionization rate high enough to stimulate cloud fragmentation and gravitational collapse of the fragments, and it will cat-

TABLE I
Spectral Lines to be Observed

	Species	Transition	Frequency (GHz)
Phase I	[CI]	$^3P_1 - {}^3P_0$	492.1612
		$^3P_2 - {}^3P_1$	809.3432
Phase II	[CII]	$^2P_{3/2} - {}^2P_{1/2}$	1900.54
	[OI]	$^3P_0 - {}^3P_1$	2060.06

alyze a rapid phase of massive star formation. This, too, is subject to observational verification once observations at high angular resolution are possible.

3. Concept of the Investigation

3.1. SPECTRAL LINES TO BE OBSERVED

The propagation of UV radiation through interstellar molecular clouds has as its principal effects the dissociation of H_2 and CO, the ionization of atomic carbon and the heating of dust grains. HISAT will image these products: the (forbidden) ground-state fine-structure lines of neutral atomic carbon, CI, in Phase I and the lines of ionized carbon, CII, and neutral oxygen, OI, in Phase II. The second phase of the investigation is paced by development of low-noise receivers for frequencies above 1 THz. (See Table I.)

Note that the isotopic spectral line of ^{13}C will appear in the same spectra as the [CI] and [CII] main lines and will also be imaged.

In Phase I, simultaneous measurement of both transitions of CI is important: the two lines together provide a complete characterization of the 3-level ground-state of atomic carbon and hence of the physical conditions in the region of excitation.

3.2. TARGETS TO BE OBSERVED

Observations will be made of a large sample of molecular clouds found near Galactic HII regions. Since a primary goal of the mission is to investigate Galactic abundance gradients and spatial inhomogeneity of Galactic chemical abundances, the targets are distributed throughout the Galactic disk.

3.3. ANGULAR RESOLUTION TO BE ACHIEVED

HISAT will image the fine-scale spatial structure, filamentation, clumping and granularity in molecular clouds. A gravitational condensation of stellar mass – a protostar – has a size of 0.1 pc. In the inner Galaxy, 8 kpc from the Sun, this corresponds to 2″. HISAT provides this resolution (see Table II).

TABLE II

Interferometer Resolution (arcseconds)

B (meters)	Species Transition Frequency			
	492 GHz	809 GHz	1900 GHz	2060 GHz
5	12.6	7.7	3.3	3.0
10	6.3	3.8	1.6	1.5
15	4.2	2.6	1.1	1.0
20	3.1	1.9	0.81	0.75
30	2.1	1.3	0.54	0.50
40	1.6	0.96	0.41	0.38
50	1.3	0.76	0.33	0.30

4. High-Fidelity Imaging on Space Station Freedom

The fidelity of an interferometric image is directly related to the completeness with which the interferometer samples the range of spatial frequencies represented by the parent aperture which the interferometer "synthesizes". Ground-based interferometers, such as the VLA, fill in the aperture-synthesis plane by tracking a source in hour-angle: rotation of the earth gradually changes the vector orientation and separation of each pair of antennas. A similar technique can be exploited by HISAT on the space station.

Of particular interest is that locus of points in the u–v plane which is traced out by the baseline vector as the orbital plane of the space station precesses in longitude. For the space station, the rate of the regression of the nodes is $\sim 7°$ of longitude per day. Over a time interval equal to the period of revolution of the nodes ($360/7 \approx 50$ days), the baseline vector $\mathbf{b}$ traces out a closed elliptical curve in the u–v plane. Its equation is $(u^2 + (v - |\mathbf{b}| \cos \delta \cos i)^2 / \sin^2 \delta = |\mathbf{b}|^2 \sin^2 i)$. An example of such an ellipse, for $\delta = 45°$ and $i \doteq 28°.5$, is shown in Figure 2, along with a similar ellipse corresponding to $-\mathbf{b}$.

Instead of a simple 2-element interferometer, one might consider a multi-element interferometer, comprising three or more elements (the more elements, the better the u–v coverage and the better the imaging fidelity and speed). For a 3-element, 3-baseline instrument the u–v coverage would consist of three pairs of ellipses. A potentially attractive option for a space station interferometer might be a 3-element instrument comprising two fixed elements and one movable element. The movable element would be re-positioned only infrequently – probably at time intervals no shorter than the internodal period of the orbit.

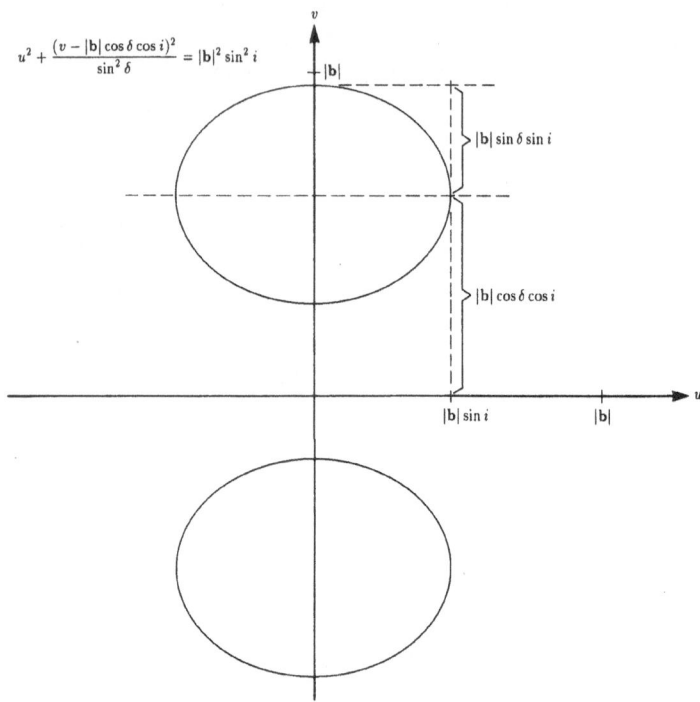

Fig. 2. The $u–v$ coverage provided by a single-baseline (i.e., a two-element) space station interferometer. In this example, $\delta = 45°$ and $i = 28°.5$. Over a time interval equal to the internodal period of the longitudinal precession of the satellite orbit the tip of the baseline vector traces out an ellipse in the $u–v$ plane.

5. HISAT Instrumentation

Technical Summary:

Antennas:	2 (or more), each of 2-meter diameter; surface accuracy $\leq 10\ \mu m$ r.m.s.;
Pointing:	3 arcseconds r.m.s.; servo star-tracker on each antenna;
Frequencies:	492 and 809 GHz (during Phase I), both frequencies observed simultaneously;
Receivers:	Low-noise, SIS, $T_{sys} < 1000$ K;
Correlator:	Analog/digital hybrid, with bandwidth ≥ 250 MHz;
Baselines:	5–30 meters, reconfigurable.

Technical Goals:

– The observations first require low-noise receivers which will allow us to achieve $T_{sys} < 1000$ K at 492 and 809 GHz. Our principal goal is to demonstrate the feasibility of such receivers at the Phase I frequencies;
– High-fidelity spectroscopic imaging will require data to be taken over the full internodal period of the space station orbit. The secondary goal of the concept

study is to define a data-taking technique, given the constraints of the space station orbit, that provides for images of the highest fidelity.

Technical Evolution:

− As the technology of low-noise heterodyne receivers matures at still higher frequencies, we plan to complement the Phase I receivers with Phase II instrumentation and extend the observational program.

− An imaging interferometric array is an instrument that could grow as the space station expands, allowing for longer antenna spacings to provide even higher-resolution images. The possibility might also exist of increasing the number of antennas, thereby enhancing the sensitivity and speed of imaging. Thus although the present study is scientifically complete as outlined here it also provides a necessary first step toward a possibly larger and more general-purpose imaging array. Such an array could be used for a wide range of remote sensing problems, including planetary and extragalactic astronomy.

SCIENCE OBSERVATIONS WITH THE IUE USING

THE ONE GYRO MODE

C. IMHOFF, R. PITTS, R. ARQUILLA, C. SHRADER, M. PEREZ

IUE Observatory, NASA Goddard Space Flight Center, Greenbelt, MD

and

J. WEBB

*Astronomy Programs, Computer Sciences Corporation, 10000 A Aerospace Road,
Lanham-Seabrook, MD 20706*

Abstract. The International Ultraviolet Explorer (IUE) is a geosynchronous orbiting
telescope launched by the National Aeronautics and Space Administration (NASA) on
January 26, 1978, and operated jointly by NASA and the European Space Agency. The
science instrument consists of two spectrographs which span the wavelength range of 1150
to 3200 Å and offer two dispersions with resolutions of 6 Å and 0.2 Å. The spacecraft's
attitude control system originally included an inertial reference package containing 6 gyroscopes for 3-axis stabilization. The science instrument includes a prime and redundant
Field Error Sensor (FES) camera for target aquisition and offset guiding. Since launch, 4
of the 6 gyroscopes have failed. The current attitude control system utilizes the remaining
2 gyros and a Fine Sun Sensor (FSS) for 3-axis stabilization. When the next gyro fails, a
new attitude control system will be uplinked which will rely on the remaining gyro and
the FSS for general 3-axis stabilzation. In addition to the FSS, the FES cameras will be
required to assist in maintaining fine attitude control during target aquisition. This has
required thoroughly determining the characteristics of the FES cameras and the spectrograph aperture plate as well as devising new target acquisition procedures. The results of
this work are presented.

1. Introduction

The International Ultraviolet Explorer (IUE) satellite was launched on January
26, 1978, by the National Aeronautics and Space Administration (NASA) with an
expected 3–5 year lifetime.

It is a collaborative project supported by NASA, the European Space Agency
(ESA), and the United Kingdom's Science and Engineering Council (SERC). The
satellite, located in synchronous orbit, is operated in real-time from the NASA
Goddard Space Flight Center in Greenbelt, Maryland, for 16 hours a day and from
the ESA Villafranca del Castillo station near Madrid, Spain, for 8 hours a day. The
science instrument consists of a 45-cm diameter f/15 Cassegrain telescope, prime
and redundant Fine Error Sensors (FES) for target acquisition and offset guiding,
and two echelle spectrographs with SEC Vidicon cameras. The spectrographs cover
the spectral range from 1150 to 3350 Å, with low (6 Å) and high (0.2 Å) resolution modes. The satellite continues to function well, recently passing its twelfth
anniversary on-station.

Y. Kondo (ed.), Observatories in Earth Orbit and Beyond, 517–520.

2. Attitude Control

The original attitude control mode employed gyros (three prime, three redundant) to sense motion in 3 axes. For slews and target acquisition, the On-Board Computer (OBC) used only the gyro information to maintain pointing by commanding 3 reaction wheels. During exposure integrations on target, the offset guiding information provided by an FES camera was also used in order to provide fine pointing control to a stability of better than 0.5 arcseconds.

By 1982, 2 of the 6 gyros had failed. A new attitude control mode was developed which employed information provided by the Fine Sun Sensor (FSS). The FSS is a solid state, two-dimensional device which can provide pitch and roll information with a resolution of ~ 15 arcseconds. Although the device had not originally been intended for use in attitude control, the interconnections of the IUE design allow the OBC to access and use the sensor information. Various combinations of gyro, FSS, and FES information are used to maintain control while slewing, for acquisition, and for offset guiding during exposures. A prominent change from the previous system was the use of FSS information to monitor roll axis motion. When the fourth gyro failure occurred on August 17, 1985, the two-gyro/FSS mode was implemented on the satellite. This mode has proved to be very effective, with essentially no degradation in the capabilities and efficiency of obtaining science observations.

In anticipation of another gyro failure, a one-gyro/FSS control mode has been developed. This mode makes more extensive use of the pitch information provided by the FSS. By using the remaining gyro to sense yaw motion and the FSS for pitch and roll, the mode fully maintains the current capability to slew around the sky with no noticable degradation. However the 15 arcsecond resolution of the FSS makes this mode unsuitable for the fine control required for target acquisition and exposure integrations. To provide fine pointing control, plans are to utilize the prime and redundant FES cameras. A new element of the one-gyro mode is the ability to bias the FES x and y position values. With the FES in track mode, the OBC normally uses the FES information to command the wheels so that a star maintains a constant position within the FES field of view to ~ 0.5 arcseconds. By biasing the position values, it will be possible to slowly move a star to any desired position in the FES field of view. This allows the fine control needed for target acquisition and exposures.

3. Operations with the One-Gyro/FSS Mode

Under the present two-gyro/FSS mode, an FES image taken in field camera mode is used to identify the target. The FES field of view is a 15.8 arcminute square image obtained from the reflection of the star field from the aperture plate. The plate contains the apertures of the spectrographs as well as various focus slots and other openings. Once located, the target is brought to a reference point in the FES field of view by performing a small fixed-rate slew. From the reference point, calibrated fixed-rate slews are used to place the target in any one of four spectrograph apertures.

The FES is then "primed" on a convenient field star for offset guiding and the exposure is begun. In the present mode, it is not a concern if the FES briefly loses star presence as the target crosses an aperture or focus slot during the small fixed-rate slews since they are being made under gyro control. It also does not matter if the exact relative positions of the guide star and target star are known prior to target acquisition since any convenient field star can serve as a guide star and the target is placed in an aperture under gyro control.

With the one-gyro/FSS control system, there will be three major differences. The telescope pointing under the remaining gyro and FSS control will not be steady enough for an FES image of the target area in field camera mode without significant blurring and degradation. To overcome this, an FES camera will first be used in a raster scan search pattern to locate any available field star. Once located, the FES will be commanded to hold the position of the star fixed within its field of view. With the telescope pointing now fixed to within ~0.5 arcseconds, the second FES can be used to obtain an image of the target area.

Secondly, since the target star will be moved about the FES field of view by biasing of the FES tracking position, the target cannot temporarily disappear in a focus slot or aperture without loss of fine attitude control. The locations and effective sizes of the aperture plate openings must be known to a high degree of accuracy and the small fixed-rate slews selected to move a target star within the field of view must avoid crossing any of these openings.

Finally, the target cannot be placed in the aperture by biasing of the target position since the FES will lose star presence at the edge of the camera aperture. Instead, the FES will have to be tracking on a selected guide star. The guide star will then be positioned such that the target is centered in the appropriate aperture. This requires not only that the relative positions of the guide and target stars be known to a high degree of accuracy, but that any geometric distortions of the FES cameras, which are magnetically focused image disectors, be taken into account.

Over the last several years extensive studies have been made by IUE Observatory staff members to measure both the locations and effective sizes of the aperture openings and the geometric distortions in the FES cameras. From initial testing of the two-gyro/FSS control mode, it was found that small fixed-rate slews done under gyro control within the FES's field of view were both repeatable and generally accurate to ~0.25 arcseconds.

Thus by performing very slow fixed-rate slews across aperture openings at several angles, both the locations and effective areas of the openings have been determined. Determination of FES geometric distortions required a more extensive series of measurements. Using the FES coordinate system, a grid of points was selected which covered the FES field of view. A target star was then placed at the FES reference point. Next a small fixed-rate slew was calculated and uplinked to the spacecraft to move the target star to a selected grid point. Upon completion of the slew, the FES was commanded to measure the target's position. The difference found by subtracting the predicted geometric position from the found position then constituted the measured geometric distortion for the grid point. As a by-product of these measurements, the surface brightness of the aperture plate with its effective boundary as seen by the FES cameras was also obtained.

The spacing of the measured grid points was varied as a function of the rate of change of the measured distortion. For lookup table purposes, a uniform grid was created with intermediate points being calculated graphically from the measured data to form a smoothly varying distortion pattern. The lookup tables have already been successfully utilized in software which calculates accurate guide star positions in the FES field of view based on information from previous observations.

4. Conclusions

Development of the calibrations of the FES cameras and of the one-gyro/FSS control system is essentially complete. Current efforts by the Observatory staff are to prepare software for the ground system computers which will be needed for the one-gyro/FSS control mode. Experience with the IUE spacecraft and the various attitude control modes being developed to extend its useful mission lifetime may be applicable to other future missions and help prevent premature ending of missions due to problems with gyros and spacecraft inertial reference systems.

Acknowledgements

This work is supported by NASA contract NAS 5-29375 with CSC. We gratefully acknowledge the significant contributions of the designers, engineers, and software specialists at GSFC, OAO, and Bendix who have developed the one- gyro/FSS attitude control mode.

NEW METHODS OF DETERMINING SPACECRAFT ATTITUDE

R. PITTS

Staff Member of the IUE Observatory/NASA and Astronomy Programs,
Computer Sciences Corporation, 10000A Aerospace Road, Lanham-Seabrook, MD 20706

T. JACKSON

Staff Member of the IUE Observatory/NASA and Bendix Field Engineering Corporation,
One Bendix Road, Columbia MD 21045

and

R. GILMOZZI

Former Staff Member of the IUE Observatory/ESA
and ESA/Space Telescope Science Institute,
3700 San Martin Drive, Baltimore, MD 21218

Abstract. The IUE spacecraft was launched with prime and redundant mechanical Panoramic Attitude Sensors (PAS) to determine coarse spacecraft pointing. Attitude determination typically took at least 24 hours. After launch both systems failed. A new method was developed which required pointing the spacecraft at the antisolar position. After the failure of the 4th IUE gyro, it was no longer possible to point in the antisolar direction. A second method was developed which utilizes IUE's ability to track the sun with a solid state two-dimensional sun sensor. Attitude determination can now be completed in several hours. An hour is required for coarse position measurement and several more hours are needed, using a small 15 arc minute square finder camera, for final attitude confirmation. These methods should be of use for other spacecraft where weight is critical or there is a desire to avoid mechanical devices.

1. Introduction

The International Ultraviolet Explorer (IUE) is a geosynchronous orbiting telescope launched by the National Aeronautics and Space Administration (NASA) on January 26, 1978, and operated jointly by NASA and the European Space Agency. A description of the spacecraft and science instrument is given by Boggess *et al.*. (1978a, b). The science instrument consists of two spectrographs which span the wavelength range of 1150 to 3200 Å and offer two dispersions with resolutions of 7 Å and 0.2 Å. A discussion of the science instrument characteristics and the processing of the telescope raw data is given by Turnrose *et al.* (1981, 1984). A description of the spacecraft's mechanical and electronic systems is found in the System Design Report (1976a, b). A general history of the spacecraft has been written by Boggess *et al.*. (1987) and a description of the Observatory operations given by Fälker *et al.*. (1987). During the twelve years that the spacecraft has been on station, 4 of the 6 gyroscopes, which were originally designed to provide the attitude reference required for pointing and slewing of the telescope, have failed. Since the failure of the 4th gyro, an attitude control system has been employed which utilizes a solid state, two-dimensional Fine Sun Sensor (FSS) to provide the third axis of stabilization.

Y. Kondo (ed.), Observatories in Earth Orbit and Beyond, 521–524.

A discussion of this system is given by Femiano (1986). A third control system, which requires a single gyro and uses the second dimension of the FSS to maintain three-axis stabilization, has been developed and will be employed upon the failure of the 5th gyro.

2. Spacecraft Attitude Control

The IUE spacecraft can move about any one of three independent axes. The terminology of attitude control has been adapted from aviation. If an airplane is on course in level flight, an up/down motion of the nose of the craft is called "pitch", a movement of the plane from side to side is called "yaw", and a turning about the axis of the fuselage is called "roll". Analogously, the IUE can be considered "on course", when the telescope is pointing directly away from the sun. Pitch is toward/away from the sun, yaw is a great circle perpendicular to the pitch great circle, and roll is motion about the telescope tube axis. The north celestial pole projected onto the spacecraft roll plane provides a reference point for roll. The system is local as it moves with the spacecraft, but since it is defined with w.r.t the sun and north celestial pole, transformations using Euler angles can be made from the more familiar coordinates of right ascension (RA) and declination (DEC) to that of the spacecraft's pitch, yaw, and roll. For discussions of spacecraft attitude determination and control, see Wertz (1987).

Until recently, earthbased telescopes were built with one axis of the telescope aligned with the Earth's axis of rotation. Such telescopes can be moved in the declination axis and set to a particular value. Motion about the polar axis allows for setting the RA value and tracking a star as the Earth rotates. With the IUE, the Beta angle (*i.e.* a measurement of the pitch angle) can be set. The telescope is then rotated simultaneously in yaw and roll so that the Beta angle remains constant until the pointing position of the target is reached (*i.e.* the spacecraft attitude). One complication is that the sun, as seen by telescope, changes postion as the IUE orbits the sun. Thus emphemeride calculations are part of every maneuver calculation for the spacecraft.

3. Attitude Determination

Infrequently, an error condition in the spacecraft onboard computer or the ground system will result in a loss of spacecraft attitude. To recover attitude, coarse pointing must first be determined (*i.e.* the current position to within several degrees of arc). Next a Fine Error Sensor (FES) with a field of view of 15.8 arc minutes square is used to determine exact pointing. At launch, a Panoramic Attitude Sensor (PAS) was used to determine coarse attitude. The technique relied on the positions of sun and Earth seen by the IUE to accomplish coarse attitude determination and required 24 hours to complete.

A second method of attitude determination was developed after launch which took much less time and which did not rely on the PAS, both of which have since failed. This method, conceptualized by A. Holm, and developed by A. Holm and F. Schiffer (Holm, 1990), makes use of the relationship between the local spacecraft axes and RA and DEC. If the spacecraft is pointed in the antisolar direction, its Beta and yaw angles are zero. The RA and DEC of the antisolar position can be accurately calculated for any given time. An FES image can then be taken and the field matched to a Palomar Sky Survey (POSS) chart. From the match, the current spacecraft roll at that position (*i.e.* w.r.t. the north celestial pole) can be estimated. Assuming this roll, the spacecraft is maneuvered to a target a few degrees away. After locating the target with a second FES image, the process is repeated to obtain a refined estimate of the spacecraft roll. As a final check, the telescope is maneuvered to a target at a Beta angle of 15 to 20 degrees of arc before resuming normal slewing. The process takes two to three hours. Since the initial attitude determination is made at a Beta angle of zero, it was named the *Betazero* method.

After adoption of the two-gyro/FSS control system, the *Betazero* method could no longer be used since the FSS heads lose sun presence below a Beta angle of 15 degrees. A new method was conceptualized by R. Pitts and C. Imhoff, and developed by R. Gilmozzi. The method utilizes the solar motion as seen by the IUE. The telescope is pointed toward a star in the current FES field of view. The spacecraft is then locked to the solar motion in pitch by using the FSS sensor. As the spacecraft tracks the sun, the star slowly moves across the FES field of view while the magnitude and sign of its motion in pitch are measured. Again using the relationship between pitch, yaw, and roll, and RA and DEC, two possible solutions of the position of the spacecraft are obtained. The sign of the roll, determined from other spacecraft telemetry, allows the selection of the correct solution. This provides the coarse attitude determination to within 1 to 3 degrees with all uncertainty being in the yaw axis. Several FES fields are then taken as the spacecraft position is changed by small amounts in yaw. These are then matched to the POSS charts of the area to complete the attitude recovery. The method requires several hours to perform. Since it is based on measuring the change in Beta angle of a target due to solar motion, it was named the *Betadot* method. A complete description is given by Gilmozzi *et al.*. (1990). Recently, T. Jackson has written an on-line computer program based on an equivalent mathematical derivation using conic sections which employs two computed measurements of the antisolar position rather than directly measuring the rate of change of solar position. Again, the sign of the roll angle is determined by other spacecraft telemetry. He has also modified the *Betadot* command sequence sent to the spacecraft which has improved the S/N of the telemetry measurements so that a full attitude recovery can now be completed in from one to two hours.

Acknowledgements

This work is supported by NASA contracts NAS 5-29375 and NAS 5-31000. We gratefully acknowledge the sigificant contributions of the designers, engineers, and software specialists at GSFC and OAO which maintain the high productivity of the aging IUE spacecraft which recently celebrated its twelfth anniversary on station.

References

Boggess, A., and Wilson, R.: 1987, in The History of IUE, ed(s)., *3*, Exploring the Universe with the IUE Satellite, ed. Y. Kondo, D. Reidel, Dordrecht

Boggess, A., Carr, F., Evans, D., et al..: 1978a, *Nature* **275**, 372–377

Boggess, A., Bohlin, R., Evans, D., et al.: 1978b, *Nature* **275**, 377–385

Fälker, J., Gordon, F., and Sandford, M.: 1987, in Operation of a Multi-Year, Multi-Agency Project, ed(s)., *21*, Exploring the Universe with the IUE Satellite, ed. Y. Kondo, D. Reidel, Dordrecht

Femiano, M.: 1986, in Inflight Redesign of the IUE Attitude Control System, ed(s)., *164*, AIAA Space Systems Technology Conference, American Institute of Aeronautics and Astronautics, AIAA-86-1193-CP

Gilmozzi, R., and Pitts, R.: 1990, submitted to AIAA Journal of Spacecraft and Rockets

Holm, A.: 1990, private communication

Wertz, J. (ed.): 1978, *Spacecraft Attitude Determination and Control*, D. Reidel, Dordrecht

System Design Report: 1976a, *System Design Report of IUE, Vol. I*, NASA/GSFC IUE-401-76-099

System Design Report: 1976b, *System Design Report of IUE, Vol. II*, NASA/GSFC IUE-401-76-099

Turnrose, B., and Thompson, R.: 1984, *International Ultraviolet Explorer Image Processing Information Manual, Version 2.0*, Computer Sciences Corporation, CSC/TM-84/6058 under NASA Contract NAS 5-27295

Turnrose, B., Harvel, C., and Stone, D.: 1981, *International Ultraviolet Explorer Image Processing Information Manual, Version 1.1*, Computer Sciences Corporation, CSC/TM-81/6268 under NASA Contract NAS 5-24350

KNOWLEDGE BASED AUTOMATED SCHEDULING
AND PLANNING TOOLS FOR IUE

CHRIS R. SHRADER

Astronomy Programs, Computer Sciences Corporation,
Code 684.9, NASA/Goddard Space Flight Center, Greenbelt, Maryland 20771

Abstract. The International Ultraviolet Explorer has been successfully operated as a real time user-interactive space observatory for twelve years. It is expected to continue operation for up to five additional years, but under increasing constraints. The option to operate IUE in a more automated, non user-interactive mode is under consideration. A sophisticated software system to support such an operation is a clear requirement. The conceptual framework of such a system is described. Results of a preliminary tests are presented for which a hypothetical four day schedule of space-craft activities at a time resolutions as low as ten minutes was generated.

1. Introduction

The International Ultraviolet Explorer (IUE) has been arguably the most successful of all satellite observatories to date. Part of its legacy of success is a result of the mode in which it has been operated: a real time, user-interactive observatory. Guest observers have been assigned telescope time in units of eight hour blocks, and are given full flexibility for planning their observing strategies within that time (a basic familiarity of IUE operations on the part of the reader is assumed; for a description see (4; 4)).

It is now recognized however, that in view of increasing pointing constraints, or in the event of a major S/C subsystem failure, that it might become necessary to operate IUE in more automated manner, incorporating an *integrated* mode of planning and scheduling. Observations would be preplanned, with a detailed timeline derived from all guest observer programs, and little realtime capability. Careful preplanning could in principle offset the decrease in efficiency imposed by stricter constraints and possible onboard hardware failures.

The complexity of the problem would clearly necessitate a high degree of automation in the form of an off-line software system. (see e.g. (4; 4)). Efforts within the IUE project have been made to develop such a system. This effort has benefited by drawing conceptually from several existing systems, most notably the powerful STScI Scheduling and Planning Interactive Knowledge Environment SPIKE (e.g. (4)). Other systems which have been studied are the JPL PLANIT (4) system which supports Voyager activity scheduling, also see (4). I will describe my conceptual approach, and later present some test results.

2. Conceptual Background

The similarity between the methods proposed here, and the HST SPIKE system is use of a "suitability function" (SF) framework, as the primary means of constraint

Y. Kondo (ed.), Observatories in Earth Orbit and Beyond, 525–530.

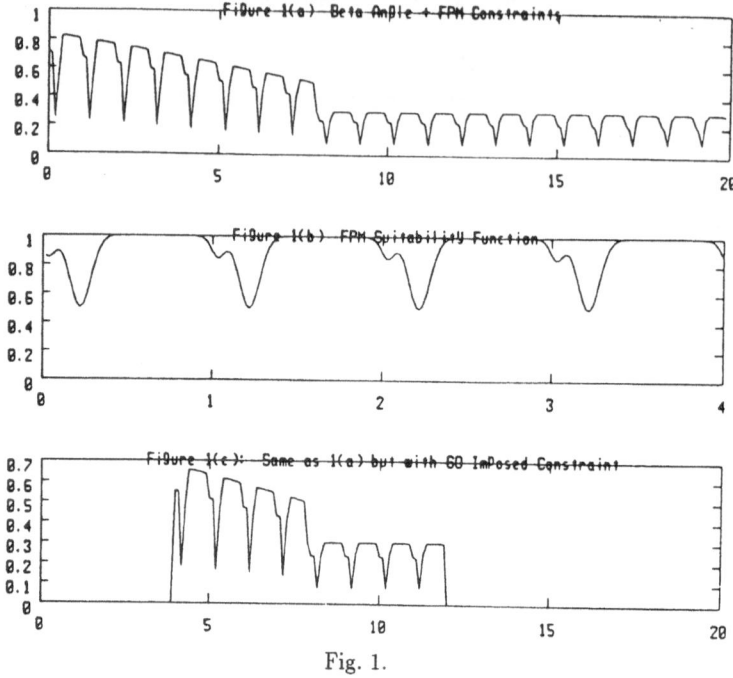

Fig. 1.

representation (4). This is an elegant and powerful technique, as should become evident. The SF for a given S/C event $S(t)$, has a probabilistic interpretation and is a function of time. The resolution of the time axis is determined by the minimum size, τ, of an activity cell to be considered. Values of $S(t)$ of order unity represent optimal scheduling situations and values tending towards zero represent undesirable or unallowed scheduling situations. Some examples are depicted in Figure 1; in Figure 1(a) only β-angles and charged particle (FPM) restrictions are considered. In Figure 1(b) the event is unconstrained with respect to β-angle; the FPM constraint is depicted at higher resolution than in (a); (c) is the same as (a) but with an additional observer imposed constraint, i.e. a window.

In addition to the SF methodology, a *Rule Based* mechanism is applied for interrelating the constraints of separate, but 'linked' events. This is where an object oriented programming language such as LISP is well suited (e.g. (4)), but out of necessity, development was done with existing resources and no LISP compiler was available. The code was written in C under VAX/VMS and run on a micro-VAX II with 5 Mb memory.

3. Detailed Description

It will now be necessary to treat consider individual **observations** rather than **observing programs**. A timeline can be represented as a series of events each consisting of a series of activities. For example, an FES image, target acquisition, followed by one or more exposures are an event in this context. Generally, one event occurs between each major spacecraft (S/C) slew. The IUE Resident Astronomers (RAs), as a part of the technical proposal review, will make entries into a knowledge base (KB) which will be used in assessing program scheduling priority and to drive a rule based inference engine for constraint handling. The RA will also estimate time and resource requirements.

The (subtle) difference between a knowledge base and a data base is that the former attempts to capture the educated intuitive response of the human expert, in this case in of a simple parameterization of constraints. These parameters are then combined algebraically to quantify the level of scheduling constraint for the event. This is then combined with a priority ranking based on scientific merit; the result determines the overall scheduling priority . The integrated SF over the balance of the scheduling period can also be folded into this calculation. If this is small, one should increase the priority, ie. this could be the last window of the scheduling period. The KB will also be used to treat linked event constraints, as will be described subsequently.

The SF for each event is computed and represented as a discrete vector. The vector dimension N, is the length of the scheduling interval divided by the size of an activity cell. M such vectors are used to construct an $N \times M$ matrix [S]. M is then the number of events to be considered for scheduling in the interval $N\tau$. Each column of [S] is the scheduling suitability vector for an event. Associated with each column are KB parameters as described above, including the time requirements of the event. The overall scheduling priority decreases from left to right. This is the order in which events will be scheduled, or at least be considered for scheduling. As is the case with a human scheduler, a schedule is typically derived from a pool of events larger than what can actually be scheduled. The following quantity is then optimized:

$$\chi(i_1, i_2) = \sum_{i_1}^{i_2} \left\{ S_{ij} - \sum_{k=j+1}^{M} \alpha_j S_{ik} \right\} \tag{1}$$

where α_j is a monotonically decreasing function of j of order $1/M$. The summation interval (i_1, i_2) is over the required time allocation for the j-th event; ie. $(i_2 - i_1)\tau$. The first term, $\sum S_{ij}$ is thus the integrated SF for event j. The second term, $\sum \sum \alpha S$ represents the net impact to all unscheduled events resulting from scheduling event j in (i_1, i_2). That α is monotonically decreasing reflects the hierarchal nature of the scheduling strategy; events of lower priority will contribute less to the net impact term than those of higher priority. χ is then optimized over all allowed intervals (i_1, i_2). An interval is allowed only if **every** suitability vector element on (i_1, i_2) is greater than a prespecified threshold. If no intervals are found to satisfy this condition, the event is flagged as unschedulable and the event of next highest priority is tried. After an event j is scheduled on (i_1, i_2), the elements S_{ik} in (i_1, i_2) and $k > j$ are set to zero (ie. no other event can be scheduled in this interval). The overall quality of the schedule is reflected in the parameter $E = \sum \chi(\Delta i)$ where the summation is over all scheduled events.

If all S/C activities were mutually independent, it would suffice to simply apply the above algorithm until all or most cells were occupied. However, the actual IUE scheduling problem is not so simple; events generally impact or are dependent on other events. For example:

1. Heavy saturation of a detector precludes long integrations on that detector for a subsequent period.

2. Certain types of events require phased or periodic scheduling. In such a case scheduling of one event predetermines the scheduling of subsequent ones. These are referred to as linked events.

3. It is undesirable to sequentially schedule observations of targets widely separated on the sky.

I will now describe my approach to handling linked-constraints of this nature.

Since S/C events are treated sequentially by decreasing priority, upon scheduling the j-th event, one needs to consider potential implications on only the subsequent (unscheduled) $M - j$ events, thus the size of the computational problem decreases with each step. Handling of linked-constraints is accomplished by applying a set of if-then type rules. A logical premise is evaluated on the basis of information in the KB. The consequence is a biasing of the appropriate suitability vector, e.g. consider the following piece of pseudo code:

Schedule event j in cells (i_1, i_2) **For** $k = j + 1$ to M

If {event j involves heavy saturation of the SWP}
And event k needs quiet SWP **then**
Bias S_{ik} on the interval $(i_2, i_2 di)$

The bias applied to S_{ik} is such that its elements now tend to zero for values of i in $(i_2, i_2 + di)$, where di is the length of time required for the SWP to recover. This is an example of an inhibitory bias. The quantity di can have a fixed canonical value or it can be weighted using the KB parameters. This type of inhibitory biasing can be applied for various types of linked-constraints; I have thus far considered: maneuver distances, heavy overexposures, thermal control, efficiency of instrument overhead and phased or periodic observations.

Spacecraft maneuvers are a special type of linked-constraint. Long maneuvers are undesireable, although it is unlikely that they can be eliminated entirely as thermal control and maneuvering efficiency tend to work against each other. Here, we represent each target in terms of its celestial, unit-position vector. Consider the following pseudo code:

Schedule { event j in the interval (i_1, i_2) } **For** { $k > j$ and $k \leq M$ }

If { R_j R_k ¡ threshold value }
Then { **Bias** { S_{ik} in $(i_1 - di, i_1)$ and in $(i_2, i_2 + di)$ } }

An inhibitory bias is applied for to activity cells preceding and succeeding event j. This discourages maneuvers longer than some prespcified threshold.

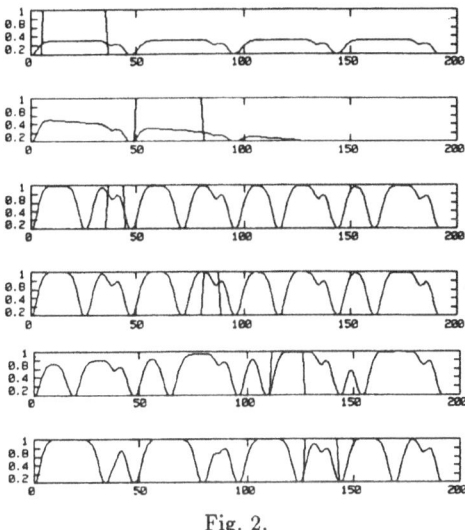

Fig. 2.

4. Test Results

I have conducted a zeroth order test employing the scheduling strategy described above, for which the following simplifications were made:

1. The smallest activity cell considered was 30 minutes.
2. Events were assumed to be mutually independent.
3. Programs were arbitrarily prioritized, except that the longest exposures were given highest priority.

Four days of hypothetical S/C events were scheduled. Events were constructed using target and exposure data from four days of actual S/C operations. The types of constraints considered were β-angles, earth occultation and FPM. Some results are depicted graphically in Figure 2. Vertical lines depict scheduled event boundaries superimposed on SF curves. Some notable features are:

1. Long exposures were set into the schedule with endpoints overlapping the wings of FPM troughs, but never centered about these troughs. Thus, FPM avoidance for long exposures was accomplished.
2. Centers of deep earth occultation troughs were avoided, e.g. Figure 2(c) and (d).
3. 95% of the activity cells were scheduled. The balance of unscheduled cells were distributed more or less uniformly over the scheduling period. Some of this could be filled by extending long integrations.
4. In cases of high scheduling priority, some situations of marginal suitability were scheduled. 18 out of the 20 highest priority events were scheduled.

A follow up, first order test, also for 4 days of S/C activities was made, this time using a more realistic 10 minute time granularity. In addition to a finer time resolution, several linked-constraints were applied. A series of periodic observations was scheduled successfully, and a heavy saturation of both primary instruments was fixed in the schedule. The SW successfully avoided following the latter by a long integration.

These initial results are encouraging and warrant further investigation.

References

Clavel, J.F. Requirements for IUE Integrated Scheduling 1989, in proc. IUE Three Agency Coordination Meeting, SERC

Eggemeyer, W.C., Grenander, S.U. 1988 proc. GSFC conf. on Space Applications of AI

Johnston, M.D., "Automated Telescope Scheduling" 1987 proc. Symposium on Coordination of Observational Projects

Johnston, M.D. and Miller, G. "Artificial Intelligence Approach to Astronomical Observation Scheduling" 1989 Proc. NASA conf. on Space Telerobotics

Johnston, M.D. Reasoning With Scheduling Constraints and Preferences 1989 SPIKE Technical Report 1989-2

McLean, D. and Yen W. Plan Specification Tools and Planning and Resources Reasoning 1989 proc. GSFC conf. on Space Applications of AI

Pitts, R., Jackson, T. and Gilmozzi R. These proceedings.

Shrader, C.R. A Proto-type Knowledge Based Integrated Scheduling and Planning System for IUE 1989 Proc. IUE Three Agency Coordination Meeting

EARTH OBSERVATION SYSTEM PLANS OF INDIA

M.G. CHANDRASHEKAR, V. JAYARAMN, C.B.S. DUTT, B. MANIKIAM

Indian Space Research Organisation (Headquarters), Bangalore, India

Abstract. Operational methodologies are available to retrieve several parameters related to the land, air and oceans from satellite data which is capable of providing well calibrated data/observations over large areas giving a synoptic view on a repetitive and reliable basis. The capability of satellites to provide data in various spectral, spatial and temporal scales is of great advantage in studying the dynamic aspects of earth atmosphere system. The present day capabilities of satellites include spatial resolutions ranging from 10 m and above and repetition of a few hours (geosynchronous Satellite) to few days. Higher spatial resolutions and all weather capabilities (through microwave sensing) are becoming available in the immediate future.Towards utilising the potentials of space based systems, India has been operating INSAT series of satellite for weather monitoring and IRS series of satellites for natural resources monitoring/management. The INSAT is a series of geostationary satellites stationed over Indian region to provide meteorological observations on a continuous basis in visible and thermal regions in addition to providing services for disaster warning related to Cyclones and remote location data collection platforms. The space based observations on meteorology over the past 5 years is proving to be a valuable data base for studies related to monsoon dynamics and tropical cyclones.

The IRS series of satellites are planned to provide Remote Sensing Observations to cater to the requirements of natural resources monitoring and management. The high resolution data obtained from IRS-IA presently is being complimented by data from contemporary satellites such as Landsat, SPOT and NOAA. A large number of operational packages have been developed to suit the user needs. A National Natural Resources Management System has been established in the country with thrust on use of remote sensing for management of resources in the country. In future, it is planned to have Remote Sensing satellite with enhanced capabilities by way of better resolutions, more spectral bands and stereo capabilities. Development of Microwave sensor on air-craft based platform is in progress towards reaching satellite payload capabilities in near future.

Y. Kondo (ed.), Observatories in Earth Orbit and Beyond, 531.
©1990 *Kluwer Academic Publishers.*

THE STABILITY OF THE PLANETARY TRIANGULAR LAGRANGE POINTS

SEPPO MIKKOLA

Turku University Observatory, Turku, Finland

and

K. A. INNANEN

Physics Department, York University, Toronto, Canada

Abstract. Numerical, self-consistent, n-body integrations of the solar system show significant indications of medium-term (i.e. several million-year) stability for the various planet-Sun L4,L5 configurations. A progress report of our computations, emphasizing the inner solar system, will be given. There exist interesting possibilities for these locations (including the Earth) as the sites for longer term scientific applications, both pure and applied.

1. Objective

The observed existence and extensive theoretical analysis of Jupiter's two Trojan asteroid families are our starting points. We ask the question of whether or not some degree of stability at their analogous triangular Lagrange points can exist for the other planets. [From a diagram, it would appear intuitively improbable for such stability to exist.]

2. Method

We use a self-consistent n-body computer simulation of the solar system by numerical integration of the equations of motion. In this simulation [emphasizing the situation for the terrestrial planets], we have included all of the planets from Mercury to Saturn. Massless test particles are introduced in the neighborhoods of the classical L4 or L5 points for each terrestrial planet. The system is integrated forward in time, following the time history for each particle in its semi-major axis, eccentricity, inclination and angular separation between particle and planet, as viewed from the Sun.

3. Results

For the terrestrial planets, during a 2 million year integration, stability clearly exists. Near the classical L4,L5 points, the test particles oscillate in "tadpole" regions typically of angular size 1.5 degrees. Beyond this, the orbits of the particles remain stable but are horseshoe-like. In Mercury's case, all longer period orbits are likely to be horseshoes. The results and trends for the first million years are shown in Figures 1–4.

Y. Kondo (ed.), Observatories in Earth Orbit and Beyond, 533–536.

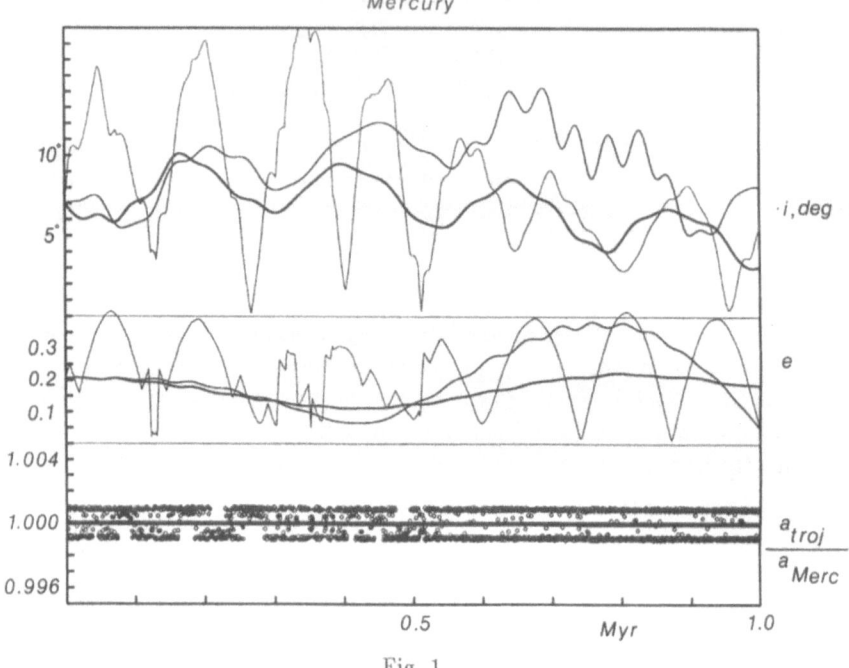

Fig. 1.

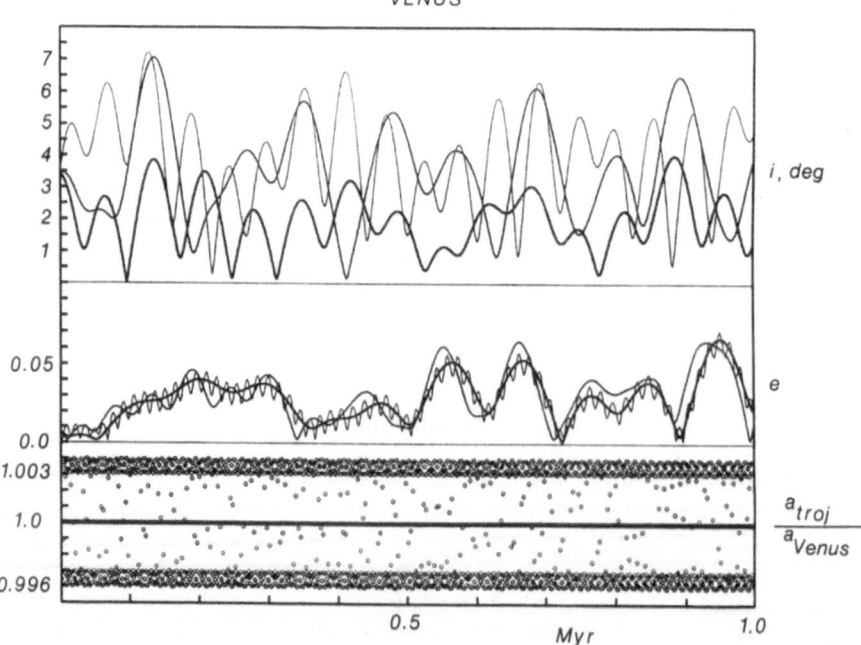

Fig. 2.

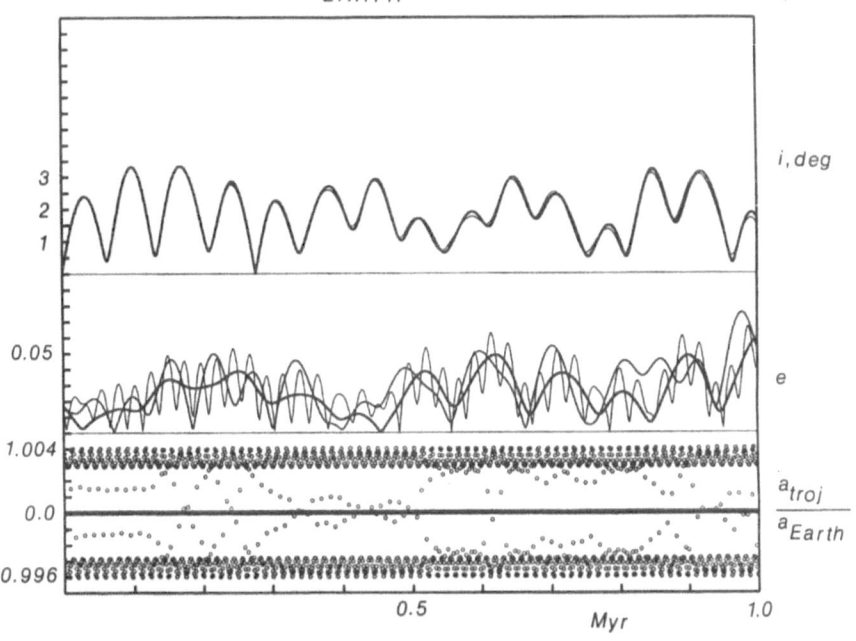

Fig. 3.

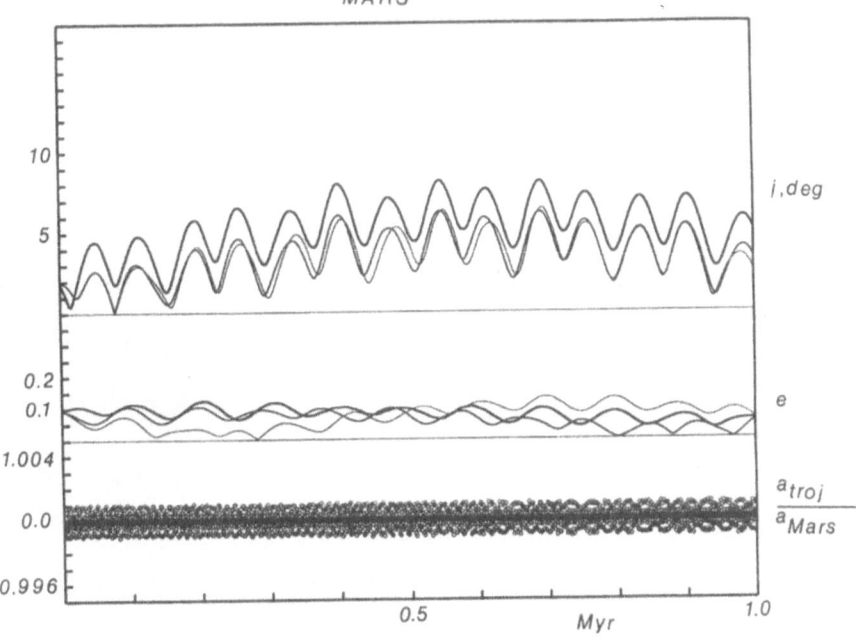

Fig. 4.

4. Conclusions

Contrary to intuition [see also the references] there is clear empirical evidence for the stability of L_4, L_5 points of all of the terrestrial planets in the several million year time frame. The subject is a fertile area for future investigation computationally, theoretically and observationally. Some of this is in progress. Interesting applications for space stations at these sites [especially the Earth] can readily be suggested, whether or not natural objects exist at them.

Note added in proof: We observe with great interest that in IAUC No. 5067 (July 28, 1990) there is the announcement of the discovery of object 1990 MB which appears to be a possible Mars "Trojan" near the Mars–Sun L5 point.

References

Innanen, K. A. and Mikkola, S.: 1989, *Astron. J.* **97**, 900
Mikkola, S. and Innanen, K. A.: 1990, *Astron. J.*, in press (July)
Weissman, P. R. and Wetherill, G. W.: 1974, *Astron. J.* **79**, 404
Zhang, S-P. and Innanen, K. A.: 1988, *Astron. J.* **96**, 1983
Zhang, S-P. and Innanen, K. A.: 1988, *Astron. J.* **96**, 1989
Zhang, S-P. and Innanen, K. A.: 1988, *Astron. J.* **96**, 1995

COSMIC RAYS AND THE DYNAMIC BALANCE IN THE LARGE MAGELLANIC CLOUD

CARL E. FICHTEL

Goddard Space Flight Center, Greenbelt, Maryland

MEHMET E. OZEL

Çukurova

and

ROBERT G. STONE

Goddard Space Flight Center, Greenbelt, Maryland

Abstract. Present and future measurement of the Large Magellanic Cloud (LMC) particularly in the radio and high energy gamma ray range offer the possibility of understanding the density and distribution of the cosmic rays in a galaxy other than our own and the role that they play in galactic dynamic balance. After a study of the consistency of the measurements and interpretation of the synchrotron radiation from our own galaxy, the cosmic ray distribution for the LMC is calculated under the assumption that the cosmic ray nucleon to electron ratio is the same and the relation to the magnetic fields are the same, although the implications of alternatives are discussed. It is seen that the cosmic ray density level appears to be similar to that in our own galaxy, but varying in position in a manner generally consistent with the concept of correlation with the matter on a broad scale.

1. Introduction

Through observations at many frequencies including the radio and the high energy gamma ray region, studies of the Large Magellanic Cloud (LMC) offer the possibilities of learning about the nature and role of cosmic rays in a galaxy other than our own. Although the greater distance is a significant handicap for some of the measurements, there is the clear advantage of being able to look at the galaxy from the outside rather than looking at the superposition of the radiation from several arms as in the case of our galaxy. The combination of the 21 cm and CO radio measurements and future medium and high energy gamma ray measurements provide information on the density and energy spectrum of the cosmic ray nucleons and at the same time a better normalization for the molecular hydrogen density estimates deduced from the CO measurements (Ozel and Fichtel, 1988). The radio continuum synchrotron observations provide information on the cosmic-ray electron spectrum as a function of position in the galaxy. The understanding of the latter is complicated by the lack of knowledge of the magnetic field strength and distribution and by some uncertainty in the interpretation of the synchrotron radiation in our own galaxy. There are, nonetheless, reasonable consistency and similarity arguments which can be used to generate a self consistent model of the cosmic rays, magnetic fields, and matter.

Y. Kondo (ed.), Observatories in Earth Orbit and Beyond, 537–541.

2. Cosmic Rays, Matter, Magnetic Fields, and Dynamic Balance

In order to attempt to understand the matters relevant to cosmic rays in the LMC, it is valuable to look first at the situation in our own galaxy and to remember that the bulk of the energy is in the cosmic ray nucleons (by a factor of about 10^2) and not the cosmic ray electrons, which produce the synchrotron radiation. In our galaxy, the local properties of the cosmic rays are well known (e.g., Simpson, 1983, Webber, 1983), but the nature of the solar modulation is not. The cosmic rays are depressed in intensity in an energy dependent manner in the vicinity of the Sun by a combination of effects including convection, diffusion, and energy changes due to the expansion of the solar wind. The exact nature of the energy dependence is not well understood, in spite of considerable theoretical effort (e.g., Parker, 1965; Gleason and Axford, 1968; Fisk, 1971). There does, however, seem to be agreement, based to a large degree on the observation of the changes in the cosmic ray intensity as a function over the solar cycle, that the degree of modulation decreases markedly with increasing energy and is quite small above about 10^2 GeV. This last deduction together with the galactic synchroton and gamma ray measurements was used by Fichtel, Stone, and Ozel (1990) to reconstruct an interstellar electron spectrum. The relationship between the electron spectrum in our galaxy and that in the LMC can be deduced from the synchrotron measurements. The solar modulation effect on the nucleonic component is less severe because of the much larger particle mass.

The matter density in the LMC, or at least the column density, is a much more straight-forward subject to discuss. The atomic hydrogen density may, of course, be obtained directly from the 21 centimeter data. The 21 centimeter map of the LMC given by Mathewson and Ford (1984) is used here for the reasons given in the paper by Ozel and Fichtel (1988). The heavier nuclei will be assumed to be present in the same relatively small abundances that they are in our galaxy. For the molecular hydrogen density, the CO map may be used to provide the distribution, although the normalization is uncertain. The recently completed full survey of the central 6 deg × 6 deg by Cohen et al. (1988) will be taken for this purpose. The normalization factor for the N_{H_2}/W_{CO} ratio of 1.7×10^{21} cm^{-2} K km s^{-1} will be used here (Cohen et al., 1988). As Ozel and Fichtel note, since the atomic and molecular hydrogen density distributions are quite different, future high energy gamma ray measurements may help to improvement the normalization value.

3. Synchroton Radiation

If the electrons have a spectrum of the form

$$N(E) = KE^{-\gamma}dE$$

where $N(E)$ is the number of electrons erg s^{-1} cm^{-3}, and where the electrons are homogenous and isotropic, then Ginzburg and Syrovatskii (1964) have shown that the intensity of the radiation is given by

$$I_\nu = 1.35 \times 10^{-22}a(\gamma)\text{LKH}^{(\gamma+1)/2}\left(\frac{6.26 \times 10^{18}}{\nu}\right)^{(\gamma-1/2)} \text{ergs cm}^{-2}\text{ s}^{-1}\text{ ster}^{-1}\text{ Hz}^{-1}$$

in the presence of random magnetic fields. In this expression, a is a slowly varying function of q with a value near 0.1 for the range of interest here, L is the length over which the electrons and magnetic fields are present, and H is the magnetic field strength. From equation (2), it is seen that, if the synchrotron spectrum is known, giving I as a function of over a reasonable frequency range, and L can be estimated, then K may be determined if H is known.

It is also important to know the relationship between the maximum in the synchrotron radiation and a given electron energy, which is (Webber, 1983).

$$E_{\text{eff}} = 2.5 \times 10^2 \left(\frac{\nu}{H_\perp} \right)^{1/2} \text{eV}$$

The frequency range of interest in the LMC study here is from about 20 to 1400 MHz., corresponding then to electrons in the energy range from approximately 0.50 to 4.2 GeV for a 5 μG field, or in fact a somewhat broader range when the distribution functions are considered, and 0.30 to 2.5 GeV for a 14 μG field.

It should be mentioned at this point, that historically there had been a concern regarding the synchrotron radiation observed in our galaxy. This was that the level appeared to be higher than would have been expected on the basis of the deduced electron spectrum and the magnetic field thought to exist. Recently, however, Fichtel, Ozel, and Stone (1990) have shown that with the interstellar cosmic ray electron spectrum now believed to exist based in part on the SAS-2 and COS-B gamma ray data at high latitudes, and the range of values for the total magnetic field including the random part now estimated from several sources, agreement can be obtained. These authors deduce a random magnetic field of about 11 μG for the local region, consistent with the currently estimated range of 5 to 14 μG, deduced in other ways.

4. The Large Magellanic Cloud

Fichtel, Ozel, and Stone further show that within the uncertainties of existing data the nonthermal radiation from the LMC and our galaxy have the same spectral shape. They then assume that the magnetic field pressure density in the LMC has the same relationship to the cosmic ray pressure as that in our galaxy as well as the relationship between the cosmic ray energy density and the electron spectrum being the same in the LMC as in our galaxy. Since the ratios of the cosmic ray electron intensity to the cosmic ray nucleon density and that to the magnetic field are assumed to be the same as in our galaxy, there is a fixed relation between K and H in Equation (2). Specifically, since the magnetic field pressure is proportional to H_2 if K_0 and H_0 are the local values of K and H in our galaxy and $w(x_i)K_0$ is a value in a local region of the LMC, then the corresponding H value in the LMC is $[w(x_i)]^{0.5}H_0$ to maintain the relationship between the cosmic-ray and magnetic fields described earlier.

If L is known, $w(x_i)$ may be determined from the knowledge of the synchrotron radiation. Following Klein et al. (1989), Fichtel, Ozel, and Stone used an effective disc thickness for L of 1 kpc, the same as in our galaxy. (Remember L for the LMC is the full thickness, not the half thickness as in our galaxy where the Sun is in the

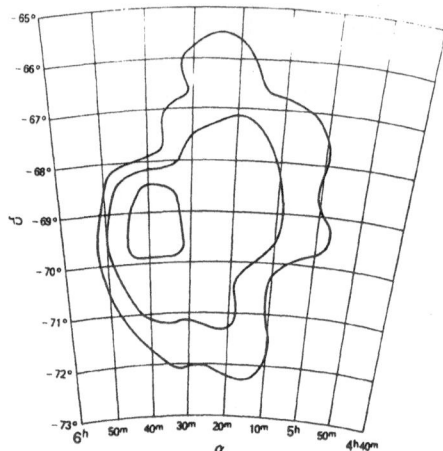

Fig. 1. The contour levels for the predicted cosmic ray energy density levels relative to those in our galaxy based on the 1400 MHz data of Haynes et al. (1986). The contour levels shown et al. (1986). The contour correspond to w equal to 0.59, 0.78, and 1.01. The contents of the figure are from Fichtel, Ozel, and Stone (1990).

middle of the plane.) They then proceed to calculate was a function of position for the LMC.

They studied three frequencies 45 MHz, 408 MHz, and 1.4 GHz. There are different considerations at each frequency and for each measurement including the degree of the thermal resolution, the beam size, and uncertainties.

For purposes of illustration, the results of their study at 1.4 GHz based on the measurements of Klein et al. (1989) are shown in Figure 1. It is seen that the cosmic ray density level on the average is similar to our own on the average, although a bit lower if the assumptions stated at the beginning of this section are valid. Notice also that a nonuniform cosmic ray level is predicted.

When this figure is compared to that of the matter density distribution and the cosmic ray density, Fichtel, Ozel, and Stone (1990) conclude the two are compatible within an uncertainty of about 1.5 and a scale of coupling between the cosmic rays and metter of about 1 kpc or greater. However, the scale of coupling is different on the two sides of the LMC.

5. Summary

The cosmic ray density in the LMC appears to be of the order of that in the local region of our own galaxy, but varying slowly with position. Further, a study of the synchrotron radiation and the matter column density in the LMC seems to indicate that it is possible to construct a consistent picture of the LMC based on the dynamic balance between the cosmic rays, the magnetic fields, and the kinetic motion of matter on the one side and gravitational attraction on the other and the additional feature that the magnetic field must be strong enough to contain the cosmic rays. A comparison of the contours related to the matter density and those related to the cosmic ray density predicted by the synchrotron radiation suggests that the scale of the coupling between the cosmic rays and the matter is probably of the order of a kiloparsec or larger since there is a smoothing required relative to the matter density controus to make the cosmic ray controus approximately consistent with those of the synchrotron radiation.

References

Cohen, R. S., T. M Dame, G. Garay, J. Montani, M. Rubio, and P. Thaddeus: 1988, *Astroph. J. (Letters)* **331**, L95

Fichtel, C. E., Ozel, M. E., and Stone, R., 1990, *Cosmic Rays and the Dynamic Balance in the Large Magellanic Cloud*, submitted for publication.

Fisk, L. A.: 1971, *J. Geophys. Res.* **76**, 221

Ginzburg, V. L. and Syrovatskii, S. I., 1964, *The Origin of Cosmic Rays*, New York: Macmillan.

Gleason, L. J. and Axford, W. I.: 1968, *Astrophys. J.* **154**, 1011

Haynes, R. F., Klein, U., Wielebinski, R., Murray, J. D.: 1986, *Astron. Astrophys.* **159**, 22

Klein, U., Wielebinski, R., Haynes, R. F. Malin, D. F.: 1989, *Astron. Astrophys.* **211**, 280

Mathewson, D. J. and Ford, V. L., 1984, in *Structure and Evolution of the Magellanic Clouds*, eds.: S. van den Berg and K. S. de Boer, Dordrecht:Reidel, 125.

Ozel, M. E. and Fichtel, C. E.: 1988, *Astroph. J.* **335**, 135

Parker, E. N.: 1965, *Planet Space Sci.* **13**, 9

Simpson, J. A.: 1983, *Ann. Rev. Nucl. Part. Sci.* **33**, 323–381

Webber, W. R., 1983, in *Composition and Origin of Cosmic Rays*, ed.: M. M. Shapiro, Dordrecht: Reidel, 83–100.

LIMITATIONS OF OBSERVATIONAL COSMOLOGY

MENAS KAFATOS

Department of Physics, George Mason University, Fairfax, VA 22030

Abstract. Unlike the usual situation with theoretical physics which is testable in the laboratory, in cosmological theories of the universe one faces the following problems: The observer is part of the system, the universe, and this system cannot be altered to test physical theory. Even though one can in principle consider any part of the observable universe as separate from the acts of observation, the very hypothesis of big bang implies that in the distant past, space-time regions containing current observers were part of the same system. One, therefore, faces a situation where the observer has to be considered as inherently a part of the entire system. The existence of horizons of knowledge in any cosmological view of the universe is then tantamount to inherent observational limits imposed by acts of observation and theory itself. For example, in the big bang cosmology the universe becomes opaque to radiation early on, and the images of extended distant galaxies merge for redshifts, z, of the order of a few. Moreover, in order to measure the distance of a remote galaxy to test any cosmological theory, one has to disperse its light to form a spectrum which would cause confusion with other background galaxies. Since the early universe should be described in quantum terms, it follows that the same problems regarding quantum reality and the role of the observer apply to the universe as a whole. One of the most fundamental properties of quantum theory, non-locality, may then apply equally well to the universe. Some of the problems facing big bang cosmology, like the horizon and flatness problems, may not then be preconditions on theoretical models but may instead be the manifestations of the quantum nature of the universe.

1. Introduction

With the theoretical framework of general relativity in place (Kafatos, 1989), Lemaitre and Friedmann postulated in the early 20's a dynamic, expanding and evolving universe. The Friedmann models obeyed the cosmological principle, which states that the universe is isotropic – the same in all directions – and homogeneous – of equal density, on the average, everywhere. On the other hand, to accommodate the obvious observational picture of Hubbles' expanding universe with a framework of an eternal universe, Herman Bondi, Thomas Gold and Fred Hoyle proposed in the late 40's the steady state theory, describing a universe which although expanding would obey the perfect cosmological principle: the universe appears the same to all observers at all times.

In the early 50's George Gamow extended Lemaitre's and Friedmann's original ideas. Nuclear physics had progressed as a branch of physics and cosmologists could now use it to answer what might have happened in the early lifetime of the universe when it achieved energies and temperatures appropriate to nuclear physics. Soon observational astronomy provided strong evidence in favor of the big bang model, the existence of the 3 black body radiation and the existence of quarsars. The 3 K black body radiation consists of a microwave background of radiation that fills all

Y. Kondo (ed.), Observatories in Earth Orbit and Beyond, 543–550.
© 1990 *Kluwer Academic Publishers.*

space. In the big bang picture, this microwave background is the relic radiation from the initial big bang. It was predicted by Gamow, more than 10 years before it was accidentally discovered by Bell Telephone physicists Penzias and Wilson. The steady state theory cannot easily account for the nature of the background. The microwave radiation, it was believed, is a natural consequence of the big bang theory and no extra assumptions are needed. The second cosmological observation (Berry, 1976) that challenges steady state is the existence of the quasars. Discovered accidentally in the early sixties these objects appear to be very distant. In relativistic cosmology, as the universe expands, galaxies recede from each other and light from a distant galaxy would then be redshifted more to the red part of the spectrum than a nearby galaxy. Quasars have their spectral lines shifted so much to lower frequencies that must be at the edges of the observable universe, some of them receding away from us at speeds exceeding 90% the speed of light. According to big bang theory, at the billions of light years that a quasar is located, the galaxy is so faint that it cannot be observed. Only the brilliant star-like nucleus can be seen. These bright nuclei were very brilliant in the past compared to now, indicating that sources evolved as time went on. If this interpretation is correct, quasars violate the perfect cosmological principle because the universe does not look the same at all times – galaxies were much brighter in the past than today. Recently, though, it has become obvious that the big bang theory itself faces theoretical challenges not known in the early sixties. Yet, the vast majority of astronomers, cosmologists, and particle physicists still adhere to the big bang theory.

In the first phase after the big bang singularity the space-time description breaks down entirely, and this is followed by the so-called "inflationary era" at about 10^{-35} sec (Guth and Steinhardt, 1984). By the time the universe had undergone inflation, it had expanded in size by a staggering factor of 10^{50} or more. When the expanding universe had cooled down to a temperature of about 10^{27} K, the universe underwent a phase transition from the false vacuum where all the Higgs fields were zero to a less energetic phase which is the true vacuum of quantum theory (Barrow, 1988).

The inflationary model was originally proposed not because of a compelling theoretical reason but to solve some observational problems faced by the standard big bang theory of the universe. For example, the horizon of the universe, within which parts of the expanding primordial matter were in contact among themselves, expanded to a huge size becoming much larger than the rradius of the observable universe. This fact is critical to resolve the observational problems of the standard big bang without inflation. Unfortunately the details of the inflationary scenario are not uniquely determined today and can only presumably become known once the quantum field theory of unification of strong, electromagnetic and weak interactions – the GUT – is firmly in place.

2. Observational Constraints of Cosmological Theory

The original big bang theory of the universe could not account for a number of features revealed by observational cosmology. One is the observed curious property of the universe that the ratio of the observed density of matter in the universe is very close to the closure value i.e. $\Omega \simeq 1$ where Ω is the density ratio $\Omega = \rho/\rho_{crit}$ and (Barrow, 1984)

$$\rho_{crit} = 5 \times 10^{-30}(H_0/50\text{km/sec/Mpc})^2 \text{ gr cm}^{-3}$$

where H_0 is the Hubble constant. Were Ω turned out to be precisely equal to unity the geometry of the universe would be exactly flat and it would expand forever.

Current observations cannot unequivocally distinguish the type of the universe we live in. Values of for luminous matter are in the range 0.1 to 2 although most observers favor values close to 0.1. If the only type of matter that there is in the universe is luminous matter found in stars and nebulae, this result would indicate an open universe. Even though present observations only indicate an approximate range of the mean density of the universe, this range is so close to the value of the critical density required for a flat geometry that many astronomers assume as a working hypothesis that the universe is exactly flat. If it turns out that the universe is not quite flat today, going back to the early times the universe must have been incredibly flat (Schramm, 1983) to one part in 10^{50}. This coincidence is known as the flatness problem and indicates an incredible fine tuning in the initial conditions of the standard big bang universe.

The second problem facing big bang cosmology has to do with the uniformity of the 3 K black body radiation. Recent COBE observations indicate a remarkable uniformity of this radiation at all parts of the sky and an almost perfect correspondence with a black body radiation at 2.735 K. In the hot big bang, though, opposite parts in the sky at the time that the microwave background formed 105 years from the beginning, were separated by distances of 107 light years (Schramm, 1983). Given the identity of temperatures from all parts of the sky and presuming that classical physics holds, one would conclude that opposite parts of the sky had to be in causal contact. This is known as the horizon problem. Again, it represents a problem of incredibly fine tuning in the conditions prevailing in the early universe or else a violation of classical causality.

Quantitative calculations show that slight anisotropies would not die away but on the contrary would get amplified. This is known as the isotropy problem. The problem becomes more severe in the steady state scenario which requires that whatever fluctuations in isotropy exist should be present at all levels. To be sure whether the universe turns out to be isotropic and homogeneous is an observational question. As we saw, the microwave radiation is highly isotropic. The universe is presumed to be expanding the same way in all directions, i.e. to be isotropic in matter as it is in background radiation. Moreover, the distribution of matter is presumed to reach homogeneity beyond the largest structures seen in the universe – the superclusters – i.e. beyond hundreds of millions of light years.

Recent observations challenge both these pillars of tradtional cosmological thinking: The universe may not be homogeneous, larger and larger structures have been found as the astronomer looks further and further into space (e.g. Burns *et al.*, 1988). Galaxies seemingly cluster themselves to increasing hierarchies of clusters, superclusters and maybe even super- superclusters. Often these structures assume the form of filaments on the surface of very large bubbles with large "voids" in between. The universe may also not be isotropic in the motion of matter, galaxies have been found which do not move in the isotropic fashion of following the Hubble flow (*Physics Today*, 1989). Yet, the interesting question remains why the universe is not even more anisotropic than observed.

Taking all the evidence at face value, one would conclude that the universe requires incredibly fine tuning at the beginning, i.e. the universe represents a very unlikely "accident". Paul A.M. Dirac (1937) first noticed in the 30's that certain ratios involving fundamental constants of nature and physical parameters which at face value should not be related to them obey simple numerical relations. Some of these coincidences yield very large numbers which are not random as one might have expected, e.g.

$$\text{Specific entropy} \simeq \frac{\text{number of photons}}{\text{number of baryons}} \simeq 10^{10}$$

$$\frac{\text{Size of elementary particle}}{\text{Planck length}} \simeq 10^{20}$$

$$\frac{\text{Electric force}}{\text{Gravitational force}} \simeq 10^{40}$$

$$\frac{\text{Radius of observable universe}}{\text{Size of elementary particle}} \simeq 10^{40}$$

Why the last, seemingly unrelated ratio, should be similar in value, which is a very large number, remains one of the most fundamental theoretical challenges facing physics today. Dirac believed this could not be a coincidence and formulated his large number hypothesis. Simply put, he reasoned that since as the universe expands its radius changes in value, in order for the ratios to be equal today one of the quantitites in the third ratio also has to change in time. He postulated that Newton's gravitational constant changes in time. Attempts to verify Dirac's hypothesis have so far failed.

Today some physicists have postulated that our existence as observers requires this fine tuning and that the seemingly unrelated ratios numerically giving such improbable values point to our existence as necessitating the kind of universe we live in. Put differently, the universe is unique because it contains conscious observers. This is known as the anthropic principle (Barrow and Tipler, 1986). Various flavors of anthropic principles exist, some of them even requiring a specific future evolution in order to preserve the transfer of information between observers.

3. Horizons of Knowledge in Cosmology

As we study the observations pertaining to the early universe, we encounter a number of observational horizons of knowledge (Kafatos, 1989). These observational horizons have to do with the quantum nature of light. For example, attempting to obtain the distance of a faint galaxy requires that we obtain its spectrum. To obtain an accurate spectrum requires that we disperse the light. This means isolating the light from the galaxy, for example by means of a narrow slit. When though few photons are involved, one cannot disperse the light without limit. Attempting to obtain more photons by decreasing the dispersion would, on the other hand, cause an observational confusion as light from neighboring galaxies in that part of the sky would also fall on the spectrograph. There is then a complementary inverse relationship between dispersion and brightness which does not permit accurate spectra of faint galaxies to be obtained.

The predictions of various competing models of the universe indicate that the observational horizons complicate the theoretical picture. Moreover, theoretical models present us with their own limitations, what one may call theoretical horizons of knowledge (Kafatos, 1989).

For example, big bang cosmology itself imposes a fundamental limit on the observability of the early universe. What we have is a situation in which direct observation of the early universe based on photons provides information only after a timescale of roughly 100,000 years after the beginning. We can express the age of the universe as a function of redshift z. One hundred thousand years after the beginning corresponds to z of 1000, i.e. when the universe was only 0.1% of 1% of its present age. On the other hand, the most distant quasars are seen at a redshift of ~ 4 and emitted their light received by us today when the universe was about 10% of its present age. Radiation can in principle tell us much more about the early universe than matter. However, the opaqueness of the universe prior to $z = 1000$ simply does not allow us to trace or confirm big-bang cosmology based on photons. At $z = 1000$ we encounter the first theoretical horizon of knowledge about our universe and as long as we are constrained to observe photons, that horizon is impregnable (Kafatos, 1989).

It is unlikely that any other observational means will provide as clear-cut evidence as light about the universe we live in. In principle, primordial neutrinos emitted a few seconds after the beginning of the universe or at a redshift $z \approx 10^9$, may one day be observable. Their numbers would be much greater than the supernova neutrinos but their energies billions of times less. Ignoring for the time being the great difficulty in detecting these neutrinos, even if one day we did, they would still not yield any information about the very early universe. The observational horizon of knowledge at $z \approx 10^9$ presents, therefore, the ultimate horizon from which we can access direct information about the universe (Kafatos, 1989). Whatever problems of interpretation we are facing today with regards to the background photons will not go away with the neutrinos. The problem of detection and interpretation will only be much worse.

Perhaps considerations from classical cosmological theory can shed more light as to the type of the universe we live in. To test the type of geometry of the universe one studies the Hubble Diagram for distant galaxies, i.e. a diagram of magnitude versus redshift. Quasars can be seen as far away as $z = 4$ but galaxies will only begin to be observable at that redshift when the Sapce Telescope starts looking at their spectra in 1990. Redshifts in the range 5–10 would be particularly important to study since astronomers suspect that at that redshift galaxies began to form more than 13 billion years ago. Unfortunately (Kafatos, 1989) galaxy images begin to blend together at those redshifts and it would be virtually impossible to obtain accurate spectra to study the geometry of the universe. The reason for this is because in a general relativistic model of the universe, a finite-size object has an apparent size that decreases until it reaches a minimum (Narlikar, 1983). The curvature of the universe causes the image to decrease for redshifts greater than the minimum value which occurs for z near unity. Obviously, when the apparent size of a galaxy becomes comparable to the mean distance between galaxies, the images blend. For most cosmological models and most "standard" sources this occurs in the approximate redshift range of 1–5. This "galaxy image" theoretical horizon is unfortunately much too close to us. It is in the range where the Space Telescope can obtain spectra of distant galaxies. It is hard to see how the traditional experimental way which revealed the hypothesis of the expanding universe can be carried out to a regime where that hypothesis can be better tested.

As we saw, these horizons which occur inherently in the particular theoretical picture used become worse when the quantum nature of light is considered: Spectra of different sources in the sky would themselves be blended together as one looks at fainter and fainter sources. Eventually, the background from different faint galaxies would dominate the spectrum from a single distant galaxy and reliable spectra could not be obtained.

For all these reasons, in cosmology we encounter horizons of knowledge which prevent us from deciding unequivocally how these tests confirm or reject particular theoretical models (Kafatos, 1989). This is precisely the case where, Bohr insisted, complementarity acquires great importance. One then begins to view the various cosmological models not as rival theories of which one day only one will emerge as the theory of the universe, but as competing complementary constructs. Coupled with the fact that the early universe should be described in quantum terms, one would conclude that these emergent complementary models and the implied under-lying wholeness are not an a-priori philosophical preference (Kafatos, 1989) but the very outcome of the observing process for the early universe.

As such, the flatness, horizon and isotropy problems should be tied to the ob-serving process itself rather than as preconditions for theory. As we look at more and more distant galaxies, the universe may be appearing as Euclidean not because of inflation or any other theoretical scheme devised after the facts of specific ob-servational results; rather, because such a universe would naturally emerge as the boundary between complementary constructs. A complementary pair of models, for example, is the open versus the closed universe.

4. The Quantum Universe

Modern cosmology faces a fundamental theoretical challenge, how a general rela-tivistic description of the expanding universe rises out of a fundamentally different description of the early cosmos, which must be quantum mechanical in nature. The assumption of classical causality then led to the presumption that subsequent events in larger and larger systems could be known after an observation disclosed the initial conditions in the systems. It was, therefore, reasonable to assume that if the description of the initial conditions in the largest system that can be known, the universe, was sufficiently complete, then we could presume knowledge of all subsequent events in the evolution of the universe.

Quantum effects in the very early stages of the life of the cosmos were quite large and the most modern of all cosmological theories – a big bang model with inflation – speculates that the universe first came into existence as a result of a quantum transition (Guth and Steinhardt, 1984). The situation will not be resolved until we have at hand a quantized theory of the universe. The subsequent evolution of the universe is described primarily in terms of a theory that we have tended not to recognize as classical in its overall philosophical outlook – the general theory of relativity. And yet it is that theory that will presumably be modified or displaced with a quantum theory of gravity. The seeming universal validity of the general theory has, in other words, contributed to the presumption that the subsequent evolution of the universe after the early initial phase could be understood in terms of the assumption of causality and, therefore, the indeterminacy principle, for all practical purposes, could be ignored. Consequently, most cosmologists tend to view the universe as an idealized closed system that is isolated from the observer and his observing apparatus.

The argument here is simply that the universe is more like a quantum rather than a classical system. It certainly cannot be viewed as a closed system in the sense of being separate from the observer. Quanta in the early universe were tangled together and one expects a similar situation as in the Bell-type experiments (Narlikar, 1983). One would then expect Bell-type correlations to be prevalent in the early universe. For example, in the electron-positron annihilation, the resultant gamma- rays would be polarized perpendicularly to each other. The very physical process of scattering provides then an "observational" or "experimental" choice: The other photon's polarization would not then be presumed to be independent. It would have a well-defined polarization and for the subsequent scattering of the second photon, an isotropic distribution of polarization could not be assumed. We have here a quantum non-local correlation which in principle could stretch across the universe. In other words, quantum effects, however small they might be on the macro level, are pervasive throughout the history of the universe. One cannot in theory at least ever presume a categorical distinction between acts of observation and the observed system, the entire universe. One would then conclude that complementarity will have to be invoked in our efforts to understand the early life of the universe based on observations involving few light quanta. In cosmological observations, the "choice" of whether to record the particle or wave aspect of light from faint sources would have appreciable consequences for our views of the type of the universe we live in.

References

Barrow, J.D.: 1988, *Q. Jl. R. Astr. Soc.*,**29**, 101–117

Barrow, J. D. and Tipler, F. J.: 1986, *The Anthropic Cosmological Principle*, Oxford Univ. Press, Oxford

Berry, M.: 1976, *Principles of Cosmology and Gravitation*, Cambridge, Univ. Press, Cambridge

Burns, J. O., Moody, J. W., Brodie, J. P. and Batuski, D. J.: 1988, *Ap. J.*, **355**, 542

Dirac, P. A. M.: 1937, *Nature*, **139**, 323

Guth, A. and Steinhardt, P.J.: 1984, *Scientific American*, June 1984, 116–128

Kafatos, M.: 1989, in *Bell's Theorem, Quantum Theory and Conceptions of the Universe*, ed. M. Kafatos, Kluwer Acad. Publ., Dordrecht, p. 195
Narlikar, J.: 1983, *Introduction to Cosmology*, Jones and Bartlett, New York
Physics Today: Jan. 1989, s. 11
Schramm, D.N.: 1983, *Physics Today*, April 1983, 27–33

STRUCTURE OF RADIATIVELY COOLED JETS

MASA-AKI KONDO

Senshu University, 3-8-1 Higashi-mita, Tama-ku, Kawasaki-shi, Kanagawa, 214 Japan

Abstract. Radiative cooling strongly affects the thermal structure of dense jet, such as in SS433, through free-free emission. From the dynamical aspect, the beam width of a cooled jet does not expand, unlike from an adiabatic jet. From the thermal aspect, cooling efficiency determines the ratio of X-ray region of high temperature to optical one of low temperature. However, this ratio is influenced by the heating due to contained high-energy particles, which produce synchrotron radiation in the tail of the jet.

Extragalactic jets can also be considered in a similar way due to other energy loss mechanisms.

1. Introduction

The astrophysical jets are so slender that a mechanism of collimation would be working. Various mechanisms have been investigated until now; cocoon model of confinement by ambient gas, turbulent magnetic effect to cocoon model, line locking, coupling of magnetic field and rotation, and others (cf. Begelmann et al, 1984).

On the other hand, with respect to the thermal structure, astrophysical jets are composed of representative three regions; X-ray region of high temperature region, optical one of lower temperature, and radio tail of synchrotron emission. Therefore, it seems collimation mechnisms should act in any thermal situation, so that collimation mechanisms might correlate with thermal situation.

Then, from the aspect of formation of filaments or knots, thermal instability has been considered such that cooled filaments will appear in hot medium (Bode et al, 1985; Brinkmann et al, 1988; dal Pino et al, 1989). If a jet pass through vacume, however, we can expect contraction of cross section due to radiative energy loss even in thermally stable state.

2. Non-adiabaticity of Astrophysical Jets

In the case of SS433, the jet is relatively dense because no forbidden line is observed. Since free-bound emissions of hydrogen or line ones of oxygen are optically thick, dominant radiative cooling is free-free emissions. Then, the efficiency factor η of energy loss can be defined as the ratio of two time scales. One is the time scale t_p of passing a characteristic distance r_0 with the jet velocity V: $t_p = r_0/V$, where r_0 is likely a radius of an orifice. Another one is the cooling time scale t_c, given by $c_p T/Q$, where c_p and Q are the specific heat capacity and energy loss.

$$\eta = t_p/t_c = r_0 Q/c_p T V$$

Since $Q \sim 5 \cdot 10^{20} \rho T^{1/2}$ for free-free emission and $V \sim c/4$, we get $\eta \sim 8.6 \cdot 10^{12} \dot{M}_{-6}/r_0 T^{1/2}$, where $\dot{M}_{-6}$ means mass flux in unit of $10^{-6} M_\odot$/year. Hence, η

Y. Kondo (ed.), Observatories in Earth Orbit and Beyond, 551–554.

of SS433 is around of 0.1, refered to the observational constraint (cf. Margon, 1984; Zwitter et al, 1990).

On the other hand, in the extragalactic case, radiative cooling is ineffective, because density is very thin. Then, synchrotron emission is possiblly effective(cf. Achterberg, 1989). However, heat conduction acts in effectiveness of $\eta \sim 1$. Moreover, relativistic nature has to play an essential role. Accordingly, the extragalactic case is not considered here.

3. Contraction of Radiatively Cooled Jets

To examine the cooling effect on jets, we set the situation where a jet flowes out vertically from horizontally layerd disk, the pessure of which steaply decreases, for example, as $exp(-5z/r_0)$. To compare a cooled jet with an adiabatic one, it is preferable to consider the steady state. Then, we will treat the steady hydrodynamical equtions, where all physical quantities are normalized by the values at the orifice, and vertical/radial distance by r_0. these equations include two basic parameters of the Mach number M and the cooling efficiency η.

We calculated the steady configuration of $M = 14.2$ and $\eta = 0.1$, by the method of characteristics (cf. Sanders, 1983). The results of two cases of adiabatic and cooled jets are shown in the figure.

3.1. THE ADIABATIC JET

The so-called potential core is formed in the cone with the base of orifice, where all quantities keep the values of orifice, and stream lines keep vertically straight. Leaving the core, the jet expands rapidly, so that quantities decrease adiabatically. Although the core with high temperatureis observed as jet beam, the optical region of low-temperature tail would be not observed in the beam form.

3.2. THE JET COOLED BY FREE-FREE EMISSION

The potential core disappears because cooling let the temperature and pressure decrease. Consequently, the jet does not expand, keeping the high density near the orifice, as shown in the figure of density profile. From the observational aspect, the X-ray region is diminished and the optical one of low temperature is observable in the beam form.

In reality, there should be heating due to high energy particles, which collisionally ionize thermal gas. Accordingly, the fraction of high energy particles to normal gas determines the ratio of optical part to X-ray one. Then, these high energy particles emittes synchrotron radiation in the tail.

One of noted points is the horizontal stream, that appeared in both cases, which keeps the high velocity and contact tangentially to the static atmosphere of the disk. This stream should be unstable for the Kelvin-Helmholtz instability, in reality. Hence, turbulent gas would be blown off by this shear instability, emitting Balmer lines with broad line widths similar to the central lines of SS433. Nevertheless, this phenomena would not affect the central jetstream, differing from the case of coccon jet.

Adiabatic case (η = 0)   Cooled case (η = 0. 1)

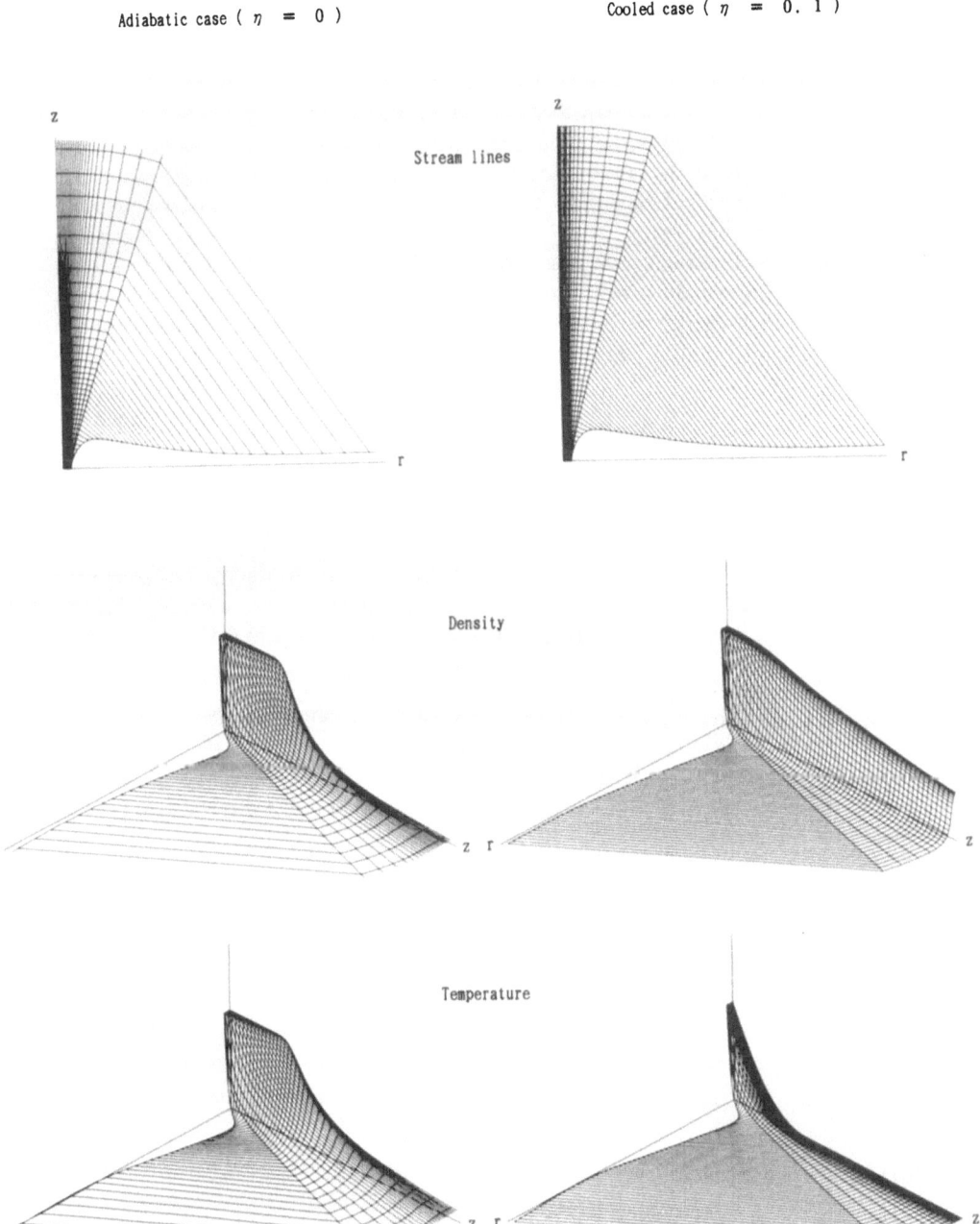

Fig. 1. Comparison between the adiabatic and the cooled case of $M = 14.2, \eta = 0.1$. Stream lines, density and temperature are shown. The scales of vertical z and radial r are in units of orifice.

4. Conclusion

The cooling effect of free-free emission inhibits expansion of jets in such a way that the pressure of the potential core, that causes expansion, is diminished. Hence, any artificial mechanisms including magnetic or rotational effects are unnecessary for collimating jets. However, another effect of heating due to high energy particles should be considered, because it determines the state of the lower temperature region through collisional ionization. Hence, the relation between cooling and heating affects the nature of jets.

In the case of extragalactic jets, another mechanism of energy loss has to act for collimation. To explain synchrotron radiation, we should consider gas containing relativistic, high energy particles, whether they originate from a central engine or have been accelerated in jets.

References

Achterberg, A.: 1989, *Astron. Astrophys.* **221**, 364

Begelmann, M. C., Blandford, R. D., and Rees, M. J.: 1984, *Rev. Mod. Phys.* **56**, 255

Bodo, G., Ferrari, A., Massaglia, S., and Tsinganos, K.: 1985, *Astron. Astrophys.* **149**, 146

Brinkmann, W., Fink, H. H., Massaglia, S.,Bodo, C., and Ferrari, A.: 1988, *Astron. Astrophys.* **196**, 313

de Gouveia Dal Pino, E. M., and Opher, R.: 1989, *Astrophys. J.* **342**, 686

Margon, A.: 1984, *Ann. Rev. Astron. Astrophys.* **22**, 507

Sanders, R. H.: 1983, *Astrophys. J.* **266**, 73

Zwitter, T., Calvani, M., Bodo, G., and Massaglia, S.: 1990, *Fundamentals of Cosmic Physics*, in press

A NEW WAY FOR TESTING OF LIGHT DEFLECTION IN EARTH ORBIT OR BEYOND

QIN YI-PING

Dept. of Astronomy, Beijing Normal University, Bejing, China

Abstract. According to the principles of light deflection, different occulting times of an occultation respond to different basic theories of physics. An occulting way for testing of light deflection in earth orbit or beyond may be possible and is proposed. An application is studied and some suggestions are presented.

1. Introduction

As is well known, the testing of light deflection is one of the three "classical" tests of relativistic gravity (Will 1981). The conclusion of light deflection can also be derived from other theories, with the results being different from that from general relativity. Generally, the testing of light deflection is operated by observing the deflection angle of light (Will 1981). This kind of operation lasts a long time and brings some observational difficulties.

This work presents an occulting way for testing of light deflection. It avoids the long time observing difficulties.

2. The Principles and Basic Formulas of Occulting

Shown in Fig. 1, P presents the occultation while E presents the observing body and S presents the centre of the deflection source. Q_1 and Q_2 stand for the two points where the two observed lights pass and the distances to S are the shortest.

Suppose P and S be motionless, and E travel round S. When E moves into the opposite side of S relative to P (see Fig. 1), the occulting happens. Let α be the angle, within which P can not be seen by the observer in E (we call that an occulting angle). For a known motion of E traveling round S, the occulting angle and occulting time are well related.

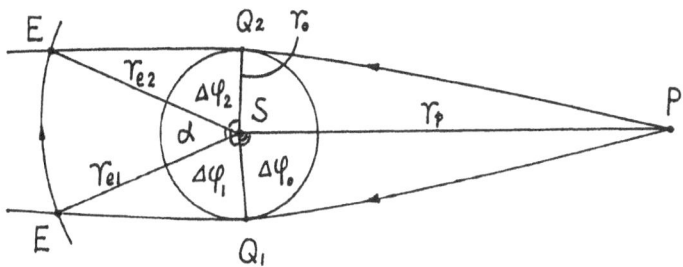

Fig. 1. The diagram of an occultation.

Y. Kondo (ed.), Observatories in Earth Orbit and Beyond, 555–558.
© 1990 *Kluwer Academic Publishers.*

From Fig. 1, we get

$$\alpha = 2\pi - 2\Delta\varphi_0 - \Delta\varphi_1 - \Delta\varphi_2 \qquad (1)$$

The angles $\Delta\varphi_0$, $\Delta\varphi_1$ and $\Delta\varphi_2$ can be calculated with the theories of general relativity and Newtonian mechanics. This kind of work is often presented in literature (Carmeli 1982).

2.1. THE ANGULAR FORMULA FROM GENERAL RELATIVITY

In general relativity, the solutions of free fall equations are (Weinberg 1972)

$$\begin{aligned} r^2(d\varphi/dp) &= J, \\ dt/dp &= b/(1 - r_s/r), \end{aligned} \qquad (2)$$

where J and b are constants, while r_s is the Schwarzschild radius, and p is a parameter describing the trajectory.

Applying Eqs. (2) to a photon and using the condition of $(dr/d\varphi)$ $(r = r_0) = 0$ (see Fig. 1), we have (for isotropic orbits, it is enough to retain the positive sign)

$$d\varphi = (1/r)((1 - r_s/r_0)(r/r_0)^2 - 1 + r_s/r)^{-\frac{1}{2}} dr. \qquad (3)$$

Expanding Eq. (3) to first order in r_s/r_0, and then integrating it, we have

$$\begin{aligned} \varphi(r_2) - \varphi(r_1) &= [-\sin^{-1}(r_0/r) + (1/2)(r_s/r_0)(1 - r_0^2/r^2)^{\frac{1}{2}} \\ &\quad + (1/2)(r_s/r_0)(r - r_0)^{\frac{1}{2}}(r + r_0)^{-\frac{1}{2}}]_{r_1}^{r_2}. \end{aligned} \qquad (4)$$

With equation (4), $\Delta\varphi_0$, $\Delta\varphi_1$ and $\Delta\varphi_2$ in general relativity can be obtained.

2.2. THE ANGULAR FORMULA FROM NEWTONIAN MECHANICS

In Newtonian mechanics, the angular momentum integral and energy conservation equation of two bodies sytem are (Roy 1978)

$$\begin{aligned} r^2(d\varphi/dt) &= h, \\ (1/2)v^2 - \mu/r &= C, \end{aligned} \qquad (5)$$

where h and C are constants, while $\mu = G(M = m)$.

Applying Eqs. (5) to a photon and using the conditions $v(r = r_p) = 1$ and $(dr/d\varphi)$ $(r = r_0) = 0$, we have

$$\begin{aligned} d\varphi &= (r_0/r)(1 + r_s/r_0 - r_s/r_p)^{\frac{1}{2}} \\ &\quad \cdot (r^2(1 - r_s/r_p) + rr_s - r_0^2(1 + r_s/r_0 - r_s/r_p))^{-\frac{1}{2}} dr. \end{aligned} \qquad (6)$$

Let $k = (r_s/r_0)/(1 + r_s/r_0 - r_s/r_p)$. Expaning Eq. (6) to first order in k and integrating it, we have

$$\begin{aligned} \varphi(r_2) - \varphi(r_1) &= [-\sin^{-1}(r_0/r) \\ &\quad + (1/2)(r_s/r_0)(1 + r_s/r_0 - r_s/r_p)^{-1} \\ &\quad (r - r_0)^{\frac{1}{2}}(r + r_0)^{-\frac{1}{2}}]_{r_1}^{r_2}. \end{aligned} \qquad (7)$$

With Eq. (7), $\Delta\varphi_0$, $\Delta\varphi_1$ and $\Delta\varphi_2$ in Newtonian mechanics can be obtained.

3. An Application

Take the sun as S in Fig. 1, and a far distance celestical body as P, and the earth as E. Considering an average motion, we take $r_0 = R_\odot = 6.9599 \cdot 10^5$ km, $r_{e1} = r_{e2} = r_e = r_\oplus = 1.4946 \cdot 10^8$ km, $r_s = 2.95$ km, $r_s/r_0 = 4.239 \cdot 10^{-6}$, $r_0/r_e = 4.656697 \cdot 10^{-3}$.

(i) From Eq. (4), to the accuracy of 10^{-6}, we then have

$$\Delta\varphi_0 = \pi/2 + r_s/r_0 \,,$$
$$\Delta\varphi_1 = \Delta\varphi_2 = \pi/2 - r_0/r_e + r_s/r_0 \,.$$

So we get the following from Eq. (1):

$$\alpha = 2r_0/r_e - 4r_s/r_0 \,. \tag{8}$$

This is the formula of occulting angle in general relativity, where in the right hand side, the first term stands for the motion that a photon travels in a straight line while the second term stands for the correction from general relativity.

(ii) In the same way, Eq. (7) gives (to the accuracy of 10)

$$\Delta\varphi_0 = \pi/2 + (1/2)r_s/r_0 \,,$$
$$\Delta\varphi_1 = \Delta\varphi_2 = \pi/2 - r_0/r_e + (1/2)r_s/r_0 \,.$$

So the corresponding occulting angle is

$$\alpha' = 2r_0/r_e - 2r_s/r_0 \,. \tag{9}$$

This is the formula of occulting angle in Newtonian mechanics, where in the right hand side, the second term is the term standing for the correction from Newtonian mechanics.

The time taken by the earth round the sun for a circle is $T = 3.1558149984 \cdot 10^7$ seconds. Then, the occulting time in general relativity is obtained as

$$\Delta T = T\alpha/(2\pi) = (T/\pi)(r_0/r_e - 2r_s/r_0) = 12^h 58^m 12^s.66 \,,$$

while the occulting time in Newtonian mechanics is

$$\Delta T' = T\alpha'/(2\pi) = (T/\pi)(r_0/r_e - r_s/r_0) = 12^h 58^m 55^s.24 \,.$$

The marks h, m and s denote hours, minutes and seconds.

The difference between the occulting times in Newtonian mechanics and general relativity is

$$\Delta T' - \Delta T = 42.58 \text{ seconds} \,.$$

4. Discussions and Suggestions

The conclusions of last part show, the occulting time of a star occulted by the sun is about 13 hours. Comparing with the observation of the deflection angle of light, the long time observing difficulties are avoided. It is also shown, the difference of occulting times between the theories of Newtonian mechanics and general relativity is about 43 seconds. It is sensitive enough to be detected by instruments if there are no other problems of certainty. We also see that the precise operation of observation is only required at the beginning and the end of an occultation. If an observation is taken in an earth orbit satellite or a vehicle beyond the earth orbit, it occupies only a little time.

On the other side, the observation may produce some difficulties which should be considered seriously. In the following, we present some problems and give some suggestions.

(i) The occulting time involves the geometry of the occulting body. In the case of the sun, if the thickness of the solar chromosphere causes an uncertainty of radius of the sun, the geometrical uncertainty will be in the same order of magnitude of the effect of light deflection. It is depended on the extent of the extinction and the brightness of the solar chromosphere and the magnitude of the observed star, whether the effect of the solar chromosphere should be taken into account.

If the observed star is occulted by the chromosphere, there are some ways tending to solve it. We may use a chromospheric telescope for observing. The distance from the centre of the sun to the edge of the chromosphere is taken as the radius of the occulting body in the theoretical calculation. If the uncertainty of the radius is less than 1/16 of the chromospheric thickness (about 2 seconds of angle), the observation will be valid. Or, when observing an occultation with an ordinary telescope, we observe the chromosphere in the same time. By analysing occulation pictures and the solar chromosphere pictures in different cases, we are supposed to find the effect of the chromosphere because the different thicknesses of the chromosphere may cause different effects (if there is any). With these two kinds of pictures, the radius of the occulting body is also obtained.

(ii) The scattering of rays from atmosphere will strongly effect the optical observation on the ground. To avoid it, a satellite-based observation is needed.

We come to a conclusion that by using an earth orbit or beyond satellite, the observation of an occultation may become a new way for testing of light deflection.

This theory may also be applied to Jupiter or other celestial bodies.

Acknowledgements

The author thanks Dr. Wu Shi-Min and Dr. Zheng Xue-Tang for all their helps and advices. The author also thanks Dr. Du Sheng-Yun, Dr. Ma Wen-Zhang and Dr. Du Jin-Sheng for their helpful discussions and suggestions.

References

Carmeli, M.: 1982, *Classical Fields*, New York: John Wiley and Sons, Inc.2

Roy, A.E.: 1978, *Orbital Motion*, Bristol: Adam Hilger Ltd)

Weinberg, S.: 1972, *Gravitation and Cosmology*, (New York: John Wiley and Sons, Inc.

Will, C.M.: 1981a, *Theory and Experiment in Gravitational Physics*, Cambridge: Cambridge University Press, 166

Will, C.M.: 1981b, *Theory and Experiment in Gravitational Physics*, Cambridge: Cambridge University Press, 172

STUDYING THE GALACTIC CENTRAL ENGINE
FROM SPACE OBSERVATORIES

HOWARD D. GREYBER

Greyber Associates, 10123 Falls Road, Potomac, MD 20854, U.S.A.

1. Introduction

Three general models have been constructed for the fantastically powerful "central engine" that powers the enormous energy output from quasars and active galactic nuclei (AGN). One model assumes a rapidly rotating accretion disk around a central black hole (however the disks, thick or thin, are subject to violent instabilities). Another assumes that in some postulated circuitry energy is extracted from the rotational portion of the deepest potential hole known, a black hole. Both models appear implausible.

The third model is the *STRONG MAGNETIC FIELD MODEL* (SMF) in which an extremely strong gravitationally bound current loop (GBCL) is formed during the gravitational collapse that forms the galaxy or quasar, producing a very intense dipole magnetic field anchored in the nucleus. SMF, first published in 1962, thus predicted the vertical magnetic field configuration seen today at our own galactic nucleus; to some the radio arcs observed suggest a dipole magnetic field there, just as SMF predicts.

When accretion occurs the diamagnetic effect of the bulk plasma produces topological magnetic mirrors as seen in Figure 1. The explanation for K. Kellerman's 1985 comment, "Typical morphology shows a compact core with one or more blobs moving away in a single direction" is simply that the throat must open when the plasma pressure locally exceeds the magnetic field pressure, so that a radio blob is expelled. The succession of blobs produced as accretion continues constitutes the radio jets observed.

SMF does not assume equipartition and thus does not need the exotic acceleration mechanisms required by other models. The two other models choose rotation as the important fundamental axial vector. SMF is different, choosing magnetic field.

In SMF the general morphology, energy production and dynamics of objects of galactic dimension are determined by the ratio of magnetic field energy to rotational energy. This ratio is extremely high for quasars and BL Lac objects, then decreasing as one goes to radio galaxies, Seyferts, Markarians, ordinary spirals, until it is close to zero for the ordinary elliptical galaxies. However, the activity seen is a function of the matter accretion rate at the time of observation! A great number of "quiet engines" may exist in the Universe.

Y. Kondo (ed.), Observatories in Earth Orbit and Beyond, 559–561.

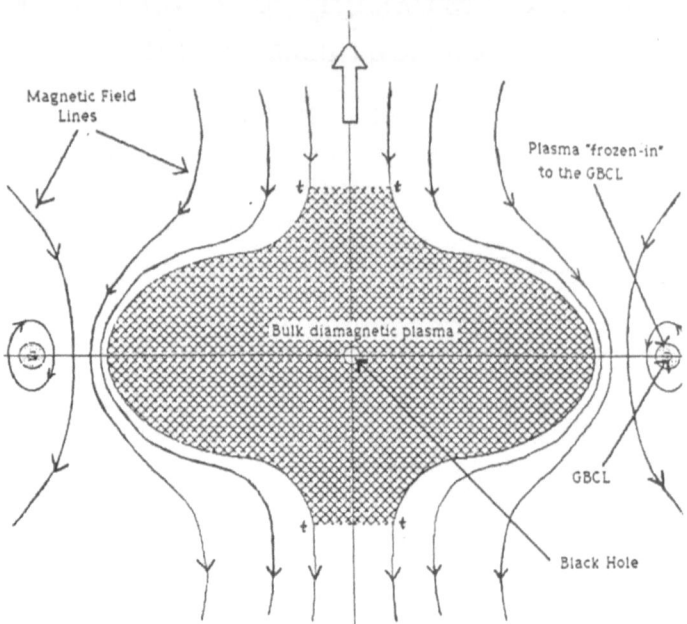

Fig. 1. SMF model for a quasar, BL Lac object or AGN when a very large accretion of matter around the black hole has occurred. t-t is the throat of the magnetic mirrors. The arrow shows the direction of the motion of the explosively ejected hot plasma blob after the upper throat suddenly opens. As accretion continues, the process repeats every few years.

2. Suggested Crucial Observations

Since, unlike other models, SMF, predicts magnetic fields of 10,000 megagauss and higher at the center of galaxies and quasars, it would be desirable to measure the Zeeman splitting of some line. This may be easier from space observatories. SMF predicts that with better radio resolution (below the microarcsecond level), possibly gained from coupling an antenna in space to the VLBI networks on Earth, one should find evidence for a relatively weak GBCL around Sag A*, which is at the center of our Milky Way galaxy. An infrared interferometer located in space or on the Moon would also aid in the detection of a GBCL. Other evidence for the presence of a GBCL would be rapid brightness changes, or rapid changes of polarization intensity and/or direction in narrow regions located symmetrically on each side of the central object.

Another method is to search for rapid fluctuations in intensity or polarization from the quasar core or AGN. Several fluctuations of the order of 100 seconds or less have already been reported. This is significant since black hole masses greater than 100 million solar masses are required to power the strongest radio sources – and such a black hole has a diameter of 2000 light-seconds. If space observations confirm such rapid time variations, other models apparently must be rejected. However in SMF very rapid time variations can occur as neutral matter accretes and plunges across the minor diameter of the thin GBCL toroid, ionizes and radiates in the intense magnetic field close to the loop. In SMF it is not the quasar diameter that limits the characteristic time for a significant intensity variation, but the small diameter of the extremely slender GBCL.

Other crucial observations are to verify from space observations important conclusions about galaxies and quasars drawn from hundreds of papers of mostly ground-based observations.

Another crucial test is to analyze the magnetic field configuration in spiral arms close to the galactic nucleus. SMF predicts the field lies in one direction along the spiral arm above the plane of the galaxy, and in the other direction below the plane. with a neutral magnetic field sheet delineating the plane.

3. Conclusions

The expanded use of space observatories in all wavelength ranges should enable us to understand finally the physics of the "central engine" that powers the enormous energy output we observe from quasars and AGN. This physics will probably also explain the formation of galaxies and the variety of objects of galactic dimension that are observed.

References

Greyber, H.D., 1990 *Strong Magnetic Fields, Galaxy Formation and the Galactic Engine* in 14th Texas Symposium on Relativistic Astrophysics, New York: Annual of the New York Acad. of Sciences, **571 239**, and earlier references cited therein.

TABLE 1 : COMPARISON OF THE STRONG MAGNETIC FIELD MODEL (SMF) WITH THE ROTATING ACCRETION DISK MODEL

	Strong Magnetic Field Model (SMF)	Rotating Accretion Disk Model
ENERGY SOURCE	Gravit. Collapse + accretion	Accretion
ENERGY STORAGE	Gravit. Bound Current Loop	Velocity of Accreting Material
EXPLAINS:		
Very frequent injection of high energy relativistic electrons	Yes	No
Millions of points where relat. electrons inject into strong mag. field	Yes	No
Production of successive radio blobs	Yes	No
Shape of giant double lobe radio sources	Yes	No
One-sided jets in strong sources	Yes	No
Strong magnetic field in very young (i.e. high Z) galaxies	Yes	No
Largest polarization near the innermost radio blob	Yes	No
Core-halo polarization difference in Core-Halo radio galaxies	Yes	No
Morphology of Objects of Galactic Dimension	Yes	No
Extremely rapid time variations observed (50-100 seconds)	Yes	No
Remarkable straightness of jets	Yes	?
Remarkable confinement of jet material over huge distances	Yes	No
Lack of evidence for rotation in the spectra of quasars	Yes	No
Production of jet	Yes	Yes- with assumed funnel
Radio-quiet quasars	Yes	?
Stability of configuration	Yes	No
Generation Mechanism	Yes	Yes
Existence for times of interest	Yes	?
Very few spirals in Rich Clusters	Yes	No

Morris Aizenman, National Science Foundation, Division of Astronomical Sciences, Washington, DC 20550
B. Andreev, IKI, Space Research Institute, Profsoyuznaya 88, 117810 Moscow, USSR
Manuel Angulo, IRTA, Programas Espaciales, CBA de Ajalvir P.K.Y., Torrejon, Madrid, SPAIN
R. Aptekar, USSR Ioffe Insitute, 194021 Leningrad, USSR
Halton Arp, MPI fur Astrophysics, D-8046 Garching Bei Munchen, West Germany,
David Batchelor, NASA/GSFC Code 633, Greenbelt, MD 20771,
David Blanchard, Ford Aerospace, 7375 Executive Place, Ste 400, Seabrook, MD 20706
Elihu Boldt, NASA/GSFC Code 666, Greenbelt, MD 20771,
Roger-Maurice Bonnet, European Space Agency, 8-10, Rue Mario Nikis, 75738 Paris Cedex 15, FRANCE
Stuart Bowyer, University of California, Center for EUV Astrophysics, Berkeley, CA 94720
Hale Bradt, MIT, 37-581, Cambridge, MA 02139
V. Bremenkamp, Associated Universities, Inc., 1717 Massachusetts Ave. N.W. Ste. 603, Washington, DC 20036
G. Briggs, NASA Headquarters, Code EL-4, Washington, DC 20546
A. Lyle Broadfoot, University of Arizona, 901 Gould-Simpson Bldg., Tucson, AZ 85721
Noah Brosch, Tel Aviv University, Nise Observatory, Tel Aviv 69978, ISRAEL
Robert Brown, NRAO, Edgemont Road, Charlottesville, VA 22903
Ron Browning, Ford Aerospace, 7375 Executive Place, Ste. 400, Seabrook, MD 20706
Andrea Bucconi, CARSO, Padriciano 99, 34612 Trieste, ITALY
Alan Bunner, NASA HQ Code EZ, Washington, DC 20546
Bernard Burke, MIT, 10 Bloomfield Street, Lexington, MA 02173
Chris Butler, Italian Space Agency, ASI, Via Reg Margrerriza 202, 00198, Roma, ITALY
Larry Caroff, NASA Headquarters, Code EZF, Washington, DC 20546
Ken Carpenter, NASA/GSFC Code 681, Greenbelt, MD 20771
Catherine Cesarsky, Cen-Saclay-DphG/SAp, 91191 Gif Sur Yvette FRANCE
David Cherry, 21st Century, P.O. Box 17285, Washington, DC 20041-0285
Hong-Yee Chiu, NASA/GSFC Code 610.1, Greenbelt, MD 20771,
Geff Clayton, NASA HQ Code EZ, Washington, DC 20546
Arthur D. Code, University of Wisconsin, Astronomy Dept., Madison, WI 53705
Mike Corcoran, NASA/GSFC Code 666, Greenbelt, MD 20771
Larry D'Addario, NRAO, 2015 Ivy Road, Charlottesville, VA 22903
Laura Danley, STSCI, 3700 San Martin Drive, Baltimore, MD 21218
Arthur Davidsen, Johns Hopkins University, Dept. of Physics & Astronomy, Baltimore, MD 21218
John Davies, Edinburgh Royal Observatory, Blackford Hill, Edinburgh, EH9, 3HJ, UK
Valentin Dorodnov, IKI, Space Research Institute, Profsoyuznaya 88, 117810 Moscow, USSR
David W. Dunham, Computer Sciences Corporation, 10110 Aerospace Road, Lanham, MD 20706
Freeman Dyson, Institute for Advanced Studies, 105 Battle Road Circle, Princeton, NJ 08540
Ronald Estes, STX, 4400 Forbes Blvd., Lanham, MD 20706
D. Evans, Los Alamos National Laboratory, M/S F650, Los Alamos, NM 87545
Gregory G. Fahlman, University British Columbia, 2219 Main Mall, Vancouver, BC V6T 1W5, CANADA
Robert Farquhar, NASA/GSFC Code 550, Greenbelt, MD 20771
Tom Faska, Martin Marietta, 1450 S. Rolling Rd., Baltimore, MD 21227
Giovanni Fazio, SAO, 60 Garden Street, Cambridge, MA 02138
Paul D. Feldman, Johns Hopkins University, Dept. of Physics & Astronomy, Baltimore, MD 21218
Austin Fehr, Martin Marietta, P.O. Box 179, Denver, CO 80201
Carl Fichtel, NASA/GSFC Code 660, Greenbelt, MD 20771
Richard Tresch Fienberg, Sky & Telescope, P.O. Box 9111, Belmont, MA 02178-9111
Walter Fowler, NASA/GSFC Code 683, Greenbelt, MD 20771
Urban Frisk, ESA/ESTEC P.O. Box 299, 2200AG Noordwijk, THE NETHERLANDS
Chris Ftaclas, Hughes Danbury Optical Systems, 100 Wooster Heights Rd., Danbury, CT 00810
Arkady M. Galper, Moscow Physical Eng. Institute, Kashizscoc sh. 31, 115409 Moscow, USSR
Ian Gatley, NOAO, 950 North Cherry Ave., Tucson, AZ 85719
Harold Geller, SAIC, 600 Maryland Ave., S.W., Ste. 307 W, Washington, DC 20024

Francoise Genova, CNES 2 Pl Maurice Quentin, F-75039, Paris Cedex 01, FRANCE
Alvaro Gimenez, Inst Nacl de Tecn Aerospacial, 28850 Torrejon de Ardoz, Carretera de Ajalvir, Madrid, S
Fred Gordon, NASA/GSFC Code 602, Greenbelt, MD 20771
Ian Gordon, Un of Arizona/Lunar Lab West, 901 Gould Simpson Building, Tucson, AZ 85721
Howard Greyber, World Technology Foundation, 10123 Falls Road, Washington, DC 20036
Ed Guinan, Villanova University, Dept. of Astronomy, Villanova, PA 19085
Bill Guit, Ford Aerospace, 8901 Harmony Court, Owings, MD 20736
Ted Gull, NASA/GSFC Code 680, Greenbelt, MD 20771
Herbert Gursky, Naval Research Laboratory, Code 4100, Washington, DC 20375
Kenneth Hallam, NASA/GSFC Code 684, Greenbelt, MD 20771
Michael Hauser, NASA/GSFC Code 680, Greenbelt, MD 20771
Robert Haymes, NASA HQ 2007 Old Stage Road, Alexandria, VA 22308
Hisashi Hirabayashi, ISAS 3-1-1 Yoshinodai, Sagamihara-shi, Kanagawa, 229, JAPAN
Jay Holberg, Un of Arizona/Lunar Lab West, 901 Gould-Simpson Bldg., Tucson, AZ 85745
Steve Holt, NASA/GSFC Code 600, Greenbelt, MD 20771
Colin Humphries, Royal Observatory Edinburgh, Blackford Hill, Edinburgh EH9 3HJ, Scotland
Garth Illingworth, University of California, Lick Observatory, Santa Cruz, CA 95064
Catherine Imhoff, NASA/GSFC/CSC, Code 684.9, Greenbelt, MD 20771
Kimmo Innanen, York University, 4700 Keele St., North York, Ontario, M3J 1P3 CANADA
Stuart Jordan, NASA/GSFC Code 682, Greenbelt, MD 20771
Menos Kafatos, George Mason University, Physics Department, Fairfax, VA 22030
Louis Kalezienski, NASA HQ Code EZ, Washington, DC 20546
Nick Kardashev, Astro Space Center, Profsoyusuaya 84/32, 810384, Moscow, USSR
Kenneth Kellerman, NRAO, Edgemont Road, Charlottesville, VA 22903
George A. Keyworth, Hudson Institute, 3956 Georgetown Court, N.W., Washington, DC 20007
Donald A. Kniffen, NASA/GSFC Code 662, Greenbelt, MD 20771
Masaaki Kondo, Senshu University, 3-8-1 Higashi-mita, Tama-ku, Kawasaki, Kanagawa, 214 JAPAN
Yoji Kondo, NASA/GSFC Code 684, Greenbelt, MD 20771
Gerhard Kraemer, University of Tuebingen, 7400 Tuebinger, Germany
Robert B. Krause, NASA HQ Code MCI, Washington, DC 20546
Jim Kurfess, NRL Code 4150, Washington, DC 20375
Peter B. Landecker, Hughes Aircraft Co., Space & Comm. Grp. S41/B326,, Los Angeles, CA 90009
Andrew Lange, University of California, , Dept. of Physics, Berkeley, CA 94720
Igor Lapshov, IKI, Space Research Institute, Profsoyuznaya 88, 117810 Moscow, USSR
Larry Lesyna, Grumman Aerospace, Mail Stop A01-26, Bethpage, NY 11714
Eugene Levy, University of Arizona, Lunar & Planetary Laboratory, Tucson, AZ 85721
Gerald Levy, NASA HQ Code EZ, Washington, DC 20546
Charles F. Lillie, TRW, One Space Park, Redondo Beach, CA 90293
Roger Linfield, Jet Propulsion Laboratory, Mail Stop 238-700, Pasadena, CA 91109
Malcolm Longair, Royal Observatory Edinburgh, Blackford Hill, Edinburgh, EH6 6DU United Kingdom
Duccio Macchetto, ESA/STScI, 3700 San Martin Drive, Baltimore, MD 21218
Fumiyoshi Makino, ISIS, 3-1-1, Yoshinodai, Sagamihara, Kanagawa 224, JAPAN
Steve Maran, NASA/GSFC Code 680, Greenbelt, MD 20771
Blaine Marshall, TLB, 777 Duke Street, Alexandria, VA 22314
Francis Marshall, NASA/GSFC Code 666, Greenbelt, MD 20771
Toshio Matsumoto, Nagoya University, Dept. of Astrophysics, Furo-cho, Chikusa-ku, Nagayoa, 464-01, JAPAI
Masaru Matsuoka, Institute of Physical and Chemical Research, Hirosawa 2-1, Wako, Saitama, 351-01, JAPAI
Mary McCarthy, TLB, 777 Duke Street, Alexandria, VA 22314
R.A. McCuicheon, Computer Sciences Corporation, 1100 West St., Laurel, MD 20707
Jaylee Mead, NASA/GSFC Code 630, Greenbelt, MD 20771
Gary Melnick, SAO, 60 Garden Street, Cambridge, MA 02138
Andrew Michalitsianos, NASA/GSFC Code 684, Greenbelt, MD 20771
Patricia Mitchell, TLB, 777 Duke Street, Alexandria, VA 22314
Warren Moos, Johns Hopkins University, Dept. of Physics & Astronomy, Baltimore, MD 21218

Tom Morgan, NASA HQ Code EL, Washington, DC 20546
Michael J. Mumma, NASA/GSFC Code 693, Greenbelt, MD 20771
Ravil Nazinov, IKI, Space Research Institute, Profsoyuznaya 88, 117810 Moscow, USSR
James Neff, NASA/GSFC Code 681, Greenbelt, MD 20771
Gerhard Neukum, DLR, 8031 Oberpfaffenhofen, West Germany
George Newton, NASA HQ Code EZ, Washington, DC 20546
Joy Nichols-Bohlin, Computer Sciences Corporation, 10000-A Aerospace Road, Lanham-Seabrook, MD 20706
Colin Norman, STScI, 3700 San Martin Drive, Baltimore, MD 21215
R.A. Novaria, Ball Corporation, 4509 South Meadow Drive, Boulder, CO 80301
Rolf Ockert, Dornier GMBH, D-7990 Friedrichshafen 1, Germany
Minoru Oda, RIKEN, Hirosawa, Wako, Saitama, 351-01, JAPAN
Goetz Oertel, AURA, 1625 Massachusetts Ave, N.W., Ste 701, Washington, DC 20036
D.S. Page, European Space Agency/JPL, M/S 169-506, Pasadena, CA 91109
Arvind Parmar, ESA/ESTEC, Postbus 299, 2200 AG Noordwijk, The NETHERLANDS
Paul Pashby, NASA/GSFC Code 602, Greenbelt, MD 20771
Jacques Paul, CEA, SAP/Geres Cen Saclay, 91191 GIF/Yvette FRANCE
Jeff Pedelty, NASA/GSFC Code 636, Greenbelt, MD 20771
Charles Pellerin, NASA HQ Code EZ, Washington, DC 20546
Michael A.C. Perryman, ESA/ESTEC Astrophysics Division, 2200AG Noordwijk, THE NETHERLANDS
Geraldine J. Peters, Univ of Southern California, Space Sciences Center, Los Angeles, CA 90089-1341
Robert Petre, NASA/GSFC Code 666, Greenbelt, MD 20771
Tom Phillips, CalTech, MS 320-47, Pasadena, CA 91125
Ryszard L. Pisarski, NASA/GSFC Code 636, Greenbelt, MD 20771
Ronald Pitts, NASA/GSFC/CSC, Code 684.9, Greenbelt, MD 20771
Art Poland, NASA/GSFC Code 682.1, Greenbelt, MD 20771
Ronald Polidan, NASA/GSFC Code 681, Greenbelt, MD 20771
Kenneth Pounds, University of Leicestes, Dept. of Physics & Astronomy, Leicestes, LEI 7RH UNITED KINGDOM
Ian Pryke, European Space Agency, 955 L'Enfant Plaza, Ste. 2800, Washington, DC 20024
Jurgen Rahe, NASA HQ Code EZ, Washington, DC 20854
Guenter Riegler, NASA HQ Code EZ, Washington, DC 20546
David A. Roalstad, Ball Aerospace, P.O. Box 1062, Boulder, CO 80506
Fred Roesler, University of Wisconsin, Dept. of Physics - 1150 , Madison, WI 53706
Nancy G. Roman, McDonnell Douglas, 4620 N. Park Ave., #306W, Chevy Chase, MD 20815
Janet Rountree, AFE/Technical Services Staff, Bolling Air Force Base, Washington, DC 20332
Janet Ruff, NASA/GSFC Code 130, Greenbelt, MD 20771
 ohn A. Sand, Ball Aerospace, 2200 Clarendon Blvd., Ste 1006, Arlington, VA 22201
R.T. Schilizzi, Netherlands Found. for Research in Astronomy, Postbus 2, 7990 AA Dwingeloo, The Netherlands
Herbert Schnopper, Danish Space Res Institute, Lundtoftevej 7, DK-2800 Lingby, Danmark
George A. Seielstad, NRAO, P.O. Box 2, Green Bank, WV 24944
Peter Serlemitsos, NASA/GSFC Code 666, Greenbelt, MD 20771
M.M. Shapiro, University of Maryland, 205 Yoabum Parkway, #1720, Alexandria, VA 22304
Peter Shaver, ESO, Karl-Schwarzschild-Strasse 2, D-8046 Garching bei Munchen, WEST Germany
Charles Sheffield, Earth Satellite Corp., 7222 47th Street, Chevy Chase, MD 20815
G. Sholomitskii, IKI, Space Research Institute, Profsoyuznaya 84/32, 117485 Moscow, USSR
Chris Shrader, NASA/GSFC CSC/Code 684.9, Greenbelt, MD 20771
Rein Silberberg, Naval Research Laboratory, Code 4154, Washington, DC 20375-5000
Ramesh Sinha, NASA/GSFC Code 685, Greenbelt, MD 20771
George Sonneborn, NASA/GSFC Code 681, Greenbelt, MD 20771
D.S. Spicer, NASA/GSFC Code 630.1, Greenbelt, MD 20771
Robert V. Stachnik, NASA HQ Code EZ, Washington, DC 20546
Ted Stecher, NASA/GSFC Code 680, Greenbelt, MD 20771
N.V. Steshenko, USSR Academy of Sciences, Crimean Astroph Observatory, Crimea, p/o Nauchny, USSR
Mark T. Stier, Hughes Danbury Optical Systems, 100 Wooster Heights Rd. MS 813, Danbury, CT 06810
Igor Strukov, IKI Space Research Institute, Profsoyuznaya 88, 117810 Moscow, USSR

Ernst Stuhlinger, 3106 Rowe Drive, Huntsville, AL 35801

A. Sukhanov, USSR Academy of Sciences, 117810 Profsoyuznaya 84/32, Moscow, USSR

Rashid Sunyaev, IKI, Space Research Institute, Profsoyuznaya 88, 117810 Moscow, USSR

Jean Swank, NASA/GSFC Code 666, Greenbelt, MD 20771

Yasuo Tanaka, ISAS, 3-1-1 Yoshinodai, Sagamihara, Kanagawa-Kgn, 229 JAPAN

Eugene A. Tanzi, IFC Milano, Istituto di Fisica Cosmica, Via Bassini 15, Milano, 20133 ITALY

Brian G. Taylor, European Space Agency, Astrophysics Division, ESTEC, Noordwijk, THE NETHERLANDS

Terry Teays, NASA/GSFC CSC/Code 684.9, Greenbelt, MD 20771

Harley Thronson, Royal Observatory Edinburgh, Blackford Hill, Edinburgh, EH9 3HJ, SCOTLAND

John Townsend, NASA/GSFC Code 100, Greenbelt, MD 20771

Joachim Truemper, MPI fur Extraterr.Physik, 8046 Garching/Munchen, Germany

A. Tsarev, IKI, Space Research Institute, Profsoyuznaya 88, 117810 Moscow, USSR

Zlatan Tsvetanov, University of Maryland, Astronomy Program, College Park, MD 20742

Michael Van Steenberg, NASA/GSFC Code 633, Greenbelt, MD 20771

Osmi Vilhu, University of Helsinki Observatory, Tahtitorninmaki SF-00130, Helsinki, FINLAND

Roberto Viotti, Istituto Astrofisica Spaziale, Via E. Fermi, 21 I-00044, Frascati, Roma, ITALY

Serge Volonte, European Space Agency, 8-10 rue Mario Nikkis, 75738 Paris Cedex 15, FRANCE

Willem Wamsteker, European Space Agency, P.O. Box 50727, 28080 Madrid, SPAIN

Sander Weinreb, Martin Marietta, 1450 S. Rolling Rd., Baltimore, MD 21227

Werner Weiss, Institute for Astronomy, Tuerkenschanzstrasse 17, A-1180 Vienna, AUSTRIA

Donald K. West, NASA/GSFC Code 684.1, Greenbelt, MD 20771

R.A. White, NASA/GSFC Code 936, Greenbelt, MD 20771Jim Weiss, NASA HQ Code EZ, Washington, DC 20546

Martin C. Weisskopf, NASA/MSFC ES 65, Huntsville, AL 35812

Alan Whitney, MIT Haystack Observatory, Off Route 40, Westford, MA 01886

Adolf N. Witt, University of Toledo, 2801 W. Bancroft St., Toledo, OH 43606

Kent Wood, Naval Research Laboratory, Code 4121, Washington, DC 20375

Greg Wright, Princeton University, Physics Dept./P.O. Box 708, Princeton, NJ 08544

Michael Wright, NASA/GSFC Code 440, Greenbelt, MD 20771

Chi-Chao Wu, Space Telescope Science Institute/CSC, 3700 San Martin Drive, Baltimore, MD 21218

V. Yatsenko, IKI, Space Research Institute, Profsoyuznaya 88, 117810 Moscow, USSR

Linda Zall, Department of Defense, Industrial Activities Office Pentagon, Washington, DC 20370-0440

Index of Telescopes and Instruments

NATIONAL AERONAUTICS SPACE ADMINISTRATION
GODDARD SPACE FLIGHT CENTER

IAU COLLOQUIUM NO. 123:
OBSERVATORIES in EARTH ORBIT and BEYOND

Building 8 Auditorium
GSFC, Greenbelt, MD 20771

April 24-27, 1990

AGENDA

MONDAY, APRIL 23, 1990

Registration and Reception at Holiday-Inn Greenbelt

TUESDAY, APRIL 24, 1990

Welcome
        J. Townsend
        R. Bonnet

(I)   CURRENT MISSIONS

        HST (Boggess, A.)
        COBE (Mather, J.C.)
        GRANAT (Sunyaev, R.)
        Hipparcos (Perryman, M.)
        IUE (Kondo. Y.)
        GINGA (Makino, F.)
        MIR (Sunyaev, R.)
        Voyager (Holberg, J.B.)

(II)   FUTURE MISSIONS

(A)   X-RAY AND GAMMA-RAY MISSIONS

        ROSAT (Trümper, J.)
        GRO (Kniffen, D.A.)

AXAF (Weisskopf, M.C.)
ASTRO-D (Tanaka, Y.)
XTE (Bradt, H.V.)
Spectrum-X (Sunyaev, R.)
XMM (Peacock, A.)
SAX (Scarsi, L.)
GAMMA-1 (Galper, A.)

Reception at the Goddard Library

WEDNESDAY, APRIL 25, 1990

(II)   FUTURE MISSIONS (continued from April 24)

(B)   ULTRAVIOLET MISSIONS

EUVE (Bowyer, S.)
Lyman-FUSE (Moos, W.)
SPECTRUM-UV (Steshenko, N.)
ORFEUS (Krämer, G.)

(C)   INFRARED AND SUBMILLIMETER MISSIONS

SIRTF (Fazio, G.G.)
ISO (Cesarsky, C.)
IRTS (Matsumoto, T.)
FIRST (Frisk, U.O.)
Submillimeter Explorer, LDR (Phillips, T.G.)
SWAS (Melnick, G.J.)

(D)   RADIO MISSIONS

Radioastron (Kardashev, N.)
IVS (Schilizzi, R.T.)
VSOP (Hirabayashi, H.)
VLBI with TDRSS (Linfield, R.P.)

(E)   SOLAR SYSTEM & PLANETARY SYSTEMS

SOHO (Poland, A.I.)
OSL (Spicer, D.S.)
Planeten (Neukum, G.)
AIT (Levy, E.H.)

(F)   SHUTTLE-BORNE ASTRO MISSIONS

       WUPPE (Code, A.D.)
       HUT (Davidsen, A.
       UIT (Stecher, T.P.)
       BBXRT (Serlemitsos, P.J.)

(G)   INTERPLANETARY MISSIONS

       Interplanetary Probes - An Overview (Briggs, G.A.)
       Ulysses (Page, D.E.)

(H).   DATA ANALYSIS ACTIVITIES

       NASA Data Analysis Program (Riegler, G.R.)

Banquet at the Goddard Recreation Center

THURSDAY, APRIL 26, 1990

(III)   LAUNCH VEHICLES

(A)   CURRENT & "NEAR" FUTURE

       USA (Krause, R.B.)
       USSR (Boyarchuk, M.)
       ESA (Feustel-Buechler, J.)
       Japan (Tanaka, Y.)

(B)   "LONG-TERM" FUTURE

       Earth to Near-Earth Space (Keyworth, G.A.)
       Deep Space Propulsion Systems (Stuhlinger, E.)

(IV)   RELATIVE MERITS OF VARIOUS OBSERVATORIES

(A)   VARIOUS ORBITS AND SITES

       Lagrangian-point behind the Moon (Strukov, I.)
       Moon (Smith, Harlan)
       Elsewhere (Illingworth, G.)

Panel:

     Relative Merits of Low-Earth, Eccentric, Geosynchronous and Interplanetary Orbits and Sites in Space. (Farquhar, R.; Kardashev, N.; Mumma, M.; Oda, M.; Taylor, B.G.; Newton, G.; Smith, Harlan; Volonte,S.)

(B)   ALTERNATIVE APPROACHES

     Major Observatories vs. Economy Class Observatories in Space (Dyson, F.J.)

(V)   <u>LONG TERM FUTURE ISSUES</u>

Panel:

     Long Term Plans for Future Space Observatories(Bonnet, R.; Kardashev, N.; Longair, M.; Pellerin, C.; Tanaka, Y.)

Dinner Reception at Martin's Crosswinds

<u>FRIDAY, APRIL 27, 1990</u>

(V)   <u>LONG TERM FUTURE ISSUES (continued from April 26)</u>

Panel:

     Photon-Gathering Instruments and Detectors of the Future (Pounds, K.; Smith H.; Humphries, C.; Gatley, I.; Illingworth, G.)

Panel:

     Major Unsolved Problems of Astronomy (Arp, H.; Dyson, F.; Oda, M.; Hoyle, F.; Longair, M.; Sunyaev, R.)

Adjourn

Posters were exhibited throughout the meeting